T0323935

Molecular Breeding and Nutritional Aspects of Buckwheat

Molecular Breeding and Nutritional Aspects of Buckwheat

Editors

Meiliang Zhou
Biotechnology Research Institute
Chinese Academy of Agricultural Sciences
Beijing, China

Ivan Kreft
Department of Forest Physiology and Genetics
Slovenian Forestry Institute
Ljubljana, Slovenia

Sun-Hee Woo
Department of Crop Science
Chungbuk National University
Cheong-ju, Korea

Nikhil Chrungoo
Department of Botany
North Eastern Hill University
Shillong, India

Gunilla Wieslander
Department of Medical Sciences, Uppsala University
Occupational and Environmental Medicine
Uppsala, Sweden

AMSTERDAM • BOSTON • HEIDELBERG • LONDON
NEW YORK • OXFORD • PARIS • SAN DIEGO
SAN FRANCISCO • SINGAPORE • SYDNEY • TOKYO

Academic Press is an imprint of Elsevier

Academic Press is an imprint of Elsevier
125 London Wall, London EC2Y 5AS, United Kingdom
525 B Street, Suite 1800, San Diego, CA 92101-4495, United States
50 Hampshire Street, 5th Floor, Cambridge, MA 02139, United States
The Boulevard, Langford Lane, Kidlington, Oxford OX5 1GB, UK

Notices
Knowledge and best practice in this field are constantly changing. As new research and experience
broaden our understanding, changes in research methods, professional practices, or medical treatment may become necessary.

Practitioners and researchers must always rely on their own experience and knowledge in evaluating
and using any information, methods, compounds, or experiments described herein. In using such
information or methods they should be mindful of their own safety and the safety of others, including parties for whom they have a professional responsibility.

To the fullest extent of the law, neither the Publisher nor the authors, contributors, or editors, assume
any liability for any injury and/or damage to persons or property as a matter of products liability,
negligence or otherwise, or from any use or operation of any methods, products, instructions, or
ideas contained in the material herein.

British Library Cataloguing-in-Publication Data
A catalogue record for this book is available from the British Library

Library of Congress Cataloging-in-Publication Data
A catalog record for this book is available from the Library of Congress

ISBN: 978-0-12-803692-1

For information on all Academic Press publications
visit our website at https://www.elsevier.com/

Working together
to grow libraries in
developing countries

www.elsevier.com • www.bookaid.org

Publisher: Nikki Levy
Acquisition Editor: Nancy Maragioglio
Editorial Project Manager: Billie Jean Fernandez
Production Project Manager: Caroline Johnson
Designer: Ines Cruz

Typeset by Thomson Digital

Contents

List of Contributors

J. Aii, Niigata University of Pharmacy and Applied Life Science, Faculty of Applied Life Science, Akiha-ku, Niigata, Japan

S. Archak, National Bureau of Plant Genetic Resources Regional Station, Shimla, India

Y.N. Barsukova, Primorsky Scientific Research Institute of Agriculture, Primorsky Krai, Russia

S. Bobkov, Laboratory of Plant Physiology and Biochemistry, All-Russia Research Institute of Legume and Groat Crops, Orel, Streletsky, Russia

A. Brunori, ENEA, SSPT-BIOAG Department, Laboratory of BioProducts and BioProcesses, Rome, Italy

B. Budič, Laboratory for Analitical Chemistry, National Institute of Chemistry, Hajdrihova, Ljubljana, Slovenia

P. Hlásná Čepková, Gene Bank, Crop Research Institute, Prague, Ruzyně, Czech Republic

R.K. Chahota, National Bureau of Plant Genetic Resources, Pusa Campus, New Delhi, India

R.S. Chauhan, Department of Biotechnology & Bioinformatics, Jaypee University of Information Technology, Solan, India

H. Chen, Sichuan Agriculture University, College of Life Sciences, Sichuan, People's Republic of China

U. Chettry, Plant Molecular Biology Laboratory, UGC-Centre for Advanced Studies in Botany, North-Eastern Hill University, Shillong, India

M. Chnapek, Department of Biochemistry and Biotechnology, Slovak University of Agriculture in Nitra, Faculty of Biotechnology and Food Sciences, Nitra, Slovakia

S.-W. Cho, Division of Rice Research, National Institute of Crop Science, Rural Development Administration, Suwon, Korea

J.-S. Choi, Biological Disaster Research Group, Korea Basic Science Institute; Department of Analytical Science and Technology, Graduate School of Analytical Science and Technology, Chungnam National University, Daejeon, Korea

N.K. Chrungoo, Department of Botany, North Eastern Hill University; Plant Molecular Biology Laboratory, UGC-Centre for Advanced Studies in Botany, North-Eastern Hill University, Shillong, India

K.-Y. Chung, Department of Environmental & Biological Chemistry, Chungbuk National University, Cheong-ju, Korea

F. Ahmad Dar, Department of Bioresources, University of Kashmir, Srinagar, Jammu and Kashmir, India

N. Devadasan, Department of Botany, North Eastern Hill University, Shillong, Meghalaya, India

M.-Q. Ding, Biotechnology Research Institute, Chinese Academy of Agricultural Sciences, Beijing; School of Life Sciences, Sichuan Agricultural University, Yaan, Sichuan, China

L. Dohtdong, Plant Molecular Biology Laboratory, UGC-Centre for Advanced Studies in Botany, North-Eastern Hill University, Shillong, India

S. Farooq, Department of Botany, University of Kashmir, Srinagar, Jammu and Kashmir, India

A.N. Fesenko, Laboratory of Groats Crops Breeding, All-Russia Research Institute of Legumes and Groats Crops, Orel, Streletskoe, Russia

I.N. Fesenko, Laboratory of Genetics and Biotechnology, All-Russia Research Institute of Legumes and Groats Crops, Orel, Streletskoe, Russia

N.N. Fesenko, Laboratory of Genetics and Biotechnology, All-Russia Research Institute of Legumes and Groats Crops, Orel, Streletskoe, Russia

A. Gaberščik, Department of Biology, Biotechnical Faculty, University of Ljubljana, Jamnikarjeva, Ljubljana, Slovenia

M. Germ, Department of Biology, Biotechnical Faculty, University of Ljubljana, Jamnikarjeva, Ljubljana, Slovenia

K. Ikeda, Kobe Gakuin University, Faculty of Nutrition, Kobe, Japan

S. Ikeda, Kobe Gakuin University, Faculty of Nutrition, Kobe, Japan

D. Janovská, Gene Bank, Crop Research Institute, Prague, Ruzyně, Czech Republic

T. Katsube-Tanaka, Graduate School of Agriculture, Kyoto University, Kitashirakawa, Kyoto, Japan

H.-H. Kim, Department of Food Nutrition and Cookery, Woosong College, Daejeon, Korea

A.G. Klykov, Primorsky Scientific Research Institute of Agriculture, Primorsky Krai, Russia

E. Kovačec, Department of Biology, University of Ljubljana, Biotechnical Faculty, Ljubljana, Slovenia

I. Kreft, Department of Forest Physiology and Genetics, Slovenian Forestry Institute; University of Ljubljana, Biotechnical Faculty, Ljubljana, Slovenia

P. Kump, Department of Low and Medium Energy Physics, Jožef Stefan Institute, Jamova, Ljubljana, Slovenia

S.J. Kwon, Department of Crop Science, Chungbuk National University, Cheong-ju, Korea

D.-G. Lee, Biological Disaster Research Group, Korea Basic Science Institute, Daejeon, Korea

M.-S. Lee, Department of Industrial Plant Science & Technology, Chungbuk National University, Cheong-ju, Korea

F. Leiber, FiBL, Research Institute of Organic Agriculture, Frick, Switzerland

F.-L. Li, Xichang Institute of Agricultural Science, Alpine Crop Research Station, Xichang, Sichuan, China

B. Malik, Department of Bioresources, University of Kashmir, Srinagar, Jammu and Kashmir, India

K. Matsui, National Agriculture and Food Research Organization, Kyushu Okinawa Agricultural Research Center, Suya, Koshi, Japan

L.M. Moiseenko, Primorsky Scientific Research Institute of Agriculture, Primorsky Krai, Russia

T. Morishita, National Agriculture and Food Research Organization (NARO) Hokkaido Agricultural Research Center, Hokkaido, Japan

C. Nobili, ENEA, SSPT-BIOAG Department, Laboratory of Sustainable Development and Innovation of Agro-industrial System, Rome, Italy

O. Ohnishi, Plant Germ-Plasm Institute, Graduate School of Agriculture, Kyoto University, Mozume-cho, Muko City, Japan

T. Ota, SOKENDAI (The Graduate University for Advanced Studies), School of Advanced Sciences, Department of Evolutionary Studies of Biosystems, Hayama, Japan

T.B. Pirzadah, Department of Bioresources, University of Kashmir, Srinagar, Jammu and Kashmir, India

G. Podolska, Institute of Soil Science and Plant Cultivation—State Research Institute, Puławy, Puławy, Poland

P. Pongrac, Department of Biology, University of Ljubljana, Biotechnical Faculty, Ljubljana, Slovenia

M. Potisek, Department of Biology, University of Ljubljana, Biotechnical Faculty, Ljubljana, Slovenia

S. Procacci, ENEA, SSPT-BIOAG Department, Laboratory of BioProducts and BioProcesses, Rome, Italy

J.C. Rana, National Bureau of Plant Genetic Resources Regional Station, Shimla, India

M. Regvar, Department of Biology, University of Ljubljana, Biotechnical Faculty, Ljubljana, Slovenia

R. Ul Rehman, Department of Bioresources, University of Kashmir, Srinagar, Jammu and Kashmir, India

O.I. Romanova, Department of Small Grains, N.I. Vavilov's Institute of Plant Industry, Saint-Petersburg, Bolshaya Morskaya, Russia

S.K. Roy, Department of Crop Science, Chungbuk National University, Cheong-ju, Korea

J. Ruan, Sichuan Agriculture University, College of Life Sciences, Sichuan, People's Republic of China

K. Sarker, Department of Crop Science, Chungbuk National University, Cheong-ju, Korea

S. Sato, Niigata University of Pharmacy and Applied Life Science, Faculty of Applied Life Science, Akiha-ku, Niigata, Japan

J.-R. Shao, School of Life Sciences, Sichuan Agricultural University, Yaan; Department of Food Science, Sichuan Tourism University, Chengdu, Sichuan, China

T.R. Sharma, National Bureau of Plant Genetic Resources, Pusa Campus, New Delhi, India

Mohar Singh, National Bureau of Plant Genetic Resources Regional Station, Shimla, India

V. Škrabanja, Department of Food Science and Technology, Biotechnical Faculty, Ljubljana, Slovenia

G. Suvorova, All-Russia Research Institute of Legumes and Groat Crops, Laboratory of Genetics and Biotechnology, Orel, Russia

T. Suzuki, National Agriculture and Food Research Organization (NARO), Kyushu Okinawa Agricultural Research Center, Koshi, Kumamoto, Japan

G. Taguchi, Department of Applied Biology, Faculty of Textile Science and Technology, Shinshu University, Ueda, Nagano, Japan

I. Tahir, Department of Bioresources, University of Kashmir, Srinagar, Jammu and Kashmir, India

Y. Tang, Department of Food Science, Sichuan Tourism University, Chengdu, Sichuan; Biotechnology Research Institute, Chinese Academy of Agricultural Sciences, Beijing; School of Life Sciences, Sichuan Agricultural University, Yaan, Sichuan, China

Y.-X. Tang, Biotechnology Research Institute, Chinese Academy of Agricultural Sciences, Beijing, China

L.K. Taranenko, Scientific-Production Enterprise Antaria, Kiev, Ukraine

P.P. Taranenko, Scientific-Production Enterprise Antaria, Foreign Relations Department, Kiev, Ukraine

T.P. Taranenko, Scientific-Production Enterprise Antaria, Marketing Department, Kiev, Ukraine

M. Ueno, Kyoto University, Graduate School of Agriculture, Kitashirakawa Oiwake-cho, Sakyou-ku, Kyoto, Japan

D. Urminska, Department of Biochemistry and Biotechnology, Slovak University of Agriculture in Nitra, Faculty of Biotechnology and Food Sciences, Nitra, Slovakia

K. Vogel-Mikuš, Department of Biology, University of Ljubljana, Biotechnical Faculty, Ljubljana; Department of Low and Medium Energy Physics, Jožef Stefan Institute, Jamova, Ljubljana, Slovenia

B. Vombergar, Education Centre Piramida, Maribor, Slovenia

G. Wieslander, Department of Occupational and Environmental Medicine, Uppsala University; Department of Medical Sciences, Uppsala University, Occupational and Environmental Medicine, Uppsala, Sweden

S.H. Woo, Department of Crop Science, Chungbuk National University, Cheong-ju, Korea

Y.-M. Wu, Biotechnology Research Institute, Chinese Academy of Agricultural Sciences, Beijing, China

R. Yadav, National Bureau of Plant Genetic Resources Regional Station, Shimla, India

Y. Yasui, Kyoto University, Graduate School of Agriculture, Kitashirakawa Oiwake-cho, Sakyou-ku, Kyoto, Japan

O.L. Yatsyshen, National Scientific Center "Institute of Agriculture" of the National Academy of Agricultural Sciences, Department of breeding of groat crops, Kiev, Ukrain

M.-L. Zhou, Biotechnology Research Institute, Chinese Academy of Agricultural Sciences, Beijing, China

Foreword

It is with pleasure that I have an opportunity to write a preface for the book entitled *Molecular Breeding and Nutritional Aspects of Buckwheat*, edited by Drs Meiliang Zhou, Ivan Kreft, Sun-Hee Woo, Nikhil Chrungoo, and Gunilla Weislander.

Molecular biology of plants, in particular molecular genetics of plants, rapidly and greatly progressed in the 1980s. In this context, PCR (polymerase chain reaction) technology contributed significantly to this progress through isolation and identification of genes/nucleotide sequences, which could have applications in genetic engineering/phylogenetic analyses. Buckwheat (*Fagopyrum* spp.), belonging to the family Polygonaceae, is an important crop in mountainous regions in the Himalayan countries, China, Korea, Japan, Russia, Ukraine, and parts of Eastern Europe, primarily because of its short growth span, capability to grow at high altitudes, and the high quality of protein contents of its grains. International Symposia on Buckwheat have been held every 3 years since 1980 and buckwheat scientists have exchanged information on new advances in buckwheat research. However, research into molecular biology has not kept pace with the advances, especially molecular genetics of other crops. As a result, buckwheat research into molecular analyses in such fields as genetics of economically important genes, molecular breeding of new buckwheat varieties, and analyses of nutritional elements of buckwheat are now falling behind the progress of major crops in agriculture. Furthermore, in my opinion, in many countries where buckwheat production and consumption are prominent, young buckwheat scientists have not progressed with time, that is to say, the generational change of buckwheat scientists is not being practiced well in these countries.

Now, in the 2010s, it is the time to catch up with the progress of molecular analyses in buckwheat research. PCR and other molecular techniques can contribute significantly toward the generation of information on potentially important buckwheat genes and on the molecular breeding of buckwheat. I hope that this book will be an introductory guide to molecular research in buckwheat, particularly for young buckwheat scientists. Molecular breeding in buckwheat may lead to improved production of buckwheat grains and increased consumption of buckwheat flour.

Ohmi Ohnishi
Professor emeritus, Kyoto University, Kyoto, Japan

Preface

Buckwheat (*Fagopyrum* spp.) is an ancient crop, which has long been grown in East Asia and the Himalayan region. It is a major staple food crop in high-altitude zones including the Daliang Mountain in Southwest China. It is the most important crop of mountain regions above 1800 m elevation both for grain and greens. Unlike common cereals, which are deficient in lysine, buckwheat has excellent protein quality in terms of essential amino acid composition. The upsurge in interest in buckwheat is based on its nutritional qualities including the high protein content of its grains, presence of flavonoids, ability to grow in marginal areas, and suitability to be cultivated as an organic (biological or ecological) traditional crop.

International scientific cooperation in research of buckwheat intensified after 1980, when the First International Symposium on Buckwheat was organized at the University of Ljubljana, Slovenia, from Sep. 1–3, 1980. This symposium was attended by some of the key personalities working in the field of buckwheat research. An outcome of the deliberations held during the symposium was the acceptance of the proposal of Marek Ruszkowski (Poland), Björn O. Eggum (Denmark), Toshiko Matano (Japan), Takashi Nagatomo (Japan), Taiji Adachi (Japan), and Ivan Kreft (Slovenia) to form the International Buckwheat Research Association (IBRA) for coordinating research into this important crop. It was decided to hold a symposium under the aegis of the IBRA every third year in different member countries and also to publish *Fagopyrum* as the official journal of the IBRA, with headquarters in Ljubljana. Professor Ivan Kreft was requested to coordinate the activities of the IBRA as its president until the next symposium. The journal *Fagopyrum* started publication in 1981 from Ljubljana with I. Kreft et al. as editors. Subsequently, it moved to Ina, Japan (editors T. Matano et al.) in 1995, to Kyoto (editors O. Ohnishi et al.) in 1998, and to Kobe (editors K. Ikeda et al.) in 2007.

Besides the founding members of the IBRA, other eminent scientists and professionals who contributed actively in the first steps of the IBRA (1980–1983) included O. Ohnishi, K. Ikeda, S. Ikeda, H. Namai, A. Ujihara, and R. Shiratori (Japan), and N.V. Fesenko (Russia). The subsequent symposia held under the aegis of the IBRA included those at Miyazaki (Japan, 1983, T. Nagatomo, T. Adachi et al.), Puławy (Poland, 1986, M. Ruszkowski et al.), Orel (Russia, 1989, N.V. Fesenko et al.), Taiyuan (China, 1992, Lin Rufa et al.), Ina (Japan, 1995, T. Matano, A. Ujihara et al.), Winnipeg, Manitoba (Canada, 1998, C. Campbell et al.), Chuncheon, Kangwon (Korea, 2001, C.H. Park, S.S. Ham, Y.S. Choi, N.S. Kim et al.), Prague (Czech Republic, 2004, A. Michalova, Z. Stehno et al.), Yangling (China, 2006, Chai Yan et al.), Orel (Russia, 2010,

V.I. Zotikov, G.N. Suvorova et al.), and Laško (Slovenia, 2013, B. Vombergar, M. Vogrinčič, M. Germ, I. Kreft et al.). The next symposium on buckwheat is scheduled to be held at Cheongu and Bongpyeong in Korea during 2016. The symposium will be organized by S.H. Woo et al.

Besides the aforementioned symposia, several other national, regional, and thematic meetings on buckwheat with participation of the IBRA members were also held from time to time. These included "World Soba Summit" at Togakushi, Japan, in 1992 (A. Ujihara et al.), "Marathon Soba Symposium" at Togamura, Japan, in 1992 (R. Shiratori, Z. Luthar, I. Kreft et al.), "Buckwheat in Diets" at Xichang, China, in 2005, and "Buckwheat Sprouts" at Bongpyoung, Korea, in 2009, C.H. Park et al. Several other symposia were also held in Italy (Teglio, 1995; Sondrio, Teglio, 2000, D. Filippini, A. Scotti, and G. Bonafaccia; Roma, 2013), Luxemburg (1999; 2015, C. Zewen, C. Ries, I. Kreft), Norway (Larvik, 1996), Czech Republic (Prague, 1997, A. Michalova et al.), Slovenia (Maribor, 2004, B. Vombergar et al.).

As Sep. 3 (1980) was the start of worldwide international cooperation on buckwheat research, it could be proclaimed as the "International Day of Buckwheat," to promote, develop, and utilize buckwheat and its products. The book *Molecular Breeding and Nutritional Aspects of Buckwheat* is a well-written document with chapters authored by eminent scientists who have contributed significantly to buckwheat by their research and also through international cooperation over the 35 years of activities of the IBRA.

Meiliang Zhou

Ivan Kreft

Sun-Hee Woo

Nikhil K. Chrungoo

Gunilla Wieslander

Chapter one

Molecular Taxonomy of the Genus *Fagopyrum*

O. Ohnishi

*Plant Germ-Plasm Institute, Graduate School of Agriculture,
Kyoto University, Mozume-cho, Muko City, Japan*

INTRODUCTION

When Steward (1930) classified buckwheat species he recognized two culti-vated species, *Fagopyrum esculentum* and *Fagopyrum tataricum*, and eight wild species. At that time he classified buckwheat species into a section of the genus *Polygonum* in a broad sense. However, most taxonomists later treated buck-wheat species as the species of a distinct genus *Fagopyrum*, based on chro-mosome number (Munshi and Javeid, 1986), and based on pollen morphology (Hedberg, 1946). In the 1990s I and my students found a new species of the genus *Fagopyrum* in southern China, including the wild ancestor of culti-vated common buckwheat, and tried to classify the new species and already known species (Ohnishi, 1990, 1998a; Ohnishi and Matsuoka, 1996; Yasui and Ohnishi, 1998a,b; Ohsako and Ohnishi, 1998, 2000; Ohsako et al., 2002). First, we tried a morphological classification. However, we immediately faced difficul-ty in finding key character(s) separating different groups of species or different species. For example, at that time it was very suspicious that common buckwheat was closely related to perennial buckwheat, *Fagopyrum cymosum*. Molecular taxonomic study already suggested that Tartary buckwheat, rather than common buckwheat is more closely related to *F. cymosum* (Kishima et al., 1995). But what character(s) does separate common buckwheat from the *F. cymosum*-Tartary buckwheat group? It was a very hard task to find such a character. For more un-familiar wild species or groups of wild species, classification by morphological characters is much more difficult.

On the other hand, classification by molecular markers has been difficult in mastering the technique of treating target DNAs. However, the results on phylogeny are reasonable in the sense that all the results by different sci-entists on either chloroplast DNA (cpDNA) or nuclear DNA do not differ by much (compare the results of Ohnishi and Matsuoka, 1996; Yasui and

1

Ohnishi, 1998a,b; Ohsako and Ohnishi, 1998, 2000, 2001; but also see Nishimoto et al., 2003 for the incongruence between nuclear and chloroplast DNA trees).

Molecular classification, however, has such weak points that it cannot be practiced in fieldwork, and getting results takes time. Hence morphological classification should be of primary important in such field research as finding a new species. Molecular classification is much more reliable for phylogenetic analyses; hence we should use the molecular classification for confirming new species or new group(s) of wild species.

In this review, I will show that how agreeably molecular classification solved phylogenetic issues in taxonomy of the genus *Fagopyrum*.

TWO GROUPS OF THE GENUS *FAGOPYRUM*: THE *CYMOSUM* GROUP AND THE *UROPHYLLUM* GROUP

Before discussing the two groups of *Fagopyrum*, I would like to discuss the genus *Fagopyrum*. This issue was clarified through the discussions by Ohnishi and Matsuoka (1996), namely, they dissolved this issue by taking Nakai's (1926) morphological criterion of the genus *Fagopyrum*, that is, thick plaited cotyledons lie in the center of the achene and petiole does not touch the wall of an achene. This criterion was adopted in the study of whether *Fagopyrum megacarpum*, a new species found by Hara (1966) in Nepal, should be included in *Fagopyrum* or not (Ohsako et al., 2001) (Fig. 1.1).

It was concluded that *F. megacarpum* should not be included in *Fagopyrum* by molecular classification study. In fact, Hara (1982) himself considered that this species (achene's morphology looks like a species of *Fagopyrum*) is not a species of *Fagopyrum*. This is the only molecular study where the criterion of *Fagopyrum* is the main issue to be discussed.

Let us now return to the issue of two major groups of *Fagopyrum*. Ohnishi and Matsuoka (1996) first proposed two groups in the genus *Fagopyrum* based on both morphological and molecular classifications. All later molecular studies supported this subdivision of the genus *Fagopyrum* (Yasui and Ohnishi, 1998a,b; Ohsako and Ohnishi, 1998, 2000; Nishimoto et al., 2003; Kochieva et al., 2010). Morphological subdivision of *Fagopyrum* is rather easy. Namely, the morphological criterion for this subdivision is achene morphology. The species of the *cymosum* group have such achene and cotyledon morphology that cotyledons are horizontally long and a large lusterless achene is partially covered with persistent perianths, whereas the species of the *urophyllum* group have the characteristics that cotyledons are laterally long or round, and small lustrous achenes are completely covered with persistent perianths. Thus the subdivision of *Fagopyrum* into the *cymosum* group and the *urophyllum* group was supported by both morphological studies and molecular studies. Therefore there remains no problem in subdividing *Fagopyrum* species into two groups, even in field research (Fig. 1.2).

FIGURE 1.1 *Fagopyrum (Eskemukejea) megacarpum* **Hara.** A new species found in Nepal by Hara (1966). Should this plant be classified in *Fagopyrum*? Ohsako et al. (2001) gave a conclusion that this plant should not be classified in *Fagopyrum*.

CONTROVERSIAL OPINIONS: IS *F. CYMOSUM* PHYLOGENETICALLY CLOSE TO COMMON BUCKWHEAT OR CLOSE TO TARTARY BUCKWHEAT?

It was a fact that only three buckwheat species, *F. esculentum*, *F. tataricum*, and *F. cymosum*, were known to European scientists until they began to search wild buckwheat species in China at the end of the 19th century (Bredtschneider, 1898). Hence buckwheat scientists including De Candolle, intuitively, or based on kernel morphology, considered that *F. cymosum* is closer to common buckwheat than to Tartary buckwheat. Hereafter, *F. cymosum* is the most probable candidate of the wild ancestor of cultivated common buckwheat. Is this true? What is the character common to both *F. cymosum* and common buckwheat, what is the key character separating them from each other? It seems quite a difficult problem to find such character(s). Fortunately, however, Kishima et al. (1995) showed that *F. cymosum* is closer to Tartary buckwheat than to common buckwheat by analyzing cpDNA. All later molecular studies supported Kishima's proposition (Ohnishi and Matsuoka, 1996; Yasui and Ohnishi, 1998a,b; Ohsako and Ohnishi, 2000; Nishimoto et al., 2003). Then the issue to be tackled became: what is the key character separating *F. tataricum* from *F. cymosum*? Ohnishi and

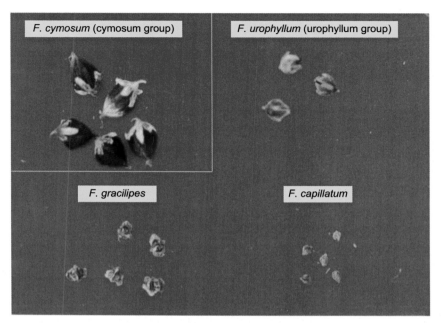

FIGURE 1.2 Examples of the achenes of two groups of *Fagopyrum*: the *cymo-sum* group and the *urophyllum* group. *F. cymosum* (*upper left*) is an example of the *cymosum* group. Large lusterless achenes are partially covered with persistent perianths. *F. urophyllum* (*upper right*), *F. gracilipes* and *F. capillatum* (*lower lane*) are the examples of the *urophyllum* group. Small lustrous achenes are completely covered with persistent perianths.

Matsuoka (1996) showed that the surface of achenes is smooth in *F. cymosum*, while the surface of achenes is rough with a canal in *F. tataricum*. This key character was valid until so-called *Fagopyrum pilus* was found in the Tibetan side of the Huongdan Mountains (Chen, 1999; Tsuji et al., 1999). It was shown that *F. pilus* is crossable with *F. cymosum*, yet this species has *F. tataricum*-like achenes with a rough surface. Since *F. cymosum* is crossable with *F. pilus*, Ohnishi (2010) classified *F. pilus* as a subspecies of *F. cymosum*. Now, the key character separating *F. cymosum* from *F. tataricum* is that the smooth or rough surface of achenes has canals in *F. tataricum* and has no canal in *F. cymosum*. The common character to both *F. cymosum* and *F. tataricum*, which separates these two species from *F. esculentum*, is the character of cotyledons in endosperm. The cotyledons in endosperm are yellowish and blade veins are transparent in *F. cymosum* and *F. tataricum*, while cotyledons are colorless and blade veins are not transparent in common buckwheat (Ohnishi and Matsuoka, 1996). By these morphological characters we could classify the members of the *cymosum* group (see Ohnishi, 2010). Newly discovered species *Fagopyrum homotropicum* is similar to *F. esculentum* ssp. *ancestrale* in general morphology, but they are distinct in the following characters: *F. esculentum* ssp. *ancestrale* is a heterostylous outbreeder, while *F. homotropicum* is a homostylous self-pollinator. So far as the members of the *cymosum* group are concerned, there remains only

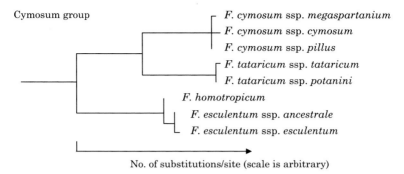

Cymosum group

F. cymosum ssp. *megaspartanium*
F. cymosum ssp. *cymosum*
F. cymosum ssp. *pillus*
F. tataricum ssp. *tataricum*
F. tataricum ssp. *potanini*
F. homotropicum
F. esculentum ssp. *ancestrale*
F. esculentum ssp. *esculentum*

No. of substitutions/site (scale is arbitrary)

FIGURE 1.3 A phylogenetic tree of the *cymosum* group constructed based on nuclear DNA data by Yasui and Ohnishi (1998b), Ohsako and Ohnishi (1998), Nishimoto et al. (2003), and other unpublished data. Relative length of blanches show DNA substitution rates.

a diploid–tetraploid problem, that is, should diploid and tetraploid *F. cymosum* and diploid and tetraploid *F. homotropicum* be treated as one species or should they be treated as different species? This issue will be discussed later. The conclusion is that in the *cymosum* group diploid and tetraploid (probably autotetraploid) are similar in morphology and diploid and tetraploid should be treated as the same species. Therefore the species in the *cymosum* group are classified as shown in Fig. 1.3.

DIFFICULTY IN MORPHOLOGICAL CLASSIFICATION OF THE MEMBERS OF THE *UROPHYLLUM* GROUP

When Ohnishi and Matsuoka (1996) classified buckwheat species, three species, *Fagopyrum lineare*, *Fagopyrum statice*, and *Fagopyrum gilesii*, which were known at that time, could not be included in the experiment, because our group has not found those three species yet. Incidentally, Ohnishi and Matsuoka (1996)'s molecular classification was immediately faced with a curious problem, that is, *F. urophyllum* has unbelievably wide molecular variation; *F. urophyllum*-Kunming (samples from Kunming, Yunnan Province) and *F. urophyllum*-Dali (samples from Dali, Yunnan Province) have so different molecular constitutions that they should be classified as different species. We could not find a key character that distinguished *F. urophyllum*-Kunming and *F. urophyllum*-Dali from each other. This difficult issue in *F. urophyllum* was tackled by Kawasaki and Ohnishi (2006). The results will be shown later.

In the molecular phylogenetic tree of the genus *Fagopyrum*, Ohnishi and Matsuoka (1996)'s classification met another difficulty of finding key character(s) separating two species or two groups of species of the *urophyllum* group. *F. lineare* looks like *Fagopyrum leptopodum* in general morphology, yet it is at the molecular level close to *F. urophyllum*, particularly to *F. urophyllum*-Dali (Ohsako and Ohnishi, 2000). *F. lineare* has species-specific thin liner blades (1–2 mm width, 20–30 mm length). No other species in *Fagopyrum* has such a character. There is no common character shared by *F. urophyllum* and *F. lineare*. Therefore finding a morphological key character is difficult.

Similarly, molecular studies on the *urophyllum* group (Ohsako and Ohnishi, 2000; Ohsako et al., 2002) showed that *Fagopyrum jinshaense* is closely related to *F. urophyllum*-Kunming; however, we could not find any morphological character suggesting similarity between *F. jinshaense* and *F. urophyllum*-Kunming. What kind of phylogenetic events are involved in this molecular similarity between morphologically different species? The phylogenetic tree of the species of the *urophyllum* group is summarized in Fig. 1.4 and allows us to speculate the early evolutionary story of *Fagopyrum* as follows. After differentiation of *F. urophyllum*-Kunming and *F. urophyllum*-Dali, two annual grass species, *F. lineare* and *F. jinshaense*, were born from each differentiated woody perennial *F. urophyllum*. All other species of the *urophyllum* group were born in the lineage of *F. urophyllum*-Dali, after the differentiation of *F. lineare* and *F. jinshaense*. This story also gives an interpretation for the reason why *F. lineare* and *F. jinshaense* have such unique morphological traits.

Three new species in the upper Min River valley of Sichuan Province are distinct in morphology from each other; however, we cannot say anything about the phylogenetic relationships among the three species. From the characteristic distribution of these three species, far away from the center of distribution of

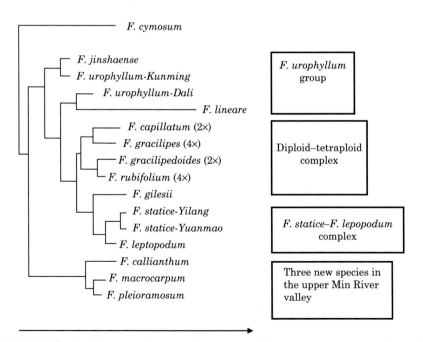

FIGURE 1.4 A phylogenetic tree of the *urophyllum* group constructed based on nuclear DNA data by Yasui and Ohnishi (1998b), Ohsako and Ohnishi (1998), and Nishimoto et al. (2003). Relative length of blanches shows DNA substitution rates. Exact scale is not shown (after Ohnishi, 2011).

Fagopyrum and far away from other species of the *urophyllum* group (Ohnishi, 2012), it is supposed that three species are closely related each other, and probably far away from other species of the *urophyllum* group. In fact, molecular data provided the expected results, that is, three new species from the upper Min River valley are far away from other species of the *urophyllum* group, as shown in Fig. 1.4 (see also Ohnishi and Matsuoka, 1996; Yasui and Ohnishi, 1998a,b; Ohsako and Ohnishi, 2000; Nishimoto et al., 2003). Later studies showed that three new species in the upper Min River valley differentiated at a very early stage in the history of the genus *Fagopyrum*, probably from *F. urophyllum* (Ohnishi, 2011).

Another difficulty in morphological classification lies in the *F. statice–F. leptopodum* group (Fig. 1.4). *F. leptopodum* has large morphological variation according to its distribution areas, southern Sichuan, western Yunnan, and eastern Yunnan. *F. statice* also has two distinct morphological types: a standard type of small perennial plant with ovate blades growing in eastern Yunnan Province, and an exceptional type of semiwoody perennial plant with sagittate blades growing in a limited area in Yanmao district in Yunnan Province. We call this later type the Yanmao type. We could not classify the samples of *F. leptopodum* and *F. statice* based only on morphological characters. However, molecular study of *F. leptopodum* and *F. statice* by Ohsako and Ohnishi (2001) revealed that geographical variations at the molecular level are parallel to morphological variations in both species and they clarified a close phylogenetic relationship between these two species.

Another example, where molecular taxonomy helped morphological taxonomy in clarifying phylogenetic relationships between newly discovered species and already known species, is found in the *F. jinshaense–F. leptopodum–F. gilesii* group studied by Ohsako et al. (2002). A newly discovered species, *F. jinshaense* has general morphology similar to that of *F. leptopodum*, but has distinct inflorescence different from the cyme of *F. leptopodum* and pod-like inflorescence of *F. gilesii* (Fig. 1.1 of Ohsako et al., 2002). So *F. jinshaense* could be easily distinct from *F. leptopodum* and *F. gilesii*. Through their molecular study it was shown that *F. jinshaense* is curiously close to *F. urophyllum*-Kunming and *F. lineare* is close to *F. urophyllum*-Dali.

Now, molecular classification of the species in the *urophyllum* group is well established (Ohnishi, 2011), hence a new morphological classification as a revised edition of Ohnishi and Matsuoka (1996) should be established in the near future, according to the current molecular phylogeny.

DIPLOID AND TETRAPLOID PLANTS BELONG TO A SAME SPECIES OR BELONG TO DIFFERENT SPECIES

It is well known that *F. cymosum* has both diploid and tetraploid plants and all plants are classified as *F. cymosum* morphologically and at the molecular level (see Yamane et al., 2003; Ohnishi, 2010). Diploid and tetraploid species, particularly when tetraploid species are autotetraploid, are often classified as the same species; on the other hand, diploid and tetraploid species, particularly when a tetraploid species is allotetraploid, are often classified as distinct species.

However, this is not always true, that is, diploid and allopolyploid species are classified as the same species, because of the dominance of a genome in allopolyploid species. We can see an example of this in the wheat *Triticum–Aegilops* complex (see, eg, Grant, 1971).

In *Fagopyrum*, tetraploid species were not clarified as autotetraploid or allotetraploid, except to *F. cymosum*, being shown as autotetraploid by molecular study (Yamane et al., 2003). Tetraploid plants of *F. cymosum* consist of various combinations of diploid genomes; the diploid genomes are not so different to each other at the molecular level. This observation is the reason for the conclusion of tetraploids of *F. cymosum* being autotetraploid. In the case of *F. homotropicum*, the story for autotetraploid is slightly different. For the establishment of tetraploid *F. homotropicum*, diploid *F. esculentum* ssp. *ancestrale* was involved as a parent of tetraploid *F. homotropicum*. This was shown by allozyme study on diploid and tetraploid *F. homotropicum* and on diploid *F. esculentum* ssp. *ancestrale* (Ohnishi and Asano, 1999). However, the genomes of diploid *F. homotropicum* and diploid *F. esculentum* ssp. *ancestrale* have not differentiated as much. These genomes of two diploid parents are similar and tetraploid plants may be called autotetraploid. In the two diploid parents, the gene of self-compatible or self-incompatible characteristic has differentiated; as a result, two diploid taxa are recognized as two distinct species. Now we arrive at the conclusion that diploid and tetraploid plants in the *cymosum* group can be classified as the same species.

On the other hand, it was shown that *Fagopyrum capillatum* is the diploid ancestor of the tetraploid species *Fagopyrum gracilipes* and *Fagopyrum gracilipedoides* is the diploid ancestor of tetraploid species *Fagopyrum rubifolium* by molecular analyses of tetraploid species and their diploid parents (Ohnishi, 2011). Although both parents of a tetraploid species have not been completely clarified, tetraploid species and their clarified diploid parents are easily distinguished morphologically as well. At present, all the species in the diploid–tetraploid complex of the *urophyllum* group are classified as distinct different species morphologically as well as at the molecular level (Ohnishi, 2011).

Cytological and molecular studies on autopolyploidy or allopolyploidy of these *Fagopyrum* species of the diploid–tetraploid complex in the *urophyllum* group should be carried out in the near future, and the conclusion given previously should be confirmed.

THE CASES OF TWO TAXA THAT ARE MORPHOLOGICALLY SIMILAR BUT DISTINCT AT THE MOLECULAR LEVEL

There are several cases where morphologically similar taxa are distinct at the molecular level. A typical example is the case of *F. urophyllum*-Dali and *F. urophyllum*-Kunming. This was extensively studied by Kawasaki and Ohnishi (2006). *F. urophyllum*-Dali and *F. urophyllum*-Kunming are not different so much in morphology, but there are minor differences in inflorescence and flowers. The flowers of *F. urophyllum*-Dali are half-open, whereas the flowers of *F. urophyllum*-Kunming are fully open in bloom (Fig. 1.1 of Kawasaki and

Ohnishi, 2006). The two groups have different geographical distribution areas, the distribution areas of each group do not overlap (Fig. 1.2 of Kawasaki and Ohnishi, 2006). *F. urophyllum*-Dali and *F. urophyllum*-Kunming are revealed to be reproductively isolated. That is, the percentage of seed set (number of flowers that set a seed/number of flowers pollinated) was only 0.51% in the crosses between *F. urophyllum*-Kunming and *F. urophyllum*-Dali, whereas it was 25.6% in the cross combinations between two populations of the same group. Astonishingly, *F. urophyllum*-Dali and *F. urophyllum*-Kunming are quite different at the molecular constitution; many species lie between two groups in molecular phylogenetic trees (see Kawasaki and Ohnishi, 2006 for more details). As a conclusion, *F. urophyllum*-Kunming and *F. urophyllum*-Dali should be treated as two distinct species.

Nishimoto et al. (2003) found two distinct genomes within a species of *F. rubifolium* during the sequence analyses of *FLO/LFY*. No morphological difference was found between genome donors. They provided a hypothesis of hybridization between the *F. rubifolium–F. gracilipedoides* clade and the *F. statice–F. leptopodum* clade and considered the molecular polymorphism caused by hybridization to be the reason for two distinct lineages of *F. rubifolium* (see Nishimoto et al., 2003 for more detailed discussion on the hypothesis).

THE GROUP OF PLANTS THAT HAVE SPECIFIC CHARACTERS, BUT HAVE NOT BEEN ANALYZED AT MOLECULAR LEVEL, HENCE HAVE NOT BEEN DECIDED AS BEING NEW SPECIES OR NOT

There are several morphologically distinct groups, but not recognized as being a distinct species. Among them, a few cases have been resolved. One is the case of *F. cymosum*. Chen (1999) proposed a new species *F. pilus* having characteristic kernel morphology like the kernel of wild Tartary buckwheat. However, *F. pilus* is not so different from ordinal typical *F. cymosum* at the molecular level (Yamane et al., 2003). Since *F. pilus* and ordinary *F. cymosum* are not reproductively isolated, Ohnishi (2010) concluded that *F. pilus* should be a subspecies of *F. cymosum*. By similar reasons, Chen (1999)'s *Fagopyrum megaspartanium* is also a subspecies of *F. cymosum*. Regarding the criteria of a species, reproductive isolation should be considered as a primary criterion.

The second resolved issue is on *F. leptopodum*. As described in Ohnishi (2012), *F. leptopodum* from southern Sichuan Province and Shimian and Hanyuan districts plants have larger blades and are taller than the plants from other places. However, at the molecular level, *F. leptopodum* from southern Sichuan and that from other places are not so different and there exists no reproductive barrier between them. Hence, a large morphological variation observed is not large enough to be the variation of different species and is considered as the variation within a species, *F. leptopodum*.

Ye and Gao (1992) described a new species, *Fagopyrum caudatum*, although the description is not complete (no Latin description, etc.). I also collected plants that look like *F. gracilipes*, but taller (taller than 30 cm) and with stems and leaves that are heavily pubescent. My collection does not directly correspond to *F. caudatum*, but apparently overlaps with *F. caudatum*. I am still doubtful of *F. caudatum* being a distinct new species. Molecular analysis and the test of cross-ability should be carried out before concluding *F. caudatum* as a new species.

Fagopyrum macrocarpum is one of the three new species distributed in the upper Min River valley. This species was first found in Putou village of Lixian district as a weed at the margin of cultivated fields. It has larger white flowers and larger achenes than other related *Fagopyrum* species, for example, *Fagopyrum pleioramosum*; hence it has been given the name "macrocarpum." After a detailed search of the distribution, this species is most frequently distributed in Maoxian district, particularly in apple orchards. The flower color is variable; some are white, but others are pink to red. Intuitively, this color variation has a genetic basis, that is, polymorphic; however, there is no reproductive isolation between color variants. Although no molecular study has been carried out, there may not be a great difference at the molecular level between flower color variants. Since there is no reproductive isolation between color variants, we may conclude that all the samples of *F. macrocarpum* belong to one species.

Dr Zhou and his coworkers found new species *Fagopyrum pugense* (Tang et al., 2010), *Fagopyrum wenchuanense*, and *Fagopyrum qiangcai* (Shao et al., 2011) in Sichuan Province and they tried to analyze genetic diversity and interspecific relationships among new species and a new species *Fagopyrum crispatofolium* (Liu et al., 2008) and several other species (*F. esculentum*, *F. tataricum*, *F. cymosum*, and *F. gracilipes*) using karyotype, inter simple sequence repeat (ISSR), and allozyme (Zhou et al., 2012).

Are those new species surely new distinct species? I am not sure. They look like the species closely related to *F. gracilipes* (probably excluding *F. wenchuanense* judging from their result). According to the distribution area of *F. wenchuanense*, *F. wenchuanense* has some relationships with Ohnishi's three new species in the upper Min River valley (see Ohnishi, 2012 and Section 1.5). I suspect a close relationship between *F. wenchuanense* and *F. pleioramosum*. The dendrograms based on isozymes and ISSR (Fig. 1.3 of Zhou et al., 2012) show this. The conclusion on new species or not of *F. wenchuanense* should be drawn after the comparison of *F. wenchuanense* and *F. pleioramosum*. As for *F. crispatofolium*, it is a tetraploid species, yet it is very close to a diploid species *F. pugense* and to a tetraploid species *F. gracilipes*. *F. crispatofolium* has unique leaf morphology, leaves like shrunken cloth, however, similar morphology was found as a mutant of common buckwheat, named crepe with gene symbol *cp* on the second chromosome (crepe, *cp* (II) in Ohnishi, 1990). This means that a great morphological difference does not immediately imply a unique new species. At present, genetic data are not enough to give a conclusion that *F. crispatofolium* and *F. pugense* are new species.

Similarly, a so-called new species *F. qiangcai* may have a genetic relationship to three new species from the upper Min River valley (Ohnishi, 2012), particularly to *Fagopyrum callianthum* judging from the figure of the *F. qiangcai* plant (Fig. 1.1

of Shao et al., 2011). We must wait for accumulation of molecular data on four new species of Zhou et al. (2012) and wait for the comparison of those new species and three new species from the upper Min River valley (see Ohnishi, 2012).

REFERENCES

Bredtschneider, F., 1898. History of European Botanical Discovery in China. Press Imp. Russ. Acad. Sci., Petersburg.

Chen, Q.F., 1999. A study of resources of *Fagopyrum* (Polygonaceae) native to China. Bot. J. Linnean Soc. 130, 53–64.

Grant, V., 1971. Plant Speciation. Columbia University Press, New York.

Hara, H., 1966. The Flora of East Himalaya. Tokyo University Press, Tokyo.

Hara, H., 1982. Polygonaceae. Hara, H., Chater, A.O., Williams, L.H.J. (Eds.), An Enumeration of the Flowering Plants of Nepal, vol. 3, pp. 172–180.

Hedberg, O., 1946. Pollen morphology in the genus *Polygonum* L. (s.l.) and its taxonomic significance. Svensk Bot. Tidskr. 40, 371–404.

Kawasaki, M., Ohnishi, O., 2006. Two distinct groups of natural populations of *Fagopyrum urophyllum* (Bur. et Franch.) Gross revealed by the nucleotide sequence of noncoding region in chloroplast DNA. Genes Genet. Syst. 81, 323–332.

Kishima, Y., Ogura, K., Mizukami, K., Mikami, T., Adachi, T., 1995. Chloroplast DNA analysis in buckwheat species: phylogenetic relationships, origin of the reproductive systems and extended inverted repeats. Plant Sci. 108, 173–179.

Kochieva, E.Z., Kadyrova, G.D., Ryzhova, N.N., 2010. Variability of plastid and mitochondrial sequences in Fagopyrum species. Procedings of Eleventh International Symposium on Buckwheat at Orel, pp. 265–267.

Liu, J.L., Tang, Y., Xia, Z.M., Shao, J.R., Cai, G.Z., Luo, Q., Sun, J.X., 2008. *Fagopyrum crispatofolium* J.L. Liu, a new species of Polygonaceae from Sichuan, China. J. Syst. Evol. 46, 929–932.

Munshi, A.H., Javeid, C.N., 1986. Systematic Studies in Polygonaceae of Kashmir Himalaya. Scientific Publishers, Jodhpur.

Nakai, T., 1926. A new classification of Linnean *Polygonum*. Rigakkai 24, 289–301, (in Japanese).

Nishimoto, Y., Ohnishi, O., Hasegawa, M., 2003. Topological incongruence between nuclear and chloroplast DNA tree suggesting hybridization in the *urophyllum* group of the genus *Fagopyrum* (Polygonaceae). Genes Genet. Syst. 78, 139–153.

Ohnishi, O., 1990. Analyses of genetic variants in common buckwheat, *Fagopyrum esculentum* Moench: a review. Fagopyrum 10, 12–22.

Ohnishi, O., 1998a. Search for the wild ancestor of buckwheat. I. Description of new *Fagopyrum* (Polygonaceae) species and their distribution in China and the Himalayan hills. Fagopyrum 15, 18–28.

Ohnishi, O., 1998b. Search for the wild ancestor of buckwheat. III. The wild ancestor of cultivated common buckwheat and of Tartary buckwheat. Econ. Bot. 52, 123–133.

Ohnishi, O., Asano, N., 1999. Genetic diversity of *Fagopyrum homotropicum*, a wild species related to common buckwheat. Genet. Resour. Crop Evol. 46, 389–398.

Ohnishi, O., Matsuoka, Y., 1996. Search for the wild ancestor of buckwheat. II. Taxonomy of *Fagopyrum* (Polygonaceae) species based on morphology, isozymes and cpDNA variability. Genes Genet. Syst. 72, 383–390.

Ohnishi, O., 2010. Distribution and classification of wild buckwheat species. 1. Cymosum group. Fagopyrum 27, 1–8.

Ohnishi, O., 2011. Distribution and classification of wild buckwheat species. 2. Urophyllum group. Fagopyrum 28, 1–8.

Ohnishi, O., 2012. On the distribution of natural populations of wild buckwheat species. Fagopyrum 29, 1–6.

Ohsako, T., Ohnishi, O., 1998. New *Fagopyrum* species revealed by morphological and molecular analyses. Genes Genet. Syst 73, 85–94.

Ohsako, T., Ohnishi, O., 2000. Intra- and interspecific phylogeny of wild *Fagopyrum* (Polygonaceae) species based on nucleotide sequences of noncoding regions in chloroplast DNA. Am. J. Bot. 87, 573–582.

Ohsako, T., Ohnishi, O., 2001. Nucleotide sequence variation of the chloroplast *trnK/matK* region in two wild *Fagopyrum* (Polygonaceae) species, *F. leptopodum* and *F. statice*. Genes Genet. Syst. 76, 39–46.

Ohsako, T., Fukuoka, S., Bimb, H., Banya, B.K., Yasui, Y., Ohnishi, O., 2001. Phyloenetic analysis of the genus *Fagopyrum* (Polygonaceae), including the Nepali species *F. megacarpum* based on nucleotide sequence of the *rbcL-accD* region in chloroplast DNA. Fagopyrum 18, 9–14.

Ohsako, T., Yamae, K., Ohnishi, O., 2002. Two new *Fagopyrum* (Polygonaceae) species, *F. gracilipedoides* and *F. jinshaense* from Yunnan, China. Genes Genet. Syst. 77, 399–408.

Shao, J.R., Zhou, M.L., Zhu, X.M., Wang, D.Z., Bai, D.Q., 2011. *Fagopyrum wenchuanense* and *Fagopyrum qingcai*, two new species of Polygonaceae from Sichuan, China. Novon 21, 256–261.

Steward, A.N., 1930. Polygoneae in eastern Asia. Contributions from the Gray Herbarium of Harvard University 88, 1–129.

Tang, Y., Zhou, M.L., Bai, D.Q., Shao, J.R., Zhu, X.M., Wang, D.Z., Tang, Y.X., 2010. *Fagopyrum pugense* (Polygonaceae), a new species from Sichuan, China. Novon 20, 239–242.

Tsuji, K., Yasui, Y., Ohnishi, O., 1999. Search for *Fagopyrum* species in eastern Tibet. Fagopyrum 16, 1–6.

Yamane, K., Yasui, Y., Ohnishi, O., 2003. Interspecific cpDNA variations of diploid and tetraploid perennial buckwheat, *Fagopyrum cymosum* (Polygonaceae). Amer. J. Bot. 90, 339–346.

Yasui, Y., Ohnishi, O., 1998a. Interspecific relationships in *Fagopyrum* (Polygonaceae) revealed by nucleotide sequences of the *rbcL* and *accD* genes and their intergenic region. Am. J. Bot. 85, 1134–1142.

Yasui, Y., Ohnishi, O., 1998b. Phylogenetic relationships among *Fagopyrum* aspecies revealed by the nucleotide sequences of the ITS region of the nuclear rRNA gene. Genes Genet. Syst. 73, 201–210.

Ye, N.G., Guo, G.Q., 1992. Classification, origin and evolution of genus *Fagopyrum* in China. Proceedings of the Fifth International Symposium on Buckwheat at Taiyuan. pp. 19–28.

Zhou, M.L., Bai, D.Q., Tang, Y., Zhu, X.M., Shao, J.R., 2012. Genetic diversity of four new species related to southwestern Sichuan buckwheat as revealed by karyotype, ISSR and allozyme characterization. Plant Syst. Evol. 298, 51–75.

Chapter two

Germplasm Resources of Buckwheat in China

Y. Tang*,**, M.-Q. Ding**,†, Y.-X. Tang**, Y.-M. Wu**, J.-R. Shao†, M.-L. Zhou**

*Department of Food Science, Sichuan Tourism University, Chengdu, Sichuan, China; **Biotechnology Research Institute, Chinese Academy of Agricultural Sciences, Beijing, China; †School of Life Sciences, Sichuan Agricultural University, Yaan, Sichuan, China

INTRODUCTION

The history of Chinese buckwheat cultivation goes back to 1st and 2nd centuries BC. After thousands of years of cultivation and evolution, cultivated buckwheat is not only widely spread in China, but has also formed a number of varieties. Southwest China especially has plentiful resources of wild-type buckwheat, which has drawn the attention of the wider world.

THE ACREAGE, PRODUCTION, AND DISTRIBUTION OF CULTIVATED BUCKWHEAT IN CHINA

China is one of the main producing countries of buckwheat, while acreage and production are ranked second in the world, trailed only by Russia. Every year the area planted to buckwheat in China is $70–100 \times 10^4$ hm^2. In this area, common buckwheat (CB) accounts for $60–70 \times 10^4$ hm^2, of which average production is $0.5–0.7$ t/hm^2 and total production is roughly 50×10^4 t. Tartary buckwheat (TB) accounts for the rest at $20–30 \times 10^4$ hm^2; its unit production is higher than CB, which generally can reach 0.9 t/hm^2, and the total production of TB is 30×10 t. Generally speaking, the production of buckwheat has a current annual output of nearly $75–150 \times 10^4$ t in China.

CB is distributed throughout the whole of China, which can expand as far as Heilongjiang Province in the north, Sanya City of Hainan Province in the south, coastal areas in the east, and Tacheng County or Hetian County of Xinjiang and Zhada County of Tibet in the west. With regard to altitude, there are a number of planting rules: CB mainly distributes at 600–1500 m, the maximum height is 4100 m and minimum height is 100 m; meanwhile, TB mainly distributes at 1200–3000 m, with a high limit of 4400 m and a low limit of 400 m.

Molecular Breeding and Nutritional Aspects of Buckwheat. http://dx.doi.org/10.1016/B978-0-12-803692-1.00002-X

Chifeng district of Inner Mongolia, Yulin district of Shaanxi, Datong district of Shanxi, Pingliang district of Gansu, Guyuan district of Ningxia, and Qujing district of Yunnan are the major producing areas of CB. Three of the largest districts are: the CB (white perianth) region in the east of Inner Mongolia, including Hure Banner, Naiman Banner, Aohan Banner, and Ougniud Banner; the CB (white perianth) region in the Houshan area of Inner Mongolia, including Guyang County, Wuchuan County, and Siziwang Banner; and the CB (red perianth) region in the juncture of Shaanxi, Gansu, and Ningxia, including Dingbian County, Jingbian County, Wuqi County, Zhidan County and Ansai County in Shaanxi Province, Yanchi County and Pengyang County in Ningxia Autonomous Region, and Huan County and Huachi County in Gansu Province. The production level of CB is relatively low; average production is 200–700 kg/hm², and the highest is 2000 kg/hm².

TB is mainly planted in Yunnan, Sichuan, Guizhou, Hunan, Hubei, Jiangxi, Shaanxi, Shanxi, and Gansu, and the major producing areas are located in Southwest China, including Zhaotong and Chuxiong district in Yunnan, Liangshan district in Sichuan, and Bijie district in Guizhou; the average production of TB is 900–2250 kg/hm², and the highest is 2900kg/hm².

BUCKWHEAT GERMPLASM RESOURCES IN CHINA

Research has shown that more than 20 buckwheat species have been named and reported, and most of them have already been found in China. Among all these species, only CB (*Fagopyrum esculentum*) and TB (*Fagopyrum tataricum*) are cultivated species and the others are wild species. What is more, CB and TB have been widely cultivated in China and a number of varieties have been further developed.

Germplasm Resources of Cultivated Buckwheat

Buckwheat has been cultured for thousands of years in China, mostly CB and TB; hence the germplasm resources are extremely rich and the varieties are numerous. However, until the 1980s, these resources had not been carefully collected and stored, nor had their characters been studied. In the 1950s more than 2000 copies of buckwheat materials had been collected but most of them have been lost for some reason. In the 1980s, the Crops Genetic Resources Institute of the Chinese Academy of Agricultural Sciences (CAAS), together with other institutes in 24 provinces, began to collect buckwheat resources. In 694 investigations, about 2000 samples of buckwheat were collected. After induction, 1500 samples were arranged into the "Content of Chinese buckwheat germplasm resources," of which 964 samples were CB and the other 536 samples were TB (Table 2.1); all of them were carefully stored in the germplasm bank of CAAS (Yang, 1992). From 1986 to 1990, these resources' distributions, characters of form, ecology, and qualities were investigated and identified in detail (Yang and Lu, 1992); the results proved a valuable reference for breeding and in further study of buckwheat.

TABLE 2.1 The Statistics of the Content of Chinese Buckwheat Germplasm Resources

Province	Variety	Common Buckwheat	Tartary Buckwheat
Heilongjiang	24	24	
Jilin	15	15	
Liaoning	75	74	1
Beijing	1	1	
Hebei	7	7	
Inner Mongolia	185	177	8
Shanxi	217	155	62
Shaanxi	271	180	91
Gansu	128	64	64
Qinghai	49	35	14
Xinjiang	30	30	
Sichuan	156	30	126
Yunnan	161	50	111
Guizhou	47	9	38
Tibet	22	5	17
Shandong	18	18	
Anhui	40	38	2
Jiangxi	54	52	2
Total	1500	964	536

Germplasm Resources of Wild Buckwheat

THE TYPES AND DISTRIBUTIONS OF WILD BUCKWHEAT

Since the 1900s, a number of reports concerning species and variants of *Fagopyrum* Mill. have been reported by botanists. Gross (1913) first classified Chinese buckwheat systematically, and he placed some of the confirmed species into *Fagopyrum* Mill., Polygonaceae. Nakai (1926) first proposed that due to the form and position of embryos in achene, *Fagopyrum* Mill. should be isolated from other Mills in Polygonaceae. Steward (1930) classified Polygonaceae plants in Asia, and put 10 species of buckwheat from *Polygonum* into *Fagopyrum* Mill. So far more than 20 buckwheat species have been found and named in reports around the world, and two of them, CB and TB, are cultivated species (Ohnishi, 1991, 1995; Ohnishi and Matsuoka, 1996; Ohsako and Ohnishi, 1998, 2000; Ohsako et al., 2002; Li, 1998; Chen, 1999; Liu et al., 2008; Tang et al., 2010; Shao et al., 2011; Hou et al., 2015; Zhou et al., 2012, 2015); most of these species distribute in China.

The distributions of different wild types are various. The 27 wild species reported in China are mostly distributed in Southeast China, such as Sichuan, Yunnan, Guizhou, and Tibet (Table 2.2). Among these wild species, *Fagopyrum cymosum* Meisn is most widely distributed in China, in the vast area in the south of Daba Mountain, including provinces belonging to Central China, East China, South China, and Southwest China in the Yangtze River basin. The distribution of *Fagopyrum gracilipes* (Hemsl.) Dammer ex Diels and its variant *F. gracilipes* (Hemsl.) Dammer ex Diels var. *odontopterum* (Gross) Sam is slightly narrower than *F. cymosum* Meisn, which is mainly distributed in Southwest and Central China in the south of the upper and middle reaches of the Yellow River basin. Other wild species only distribute in one to three provinces. Some of their distribution is extremely narrow, for example, *Fagopyrum crispatofolium* J. L. Liu, *Fagopyrum qiangcai* D. Q. Bai, and *Fagopyrum hailuogouense* J. R. Shao, M. L. Zhou, and Q. Zhang; their acreage being no more than 100 km^2. As for vertical distribution, most wild buckwheat grows at altitudes of 1000–2000 m; some individual species can reach 3500–4000 m, such as *F. gracilipes* (Hemsl.) Dammer ex Diels, *Fagopyrum zuogongense* Q-F. Chen, and *Fagopyrum megaspartanum* Q-F. Chen, while *F. cymosum* Meisn can grow at a height lower than 100 m above sea level.

INTRODUCTION OF BUCKWHEAT RESOURCES IN THE SOUTHWEST OF CHINA

Steward (1930) classified the Polygonaceae plants in Asia in his report, 10 species were included in *Fagopyrum* Mill. and they were all distributed in the southwest of China; and the Flora of China proved and described them (Li, 1998). The 10 species consisted of two cultivated species (*F. esculentum* Moench and *F. tataricum* (L.) Gaertn) and eight wild species, which were *F. gracilipes* (Hemsl.) Dammer ex Diels, *F. cymosum* (Trev.) Meisn, *Fagopyrum lineare* (Sam.) Haraldson, *Fagopyrum urophyllum* (Bur. et Fr.) H. Gross, *Fagopyrum leptopodum* (Diels) Hedberg, *Fagopyrum statice* (Lévl.) H. Gross, *Fagopyrum caudatum* (Sam.) A. J. Li, comb. nov., and *Fagopyrum gilesii* (Hemsl.) Hedberg.

Ohnishi (1991, 1995), Ohsako and Ohnishi (1998, 2000), and Ohsako et al. (2002) proved the former buckwheat species and found a new subspecies of *F. esculentum* Moench and eight wild species, for example, *F. esculentum* ssp. *ancestrale* Ohnishi, *Fagopyrum homotropicum* Ohnishi, *Fagopyrum capillatum* Ohnishi, *Fagopyrum pleioramosum* Ohnishi, *Fagopyrum callianthum* Ohnishi, *Fagopyrum macrocarpum* Ohsako et Ohnishi, *Fagopyrum rubifolium* Ohsako et Ohnishi, *Fagopyrum jinshaense* Ohsako et Ohnishi, and

TABLE 2.2 The List of Chinese Wild Buckwheat Species and Their Distribution

No.	Species	Variant	Subspecies	Distribution
1	*F. urophyllum* (Bur. et Fr.) H. Gross			Sichuan, Yunnan, and Gansu
2	*F. statice* (Lévl.) H. Gross			Sichuan, Yunnan, and Guizhou
3	*F. cymosum* Meisn			Sichuan, Yunnan, Guizhou, Chongqing, Tibet, Shaanxi, Hunan, Hubei, Zhejiang, Anhui, Fujian, Guangdong, Guangxi, and Hainan
4	*F. leptopodum* (Diels) Hedberg	*F. leptopodum* (Diels) Hedberg var. *grossi* (Lévl.) Sam		Sichuan and Yunnan
5	*F. lineare* (Sam.) Haraldson			Sichuan and Yunnan
6	*F. gracilipes* (Hemsl.) Dammer ex Diels	*F. gracilipes* (Hemsl.) Dammer ex Diels var. *odontopterum* (Gross) Sam		Sichuan, Yunnan, Guizhou, Chongqing, Shaanxi, Hubei, Gansu, and Henan
7	*F. gilesii* (Hemsl.) Hedberg			Sichuan, Yunnan, and Tibet
8	*F. caudatum* (Sam.) A. J. Li, comb. nov			Sichuan, Yunnan, and Gansu
9	*F. zuogongense* Q-F. Chen			Tibet
10	*F. pilus* Q-F. Chen			Sichuan
11	*F. megaspartanum* Q-F. Chen			Tibet
12	*F. gracilipedoides* Ohsako et Ohnishi			Yunnan
13	*F. jinshaense* Ohsako et Ohnishi			Sichuan and Yunnan
14	*F. crispatofolium* J. L. Liu			Sichuan
15	*F. pugense* T. Yu			Sichuan
16	*F. wenchuanense* J. R. Shao			Sichuan

(Continued)

TABLE 2.2 The List of Chinese Wild Buckwheat Species and Their Distribution (*cont.*)

No.	Species	Variant	Subspecies	Distribution
17	*F. qiangcai* D. Q. Bai			Sichuan
18	*F. luojishanense* J. R. Shao			Sichuan
19	*F. hailuogouense* J. R. Shao, M. L. Zhou, and Q. Zhang			Sichuan
20	*F. capillatum* Ohnishi			Yunnan
21	*F. callianthum* Ohnishi			Sichuan
22	*F. rubifolium* Ohsaka ex Ohnishi			Sichuan
23	*F. macrocarpum* Ohsaka ex Ohnishi			Sichuan
24	*F. homotropicum* Ohnishi			Sichuan and Yunnan
25	*F. pleioramosum* Ohnishi			Sichuan
26			*F. esculentum* ssp. *ancestrale* Ohnishi	Sichuan and Yunnan
27			*F. tataricum* ssp. *potanini*	Sichuan, Yunnan, Gansu, and Tibet

Fagopyrum gracilipedoides Ohsako et Ohnishi. These species are distributed in Sichuan, Yunnan, and their surroundings. Ohnishi and Matsuoka (1996) proposed that *Fagopyrum* Mill. should be divided into two groups: the *cymosum* group including two cultivated species (*F. tataricum* [L.] Gaertn and *F. esculentum* Moench) and two wild species (*F. cymosum* Meisn and *F. homotropicum* Ohnishi), and the *urophyllum* group including *F. statice*, *F. leptopodum*, *F. urophyllum*, *F. lineare*, *F. gracilipes*, *F. gracilipes* var. *odontopterum*, *F. esculentum* ssp. *ancestrale* Ohnishi, *F. homotropicum* Ohnishi, *F. capillatum* Ohnishi, *F. pleioramosum* Ohnishi, *F. callianthum* Ohnishi, etc.

Chinese buckwheat researchers have worked tirelessly searching the resources of wild *Fagopyrum* Mill. Chen (1999) found three wild species in the Sichuan, Tibet, region: *F. zuogongense* Q-F. Chen, *Fagopyrum pilus* Q-F. Chen, and *F. megaspartanum* Q-F. Chen. Later, Liu et al. (2008) found a new species in Puge County, Liangshan Prefecture, Sichuan Province, named *F. crispato-folium* J. L. Liu, and Tang et al. (2010) found *Fagopyrum pugense* T. Yu in

Puge County, Liangshan Prefecture, Sichuan Province. Later still, Shao et al. (2011) found *Fagopyrum wenchuanense* J. R. Shao and *F. qiangcai* D. Q. Bai in Wenchuan County, Aba Prefecture, Sichuan Province. Two new species, *Fagopyrum luojishanense* J. R. Shao in Liangshan prefecture and *F. hailuogouense* J. R. Shao, M. L. Zhou, and Q. Zhang in Ganzi Prefecture, were found by Hou et al. (2015) and Zhou et al. (2015), respectively. Until now, 27 buckwheat species have been named and reported in China, including two cultivated species and 25 wild species. They are widely distributed in Sichuan, Yunnan, Guizhou, and Tibet, in the southwest of China, because of the complicated geographical environments of these regions, which have been described as the treasure of plant resources.

CONCLUDING REMARKS

Chinese buckwheat germplasm resources are so plentiful. Almost all of the proved species of *Fagopyrum* Mill. in the world can be found in China, especially in the southwest, such as Yunnan, Sichuan, and Guizhou Provinces. These provinces are the major distribution areas of these species, variants, and subspecies. Therefore buckwheat researchers are continuously focusing on these regions. In the past 30 years the results of research on the germplasm resources of buckwheat have been remarkable. The local varieties of cultivated buckwheat from the provinces have been collected and carefully stored. Especially in recent years a number of scientists have been devoted to the research of buckwheat phylogenetic botany, and during the past 10 years more than 10 new species and subspecies of wild buckwheat have been found. After relentless study of these species, the origin and relationship of cultivated buckwheat have gradually become clearer. By the work of deeply evaluating, exploring the unique character and breeding new variety using preeminent material of wild buckwheat species through modern biotechnology, great progress can be made. It is believed that with the efforts of researchers on buckwheat, the collection of buckwheat resources will intensify and research methods will be continually modernized. There will be newer species discovered and the status of buckwheat in phylogeny will be confirmed. In the future, the relationship between buckwheat species and their genetic diversity will be clarified, and there will be a sizable breakthrough in the area of breeding new varieties and exploiting new buckwheat resources.

ACKNOWLEDGMENTS

This research was supported by the Key Project of Science and Technology of Sichuan, China (grant no. 04NG001-015, "Protection and exploitation of wild-type buckwheat germplasm resource").

REFERENCES

Chen, Q.F., 1999. A study of resources of *Fagopyrum* (Polygonaceae) native to China. Bot. J. Linn. Soc. 130, 53–64.

Gross, M.H., 1913. Remarques sur les Polygonees de l'Asie Orientale. Bulletin de l'Académie internationale de Géographie Botanique 23, 7–32.

Hou, L.L., Zhou, M.L., Zhang, Q., Qi, L.P., Yang, X.B., Tang, Y., Zhu, X.M., Shao, J.R., 2015. *Fagopyrum luojishanense*, a new species of Polygonaceae from Sichuan, China. Novon 24 (1), 22–26.

Li, A.R., 1998. Flora of China, Polygonaceae, vol. 25, Science Press, Beijing, pp. 108–117.

Liu, J.L., Tang, Y., Xia, M.Z., Shao, J.R., Cai, G.Z., Luo, Q., Sun, J.X., 2008. *Fagopyrum crispatofolium* J. L. Liu, a new species of Polygonaceae from Sichuan, China. J. System. Evol. 46 (6), 929–932.

Nakai, T., 1926. A new classification of Linnean *Polygonum*. Rigakkai 24, 289–301, (in Japanese).

Ohnishi, O., 1991. Discovery of the wild ancestor of common buckwheat. Fagopyrum 11, 5–10.

Ohnishi, O., 1995. Discovery of new *Fagopyrum* species and its implication for the studies of evolution of *Fagopyrum* and of the origin of cultivated buckwheat. In: Matano, T., Ujihara, A. (eds.), Current Advances in Buckwheat Research, vol. I–III. Proceedings of the Sixth International Symposium on Buckwheat in Shinshu, Shinshu University Press, August 24–29. pp. 175–190.

Ohnishi, O., Matsuoka, Y., 1996. Search for the wild ancestor of buckwheat. II. Taxonomy of *Fagopyrum* (Polygonaceae) species based on morphology, isozymes and cpDNA variability. Genes Genet. Syst. 71 (6), 383–390.

Ohsako, T., Ohnishi, O., 1998. New *Fagopyrum* species revealed by morphological and molecular analyses. Genes Genet. Syst. 73 (2), 85–94.

Ohsako, T., Ohnishi, O., 2000. Intra and inter-specific phylogeny of the wild *Fagopyrum* (Polygonaceae) species based on nucleotide sequences of noncoding regions of chloroplast DNA. Amer. J. Bot. 87 (4), 573–582.

Ohsako, T., Yamane, K., Ohnishi, O., 2002. Two new *Fagopyrum* (polygonaceae) species *F. gracilipedoides* and *F. jinshaense* from Yunnan, China. Genes Genet. Syst. 77, 399–408.

Shao, J.R., Zhou, M.L., Zhu, X.M., Wang, D.Z., Bai, D.Q., 2011. *Fagopyrum wenchuanense* and *Fagopyrum qiangcai*, two new species of Polygonaceae from Sichuan, China. Novon 21, 256–261.

Steward, A.N., 1930. The Polygoneae of eastern Asia. Gray Herbarium 88, 1–129.

Tang, Y., Zhou, M.L., Bai, D.Q., Shao, J.R., Zhu, X.M., Wang, D.Z., Tang, Y.X., 2010. *Fagopyrum pugense* (Polygonaceae), a new species from Sichuan, China. Novon 20, 239–242.

Yang, K.L., 1992. Research on cultivated buckwheat germplasm resources in China. Proc. 5th Int. Symp on Buckwheat at Taiyuan, China. Agricultural Publishing House, pp. 55–59.

Yang, K.L., Lu, D.B., 1992. The quality appraisal of buckwheat germplasm resources in China. Proc. 5th Int. Symp on Buckwheat at Taiyuan, China. Agricultural Publishing House, pp. 90–97.

Zhou, M.L., Bai, D.Q., Tang, Y., Zhu, X.M., Shao, J.R., 2012. Genetic diversity of four new species related to southwestern Sichuan buckwheats as revealed by karyotype ISSR and allozyme characterization. Plant System. Evol. 298 (4), 751–759.

Zhou, M.L., Zhang, Q., Zheng, Y.D., Tang, Y., Li, F.L., Zhu, X.M., Shao, J.R., 2015. *Fagopyrum hailuogouense* (Polygonacee), one new species of Polygonaceae from Sichuan, China. Novon 24 (2), 222–224.

Chapter three

Concepts, Prospects, and Potentiality in Buckwheat (*Fagopyrum esculentum* Moench): A Research Perspective

S.-H. Woo*, S.K. Roy*, S.J. Kwon*, S.-W. Cho**, K. Sarker*,
M.-S. Lee†, K.-Y. Chung‡, H.-H. Kim§

*Department of Crop Science, Chungbuk National University, Cheong-ju, Korea; **Division of Rice Research, National Institute of Crop Science, Rural Development Administration, Suwon, Korea; †Department of Industrial Plant Science & Technology, Chungbuk National University, Cheong-ju, Korea; ‡Department of Environmental & Biological Chemistry, Chungbuk National University, Cheong-ju, Korea; §Department of Food Nutrition and Cookery, Woosong College, Daejeon, Korea

INTRODUCTION

Common buckwheat, *Fagopyrum esculentum* Moench, is one of the oldest domesticated crops of Asia. It has been a popular grain and forage crop for centuries and it has also received attention around the world for its enhanced taste (Hagels, 1999). It is an outcrossing, self/cross-incompatible species belonging to the Polygonaceae family. The origin of buckwheat domestication is thought to be the eastern Tibetan plateau, bordering the Chinese province of Yunnan (Gondola and Papp, 2010; Ohnishi, 1998a,b). Previous studies also have provided strong evidence for the hypothesis that the center of origin for buckwheat is the Jinsha River basin (Chinese western region) (An-Hu et al., 2008; Ohnishi and Yasui, 1998). However, the buckwheat plant is grown widely in Asia, the United States, Canada, Russia, and East Europe and its utilization for human consumption is similar to that of cereals. Moreover, China, Russian Federation, Ukraine, and Kazakhstan are the main producers of common buckwheat (Campbell, 1995)

Molecular Breeding and Nutritional Aspects of Buckwheat. http://dx.doi.org/10.1016/B978-0-12-803692-1.00003-1

whereas the major exporters are China, Brazil, France, the United States, and Canada. On the other hand, Japan accounts for almost all of the world's buckwheat imports.

The seeds of buckwheat are rich in high-quality proteins, containing a well-balanced amount of essential amino acids, with a lysine content above 5% (Javornik et al., 1981). In addition, common buckwheat is also important as a nectariferous and pharmaceutical plant because of its excellent nutritional value.

Buckwheat production is greatly affected by low seed set and resulting low grain yield globally. This problem is confined to the high losses of grains during the harvest and thrashing as a result of shattering (Alekseeva and Malikov, 1992; Fesenko, 1986; Oba et al., 1998; Wang and Campbell, 1998). The several attempts toward the genetic improvement of buckwheat by means of biotechnology techniques through plant tissue and cell culture revealed that hypocotyl segments of common buckwheat and *Fagopyrum homotropicum* were capable of regeneration by somatic embryogenesis and organogenesis (Lachmann, 1991; Woo and Adachi, 1997; Woo et al., 1998). However, certain research hypothesizes that the main obstacles in buckwheat breeding might include its very strong self/cross-incompatibility and its indeterminate type of growth and flowering. To this end, modern biotechnology has paved the way to address these problems in a novel way (Neskovic et al., 1995). Several attempts were performed to investigate the in vitro regeneration of buckwheat from explants such as hypocotyls (Lachmann and Adachi, 1990; Yamane, 1974); cotyledons (Miljus-Djukic et al., 1992; Srejovic and Neskovic, 1981); immature inflorescence (Takahata, 1988); and anthers (Adachi et al., 1989; Bohanec et al., 1993).

However, programs aimed at buckwheat genetic and breeding improvement are further fraught with numerous problems. The self-incompatibility phenomenon is an inherited outbreeding mechanism that governs plants to prevent self-fertilization by discriminating between the self- and nonself-pollen grains (Miljus-Đukic et al., 2004). However, these breeding barriers have in no small measures contributed to the intractable difficulty and recalcitrance of buckwheat to respond to conventional improvement techniques. For instance, the transfer of a valuable genetic trait like self-fertility and another important agronomic character like productivity have so far proved abortive because of cross-incompatibility (Neskovic et al., 1995).

Buckwheat is still produced in significant quantities in many parts of the world. Crop improvement is now being addressed in several countries through the collection and evaluation of germplasm. This chapter reviews the current knowledge regarding the basic concepts and advances in the development of biotechnological and genetic perspectives of buckwheat. This review may assist the researcher by providing the current status of buckwheat.

BASIC CONCEPT OF BUCKWHEAT

Buckwheat is a moisture-loving grain that is thought to be one of the most versatile crops for forage and food, and has several benefits for human health. Buckwheat is a plant cultivated for its grain-like seeds and it is derived from the

Anglo-Saxon *boc* (beech) and *whoet* (wheat) because the word beech was used since the fruit of the plant was similar to that of the beechnut (Edwardson, 1995). Many species of buckwheat are grown around the world. Buckwheat is basically classified as common buckwheat, perennial buckwheat, and Tartary buckwheat. Buckwheat belongs to the genus *Fagopyrum* under the family Polygonaceae, which consists of two monophyletic groups: *cymosum* and *urophyllum* (Sangma and Chrungoo, 2010).

Common buckwheat (*F. esculentum* Moench) has been cultivated in all corners of the world because it can easily survive and grows well, even under adverse environmental conditions (Niroula et al., 2006).

Origin and Domestication History of Buckwheat

Common buckwheat is one of the oldest domesticated crops from Asia. The origin of its domestication is thought to be in Central and Western China from a wild Asian species (*Fagopyrum cymosum*) (Ohnishi, 1988, 1995, 1998a; Ohnishi and Konishi, 2001). These authors also indicate that the species *F. esculentum* spp. *ancestrale* is the wild ancestor of common buckwheat. Cultivated buckwheat was introduced to Asian countries from Southern China via two routes (Murai and Ohnishi, 1996); the first route crossed the Himalayan region and Tibet, and the second route ended up in Japan via Northern China. Archeological evidence suggests that buckwheat was believed to be introduced into Japan about 3000 years ago (Nagatomo, 1984). Buckwheat had already been cultivated extensively as a catch crop when it first appeared in records in Japan in the 8th century. Previous studies (Nagatomo, 1984; Ohnishi, 1995) suggest that buckwheat was introduced to Japan via the Korean peninsula from Northern China. Buckwheat cultivation in Japan dates back to as early as 6600 BP in the early Jōmon period (7000–5000 years BP) (Tsukada et al., 1986).

Present Status of Production

Buckwheat has been cultivated in China for 1000 years and was brought to Europe during the Middle Ages (Gondola and Papp, 2010; Murai and Ohnishi, 1996). Nevertheless, cultivation in Europe became familiar only in the Early Middle Ages. However, cultivation prevailed in Europe in the 14th to 15th centuries (Gondola and Papp, 2010; Ohnishi, 1993).

The leading producers of buckwheat are China, Russian Federation, Ukraine, and Kazakhstan (Campbell, 1995) whereas the major exporters are China, Brazil, France, the United States, and Canada (Fig. 3.1). In addition, Japan is considered to be familiar exporters in the world. On a global scale, the cultivation of buckwheat lagged behind 44 times that of wheat cultivation in 1961 (FAOSTAT, 2012).

Plant Description

Buckwheat is an annual plant, characterized by large heart-shaped leaves, growing 0.6–1.3 m tall, with reddish stems and flowers ranging in color from white

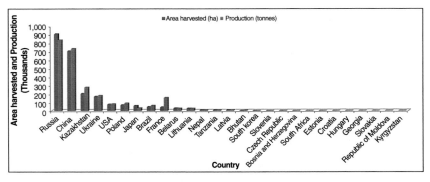

FIGURE 3.1 FAO world production estimates (2013).

to pink. Buckwheat has a shallow tap root system, with numerous laterals extending to 3–4 ft. in depth. Flowers can be white or white tinged with pink. The flowers of *F. esculentum* are perfect, but incomplete. They have no petals, but the calyx is composed of five petal-like sepals that are usually white, pink, or dark pink. The flowers are showy and densely clustered in racemes at the ends of the branches or on short pedicles that arise from the axils of the leaves (Fig. 3.2). This species is dimorphic, having plants bearing one of two flower types. The pin flowers have long pistils and short stamens while the thrum flowers have short pistils and long stamens. Flowers with pistils and stamens of similar length (Esser, 1953; Marshall, 1969) and lines with only one floral type (Fesenko and Antonov, 1973; Marshall, 1969) have been reported. The pistil consists of a one-celled superior ovary and a three-part style with a knob-like stigma and is surrounded by eight stamens.

PROGRESS OF THE BIOTECHNOLOGICAL APPROACH FOR THE IMPROVEMENT OF BUCKWHEAT

Plant tissue culture is used widely in plant science; it also has a number of commercial applications. For buckwheat, which has immense nutraceutical importance, tissue culture assumes greater significance for in vitro production of important plant metabolites.

During the last 20 years, the application of modern biotechnology in buckwheat research has attracted the concern of scientists. In this section, in vitro regeneration of buckwheat has been reported from explants such as hypocotyls, cotyledons, immature inflorescence, and anthers. Somatic embryogenesis including cultures of immature embryos of common buckwheat and of Tartary buckwheat, and plant regeneration of callus cultures from cotyledon segments of the cultivated buckwheat species *F. esculentum*, have been focused on and the several techniques have been studied related to the biotechnological approaches in buckwheat. However, in this section only a small proportion is discussed. More detailed information is listed in Table 3.1.

Common buckwheat   Tartary buckwheat

FIGURE 3.2 Morphological differences between common and Tartary buckwheat.

Advances of Callus Culture and Organogenesis

Vegetative propagation of many economic plants has so far been achieved through tissue culture methods (Murashige and Skoog, 1977). Previous results showed that isolated hypocotyls and cotyledons of buckwheat could be induced to develop calluses, with the capacity for organogenesis and restoration of plantlets (Yamane, 1974). In vitro regeneration of buckwheat has been reported from explants such as hypocotyls (Lachmann and Adachi, 1990; Yamane, 1974); cotyledons (Miljus-Djukic et al., 1992; Srejovic and Neskovic, 1981); immature inflorescence (Takahata, 1988); and anthers (Adachi et al., 1989; Bohanec et al., 1993). Moreover, research was conducted on common buckwheat (*F. esculentum* Moench) to develop a plant regeneration system for future application of genetic transformation (Woo et al., 2000b).

Progress on Somatic Embryogenesis

Somatic embryogenesis is a procedure whereby a cell or group of cells from somatic tissue forms an embryo. The expansion of somatic embryos closely

TABLE 3.1 List of Tissue Culture Studies Applied in the Buckwheat Plant

SI No.	Tissue Culture Studies	Medium Used	Species/Cultivar	Growth Conditions	Cultured Organ/Tissue	References
Shoot organogenesis						
1	An efficient protocol for shoot organogenesis and plant regeneration of buckwheat	MS medium containing 4.0 mg/L BAP	F. esculentum Moench	16-h photoperiod, 25°C	Shoot	Lee et al. (2009)
Cotyledon organogenesis						
2	Regeneration of plants from cotyledon tissue of common buckwheat	MS medium Sucrose 3%, pH 5.7	F. esculentum Moench	16-h photoperiod, 25°C	Cotyledon tissue	Woo et al. (2000)
3	Regeneration of plants from cotyledon fragments of buckwheat	High auxin (2,4-D) and low cytokinin (kinetin) content	F. esculentum Moench	16-h photoperiod, 25°C	Cotyledon fragments	Srejovic and Neskovic (1981)
Protoplast culture						
4	Plant regeneration from protoplast of common buckwheat	MS medium	F. esculentum	16-h photoperiod, 25°C	Protoplast	Adachi et al. (1989)
5	Callus regeneration from hypocotyl protoplast of Tartary buckwheat	MS medium Sucrose 3%, mannitol 0.5 M, 5 mM $CaCl_2$, pH 5.8	F. tataricum Gaertn	Dark for 4 weeks followed by 16-h photoperiod, 25°C	Hypocotyl protoplast	Lachmann and Adachi (1990)
Callus culture						
6	Induction of diploid-restored plants from callus of buckwheat	White's basal medium pH 5.6	F. esculentum Moench	Dark (transferred to fresh media after every 14–21 days), 23°C	Callus	Yamane (1974)

No.	Title	Medium	Species	Culture conditions	Explant	Reference
7	High-frequency plant regeneration of common buckwheat	MS medium pH 5.7 for culturing (1/2 MS medium containing 1.0 mg/L NAA, 3% sucrose	*F. esculentum*	16-h photoperiod, 25°C	Callus	Qin (2006)
Immature inflorescence						
8	Plant regeneration from immature inflorescence	B5 media pH 5.8	*F. esculentum* Moench	16-h photoperiod, 25°C	Immature inflorescence	Takahata (1988)
Somatic embryogenesis						
9	Plantlet regeneration via somatic embryogenesis from hypocotyls of common buckwheat (*F. esculentum* Moench)	2,4-Dichlorophenoxyacetic acid (2,4-D) 2.0 mg/L, BA 1.0 mg/L, and 3% sucrose	*F. esculentum* Moench	Diffused light (~1200 lux) at 24°C	Hypocotyl segments	Kwon et al. (2013)
10	Regeneration of plantlet via somatic embryogenesis from hypocotyls of Tartary buckwheat (*F. tataricum*)	MS supplemented with BA, 2,4-D, kinetin, 2-isopentenyl adenine, zeatin, zeatin riboside, thidiazuron	*F. tataricum*	Diffused light (~1200 lux) at 24°C	Hypocotyls	Han et al. (2011)
11	Somatic embryogenesis in common buckwheat by use of explants from hypocotyls of young seedlings	MS	*F. esculentum*	16-h photoperiod, 25°C	Hypocotyl seedling	Gumerova et al. (2001)
12	Somatic embryogenesis and bud formation from immature embryos of buckwheat (*F. esculentum* Moench)	B5 salt solution supplemented with 2,4-dichlorophenoxyacetic acid and kinetin	*F. esculentum* Moench	16-h photoperiod, 25°C	Immature embryos	Neskovic et al. (1987)

BA, benzyladenine; MS, Murashige and Skoog; BAP, 6-benzyl aminopurine.

replicates the process of zygotic embryo formation. Somatic embryogenesis mostly occurs indirectly via a superseding callus phase or directly from early explants (Han et al., 2011).

This study of somatic embryogenesis and plant regeneration of callus cultures from cotyledon segments of the cultivated buckwheat species *F. esculentum* differs from existing studies in the growth regulator combinations used. Somatic embryogenesis has previously been reported in cultures of immature embryos of common buckwheat (Neskovic et al., 1987; Rumyantseva et al., 1989) and Tartary buckwheat (Lachmann and Adachi, 1990; Rumyantseva et al., 1989). In addition, attempts have been conducted to establish a vegetable propagation system from immature inflorescence of common buckwheat (*F. esculentum*) and perennial buckwheat (*F. cymosum*) (Takahata, 1988).

Research on somatic embryogenesis and plant regeneration has been carried out using hypocotyl segments as explant of the cultivated buckwheat species *F. esculentum* to develop an efficient protocol for plant regeneration for common buckwheat, and to apply for future genetic transformation (Kwon et al., 2013).

In vitro tissue cultures of buckwheat serve as an important means for its improvement through genetic transformation as well as induced somaclonal variation. Somatic embryos have also proved to be excellent material for genetic transformation studies because of their competency in expressing incorporated DNA. However, previous research reported the effect of plant growth regulators on the culture of hypocotyls of Tartary buckwheat to develop the efficient protocol for callus induction and plantlet regeneration (Han et al., 2011). Approach towards the genetic improvement of buckwheat by means of biotechnology techniques through plant tissue and cell culture discovered that hypocotyl segments of common buckwheat and *F. homotropicum* were capable of regeneration by somatic embryogenesis and organogenesis (Lachmann, 1991; Woo and Adachi, 1997; Woo et al., 1998).

Somatic Hybridization via Protoplast Culture

Regeneration from protoplasts is one prerequisite for the successful use of somatic hybridization. Several attempts have been made to use protoplast fusion among *Fagopyrum* species, whereas the most successful attempts have been attained in plant regeneration from protoplasts in common buckwheat (*F. esculentum*) (Adachi et al., 1989; Gurnerova, 1991) but fusion experiments have been less successful (Lachmann and Adachi, 1990).

Although plant regeneration from protoplasts is still an obstacle in buckwheat research, protoplast fusion and culture of fusion products could be greatly improved. However, the isolation and culture of protoplasts from common buckwheat were described (Rumyanzeva and Lozovaya, 1988) and callus regeneration from Tartary buckwheat protoplasts (Lachmann, 1991; Lachmann and Adachi, 1990) was reported. To produce somatic hybrids between common and Tartary buckwheat, mesophyll protoplasts of *F. esculentum* were fused by polyethylene glycol-mediated fusion with hypocotyl protoplasts of *Fagopyrum tataricum*, serving as the hauler (Lachmanni et al., 1994). The hybrid nature of calli obtained was verified by restriction fragment length polymorphism analysis. The callus clones analyzed predominantly expressed one partial nuclear DNA fragment, as well as the presence of fragments of the other parent in low proportions and novel bands. All the calli analyzed carried one parental circulating tumor DNA type. The fusion experiments were aimed at the elaboration of simple procedures for overcoming breeding barriers in buckwheat (Lachmanni et al., 1994).

In Vitro Fertilization Through Protoplast Culture

Buckwheat (*F. esculentum* Moench) is a heterostylous plant displaying a sporophytic heteromorphic type of self-incompatibility. It is well known that self-incompatibility in common buckwheat (*F. esculentum* Moench) is a major barrier to the improvement of buckwheat through conventional breeding methods (Marshall and Pomeranz, 1982). Therefore in vitro fertilization could be useful to overcome the problem of self-incompatibility. Several research groups have developed different techniques for isolating female gametoplasts from a variety of plant species. In maize, fusion products of egg and sperm cells have been obtained and cultured in vitro (Kranz et al., 1991; Kranz and Lorz, 1993).

However, several studies have been conducted to develop a potential technique for the isolation of viable protoplasts from egg cells (Woo et al., 1999) so that egg cells of buckwheat can be fused with somatic protoplasts and other gametoplasts because of their biological function; a technique has also been developed for the isolation of viable protoplasts from sperm cells (Woo et al., 2000a). Consequently, artificial zygotes have been produced by in vitro fusion of male and female gametoplasts (Woo et al., 1999, 2000a).

Induction of Haploid Through Anther and Ovule Culture

During the last 20 years, haploid plants have been induced by in vitro culture techniques in many plant species, but application of this technique has not yet been developed for several agriculturally important species, though the potential of haploid (dihaploid) genotypes for buckwheat genetic and breeding studies has already been mentioned in the previously described studies (Adachi et al., 1988; Neskovic et al., 1986).

The studies on ovule culture for haploidy induction were initiated to determine the possibility of inducing a higher percentage of regenerants via gynogenesis than via androgenesis (Bohanec et al., 1993). Previous results speculated that gynogenetic regenerants will have fewer genetic aberrations than androgenetic regenerants, which was observed in *Nicotiana* (Wernsman, 1992; Kumashiro and Oinuma, 1985).

However, haploid induction via culture of unpollinated ovules was performed previously (Bohanec, 1995) whereas most of the studies were focused on determining the optimal medium for stable induction and regeneration of haploid plants.

BREEDING ADVANCES IN BUCKWHEAT

Programs aimed at buckwheat genetic and breeding improvement are further fraught with numerous problems. Prominent among them is the self-incompatibility phenomenon peculiar to the reproductive biology of this seed-propagated genus. Other factors hampering effective breeding efforts are apomixis, sterility, low seed set, and shattering, which are determined by genetic and agroecological interactive systems. These breeding barriers have in no small measures contributed to the intractable difficulty and recalcitrance of buckwheat to respond to conventional improvement techniques. However, the detailed breeding advances are listed in Tables 3.2–3.7.

Breeding Advances on Interspecific Hybridization

Interspecific hybridization has been and still is one of the useful methods for plant breeders to create new plant forms or to introduce genes from related wild species into crops of interest. Vegetative production of many economically important plants has been achieved through in vitro embryo rescue (Bhojwani and Razdan, 1983). Improvement of the agronomic characters of buckwheat could be speeded up by conventional breeding methods if interspecific hybrids were available. In this aspect, ovule culture has been very important, because hybrid seedlings in interspecific hybrids could be produced. Furthermore, the nature of breeding barriers has been reported in a cross between two diploid species, *F. esculentum* and *F. cymosum*, and developed an embryo rescue procedure to produce interspecific hybrids of self-pollinating buckwheat (Woo et al., 2002).

Improvement of Buckwheat Plant via Conventional Breeding Approaches

Self-incompatibility in buckwheat is a dimorphic, sporophytic type and therefore the production of seed is dependent on cross-pollination between "pin" (long pistil, short stamen) and "thrum" (short pistil, long stamen) flowers. In an earlier study, flower forms with reduced style length have been observed and self-fertile homomorphic lines have been developed (Marshall, 1969). However, many breeders have attempted the development of self-pollinating buckwheat as a means of increasing the ease of selection in buckwheat, and this allows for an extensive search for spontaneous recessive mutations that are normally hidden in the cross-pollinating form. The self-pollinating forms have been reported in the previous studies in buckwheat (Fesenko and Lokhatova, 1981; Marshall, 1969) whereas reports revealed that a study done on Zamyatkin's homostylous long-styled buckwheat form showed it to be a facultative cross-pollinator (Fesenko and Lokhatova, 1981). However, the homostylous form showed high self-compatibility and was capable of self-fertilization when individually isolated pollen from other plants of the same or heterostylous forms predominated. The polymorphism of buckwheat was increased using the induced mutants (Alekseeva, 1979).

TABLE 3.2 Advances of Marker-Assisted Breeding Program in Buckwheat Using Molecular Markers

Sl No.	Marker	Marker-Assisted Breeding	Objectives of the Study	Major Findings	References
1	RAPD	Characterization of interspecific hybrid between F. tataricum and F. esculentum	Study was conducted to analyze morphological and cytological characters and also DNA analysis	Hybrid was produced from interspecific crosses using ovule rescue method	Asaduzzaman et al. (2009)
2	SSR	Analysis of genetic diversity and population structure of buckwheat	To analyze the genetic variability, phylogenetic relationships, and population structure of buckwheat landraces of Korea using SSR markers	Results revealed genetic differentiation was low according to the geographic region because of outcrossing and self- incompatibility	Song et al. (2011)
3	RAPD	Population genetics of cultivated common buckwheat, F. esculentum Moench X. Diffusion routes revealed by RAPD markers	To compare RADP markers among the landraces all over the world, to construct phylogenetic trees of these landraces, and to assess the major diffusion routes of buckwheat cultivation in the world	Phylogenetic tress for landraces were constructed from RAPD variability by the unweighted pair group neighbor-joining methods	Murai and Ohnishi (1996)
4	RAPD	Genetic mapping of common buckwheat using DNA, protein, and morphological markers	Genetic mapping of F2 progeny of self-fertile hybrids between self-infertile Sobano plants with a long style and the wild-type homostylous homo pure line, which is self-fertile, to construct linkage maps of common buckwheat	Ten linkage groups were identified, involving 87 RAPD markers, 12 STS markers, four seed protein subunit (PS62/PS59, S49.8/PS51.4, PS44/ PS42.9, and PS39.9/PS37.8) markers, and three morphological alleles controlling homo/long style (H/s), shattering habit (Sht/sht), and acute/obtuse achene ridge (Ac/ac), covering a total of 655.2 cM	Pan and Chen (2010)

(Continued)

TABLE 3.2 Advances of Marker-Assisted Breeding Program in Buckwheat Using Molecular Markers (cont.)

Sl No.	Marker	Marker-Assisted Breeding	Objectives of the Study	Major Findings	References
5	RAPD	Origin of cultivated Tartary buckwheat (*F. tataricum* Gaertn.) revealed by RAPD analyses	To search for those individuals among wild subspecies that are genetically closely related to cultivated landraces by analyzing phylogenetic relationships among individuals of cultivated landraces and wild subspecies	The phylogenetic relationships among cultivated landraces and natural populations of wild subspecies of Tartary buckwheat were investigated at the individual level by constructing a phylogenetic tree based on RAPD markers. As the PCR templates, DNA of individuals rather than bulked samples was used	Tsuji and Ohnishi (2000)
6	RAPD	Identification of RAPD markers linked to the homostylar (Ho) gene in buckwheat	Attempts were made to determine molecular markers linked with the homostylar (Ho) gene to explore the basis of self-compatibility	F2 population was generated from an interspecific hybrid between *F. esculentum* and *F. homotropicum*	Aii et al. (1998)
7	RAPD	Species relationships in *Fagopyrum* revealed by PCR-based DNA fingerprinting	To study the species relationship between 28 different accessions that belong to 14 different species	Results revealed that *F. tataricum* is closer to its wild ancestor *F. tataricum* ssp. *potanini* Batalin, closely followed by *F. giganteum*-cultivated common buckwheat (*F. esculentum*) showed affinity with its putative wild ancestor *F. esculentum* ssp. *ancestrale* and the other closely related diploid species *F. homotropicum*	Sharma and Jana (2002b)
8	RAPD	RAPD variation in *F. tataricum* Gaertn. Accessions from China and Himalayan region	To determine the feasibility of using RAPD for diversity analysis in *F. tataricum* and accessions collected from different ecoregions on the basis of their genetic diversity and study the relatedness of wild ancestors *F. tataricum* ssp. *potanini* with cultivated Tartary buckwheat germplasm	The wild buckwheat accession did not group with any of the three cultivated Tartary buckwheat groups, and formed its own single-entry group	Sharma and Jana (2002a)

9	AFLP	Conversion of AFLP marker linked to the Sh allele at the S locus in buckwheat to a simple PCR-based marker	To find tightly linked markers in buckwheat homostylar locus concerned with cell compatibility	Two markers were confirmed to have been derived from a single region among the nine markers. Nucleotide sequence information from each flanking region of the two single locus markers was used to design region-specific primers for PCR amplification	Nagano et al. (2001)
10	AFLP	Identification of AFLP markers linked to nonseed shattering locus (Sht 1) in buckwheat and conversion to STS markers for marker-assisted selection	To detect molecular markers linked to Sht 1 locus, AFLP analysis was used in combination with bulked segregant analysis of segregating progeny of a cross between nonbrittle common buckwheat and brittle self-compatible line	Five AFLP markers linked to the Sht 1 locus (genes linked to brittle pedicle) in buckwheat were identified	Matsui et al. (2004a)
11	Microsatellites	Amplified fragment length polymorphism linkage analysis of common buckwheat (F. esculentum) and its wild self-pollinated relative F. homotropicum	To develop a large number of microsatellite markers in common buckwheat	An interspecific linkage map using F. esculentum and F. homotropicum was developed	Yasui et al. (2004)
12	SSR	Development of SSR markers for studies of diversity in the genus Fagopyrum	To develop 136 new SSR markers in F. esculentum ssp.	The use of SSRs showed consistent results as to using other marker systems	Ma et al. (2009)

RAPD, Random amplified polymorphic DNA; SSR, simple sequence repeat; PCR, polymer chain reaction; AFLP, amplified fragment length polymorphism.

TABLE 3.3 Advances of Interspecific Hybridization via Pollen and Anther-Related Breeding Program in Buckwheat

SI No.	Breeding Studies	Cultured Organ	Objectives	Major Research findings	References
1	Anther culture and androgenetic plant regeneration in buckwheat (*F. esculentum* Moench)	Anther	To develop a method for buckwheat anther culture, which could result in the possible induction of haploid or spontaneous diploid plants	Several abnormalities of pollen development in vitro were detected. Starch presence in pollen as a possible sign of androgenic capacity was studied. Microspores in uninucleate and early binucleate stages contained only proplastids, while in adult pollen grains a number of amyloplasts were present	Bohanec et al. (1993)
2	Gametophyte selection through pollen competition in buckwheat	Pollen	To explore the pollen competition that occurs in buckwheat, which can therefore have a significant influence on the genetic structure and vigor of populations	The results from these experiments are evidence that pollen grain competition can occur in buckwheat with benefits for progeny performance. It is also possible for selection among donors to occur through pollen competition if they differ in pollen tube growth rate. Pollen tubes from different donors did differ in growth rate. Pollen from different donors applied to different flowers on the same plant set the same proportion of seed, indicating that there was no selective abortion	Bjorkman (1995)
3	Genes outside the s supergene suppress s functions in buckwheat (*F. esculentum*)	Pollen	To clarify whether the locus controlling flower morphology and self-fertility of Pennline 10 is the same as that of KSC2, pollen tube tests and genetic analysis have been performed	The results suggest that Pennline 10 possesses the s allele as pin does, not an allele produced by the recombination in the S supergene, and that the short style length of Pennline 10 is controlled by multiple genes outside the S supergene	Matsui et al. (2004b)

#	Subject	Material	Objective	Findings	Reference
4	In vitro germination and viability of buckwheat (*F. esculentum* Moench) pollen	Pollen	The purpose of this study was to develop an in vitro germination technique for buckwheat pollen and utilize this method to determine the effect of temperature and flower age on pollen longevity	Maximum pollen viability was found 2 and 6 h after first light when plants were maintained at 25 and 20°C, respectively. Viability, as measured by germination percentage, was similar at both temperature regimes. Some pollen remained viable for approximately 34–38 h in intact flowers, but all pollen lost viability in less than an hour when stored at room temperature without humidity control	Adhikari and Campbell (1998)
5	Pollen tube behavior related to self-incompatibility in interspecific crosses of *Fagopyrum*		The purpose of this study is to assess the interspecific cross-incompatibility at the prezygotic stage with special reference to the self-incompatibility among the species of *Fagopyrum*, and also to analyze the evolution relationships between self-incompatible and self-compatible species of *Fagopyrum*	Unilateral incompatibility was observed among the self-incompatible and self-compatible species. However, the growth of the pollen tube with a longer style species was more active than that of the shorter style species. In addition, dimorphic self-incompatibility system influenced the re-cross-incompatibility and self-incompatibility	Hirose et al. (1995)
6	Inflorescence structure and control of flowering time and duration by light in buckwheat (*F. esculentum* Moench)	Pollen	The aim was to define whether the light effects could be related either to photosynthate availability or to photoperiodically regulated messages or to both. How morph type and inflorescence position on the shoot affect the flowering parameters is also reported	The number of inflorescences, and thus flowering duration, was also strongly reduced by short days. It was unaffected by light irradiance in 8-h days while, in 16-h days it was prolonged when light intensity was increased, suggesting the interaction of two different mechanisms for its regulation. Buckwheat is a distylous species, but inflorescence structure and flowering behavior were not affected by floral morph	Quinet et al. (2004)
7	Pollen–tube behavior and embryo development in interspecific crosses among the genus *Fagopyrum*	Pollen	This study was executed to identify highly compatible interspecific combinations and the optimal time for fertilization	The observed pollen tube elongation and the following embryo development showed that highly compatible pollinations were found to be crosses between *F. esculentum* × *F. cymosum* and between *F. esculentum* (thrum) × *F. homotropicum*	Woo et al. (2008)

TABLE 3.4 Advances of Interspecific Hybridization Through the Embryo Rescue Technique in Buckwheat

Sl No.	Breeding Studies	Species/Cultivar	Cultured Organ	Objectives	Major Research Findings	References
1	Interspecific buckwheat hybrid between F. esculentum and F. cymosum through embryo rescue	F. cymosum	Embryo	To explore the nature of breeding barriers in cross between two diploid species F. esculentum and F. cymosum	The comparative studies on embryo development in interspecific cross of F. esculentum × F. cymosum indicated that incompatibility caused the embryo abortion. All of the hybrid plants grew normally and produced flowers. However, almost all of the hybrid plants were male-sterile. The new hybrid plants may be useful for the improvement of self-pollinating buckwheat in future, if sterility can be overcome	Woo et al. (2002)
2	Interspecific hybrids of buckwheat (Fagopyrum spp.) regenerated through embryo rescue	Fagopyrum spp.	Embryo	To produce interspecific hybrids using Nepalese buckwheat species so as to standardize the in vitro breeding protocol and strengthen the mining of useful wild traits in Nepalese conditions	Ten interspecific hybrids, resulting from crosses between F. tataricum × F. esculentum and F. cymosum × F. esculentum were successfully produced. The highest crossability frequency (4%) was recorded in F. tataricum (h) × F. esculentum (t). Reciprocal differences and strong prefertilization barriers were recorded in some of the cross combinations. Further success in recovering interspecific hybrids needs improvement in embryo rescue technique and selection of compatible morphs	Niroula et al. (2006)
3	Possibility of interspecific hybridization by embryo rescue in the genus Fagopyrum	Fagopyrum	Embryo	To develop a new method to overcome the breeding barriers in the genus Fagopyrum by using embryo rescue and in vitro ovule culture	The genotype of the maternal plant in the cultivated species F. esculentum has relatively high capacity to allow foreign pollen penetration. The flower morphs disparity was observed in the pollen of Fagopyrum species and the endosperm deterioration in hybrid ovules was exhibited 5 days after pollination	Woo et al. (1995)

TABLE 3.5 Advances of the Studies of Self-Incompatibility in Buckwheat

Sl No.	Breeding Studies	Species/Cultivar	Objectives of the Study	Major Research Findings	References
1	A methodology for heterosis breeding of common buckwheat involving the use of the self-compatibility gene derived from *F. homotropicum*	*F. homotropicum*	To test a new methodology for heterosis breeding of buckwheat, based on inbreeding with the self-compatibility gene derived from *F. homotropicum*	This strategy successfully produced self-compatible homostyle hybrids (Shs); the percentage of hybrid plants among the progeny was more than 90%. In a greenhouse trial, the heterosis (best parent) in top dry weight, including stem and leaves, was 29.1–48.1% during the early growth stage. The seed yield of the progeny obtained in single-cross experiments was superior to that of an open-pollinated standard variety "Kitawasesoba" by 10% on average	Mukasa et al. (2010)
2	Detection of proteins possibly involved in self-incompatibility response in distylous buckwheat	*F. esculentum* Moench	An attempt to relate differences in morphological aspects of SI reaction with differences in protein profiles between the two styles	Results from one-dimensional gel electrophoresis revealed that short pistis 2 h after selfing contained a unique 50 kDa protein. In the two-dimensional electrophoresis, two distinct groups of proteins possibly involved in self-incompatibility response were detected in the short, and one in the long, pistis	Milijus-Đukic et al. (2004)
3	Heteromorphic incompatibility retained in self-compatible plants produced by a cross between common and wild buckwheat	*F. esculentum* and *F. homotropicum*	To investigate the self-compatible buckwheat derived from the wild buckwheat *F. homotropicum*, retained elements of the heterostylous incompatibility, and self-incompatibility in buckwheat	These results indicate that the self-compatibility allele, S^h, retains heteromorphic incompatibility and suggests that the S^h allele was derived from recombination in the S supergene	Matsui et al. (2003)

(Continued)

TABLE 3.5 Advances of the Studies of Self-Incompatibility in Buckwheat (cont.)

SI No.	Breeding Studies	Species/ Cultivar	Objectives of the Study	Major Research Findings	References
4	Production of self-compatible common buckwheat by ion exposure	*F. esculentum*	To overcome the self-incompatibility in common buckwheat, inducing mutation by irradiation, induced tetraploidy by colchicine treatment and interspecific crosses have been attempted; however, none of them were successful	The stable self-compatible plants were all thrum plants and this feature was dominantly inherited. Inbreeding depression occurred in selfed lines. In the local variety "Oono," albino plants segregating in the progeny of self-compatible plants were also obtained by ion exposure. However, the self-compatibility was not fixed in this instance	Nomura et al. (2002)
5	S-locus early flowering 3 is exclusively present in the genomes of short-styled buckwheat plants that exhibit heteromorphic self-incompatibility	*F. esculentum*	To determine the molecular basis of buckwheat heteromorphic self-incompatibility by identifying the primary factor(s) involved in this process. By integrating a variety of genetic and molecular approaches, including transcriptome analysis, mutagenesis screening, and evolutionary genetic analysis	By examining differentially expressed genes from the styles of the two floral morphs, a gene that is expressed only in short-styled plants was identified. The novel gene identified was completely linked to the S-locus in a linkage analysis of 1373 plants and had homology to EARLY FLOWERING 3. Furthermore, SELF3 was present in the genome of short-styled plants and absent from that of long-styled plants both in worldwide landraces of buckwheat and in two distantly related *Fagopyrum* species that exhibit heteromorphic SI	Yasui et al. (2012)
6	Treatment of isolated pistils with protease inhibitors overcomes the self-incompatibility response in buckwheat	*F. esculentum* Moench	To explore the effects of protease inhibitors on the self-incompatibility response in order to further "dissect" the self-incompatibility response in buckwheat and analyze the possible role of proteases in this complex process	Pistils were cross- or self-pollinated, and growth of pollen tubes was observed under a fluorescence microscope. Treatments with all inhibitors suppressed inhibition of self-pollen tube growth, suggesting that activity of proteases is involved in rejection of self-pollen during the SI response	Miljus-Dukic et al. (2007)

TABLE 3.6 Advances in the Studies of Heterostyle and Homostyle in BUCKWHEAT

SI No.	Breeding Studies	Species/Cultivar	Objectives	Major Research Findings	References
1	Breeding of a new autogamous buckwheat	*F. esculentum*	To investigate the inheritance pattern of the heterostyle and the homostyle genes in buckwheat using interspecific hybridization progenies in addition to heterostyle and homostyle genes, and a genetic model of self-compatible character	Homostyly appears to be controlled by a single dominant gene as F2 progeny segregated into a 3:1 (homorphic:pin type flower). Homomorphic F1 plants were selfed and all F2 experimental lines were segregated into a 3:1 ratio indicating a simple Mendelian inheritance. Only the F2 seed with the heteromorphic type segregated again in the F3 generation, with a ratio of 3:1. But the differences can be observed in nonsegregating progenies and segregating progenies among generations	Woo et al. (1998)
2	Genetic analysis of the heterostylar and the homostylar genes in buckwheat	*F. esculentum*	To explore the inheritance pattern of the heterostyle and homostyle gene in buckwheat using interspecific hybridization progenies and heterostyle and homostyle genes, marker-assisted selection of self-compatible character	The segregation pattern was studied in the F1, F2, and BCF1 generations. The F1 hybrids segregated into 1:1 (homomorphic plant:thrum-type flower) indicating a single dominant gene for homomorphism. Homomorphic F1 plants were selfed and all F2 experimental lines were segregated into a 3:1 ratio indicating a simple Mendelian inheritance. BCF1 lines segregated 1:1 ratio except 1 pin plant	Woo and Adachi (1997)

TABLE 3.7 Breeding Method Advances in Buckwheat

SI No.	Breeding Studies	Species/ Cultivar	Breeding Method	Objectives	Major Research Findings	References
1	Enhanced seed development in the progeny from the interspecific backcross (*F. esculentum* × *F. homotropicum*) × *F. esculentum*	*F. esculentum* and *F. homotropicum*	Backcross	The objective of this study is to transfer desirable agronomic traits from *F. homotropicum*, a wild annual species, into elite lines of the cultivated common buckwheat, *F. esculentum*	The F2 generation was more amenable than F1 hybrids to produce backcross progenies. Pollen tube growth of BCF1 × *F. esculentum* (thrum) and *F. homotropicum* × BCF1 was the disturbed penetration exceeded for all initial interspecific hybrids, and its requirement was proportionally lower when the common buckwheat was used as the recurrent parent and as the last parent of congruity hybrids. Growth of hybrid embryos before rescue, regeneration of mature hybrids all increased recurrent and congruity backcrosses and intercrosses between F1 plants and selected fertile plants of the second congruity backcrosses	Shin et al. (2009)
2	Cytogenetic studies on diploid and autotetraploid common buckwheat and their autotriploid and trisomics	*F. esculentum*	Cytogenetic	The main objective of this study was to develop autotriploid and trisomic common buckwheat and to report on cytogenetical findings that provide new information for buckwheat genetics and breeding	Crossing autotetraploid plants with diploid parent is an efficient way to produce autotriploid and trisomics. Siva had the highest pollen fertility (98.89%), followed by the autotetraploids Emka and Emka-1 (95.33 and 93.33%); the autotriploid hybrids had the lowest (31.28%). When autotriploid hybrids were backcrossed with diploid cultivar Sobano, 27.72% of the flowers pollinated yielding shriveled dead seeds, and only 3.08% yielded normal seeds	Ma et al. (2009)

Advances of Interspecific Hybridization via Pollen- and Anther-Related Breeding in Buckwheat

The inheritance of stylar morphology and the loss of self-incompatibility studies in the progenies of induced autotetraploid buckwheat indicated a terrible reduction of pollen fertility and decline in seed set because of the small amount of aneuploids and low frequency of multivalent formation in middle incompatibility (MI) (Adachi et al., 1983). Therefore grain yield has dramatically increased rather than the original diploid strain and some native varieties. Two types of homostyled variants (reduced style length in pin and increased style length) were observed in the progenies. Consequently, the development of pollen tube growth elongation, fertilization, and ovule in autogamous autotetraploid buckwheat revealed that all pollen tubes of illegitimate crosses did not reach the style base. Alternatively, legitimate crosses showed a high penetration ratio into the stylar base (Woo et al., 1995). Legitimate combinations in diploids result in 90% normal embryo development, both in summer and fall seasons. The results also suggest that low seed fertility is caused by sporophytic incompatibility caused by heterostyly, the defectiveness of embryo sac development, and fertilization failure and embryo abortion after fertilization (Woo et al., 1995).

Progress on Interspecific Hybridization Through the Embryo Rescue Technique in Buckwheat

Several attempts have been made to overcome breeding barriers by crossing common buckwheat with some wild relatives. Attempts have been successful in crossing (in vitro embryo rescue) between the *F. esculentum* and *F. cymosum* tetraploid levels (Hirose et al., 1993; Suvorova et al., 1994; Ujihara et al., 1990; Wang and Campbell, 1998; Woo et al., 1995). Based on pollen tube growth assay, similarity among *Fagopyrum* species is the greatest between *F. esculentum* and *F. cymosum*; *F. tataricum* is less related to *F. esculentum* and a group of six other *Fagopyrum* species is even more distant (Hirose et al., 1995). Therefore other interspecific crosses were more difficult to achieve. Such less successful study was, for instance, published previously (Samimy, 1991), which has made attempts for embryo rescue of crosses between *F. esculentum* (pin) and *F. tataricum*. But the reports on successful interspecific crosses with the help of the embryo rescue technique to solve this problem are absent in some trials (Wagatsuma and Un-no, 1995; Wang and Campbell, 1998).

Successful hybridization of Tartary buckwheat with common buckwheat has been accomplished by breeding techniques that cross the two species and then grow the resulting ovules on media at the diploid level. One of the main reasons for the utilization of *F. tataricum* as a parent in interspecific crosses has been the desire to transfer its desirable traits of higher seed yields, self-pollinating ability, frost resistance, and overall plant vigor. These traits, however, are present not only in this species, but in other species as well. The identification and classification of additional buckwheat species mentioned in an earlier study (Ohnishi, 1991) has now opened the door to increase the interspecific opportunities of buckwheat.

Several studies (Campbell, 1995; Woo and Adachi, 1997; Woo et al., 1995) have reported on the first successful interspecific hybridization of buckwheat at the diploid level in which the progeny is fertile, and further backcrosses to common buckwheat have been carried out. Conventional sexual hybridization at the diploid level has paved the way to open a new area for the improvement of common buckwheat. However, other attempts were carried out in buckwheat (*Fagopyrum* spp.) to standardize the in vitro breeding protocol by crossing three Nepalese species, namely, *F. tataricum*, *F. esculentum*, and *F. cymosum*, to produce interspecific hybrids (Niroula et al., 2006).

Advances of Self-Incompatibility Studies in Buckwheat

Self-incompatibility, caused by a condition whereby the plants have short filaments and anthers that are much below the stigmas, has been reported in *F. esculentum*. Additionally, self-incompatibility was thought to be restored by single dominant genes (Sharma and Boyes, 1961) or possibly by both single and double restorer genes (multiple genes or complex genes) (Woo and Adachi, 1997). However, self-incompatibility may help in the population improvement programs and be used to possibly exploit some of the nonadditive gene actions on characters' affective production.

Development of Molecular Marker-Assisted Breeding in Buckwheat

The development and use of molecular markers is one of the most significant issues for the detection and exploitation of DNA polymorphism in the field of molecular genetics (Cullis, 2002). Therefore the genetic research aspect is creating genetic maps that are useful to geneticists and plant breeders. In addition, DNA markers can be employed for genetic map construction, which helps in determining the chromosomal locations of genes affecting either simple or complex traits (Paterson et al., 1991); with these molecular methods, genetic maps of diploid plants can be developed more rapidly than those of polyploids. The origin of cultivated common buckwheat has been studied by the analysis of the diffusion routes of cultivated common buckwheat using random amplified polymorphic DNA (RAPD) markers (Murai and Ohnishi, 1996), by the determination of the allozyme diversity of the cultivated populations (Ohnishi, 1998b) and by the analysis of the genetic relationships between cultivated populations and natural populations of wild common buckwheat using amplified fragment length polymorphism (AFLP) markers (Konishi et al., 2005). During the course of these studies, several issues that were difficult to solve by using allozymes as well as RAPD and AFLP markers were identified, for example, the determination of the center of genetic diversity, detailed genetic relationships among cultivated populations, and gene flow between cultivated populations and natural populations of wild common buckwheat. If more powerful markers became available, it would be possible to address these issues, so that studies on the origin of cultivated common buckwheat could progress more rapidly. Using RAPD a linkage map was constructed in buckwheat (Aii et al., 1998). The RAPD map so generated is relatively dense, with a 0.6-cm distance between markers. The mapping of this gene represents a first step toward better understanding of the sexual reproduction system in buckwheat. These markers should be useful for marker-assisted selection for introgression of self-compatibility into common buckwheat. Also, using molecular markers, previous studies (Aii et al., 1998) have cloned and sequenced three RAPD fragments tightly linked to the S^h gene in buckwheat. Recent advances in molecular biology have offered a suite of molecular tools for assessing genetic diversity. Among the polymer chain reaction (PCR)-based techniques, AFLP (Vos et al., 1995) and simple sequence repeat (SSR) markers are widely used for studies of genetic diversity in crop species. Although AFLP markers have two weaknesses, a dominant mode of inheritance and the fact that homology is inferred from band comigration, they have several advantages, namely, the ability to resolve a large number of loci from a single reaction and the fact that no prior sequence information is required. The advantages of SSR markers are their codominant mode of inheritance and hypervariability, which make them ideal for a wide range of applications (Goldstein and Schlotterer, 1999). However, the development of an SSR assay requires the laborious processes of library construction, DNA sequencing, and primer synthesis. No SSR markers for common buckwheat were available previously. However, later attempts have developed using SSR markers for common buckwheat, and have observed a significant genetic differentiation among Japanese common buckwheat cultivars (Iwata et al., 2005). A study was performed (Nagano et al., 2001) using AFLP and bulked segregate analysis to develop molecular markers linked to the self-compatibility (S^h) gene. AFLP provides many genomic bands for buckwheat. Sixty-two bands specific to homomorphic flowered plants were detected. AFLP should be useful for marker-assisted selection for introgression of self-compatibility into common buckwheat and also for detailed analysis of plant genomes. To further enrich genetic resources of buckwheat, a bacterial artificial chromosome (BAC) library was constructed previously (Yasui et al., 2008). In many species a large insert genomic library, such as BAC, P1 artificial chromosome, or transformation competent artificial chromosome, has been indispensable, not only for genome analyses and functional studies of genes of interest, but also for studying the evolution of genome or genetic systems of organisms.

Progress on the Inheritance of Homostylar and Heterostylar Genes in Buckwheat

Buckwheat is a cross-pollinated, strong, self-incompatible, heterostyly plant with two flower forms, namely, pin (long styles, recessive homozygote—*ss*) and, thrum (short styles, heterozygote—*Ss*), borne on different plants; these features make the genetic improvement of buckwheat very difficult. It is thus not suitable for improvement by conventional plant-breeding methods. The majority of genotypes have an indeterminate growth habit (*DD*, *Dd*), and are governed by single genes. The gene for the determinant growth in Buckwheat has been reported earlier. The respective gene was demonstrated as monofactorial recessive inheritance (*dd*) (Fesenko, 1986). This inheritance is very good proof of a successful cross. Interspecific incompatibility often occurs unilaterally as a barrier that prevents self-incompatible species from accepting the pollen or the pollen tubes of self-compatible species. Self-incompatibility has been regarded as an important factor of interspecific cross-incompatibility in many cases. Regarding the genetic model for self-incompatibility, it was hypothesized that the self-incompatibility is controlled by the super gene *S* in *Primula* spp. (Sharma and Boyes, 1961). The authors analyzed the genetic model of loss of function of self-incompatibility in *F. esculentum*. It was already reported that interspecific cross-compatibility is closely related to the possibility of cross relationships in the genus *Fagopyrum*, and suggested that two groups (compatible and incompatible) and 5 degrees from highly compatible to completely incompatible groups had differentiated because of the development of a reproductive isolation mechanism (Woo et al., 1995). However, the genetic mechanism of the heterostyly gene in buckwheat has always remained a fascinating but unsolved puzzle. Some have even considered it too difficult for scientific research, because the direct evidence is long gone and we can only work by plausible inference. Although this is indeed a very difficult problem to approach experimentally, a number of striking observations over the years have allowed the formulation of plausible scenarios for the genetic nature of various important heterostyly genes. This strategy has been tested in homostyle common buckwheat by screening a single accession of the wild species, *F. homotropicum* Ohnishi and a cultivated common buckwheat *F. esculentum* Moench, for genes capable of increasing the yield and new homostyle appearance (Woo et al., 1998). A similar genetic explanation was reported on the development of self-compatible interspecific hybrids derived from *F. esculentum* (pin) and *F. homotropicum* (Campbell, 1995).

FUTURE PROSPECTS FOR THE IMPROVEMENT OF BUCKWHEAT

Crop improvement programs have emphasized the development of high yielding adapted cultivars containing very high amounts of the rutin and zero allergic protein. The feasibility of interrogation of desirable characteristics of other closely related species will also be addressed. The emphasis on seed yield will probably remain, and forage aspects of this crop are starting to demand more attention. Enhancing the nutritional value of the crop by reducing antinutritional factors will also receive more emphasis. Many of the crop improvement programs will add another dimension or direction to their present programs. In future research at Morden, Canada, for example, investigations will be conducted in four directions; breeding of reduced allergic protein lines with good agronomic characteristics will continue as well as crop improvement by self-pollinating and frost resistance for a longer growing season at less risk. These aspects are now increasingly being addressed by several programs.

In addition, molecular genetics have emerged as an applied research discipline during the past decade and promises to assist in the solution of many agricultural problems. Genetic engineering holds promise for the development of specific genotypes since it is based on identification characterization and transfer of specific genes into the recipient plant as compared to the mixing of two complete genomes of two parental lines followed by backcrossing for several generations to remove undesirable genes. Molecular genetic technologies can be employed to characterize the genes and identify important linkages, which could facilitate gene transfer to suitable agronomic types, genes for disease, insect resistance, frost resistance, reduced allergic protein concentrations, increased protein content, and other characters of agronomic importance that seem to be adapted to this approach. Interdisciplinary research efforts, now under way at several research establishments together with collaboration between institutes, promise to produce needed improvements in buckwheat in the near future.

REFERENCES

Adachi, T., Kawabata, K., Matsuzaki, N., Yabuya, T., Nagatomo, T., 1983. Observation of pollen tube elongation, fertilization and ovule development in autogamous autotetraploid buckwheat. Buckwheat Research, Proceedings of the 2nd International Symposium on Buckwheat, pp. 7–10.

Adachi, T., Suputtitada, S., Miike, Y., 1988. Plant regeneration from anther culture in common buckwheat (*Fagopyrum esculentum*). Fagopyrum 8, 5–9.

Adachi, T., Yamaguchi, A., Miike, Y., Hoffmann, F., 1989. Plant regeneration from protoplasts of common buckwheat (*Fagopyrum esculentum*). Plant Cell Rep. 8, 247–250.

Adhikari, K.N., Campbell, C.G., 1998. In vitro germination and viability of buckwheat (*Fagopyrum esculentum* Moench) pollen. Euphytica 102, 87–92.

Aii, J., Nagano, M., Penner, G., Campbell, C., Adachi, T., 1998. Identification of RAPD markers linked to the homostylar (Ho) gene in buckwheat [*Fagopyrum*]. Japan. J. Breed. (Japan) 48, 59–62.

Alekseeva, E., 1979. Use of induced mutants in the breeding of buckwheat. Soviet Agriculture Sciences (USA).

Alekseeva, E., Malikov, V., 1992. The results of buckwheat green floret form. Proceedings of the Fifth International Symposium on Buckwheat. Taiyuan, China, pp. 20–26.

An-Hu, W., Ming-Zhong, X., Guang-Ze, C., Ping, Y., 2008. The origin of cultivating buckwheat and the genetic analysis of the kindred species. Southwest China J. Agric. Sci. 21, 282–285.

Asaduzzaman, M., Minami, M., Matsushima, K., Nemoto, K., 2009. Characterization of interspecific hybrid between *F. tataricum* and *F. esculentum*. J. Biol. Sci. 9, 137–144.

Bhojwani, S.S., Razdan, M.K., 1983. Protoplast isolation and culture. Plant tissue culture: theory and practice. Elsevier, Amsterdam, pp. 237–260.

Bjorkman, T., 1995. Gametophyte selection through pollen competition in buckwheat. Curr. Adv. Buckwheat Res., 443–451.

Bohanec, B., 1995. Progress of buckwheat in vitro culture techniques with special aspect on induction of haploid plants. Curr. Adv. Buckwheat Res. 1, 205–209.

Bohanec, B., Nešković, M., Vujičić, R., 1993. Anther culture and androgenetic plant regeneration in buckwheat (*Fagopyrum esculentum* Moench). Plant Cell Tiss. Org. 35, 259–266.

Campbell, C., 1995. Inter-specific hybridization in the genus *Fagopyrum*. Proceedings of the 6th International Symposium on Buckwheat, pp. 255–263.

Cullis, C.A., 2002. The use of DNA polymorphisms in genetic mapping. Genetic EngineeringSpringer, New York, pp. 179–189.

Edwardson, S.E., 1995. Using growing degree days to estimate optimum windrowing time in Buckwheat. Curr. Adv. Buckwheat Res., 509–514.

Esser, K., 1953. Genome doubling and pollen tube growth in heterostylous plants. Z. für ind. Abstammungs und Vererbungslehre 85, 25–50.

FAOSTAT data, 2012. Online database. Available online: http://faostat.fao.org.

Fesenko, N., 1986. Buckwheat breeding for stable high yielding. Proceedings of the Third International Symposium on Buckwheat. pp. 99–107.

Fesenko, N., Antonov, V., 1973. New homostylous form of buckwheat. Plant Breed. Abstr., 10172.

Fesenko, N., Lokhatova, V., 1981. Self-compatibility of Zamyatkin's homostylous long-styled buckwheat form. Doklady Vsesoyuznoi Ordena Lenina Akademii Sel'skokhozyaistvennykh Nauk Imeni VI Lenina 5, 16–17.

Goldstein, D.B., Schlotterer, C., 1999. Microsatellites: Evolution and Applications. Oxford University Press, New York.

Gondola, I., Papp, P., 2010. Origin, geographical distribution and polygenic relationship of common buckwheat (*Fagopyrum esculentum* Moench). Eur. Plant Sci. Biotechnol. 4, 17–33.

Gumerova, E., Gatina, E., Chuenkova, S., Rumyantseva, N., 2001. Somatic Embryogenesis in Common Buckwheat *Fagopyrum esculentum* Moench. Proceedings of the Eighth International Conference on Systems Biology, pp. 377–381.

Gurnerova, E., 1991. Regeneration of plantlets from protoplasts of Fagopyrum esculentum Moench. Proceedings of the 8th International Protoplast Symposium. Uppsala, Sweden, Physiol. Plant, pp. pA19.

Hagels, H., 1999. *Fagopyrum esculentum* Moench. chemical review. Zbornik Biotehniske fakultete Univerze v Ljubljani, Slovenia.

Han, M.-H., Kamal, A.H.M., Huh, Y.-S., Jeon, A., Bae, J.S., Chung, K.-Y., Lee, M.-S., Park, S.-U., Jeong, H.-S., Woo, S.-H., 2011. Regeneration of plantlet via somatic embryogenesis from hypocotyls of Tartary buckwheat ("*Fagopyrum tataricum*"). Aust. J. Crop Sci. 5, 865–869.

Hirose, T., Ujihara, A., Kitabayashi, H., Minami, M., 1993. Morphology and identification by isozyme analysis of interspecific hybrids in buckwheats. Fagopyrum 13, 25–30.

Hirose, T., Ujihara, A., Kitabayashi, H., Minami, M., 1995. Pollen tube behavior related to self-incompatibility in interspecific crosses of *Fagopyrum*. Breed. Sci. 45, 65–70.

Iwata, H., Imon, K., Tsumura, Y., Ohsawa, R., 2005. Genetic diversity among Japanese indigenous common buckwheat (*Fagopyrum esculentum*) cultivars as determined from amplified fragment length polymorphism and simple sequence repeat markers and quantitative agronomic traits. Genome 48, 367–377.

Javornik, B., Eggum, B.O., Kreft, I., 1981. Studies on protein fractions and protein quality of buckwheat. Genetika 13, 115–121.

Konishi, T., Yasui, Y., Ohnishi, O., 2005. Original birthplace of cultivated common buckwheat inferred from genetic relationships among cultivated populations and natural populations of wild common buckwheat revealed by AFLP analysis. Genes Genet. Syst. 80, 113–119.

Kranz, E., Lorz, H., 1993. In vitro fertilization with isolated, single gametes results in zygotic embryogenesis and fertile maize plants. Plant Cell 5, 739–746.

Kranz, E., Bautor, J., Lörz, H., 1991. In vitro fertilization of single, isolated gametes of maize mediated by electrofusion. Sex. Plant Reprod. 4, 12–16.

Kumashiro, T., Oinuma, T., 1985. Comparison of genetic variability among anther-derived and ovule-derived doubled haploid lines of tobacco. Breed. Sci. 35, 301–310.

Kwon, S.-J., Han, M.-H., Huh, Y.-S., Roy, S.K., Lee, C.-W., Woo, S.-H., 2013. Plantlet regeneration via somatic embryogenesis from hypocotyls of common buckwheat (*Fagopyrum esculentum* Moench). Korean J. Crop Sci. 58, 331–335.

Lachmann, S., 1991. Plant cell and tissue culture in buckwheat: an approach towards genetic improvements by means of unconventional breeding techniques. Overcoming breeding barriers by means of plant biotechnology. Proceedings of the International College, Miyazaki, Japan 145, 154.

Lachmann, S., Adachi, T., 1990. Callus regeneration from hypocotyl protoplasts of Tartary buckwheat (*Fagopyrum tataricum* Gaertn). Fagopyrum 10, 62–64.

Lachmanni, S., Kishima, Y., Adachi, T., 1994. Protoplast fusion in buckwheat: preliminary results on somatic hybridization. Fagopyrum 14, 7–12.

Lee, S.Y., Kim, Y.K., Uddin, M.R., Park, N., Park, S.U., 2009. An efficient protocol for shoot organogenesis and plant regeneration of buckwheat (*Fagopyrum esculentum* Moench). Rom. Biotech. Lett. 14, 4524–4529.

Ma, K.-H., Kim, N.-S., Lee, G.-A., Lee, S.-Y., Lee, J.K., Yi, J.Y., Park, Y.-J., Kim, T.-S., Gwag, J.-G., Kwon, S.-J., 2009. Development of SSR markers for studies of diversity in the genus *Fagopyrum*. Theor. Appl. Genet. 119, 1247–1254.

Marshall, H., 1969. Isolation of self-fertile, homomorphic forms in buckwheat *Fagopyrum sagittatum* Gilib. Crop Sci. 9, 651–653.

Marshall, H., Pomeranz, Y., 1982. Buckwheat: description, breeding, production, and utilization. In: Pomeranz, Y. (ed), Advances in Cereal Science and Technology (USA). V St. Paul, Minnesota, p. 65.

Matsui, K., Tetsuka, T., Nishio, T., Hara, T., 2003. Heteromorphic incompatibility retained in self-compatible plants produced by a cross between common and wild buckwheat. New Phytol. 159, 701–708.

Matsui, K., Kiryu, Y., Komatsuda, T., Kurauchi, N., Ohtani, T., Tetsuka, T., 2004a. Identification of AFLP makers linked to non-seed shattering locus (sht1) in buckwheat and conversion to STS markers for marker-assisted selection. Genome 47, 469–474.

Matsui, K., Nishio, T., Tetsuka, T., 2004b. Genes outside the S supergene suppress S functions in buckwheat (*Fagopyrum esculentum*). Ann. Bot. 94, 805–809.

Miljus-Djukic, J., Neskovic, M., Ninkovic, S., Crkvenjakov, R., 1992. Agrobacterium-mediated transformation and plant regeneration of buckwheat (*Fagopyrum esculentum* Moench.). Plant Cell Tiss. Org. 29, 101–108.

Miljus-Đukic, J., Ninkovic, S., Radovic, S., Maksimovic, V., Brkljacic, J., Neskovic, M., 2004. Detection of proteins possibly involved in self-incompatibility response in distylous buckwheat. Biol. Plant. 48, 293–296.

Miljus-Dukic, J.D., Radovic, S.R., Maksimovic, V.R., 2007. Treatment of isolated pistils with protease inhibitors overcomes the self-incompatibility response in Buckwheat. Arch. Biol. Sci. 59 (1), 45–49.

Mukasa, Y., Suzuki, T., Honda, Y., 2010. A methodology for heterosis breeding of common buckwheat involving the use of the self-compatibility gene derived from *Fagopyrum homotropicum*. Euphytica 172, 207–214.

Murai, M., Ohnishi, O., 1996. Population genetics of cultivated common buckwheat *Fagopyrum esculentum* Moench. X. Diffusion routes revealed by RAPD markers. Genes Genet. Syst. 71, 211–218.

Murashige, T., Skoog, F., 1977. Manipulation of organ initiation in plant tissue cultures. Bot. Bull. Acad. Sin. 18, 1–24.

Nagano, M., Aii, J., Kuroda, M., Campbell, C., Adachi, T., 2001. Conversion of AFLP markers linked to the Sh allele at the S locus in buckwheat to a simple PCR based marker form. Plant Biotechnol. 18, 191–196.

Nagatomo, T., 1984. Soba-no-kagaku (Scientific considerations on buckwheat). Shincho-sha, Tokyo (in Japanese).

Neskovic, M., Srejovic, V., Vujicic, R., 1986. Buckwheat (*Fagopyrum esculentum* Moench.). Crops I. Springer, pp. 579–602.

Neskovic, M., Vujicic, R., Budimir, S., 1987. Somatic embryogenesis and bud formation from immature embryos of buckwheat (*Fagopyrum esculentum* Moench). Plant Cell Rep. 6, 423–426.

Neskovic, M., Ćulafić, L., Vujicic, R., 1995. Somatic embryogenesis in buckwheat (*Fagopyrum* Mill.) and sorrel (*Rumex* L.), Polygonaceae. Somatic Embryogenesis and Synthetic Seed II. Springer, pp. 412–427.

Niroula, R., Bimb, H., Sah, B., 2006. Interspecific hybrids of buckwheat (*Fagopyrum* spp.) regenerated through embryo rescue. Sci. World 4, 74.

Nomura, Y., Hatashita, M., Inoue, M., 2002. Production of self-compatible common buckwheat by ion exposure. Fagopyrum 19, 43–48.

Oba, S., Ohta, A., Fujimoto, F., 1998. Grain shattering habit of buckwheat. Proceedings of the VIIth International Symposium Buckwheat at Winnipeg, Canada, pp. 70–75.

Ohnishi, O., 1988. Population genetics of cultivated common buckwheat, *Fagopyrum esculentum* Moench. VII. Allozyme variability in Japan, Korea, and China. Japan. J. Genet. 63, 507–522.

Ohnishi, O., 1991. Discovery of the wild ancestor of common buckwheat. Fagopyrum 11, 5–10.

Ohnishi, O., 1993. Population genetics of cultivated common buckwheat *Fagopyrum esculentum* Moench. VIII. Local differentiation of land races in Europe and the silk road. Japan. J. Genet. 68, 303–316.

Ohnishi, O., 1995. Discovery of new *Fagopyrum* species and its implication for the studies of evolution of *Fagopyrum* and of the origin of cultivated buckwheat. Proceedings of the Sixth International Symposium on Buckwheat at Ina, pp. 175–190.

Ohnishi, O., 1998a. Search for the wild ancestor of buckwheat. III. The wild ancestor of cultivated common buckwheat, and of Tartary buckwheat. Econ. Bot. 52, 123–133.

Ohnishi, O., 1998b. Search for the wild ancestor of buckwheat. I. Description of new *Fagopyrum* (Polygonaceae) species and their distribution in China and Himalayan hills. Fagopyrum 15, 18–28.

Ohnishi, O., Konishi, T., 2001. Cultivated and wild buckwheat species in eastern Tibet. Fagopyrum 18, 3–8.

Ohnishi, O., Yasui, Y., 1998. Search for wild buckwheat species in high mountain regions of Yunnan and Sichuan provinces of China. Fagopyrum 15, 8–17.

Pan, S.J., Chen, Q.F., 2010. Genetic mapping of common buckwheat using DNA, protein and morphological markers. Hereditas 147, 27–33.

Paterson, A.H., Tanksley, S.D., Sorrells, M.E., 1991. DNA markers in plant improvement. Adv. Agron. 46, 39–90.

Qin, C.L.H.X.Z., 2006. Tissue culture and high-frequency plant regeneration of buckwheat (*Fagopyrum esculentum* Moench) [J]. J. Mol. Cell Biol. 5, 008.

Quinet, M., Cawoy, V., Lefèvre, I., Van Miegroet, F., Jacquemart, A.-L., Kinet, J.-M., 2004. Inflorescence structure and control of flowering time and duration by light in buckwheat (*Fagopyrum esculentum* Moench). J. Exp. Bot. 55, 1509–1517.

Rumyantseva, N., Sergeeva, N., Khakimova, L., Salnikov, V., Gumerova, E., Lozovaya, V., 1989. Organogenesis and somatic embryogenesis in tissue culture of two buckwheat species. Fiziol. Rast. 36, 187–194.

Rumyanzeva, N., Lozovaya, V., 1988. Isolation and culture of buckwheat (*Fagopyrum esculentum* Moench.) callus protoplasts. Prog. Plant Prot. Res. Springer, 45–46.

Samimy, C., 1991. Barrier to interspecific crossing of *Fagopyrum esculentum* with *Fagopyrum tataricum*. I. Site of pollen-tube arrest. II. Organogenesis from immature embryos of *F. tataricum*. Euphytica 54, 215–219.

Sangma, S., Chrungoo, N., 2010. Buckwheat gene pool: potentialities and drawbacks for use in crop improvement programmes. Eur. J. Plant Sci. Biotechnol. 4, 45–50.

Sharma, K., Boyes, J., 1961. Modified incompatibility of buckwheat following irradiation. Canadian J. Bot. 39, 1241–1246.

Sharma, T., Jana, S., 2002a. Random amplified polymorphic DNA (RAPD) variation in *Fagopyrum tataricum* Gaertn. accessions from China and the Himalayan region. Euphytica 127, 327–333.

Sharma, T., Jana, S., 2002b. Species relationships in *Fagopyrum* revealed by PCR-based DNA fingerprinting. Theor. Appl. Genet. 105, 306–312.

Shin, D.-H., Kamal, A., Yun, Y.-H., Bae, J.-S., Lee, Y.-S., Lee, M.-S., Chung, K.-Y., Woo, S.-H., 2009. Enhanced seed development in the progeny from the interspecific backcross (*Fagopyrum esculentum* × *F. homotropicum*) × *F. esculentum*. Korean J. Plant Resour. 22, 209–214.

Song, J.Y., Lee, G.-A., Yoon, M.-S., Ma, K.-H., Choi, Y.-M., Lee, J.-R., Jung, Y., Park, H.-J., Kim, C.-K., Lee, M.-C., 2011. Analysis of genetic diversity and population structure of buckwheat (*Fagopyrum esculentum* Moench) landraces of Korea using SSR markers. Korean J. Plant Resour. 24, 702–711.

Srejovic, V., Neskovic, M., 1981. Regeneration of plants from cotyledon fragments of buckwheat (*Fagopyrum esculentum* Moench). Z. Pflanzenphysiol. 104, 37–42.

Suvorova, G., Fesenko, N., Kostrubin, M., 1994. Obtaining of interspecific buckwheat hybrid (*Fagopyrum esculentum* Moench × *Fagopyrum cymosum* Meissn). Fagopyrum 14, 13–16.

Takahata, Y, 1988. Plant regeneration from cultured immature inflorescence of common buckwheat (*Fagopyrum esculentum* Moench) and perennial buckwheat (*F. cymosum* Meisn). Japan. J. Genet. 38, 409–413.

Tsuji, K., Ohnishi, O., 2000. Origin of cultivated Tartary buckwheat (*Fagopyrum tataricum* Gaertn) revealed by RAPD analyses. Genet. Resour. Crop Evol. 47, 431–438.

Tsukada, M., Sugita, S., Tsukada, Y., 1986. Oldest primitive agriculture and vegetational environments in Japan. Nature 322, 632–634.

Ujihara, A., Nakamura, Y., Minami, M., 1990. Interspecific hybridization in genus *Fagopyrum*—properties of hybrids (*F. esculentum* Moench × *F. cymosum* Meissner) through ovule culture. Gamma Field Symposia, pp. 45–53.

Vos, P., Hogers, R., Bleeker, M., Reijans, M., Van de Lee, T., Hornes, M., Friters, A., Pot, J., Paleman, J., Kuiper, M., 1995. AFLP: a new technique for DNA fingerprinting. Nucleic Acids Res. 23, 4407–4414.

Wagatsuma, T., Un-no, Y., 1995. In vitro culture of interspecific ovule between buckwheat (*F. esculentum*) and Tartary (*F. tataricum*). Breed. Sci. 45, 312.

Wang, Y., Campbell, C., 1998. Interspecific hybridization in buckwheat among *Fagopyrum esculentum*, *F. homotropicum*, and *F. tataricum* p. I: 1–13. Proceedings of Seventh International Symposium Buckwheat, Winnipeg, Canada, pp. 12–14.

Wernsman, E., 1992. Varied roles for the haploid sporophyte in plant improvement. Plant breeding in the 1990s. CAB International, pp. 461–484.

Woo, S., Adachi, T., 1997. Genetic analysis of the heterostylar and the homostylar genes in buckwheat. Korean J. Crop Sci. 29, 297–298.

Woo, S., Tsai, Q., Adachi, T., 1995. Possibility of interspecific hybridization by embryo rescue in the genus *Fagopyrum*. Curr. Adv. Buckwheat Res. 6, 225–237.

Woo, S.H., Adachi, T., Park, S.I., 1998. Breeding of a new autogamous buckwheat: 2 Seed protein analysis and identification of RAPD markers linked to the Ho (Sh) gene. Koean J. Plant Resour. 30, 144–145.

Woo, S.H., Adachi, T., Jong, S.K., Campbell, C.G., 1999. Isolation of protoplasts from viable egg cells of common buckwheat (*Fagopyrum esculentum* Moench). Canadian J. Plant Sci. 79, 593–595.

Woo, S.H., Adachi, T., Jong, S.K., Campbell, C.G., 2000a. Isolation of protoplasts from viable sperm cells of common buckwheat (*Fagopyrum esculentum* Moench). Canadian J. Plant Sci. 80, 583–585.

Woo, S.H., Nair, A., Adachi, T., Campbell, C.G., 2000b. Plant regeneration from cotyledon tissues of common buckwheat (*Fagopyrum esculentum* Moench). In Vitro Cell. Dev.-Pl. 36, 358–361.

Woo, S.H., Campbell, C.G., Jong, S.K., 2002. Interspecific buckwheat hybrid between *Fagopyrum esculentum* and *F. cymosum* through embryo rescue. Korean J. Breed. 34, 322–327.

Woo, S.-H., Kim, S.-H., Tsai, K.S., Chung, K.-Y., Jong, S.-K., Adachi, T., Choi, J.-S., 2008. Pollen-tube behavior and embryo development in interspecific crosses among the genus *Fagopyrum*. J. Plant Biol. 51, 302–310.

Yamane, Y., 1974. Induced differentiation of buckwheat plants from subcultured calluses in vitro. Japan. J. Genet 49, 139–146.

Yasui, Y., Wang, Y., Ohnishi, O., Campbell, C.G., 2004. Amplified fragment length polymorphism linkage analysis of common buckwheat (*Fagopyrum esculentum*) and its wild self-pollinated relative *Fagopyrum homotropicum*. Genome 47, 345–351.

Yasui, Y., Mori, M., Matsumoto, D., Ohnishi, O., Campbell, C.G., Ota, T., 2008. Construction of a BAC library for buckwheat genome research—an application to positional cloning of agriculturally valuable traits. Genes Genet. Syst. 83, 393–401.

Yasui, Y., Mori, M., Aii, J., Abe, T., Matsumoto, D., Sato, S., Hayashi, Y., Ohnishi, O., Ota, T., 2012. S-LOCUS EARLY FLOWERING 3 is exclusively present in the genomes of short-styled buckwheat plants that exhibit heteromorphic self-incompatibility. PLoS One 7, e31264.

Chapter four

Biological Resources and Selection Value of Species of *Fagopyrum* Mill. Genus in the Far East of Russia

A.G. Klykov, L.M. Moiseenko, Y.N. Barsukova

Primorsky Scientific Research Institute of Agriculture, Primorsky Krai, Russia

INTRODUCTION

Common buckwheat (*Fagopyrum esculentum*) is a cereal and melliferous crop, which is widely cultivated in many countries of the world. The main producers of buckwheat are China, Russia, and Ukraine. In some countries of Southeastern Asia (China, India), *Fagopyrum tataricum* and *Fagopyrum cymosum* are used as an edible and medicinal crop. Plants of species of *F. esculentum* are widely used in popular medicine. For medicine they use leaves and tops of shoots in blooming (Hinneburg and Neubert, 2005; Kreft et al., 2006). Representatives of the *Fagopyrum* genus are prospective resources of flavonoids. The main flavonoid is 3-O-rutinozid quercetin (rutin or vitamin P), which has antioxidant, angioprotective, antibacterial, and hepatoprotective traits.

Flavonoids are very important in plant resistance to changing environmental conditions, growth, development, reproduction, energy metabolism, disease and virus protection, and other vital functions of plants (Harborne, 1972; Zaprometov, 1993).

Modern living organisms and their habitats are under steady anthropogenic pressure.

Among many factors influencing populations, biocenosis and biota in general, it is necessary to distinguish the so-called "pollutants" of the environment.

Molecular Breeding and Nutritional Aspects of Buckwheat. http://dx.doi.org/10.1016/B978-0-12-803692-1.00004-3

Some of them are heavy metals (HMs). Their powerful effect on physiological processes is because they are often enzyme activators (Pleshkov, 1987). In response to the stress, caused by HMs, cells can activate DNA repair, an antioxidant defense system, and synthesis of low-molecular-weight substances with stress-protective properties (Chernykh et al., 1999; Goncharova et al., 2009).

The metal ions' presence in a large numbers in soil, having a toxic effect on plants, is the cause of their ionic (mineral) stress. Stress in plants can be induced by ions of heavy metals such as zinc, cadmium, copper, and mercury (Zhuchenko, 2003; Gladkov, 2006). There is an opinion that copper has no direct influence on DNA and its injuring effect has an indirect nature: the transition reaction of copper from bivalent into monovalent state is accompanied by formation of active oxygen forms in the cell. These active oxygen forms, in turn, have a destabilizing effect on chromatin (Maksymiec, 1997; Schuetzenduebel and Polle, 2002).There is evidence of an increase in chromosomal aberrations in the presence of high concentrations of HMs, including zinc (Gebhart, 1984). Thus, while not actually mutagens, heavy metal ions promote mutations (Bessonova, 1992; Barsukova, 1997).

Therefore improvement of existing varieties of *F. esculentum* and development of new varieties with high rutin content, adapted to the growth conditions, pests, and diseases, make it necessary to engage in selection of a process-rich genofund of cultivated and wild species. The search for new, efficient, economic, and profitable sources of rutin is one of the main difficulties of the pharmaceutical industry.

MATERIALS AND METHODS

Studies were conducted on the basis of the laboratory of agricultural biotechnology, and the field experiments were carried out at Primorsky Scientific Research Institute of Agriculture (44.34°N, 131.58°E) and the Pacific Institute of Bioorganic Chemistry. The object of research was members of the family Polygonaceae Juss.: cultivated species of the *Fagopyrum* Mill. genus [*F. esculentum* Moench, *F. tataricum* (L.) Gaertn., *F. cymosum* Meissn.] and wild species of the genera: *Aconogonon* (Meissn.) Reichenb. [*Aconogonon weyrichii* (Fr. Schmidt) Hara], *Fallopia* Adans. [*Fallopia convolvulus* (L.) A. Love], and *Reynoutria* Houtt. [*Reynoutria sachalinense* (Fr. Schmidt) Nakai]. *A. weyrichii* and *R. sachalinense* plants for chemical research were collected in the area of Yuzhno-Sakhalinsk (Sakhalin region). Samples of plants *F. convolvulus* were gathered in the outskirts of Timiryazevskaya, Ussuriysky district of Primorsky Krai.

To identify rutin spectra of NMR-^{1}H were used, recorded on a Bruker AC-250 spectrometer (250.13 MHz for ^{1}H) (Germany) in CDCl3 (deuterochloroform) and in acetone-d$_6$, and compared with a particularly pure rutin ("Chemopol," Czech Republic).

In order to study effect and after-effect (in soil) of ions of copper and zinc on the plant growth of buckwheat varieties Izumrud and Cheremshanka, 20 seeds were placed 5 times into Petri dishes. Depending on the experimental variant, either 10 mL of salt solution of copper sulfate ($CuSO_4 \times 5H_2O$),

11 concentrations—345–5750 mg/L [30–500 MPC (maximum permissible concentration)]—or a salt of zinc sulfate ($ZnSO_4 \times 7H_2O$) with 2, 4, 6, or 8 MPC were added (1 MPC Zn^{2+} is equal to 23 mg/L, ie, 1 MPC $ZnSO_4 \times 7H_2 O = 101$ mg/L and 1 MPC Cu^{2+} is equal to 3 mg/L, ie, 1 MPC $CuSO_4 \times 5H_2O = 11.5$ mg/L). Distilled water was used as control (Barsukova, 2009).

The effect of selective media, containing $CuSO_4$ from 6 to 230 mg/L, upon growth and proliferation of shoots of the callus cultures was studied in five genotypes of common buckwheat varieties: Izumrud, Cheremshanka, Kitavase; hybrids—Izumrud × Natasha, Izumrud × Kitavase (Barsukova, 2013). The effect of zinc ions in the selective culture medium was studied on the callus cultures of five samples of buckwheat (Moiseenko et al., 2010).

The effect and after-effect of nine concentrations of $CuSO_4 \times 5H_2O$, 1 (11.5 mg/L), 2, 6, 8, 10, 12, 14, and 16 MPC and eight concentrations of $ZnSO_4 \times 7H_2O$, 1 (101.0 mg/L), 2, 4, 6, 8, 9, 10, and 11 MPC on survivability, growth, and development of microcuttings of buckwheat (areas of aseptic plants with auxiliary bud) were studied on the Japanese variety Kitavase and the hybrid Izumrud × Kitavase. The cultivation period of microcuttings on selective medium with zinc and copper ions was 30 days. Gamborg medium (B_5) with copper sulfate 0.025 mg/L and zinc sulfate 2.0 mg/L was used as control (Gamborg et al., 1968).

RESULTS AND DISCUSSION

Species of the Polygonaceae family as sources of rutin. Rutin content depends on genus, species, and variety. We investigated rutin content and productivity of the aboveground part of different samples of three *Fagopyrum* species. The data show that high rutin content was in the aboveground part of *F. esculentum* (Izumrud variety), 38.1 mg/g; *F. tataricum* (sample k-62 from Canada), 44.5 mg/g; and *F. cymosum* (k-4231 from India), 41.2 mg/g.

On the productivity of aboveground mass and dry matter, samples of *F. tataricum* k-17 (China), k-62 (Canada), *F. esculentum* variety Izumrud (Russia, Primorsky Krai), and *F. cymosum* k-4231 (India) were distinguished. Rutin content was studied in species of the family Polygonaceae (Table 4.1).

High rutin content was observed in flowers of the studied species (63.1 and 8.5 mg/g). Less rutin content was found in the leaves (48.1 and 7.4 mg/g). Minimum rutin content was defined in the stems (12.6 and 2.2 mg/g). Maximum rutin content in all parts of the plant was observed in *F. tataricum.* Comparison of the data on rutin content in *F. tataricum, F. cymosum,* and *F. esculentum* species with those of *A. weyrichii, R. sachalinense,* and *F. convolvulus* found that *Fagopyrum* species are likely sources of this flavonoid.

Intraspecific and intravariety polymorphism of F. esculentum and prospects of rutin usage as a diagnostic trait in selection. At present, almost all varieties of *F. esculentum* represent complex heterozygotic populations with a wide genofund of traits. Our research has shown that intravariety changes of the plants' color have a wide spectrum (red, red-green, green-red, and green) and are

TABLE 4.1 Rutin Content in Species of the Family Polygonaceae in the Flowering Phase (mg/g)

		Plant Part		
Species	Flowers	Leaves	Stems	Aboveground Part
A. weyrichii	16.1 ± 0.2	12.3 ± 0.2	4.5 ± 0.1	11.1 ± 0.2
F. convolvulus	13.4 ± 0.2	13.2 ± 0.2	4.3 ± 0.1	6.1 ± 0.1
R. sachalinense	8.5 ± 0.1	7.4 ± 0.2	2.2 ± 0.1	4.3 ± 0.1
F. esculentum	57.1 ± 0.2	43.2 ± 0.2	12.4 ± 0.2	38.2 ± 0.2
F. tataricum	63.3 ± 0.3	48.1 ± 0.3	12.6 ± 0.2	41.5 ± 0.3
F. cymosum	51.4 ± 0.3	45.3 ± 0.3	11.7 ± 0.2	37.4 ± 0.3

affected not only by the variety of genotype but largely by variability, the effect of which depends on various factors: sowing date, mineral fertilizers, seeding rate, and method of sowing (Klykov and Moiseenko, 2010). As a result of the chemical analysis of buckwheat plants selected for color, it was found that plants with red coloring of stems contain more rutin in comparison with green, green-red, and red-green plants. Based on the data, we have developed a selection method of buckwheat plants with high rutin content in the aboveground part (Klykov and Moiseenko, 2005). Buckwheat plants are selected according to the stalk coloring in a fruit formation stage, choosing plants with a dark red (anthocyanin) color.

F. esculentum has white, green, pink, and red flowers. The presence of variously colored flowers in the same species indicates hereditary peculiarities of metabolism in plants, which is important in the new varieties' development, because the flower coloring is one of the basic varietal characteristics. The research found that the flower's coloring depends on the varietal characteristics of rutin content (Table 4.2).

Most of this flavonoid was found in the plants with red flowers (42.8 mg/g) compared to the plants with white flowers (34.0 mg/g). Anthocyanin coloring of the flowers and stalks is a good diagnostic indicator that can be used as a criterion in buckwheat plant selection with high rutin content. At the present time, rutin content in the roots of *Fagopyrum* Mill. species is seldom studied, as well as its role in selection on lodging resistance. Therefore we studied growth dynamics of aboveground and root mass, root system maintenance, and rutin content in roots of *F. esculentum* and *F. tataricum*.

It was found that *F. esculentum* plants, resistant to lodging, contain 6.8 mg/g of rutin in the roots (bright coloring of the root system), and sensitive plants contain −3.0 mg/g (dark brown coloring of the roots). Apparently, resistance

TABLE 4.2 Rutin Content Depending on the Coloring of Different Parts of the *F. esculentum* Plant

Plant Parts Coloring	Rutin (mg/g) (Dry Matter)		
	lim	$\overline{X} \pm S\overline{x}$	V (%)
Flowers[a]			
White	31.2–37.5	34.0 ± 0.1	8.5
Green	33.3–35.6	34.3 ± 0.2	2.7
Pink	34.1–38.9	35.6 ± 0.2	3.5
Red	39.5–46.7	42.8 ± 0.2	7.8
The root system[b]			
Light	5.1–7.9	6.8 ± 0.2	2.1
Light brown	4.7–5.6	5.1 ± 0.1	1.2
Dark brown	2.7–3.3	3.0 ± 0.1	0.9

[a]Note: Rutin content in the plant aboveground part in the mass flowering phase.
[b]In the roots in the phase of fruit formation.

to lodging plants at the fruit formation stage has a physiologically active root system (viable), which directly influences the intense accumulation of rutin in comparison with sensitive ones. The identified relation between rutin content, the roots' coloring, and the root mass became the basis for development of a selection method for buckwheat plants on lodging resistance (Klykov and Moiseenko, 2003).

The study showed that rutin content in buckwheat groats, depending on genotype, varies from 0.07 mg/g (*F. esculentum*) to 24.0 mg/g (*F. tataricum*). It was observed that buckwheat groats were colored light brown, light green, and green. Of particular interest is the possibility that rutin content has an effect on the coloring of buckwheat groats (green color implies increased rutin content). Varieties of *F. esculentum* with green coloring of buckwheat groats had the largest rutin content (0.15–0.07 mg/g). Rutin content was significantly lower in *F. esculentum* than that of *F. tataricum* and *F. cymosum.* Wild species of *F. tataricum* and *F. cymosum* are valuable in breeding because of their genetic rutin source. The studied species had different buckwheat groat colors. Thus buckwheat groats of the *F. esculentum* variety Pri7 were light brown, with variety Primorochka light green, variety Izumrud green, *F. tataricum* yellow-green, and *F. cymosum* bright yellow-green. The coloring of buckwheat groats can serve as a diagnostic sign of visual selection of forms with high rutin content.

The viability of common buckwheat plants under the influence of copper and zinc. Plant diversity in nature makes it possible to reveal forms tolerant to ion stress. We have not found information on the sensitivity and resistance of buckwheat to copper and zinc. Therefore in this study we aimed to define buckwheat plant tolerance to heavy metal ions.

In laboratory experiments with variety Izumrud seeds, it was found that germination in the variants, where concentration of copper ions was more than 70 MPC (805 mg/L), was significantly lower than control, with the exception of the experimental variants with 200 and 300 MPC of copper. Negative effect of copper ions was especially evident on further root growth of the sprout with 345 mg/L of copper salts (30 MPC). The average length of root compared with the control was 3–30 times less, and the total length of the sprout with the root in copper was 1.7–15.8 times less than in the control.

Despite the evident inhibitory effect of high concentrations of copper ions upon the seedlings of buckwheat variety Izumrud, lethal concentrations in the laboratory experiments were not defined. Thus another laboratory-and-field experiment was carried out with buckwheat varieties Izumrud and Cheremshanka. Germination of seeds, after treatment with solutions of salts of copper sulfate and zinc sulfate, after sowing in the field ranged from 0% to 77.7%. Field germination of Cheremshanka variety plants under the influence of zinc ions was higher (40.0–71.4 %) than that of Izumrud varieties (30.0–62.5%). The studied buckwheat genotypes showed a selective degree of resistance to copper ions. A concentration of 500 MPC for Cheremshanka variety plants was lethal because for plants of Izumrud variety they had 62.5% germination. Maximum germination of buckwheat was observed under treatment with copper ions at a concentration of 100 MPC—77.7% of Izumrud variety plants and 57.1% of Cheremshanka variety plants.

An account of plants that survived before harvesting showed that Izumrud variety plants were more tolerant to the after-effect of copper and zinc ions than Cheremshanka variety plants. Despite the fact that germination of the last in variants with zinc was higher at harvesting time, death of Cheremshanka variety plants with zinc content of 4 and 8 MPC was noted.

The after-effect of heavy metal ions affected not only the growth and development of the plants, but also fruit formation. Buckwheat plants of Cheremshanka variety, whose seeds were treated with zinc ions, were low and bloomed, but did not form the seed buds.

Izumrud variety plants formed maximum normally obtained seeds in the variant with 6 MPC (zinc) and 50 MPC (copper). Cheremshanka variety plants formed seeds only in the variant with 6 MPC zinc and 30 and 200 MPC copper.

Tolerance of buckwheat cell cultures and microcuttings in vitro to the effect of sulfate of copper and zinc. Observations on the callus, placed on a selective medium with copper ions, showed that growth of the callus mass was held for all variants of the medium. Insignificant lag in growth of the callus occurred when cultured on media containing from 30 to 230 mg/L of copper sulfate salt. The studied copper sulfate concentrations (6.0–230 mg/L) had no adverse or lethal effect on growth speed of buckwheat callus. However, not all calluses have retained the ability for further regeneration. Buckwheat genotypes resistant to the effect of copper ions were found. They were varieties Cheremshanka and Kitavase, characterized by the maximum number of calluses that formed an organogenic structure. In 97.4% of variety Cheremshanka callus cultures the formation of somatic embryoids was observed, but regenerative ability was kept by only 33.3% (Fig. 4.1). Part of the morphogenic calluses of buckwheat variety Izumrud consisted of 12.2%, and the hybrid of Izumrud × Kitavase at 14.3%, but there was no plant regeneration. Callus cultures of hybrid combination Izumrud × Natasha after cultivation in media with copper lost their morphogenetic potential.

Regeneration of 15 plantlets from the callus of buckwheat variety Kitavase in medium with 60 mg/L of copper salt and 29 regenerants from the callus of buckwheat variety Cheremshanka in medium with copper salt from 29 to 96 mg/L was performed. The seed progeny of regenerants was developed.

Cultivation on the media with zinc ions revealed inverse dependence between zinc content in the medium and growth of callus cells. The minimum number of actively growing calluses was observed with maximum content of zinc sulfate (460 mg/L). Calluses of buckwheat variety Pri7 were characterized by minimum growth potential. The negative after-effect of zinc salts upon growth of buckwheat callus cells was not observed in 260 days after growing on selective media.

FIGURE 4.1 Induction of somatic embryoids of callus tissue of *F. esculentum*.

Calluses of varieties Pri7 and Izumrud × Kitavase grew and actively went to organogenesis. In most cases, despite the kidney formation on calluses, further proliferation of shoots with typical morphology of stems, leaves, and roots was difficult. Numerous shoots formed flower buds that made further micropropagation difficult. Getting the seeds of regenerants developed from calluses, cultivated in the media with high zinc content, was not successful.

There was a defined negative effect of HMs upon buckwheat posterity. The negative mutagenic effect of copper ions (24–60 mg/L) was observed in the second generation of regenerated plant progeny developed through regeneration from callus, in the form of plant chlorosis and nongreen mutants. Albino plants could be seen in the phase of germination within 1–2 days, but then died because of the lack of chlorophyll (Fig. 4.2). The frequency of this lethal mutation was within 1.5–9.6%.

The studied concentrations of copper and zinc in nutrient medium under cultivation of microcuttings were very significant compared with control. Copper ion content in a selective medium was higher than in medium B5 in 460–7360 of cases and zinc ions in 51–556 of cases depending on the variant of the experiment. The most sensitive to high concentrations of zinc ions were microcuttings of buckwheat variety Kitavase, because they died in a zinc salt concentration of 1111 mg/L (11 MPC). The effect of copper ions for these cuttings was less toxic: 7.1% of them were viable, even after the 30-day cultivation of microcuttings on the medium with the copper sulfate concentration of 184 mg/L (16 MPC). However, the stress borne under the effect of HMs was reflected further:

FIGURE 4.2 Lethal mutation (albinism) of buckwheat at the cotyledon stage.

the shoots fell behind in growth and had little regenerative ability for two subsequent passages on the control medium. The plantlets' explants of buckwheat hybrid Izumrud × Kitavase presented a higher resistance to the effect and aftereffect of copper and zinc ions. Death of microcuttings cultivated on nutrient media with high concentrations of the studied HMs was not observed.

In vitro regeneration of microcuttings was better in media with a high content of copper sulfate. This regularity was confirmed in additional experiments. There was a defined curtain interval of copper sulfate concentrations in nutrient medium (9.2–23.0 mg/L) that stimulated tissue regeneration and maximum output of buckwheat-regenerated plants. On the basis of the research a patent for invention has been taken out No. 2538167 (26.07.2013) "Method for buckwheat reproduction in vitro" (Barsukova, 2015).

CONCLUSIONS

As a result of the study of cultivated and wild species of *Fagopyrum* Mill. and *Polygonum* L. there have been distinguished new sources with high rutin content, which has indicated the possibility of their usage in medicine. As a promising source of rutin we recommend: *F. esculentum* (variety Izumrud, Russia—38.1 mg/g) and *F. tataricum* (sample k-62 from Canada—44.5 mg/g). The study showed intraspecies variability of rutin content in the aboveground part of *F. esculentum* depending on color of stems, flowers, and fruits.

The research defined that rutin content in the aboveground part of *F. esculentum* depends on ecological and geographical conditions. Based on the analysis a complex of traits has been developed that should be followed in evaluating the initial material in selection and seed breeding of *F. esculentum*. These traits include morphological (coloring of grain, stem, leaves, and flowers) and chemical (rutin content). These indicators serve as diagnostic traits for selection of buckwheat with high rutin and resistance to adverse abiotic and biotic environmental factors.

As a result of the experiments the seed posterity of 50 buckwheat samples tolerant to the effects of HMs has been developed, and is under study in the selection nursery. The largest number of samples was developed using copper (71.1%), zinc (26.7%), and cadmium (2.2%) ions. Evaluation of the material in the breeding nursery in 2014 showed that 50% of samples had gross grain developed under the influence of copper and 20–22% under zinc. This is a positive result of HM usage because gross grain is one of the main signs influencing the technological properties of buckwheat varieties. For further breeding, three samples were selected, obtained as a result of copper effect, with signs of gross grains (weight of 1000 grains 37.0–38.6 g) and weight grains of 7.1–9.3 g per plant. It has been shown that the distinguished forms of the plant have an interrelation between rutin and morphological and economically valuable traits.

The results of the *F. esculentum* study as a raw material for rutin as well as data on the production maintenance with raw materials can be used in practical applications in the pharmaceutical industry. New data have been obtained that may become essential for understanding the functional role of rutin in buckwheat, as well as for identifying mechanisms of rutin accumulation in plants depending on various factors. This will allow the most promising sources to be defined and practical recommendations for their usage in agriculture, the food industry, and medicine to be developed.

REFERENCES

Barsukova,V.S., 1997. The Physiology-Genetic Aspects of Plant Resistance to Heavy Metals. Analytical Review. GPNTB, Novosibirsk.

Barsukova, Y.N., 2009. The viability of *Fagopyrum esculentum* plants under effect of copper and zinc ions In : Chaika, A.K., Vashchenko, A.P., (Eds.), Actual of development of agrarian science in the Far-Eastern region, Russia. Dalnauka, Vladivostok, pp. 129–136.

Barsukova, Y.N., 2013. Cell selection of *Fagopyrum esculentum* in the conditions of the ion stress. Agrarian Russia 10, 2–4.

Barsukova, Y.N., 2015. Patent 2538167 RU, МПКА01 4/00. Method for buckwheat reproduction in vitro. The patent owner SSI Primorsky SRIA RAAS. No. 2013135431; declaration 26.07.2013.

Bessonova, V.P., 1992. The pollen state as indicator of contamination by heavy metals. Ecology 4, 45–50.

Chernykh, N.A, Milashchenko, T.R., Ladonin, V.F., 1999. Eco-Toxicological Aspects of Soil Pollution with Heavy Metals. Agrokonsalt, Moscow.

Gamborg, O.L., Miller, R.A., Ojima, K., 1968. Nutrient requirements of suspension culture of soybean root cells. Exp. Cell. Res. 50 (1), 151–158.

Gebhart, E., 1984. Chromosome damage in individuals exposed to heavy metals. Toxicol. Environ. Chem. 8, 253–266.

Gladkov, Y.A., 2006. Methods of biotechnology for the development of plants resistant to heavy metals. 1. Comparative evaluation of the heavy metals' toxicity for callus cells and whole plants. Biotechnology 3, 79–82.

Goncharova, L.I., Selezneva, Y.M., Belova, N.V., 2009. Changes in the intensity of lipid peroxidation and free proline accumulation in leaves of barley under soil contamination by copper and zinc. Rep. RAAS 2, 12–14.

Harborne, J.B., 1972. Evolution and function of flavonoids in plants. Rec. Adv. Phytochem. 4, 107–441.

Hinneburg, I., Neubert, R.H., 2005. Influence of extraction parameters on the phytochemical characteristics of extracts from buckwheat (*Fagopyrum esculentum*) herb. J. Agric. Food Chem. 53, 3–7.

Klykov, A.G., Moiseenko, L.M., 2003. Patent 2201075 RU; IPC[7] A 01 H 1/04. The choice of method for *F. esculentum* plants with lodging resistance (in Russian).

Klykov, A.G., Moiseenko, L.M., 2005. Patent 2255466 RU; IPC[7] A 01 H 1/04. Method of choice of *F. esculentum* plants with high rutin content in the overground mass (in Russian).

Klykov, A.G., Moiseenko, L.M., 2010. Influence of buckwheat cultivation conditions upon rutin content. Advances in Buckwheat Research. Proceedings of the 11th International Symposium on Buckwheat, July 19–23, Russia. Orel., pp. 475–483.

Kreft, I., Fabjan, N., Yasumoto, K., 2006. Rutin content in buckwheat (*Fagopyrum esculentum* Moench) food materials and products. Food Chem. 98 (3), 508–512.

Maksymiec, W., 1997. Effect of copper on cellular processes in higher plants. Photosynthetica 34, 321–342.

Moiseenko, A.A., Moiseenko, L.M., Klykov, A.G., Barsukova, Y.N., 2010. Buckwheat in the Far East. Rosinformagroteh, Moscow.

Pleshkov, B.P., 1987. Biochemistry of Agricultural Plants, fifth ed., Agropromizdat, Moscow.

Schuetzenduebel, A., Polle, A., 2002. Plant responses to abiotic stresses: heavy metal-induced oxidative stress and protection by mycorrhization. J. Exp. Bot. 53, 1351–1365.

Zaprometov, M.N., 1993. Phenolic Compounds: Distribution, Metabolism, and Functions in Plants. Nauka, Moscow.

Zhuchenko, A.A., 2003. Ecological Genetics of Crops. Agrorus, Samara.

Chapter five

Buckwheat Production, Consumption, and Genetic Resources in Japan

T. Katsube-Tanaka

Graduate School of Agriculture, Kyoto University, Kitashirakawa, Kyoto, Japan

PRODUCTION

Production and the Past of Buckwheat

Annually, Japan consumes 130,000–150,000 tons and produces 30,000 tons of buckwheat (Takahashi, 2011). Japanese production is the eighth largest in the world and represents 2% of world production (FAOSTAT: http://faostat.fao.org/). Buckwheat has been important as an emergency crop or a catch crop because of various superior characteristics, such as rapid growth, low labor requirements, and high adaptability to infertile land. In Japan, the production of buckwheat was relatively large, approximately 140,000 tons, dating back to the late 19th century when rice production was low, at approximately 5 million tons (crop statistics: http://www.maff.go.jp/j/tokei/kouhyou/sakumotu/) (Fig. 5.1A). During the steady increase in rice production into the 1970s, when many resources and efforts were concentrated on rice production, and self-sufficiency in rice reached 100% for the first time in recorded history, buckwheat production decreased rapidly to approximately 20,000 tons. Then the production of rice decreased as rice consumption per capita decreased from 120 kg/year in the 1960s to 60 kg/year in the 2000s (food balance sheet: http://www.maff.go.jp/j/tokei/kouhyou/zyukyu/index.html), and an acreage reduction policy for rice started to sustain the price of rice, which was under a food control system by the government. The policy with the subsidy drove farmers to grow other crops, such as wheat, soybean, and buckwheat, resulting in a firm upturn in cropping acreage of buckwheat in the last several decades (Fig. 5.2A) and a gradual increase in

61

Molecular Breeding and Nutritional Aspects of Buckwheat. http://dx.doi.org/10.1016/B978-0-12-803692-1.00005-5

(A)

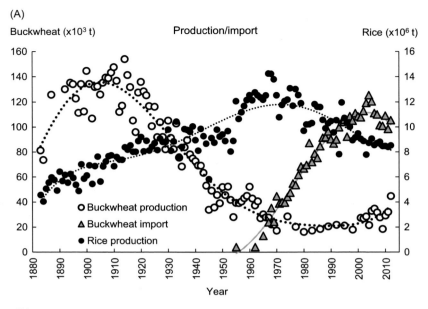

(B)

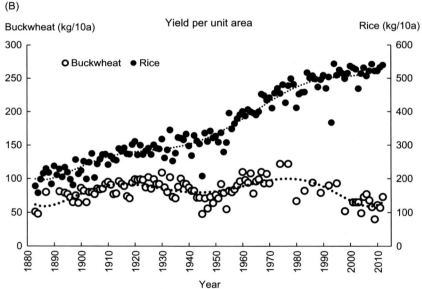

FIGURE 5.1 Interannual changes in (A) production of buckwheat and rice and buckwheat import and (B) yield per unit area of buckwheat and rice from 1883 to 2012 in Japan. *Statistical data were retrieved from crop statistics edited by the Ministry of Agriculture, Forestry and Fisheries of Japan (http://www.maff.go.jp/j/tokei/kouhyou/sakumotu/) and from foreign trade statistics edited by the Ministry of Finance of Japan (http://www.customs.go.jp/toukei/info/index.htm).*

production (Fig. 5.1A). On the other hand, the long-term decline of cropping acreage through the 1970s was also common in major upland crops, such as wheat, soybean, barley, and naked barley, with the exception of some temporal increases (Fig. 5.2B). This decrease was most likely caused by economic development, especially the rapid economic growth from 1954 to 1973, shifting labor

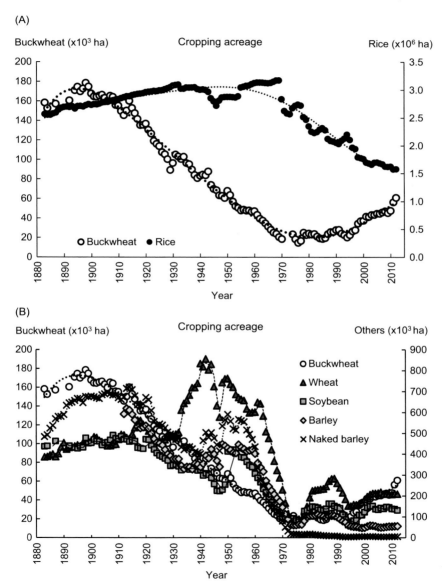

FIGURE 5.2 Interannual changes in cropping acreage of (A) buckwheat and rice and (B) buckwheat and upland crops (wheat, soybean, barley, naked barley) from 1883 to 2012 in Japan. *Statistical data were retrieved from crop statistics edited by the Ministry of Agriculture, Forestry and Fisheries of Japan (http://www.maff.go.jp/j/tokei/kouhyou/sakumotu/).*

power from primary industries (agriculture, forestry, and fisheries) to second and the tertiary industries. It is noteworthy that the import of unhulled buckwheat grains was initiated in 1952 (Takahashi, 2011), and this import rapidly increased to match the increasing consumption demands (Foreign Trade Statistics: http://www.customs.go.jp/toukei/info/index.htm) (Fig. 5.1A). The import of hulled grains or flour has also increased recently. The growing import of buckwheat appeared to inhibit domestic production. In conclusion, domestic buckwheat production in Japan has been historically affected by rice production, economic development, and is currently affected by the import of foreign buckwheat.

Productivity of Buckwheat

The productivity of rice, indicated by yield per unit area, has constantly increased over the last 130 years from 200 kg/10a to over 500 kg/10a (Fig. 5.1B). In contrast the productivity of buckwheat stagnated between 50 and 100 kg/10a during the same period. The productivity of buckwheat has declined rather than improved in the last two decades (Fig. 5.1B), resulting in slowly expanding production (Fig. 5.1A) despite the firm upturn in cropping acreage (Fig. 5.2A), as mentioned earlier. It has been noted that the recent reduction in the productivity of buckwheat might be related to the increase in paddy field ratio: the percentage the field that was converted from a paddy against the total field used for buckwheat production. The converted field is originally suitable for lowland (flooded) rice production and presents a higher risk for excess soil moisture damage to buckwheat seedlings. The essentially continued production adjustment policy for edible rice (not for forage or processing uses) has led to a greater usage of the converted field for buckwheat production, and the prevention of excess soil moisture damage is becoming more important. Sugimoto and Sato (2000) demonstrated that seed yield was remarkably reduced by 1 day of flooding at the first leaf emergence stage and by more than 3 days flooding at the flowering stage. Sakata and Ohsawa (2005) showed that the emergence rate of seeds with a 30–70-mm root length was reduced rapidly (severely) by 1–4 days of flooding at 25°C and slowly (less severely) at 15°C. However, the emergence rate of seeds with 3-mm root length severely decreased at even 15°C. Therefore it is important to avoid flood damage, especially before the seedling stage. Matsuura et al. (2005) demonstrated that common buckwheat, which produced more adventitious roots, had a greater growth response to flooding 12–36 days after sowing than Tartary buckwheat. Sakata and Ohsawa (2006) examined the 1-day flooding tolerance of germinated seeds in 17 Japanese varieties and identified a promising landrace with high and stable emergence, termed "Kitouzairai."

Matano (1990) reviewed recent research on buckwheat and noted that low seed-setting ratio as well as lodging and shattering habitats are the main causes of low productivity. Sugimoto and Sato (1999) demonstrated that the seed-setting ratio of summer ecotype (photoperiod-insensitive) buckwheat was greatly decreased when the average minimum temperature was over 17.5°C during the grain-filling period. Michiyama et al. (1998) showed that the summer ecotype grown in summer and the autumn ecotype (photoperiod-sensitive) grown in autumn produced similar numbers of flowers and seed settings, whereas the autumn ecotype grown in the summer produced a larger number of flowers with a quite low seed-setting ratio through a prolonged flowering period up to 4 months. A 15-h day-length before anthesis delayed the onset of flowering and increased the number of nodes more severely in the autumn ecotype (Michiyama et al., 2003). Hara and Ohsawa (2013) elucidated that under controlled environmental conditions with a 15-h day-length, an autumn ecotype cultivar had a larger mean value and larger variation in days-to-flowering than a summer ecotype cultivar, even though the two cultivars had similar smaller mean values and variation under a day-length of 12 h. In conclusion, appropriate ecotypes are cultivated in various environments/locations with different latitudes and altitudes to avoid frost and heat stress and to maximize biomass and yield accumulation in Japan, resulting in various cropping types that were developed as exemplified later. Note that the effect of fertilizing on the growth and yield of buckwheat have also intensively been studied to increase productivity such as using delayed release fertilizer (Sugimoto, 2004a), green manure of Chinese milk vetch (Sugimoto et al., 2000, 2007; Sugimoto, 2004b), soybeans (Allotey et al., 1997a,b), and compost manure (Hara et al., 2011, 2014).

Regional Variation in Buckwheat Production

The cropping acreage of buckwheat has firmly increased in the last two decades. This increase is especially true in Hokkaido, the most northern area of Japan, which increased cropping acreage eightfold from 2,710 ha in 1985 to 21,700 ha in 2012, and currently exceeds more than one-third of the total cropping acreage throughout the country (Fig. 5.3). When buckwheat production in

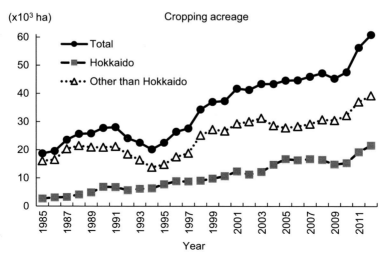

FIGURE 5.3 Interannual changes in cropping acreage of buckwheat for the whole country, Hokkaido, and other than Hokkaido from 1985 to 2012. *Statistical data were retrieved from crop statistics edited by the Ministry of Agriculture, Forestry and Fisheries of Japan (http://www.maff.go.jp/j/tokei/kouhyou/sakumotu/).*

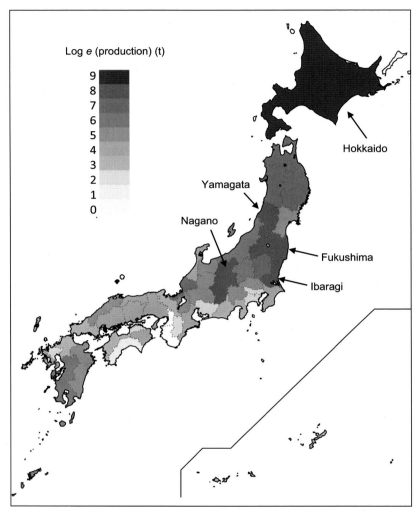

FIGURE 5.4 Regional variation of buckwheat production in 2014. Yield data (metric ton) were natural log-transformed, classified, and displayed in gray scale by prefecture. *Statistical data were retrieved from crop statistics edited by the Ministry of Agriculture, Forestry and Fisheries of Japan (http://www.maff.go.jp/j/tokei/kouhyou/sakumotu/).*

2014 was compared by prefecture, Hokkaido was followed by prefectures in the northeastern and eastern district (Yamagata, Fukushima, and Ibaragi) and mountainous central district (Nagano) that are also ranked as high yield producers (Fig. 5.4). Some prefectures in the southwestern district also produced relatively high yields.

The average yield per unit area in the recent 7 years was the highest in the Gunma and Saitama prefectures, followed by the Tochigi and Ibaragi all in the north Kanto area and by Kagoshima in the southwestern district (Fig. 5.5). The prefectures with higher productivity in the north Kanto area had lower paddy

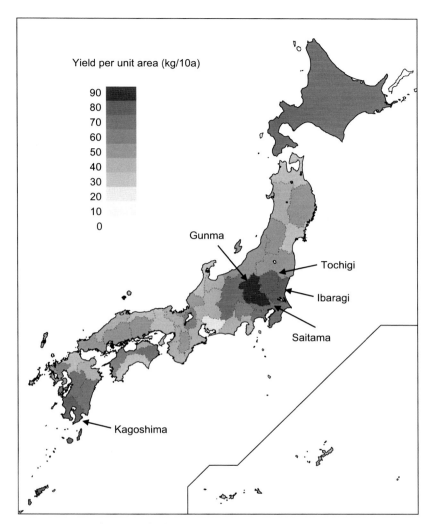

FIGURE 5.5 Regional variation of buckwheat average yield per unit area in 2014. Average yield per unit area data (kg/10a) were classified and displayed in gray scale by prefecture. *Statistical data were retrieved from crop statistics edited by the Ministry of Agriculture, Forestry and Fisheries of Japan (http://www.maff.go.jp/j/tokei/kouhyou/sakumotu/).*

field ratios (Fig. 5.6). By contrast, prefectures with higher paddy field ratios, such as Hyogo, Kyoto, Shiga, Fukui, and Toyama in the Kinki and Hokuriku areas, had relatively lower productivity (Fig. 5.5). The challenge of excess soil moisture damage is also true for soybean production grown in a converted field in Japan. Therefore some effective measures against excess soil moisture damage have been proposed for soybean and buckwheat, such as trenching, subsoil breaking, installing open/blind drainage ditch and/or groundwater level controlling system [eg, farm-oriented enhanced aquatic system (FOEAS)], and use of simultaneously operating machines for tillage-list-seeding (Hosokawa, 2011).

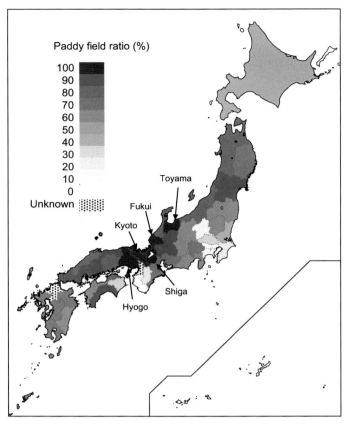

FIGURE 5.6 Regional variation in paddy field ratio of buckwheat in 2014. Paddy field ratio (%) was calculated as percentage of converted field from a paddy against the total field used for buckwheat production. The ratio was classified and displayed in gray scale by prefecture. *Statistical data were retrieved from crop statistics edited by the Ministry of Agriculture, Forestry and Fisheries of Japan (http://www.maff.go.jp/j/tokei/kouhyou/sakumotu/).*

GENETIC RESOURCES

Genetic Resources and Conservation of Landraces

The National Institute of Agrobiological Sciences (NIAS) in Japan is the largest agricultural research institute for basic life sciences, and its Genebank is the main repository of agronomically important genetic resources of plants, animals, and microorganisms. The NIAS Genebank conserved landrace accessions of 218 common buckwheat and 8 Tartary buckwheat varieties originating from Hokkaido in the north to Kagoshima in the south (Fig. 5.7) (http://www. gene.affrc.go.jp/databases_en.php). They also disclose detailed information on various characteristics of the accessions, some of which were evaluated in multiple environments. The distribution of plant height, node number, branch number, and lodging resistance of the conserved and evaluated landraces had

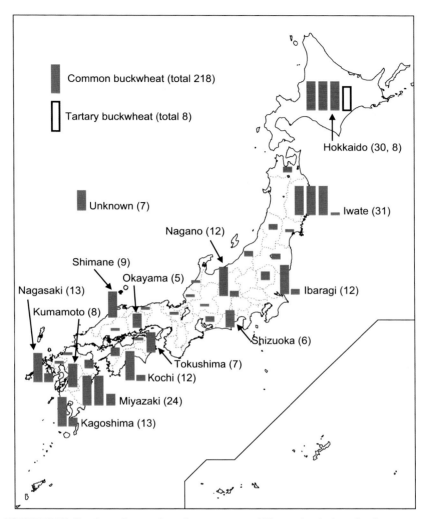

FIGURE 5.7 Number of accessions for common and Tartary buckwheat landraces conserved in the National Institute of Agrobiological Sciences Genebank (http://www.gene.affrc.go.jp/databases_en.php). The number of accessions is schematically displayed by bar chart by their origin. Major origins are also shown by name and the number of accessions.

a broad range and peaked at 90 cm height, 11 nodes, 4 branches, and medium resistance, respectively (Fig. 5.8). The flowering date, days to flowering, 1000 grain weight, and yield per unit area of the evaluated landraces were also widely distributed and peaked at mid-Jul. with late-Aug. to mid-Sep. flowering, 26 days to flowering, 30 g per 1000 grains, and 12 kg/a yield, respectively (Fig. 5.9).

Ohsawa (2011) noted several important breeding objectives for stable high yield and superior seed quality, such as defined (controlled) photoperiod and temperature susceptibility, lodging resistance, determinate growth habitat, less shattering habitat, excess soil moisture resistance as well as high rutin content,

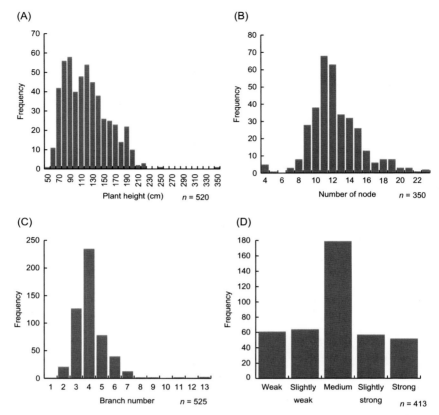

FIGURE 5.8 Histogram of (A) plant height, (B) number of main stem node, (C) branch number, and (D) lodging resistance of common buckwheat landraces recorded in the National Institute of Agrobiological Sciences Genebank (http://www.gene.affrc.go.jp/databases_en.php). Note that the data were unorganized ones from multiple varieties, locations, seasons, and years.

lowered allergenic protein content, good aroma, preharvest sprouting resistance, and enhanced palatability with established evaluation method. Recently, Funatsuki et al. (1996) and Wagatsuma (2004) independently isolated determinate-type variants from the current local leading variety "Kitawasesoba" in Hokkaido and developed determinate-type cultivars "Kitanomashu" and "Horominori (Horokei 3)," respectively (Honda et al., 2009). Matsui et al. (2008) developed an autogamous (self-pollinating) line "Buckwheat Norin-PL1" by crossing the wild relative *Fagopyrum homotropicum* and the former local leading variety "Botansoba," which was isolated from a landrace in Hokkaido. Yasui et al. (2012) identified the heteromorphic self-incompatibility associating gene, *S-ELF3*, which was confirmed to be mutated and malfunctioning in self-compatible species *F. homotropicum* and *Fagopyrum tataricum*. Matsui et al. (2004) produced STS markers for one of two complementary brittle pedicel genes (*Sht1* and *Sht2*), the trait of which is usually observed in wild species and is different from a weak

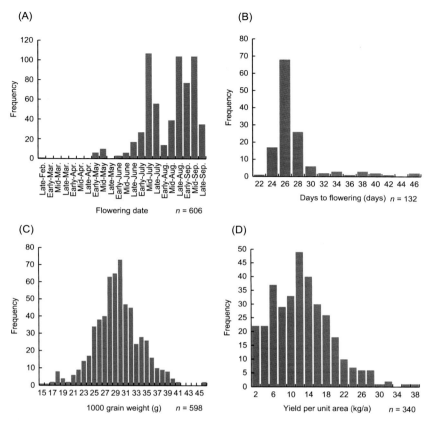

FIGURE 5.9 Histogram of (A) flowering date, (B) days to flowering, (C) 1000 grain weight, and (D) yield per unit area of common buckwheat landraces recorded in the National Institute of Agrobiological Sciences Genebank (http://www.gene.affrc.go.jp/databases_en.php). Note that the data were unorganized ones from multiple varieties, locations, seasons, and years.

pedicel trait observed in cultivated buckwheat. Oba et al. (1998, 1999, 2004) and Fujimura et al. (2001) demonstrated that common buckwheat and autotetraploids were more resistant in grain shattering than Tartary buckwheat and diploids, respectively. Suzuki et al. (2012) confirmed greater shattering-resistant traits of "W/SK86GF," which were developed by crossing Russian green flower variety "Skorosperaya 86" and "Kitawasesoba." Mukasa (2011) reviewed studies on breeding methodologies and variety developments at the National Agriculture and Food Research Organization/Hokkaido Agricultural Research Center. The review includes self-compatibility with nonshattering, estimation of outcrossing rate of self-compatible lines (ie, ~10%), the most efficient hot-water emasculation at 42°C for 5 min, heterosis of single-cross hybrid with homostyle (S^hS^h) and pin (ss) lines, and measurements of breaking tensile strength of pedicels in green flower lines, for common buckwheat. The review also contains the most efficient hot-water emasculation at 44°C for 3 min, nonadhering hull trait for dehulled whole seeds to minimize rutin degradation during flour-basis processing, two semidwarf genes derived from gamma ray irradiation, and dark red cotyledonal trait for anthocyanin-rich sprouts, for Tartary buckwheat. Tetsuka and Uchino (2002, 2005) examined plant morphological variations and seed shape/color variations in 56 Japanese landraces when cultivated in Kumamoto in the southwestern district. They found that growth duration positively correlated with "plant size"-related characteristics such as main stem length, and landraces from western Japan showed larger variation in "plant shape"-related characteristics. The growth duration also positively correlated with lightness of husk color, which changed from deep black to grayish brown as the growth duration increased.

Hayashi (2011) summarized inventory of 32 varieties/lines of common buckwheat that were developed and/or registered as a variety from 1919 to 2010 in Japan. The varieties/lines include traits such as colchicine-treated tetraploids, red flower/pericarp, high rutin content, early maturing, and preharvest sprouting resistance. Therefore the aforementioned breeding challenges indicated by Ohsawa (2011) might be overcome one by one by exploiting the diversified landrace accessions conserved at the NIAS Genebank, coupled with superior traits from the improved varieties, wild species (Ohsako and Ohnishi, 1998, 2000), and foreign genetic resources. Note that shattering- and lodging-resistant traits are more and more important to mechanical harvesting for time/labor saving. Processing suitability (higher milling ratio, preharvest sprouting resistance, good noodle-making properties) is an important target to be ameliorated. Red flower/pericarp traits are preferred for a scenic landscape formation attracting tourists in some areas planning to boost economic development. Tetraploid traits of the cultivars "Miyazaki-Ootsubu" (Nagatomo, 1984), "Shinshu-Osoba" (Ujihara et al., 1977), and "Hokkai No. 3" (Morishita, 2011) are also marked for conservation of the genetic properties of diploid landraces and breeding lines because of reproductive isolation.

Spring-Sowing Cultivation and Double Cropping

Japan covers a wide range of latitudes from 43°03′ N at Sapporo in Hokkaido to 26°12′ N at Naha in Okinawa. In Hokkaido, summer ecotype buckwheat with weak photoperiod sensitivity is sown in Jun. and harvested from Aug. to Sep. The summer ecotype buckwheat is also cultivated in mountainous Nagano from May to Jul. Meanwhile, the summer ecotype cultivar "Kitawasesoba" has been examined for a spring-sowing cultivation (seeded in Mar./Apr. and harvested in Jun.) in southwestern warm-temperature regions (Sugimoto et al., 2000, 2007; Sugimoto, 2004a,b). In Kyushu, for example, in Kagoshima at 31°33′ N, an autumn ecotype with strong photoperiod sensitivity is traditionally grown from Sep. to Nov. Recently, however, the intermediate-summer ecotype cultivar with preharvest sprouting resistance "Harunoibuki" has been developed (Hara et al., 2009a) and is grown in the spring-sowing cultivation. The preharvest sprouting, which affects flour pasting viscosity (Hara et al., 2007) and the texture of the cooked noodles (Hara et al., 2009b), is negatively correlated with a main stem length (Hara et al., 2008b). It must be an important issue especially when seed maturation proceeds in a rainy season in Jun. and Jul. in Japan except in Hokkaido.

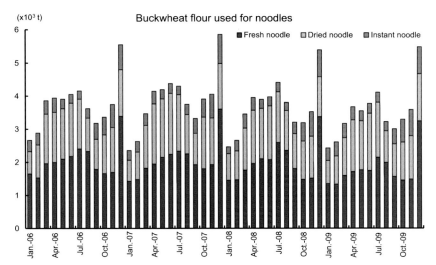

FIGURE 5.10 Monthly change of buckwheat noodle production (fresh, dried, and instant noodles) in 2006–09. *Statistical data were retrieved from the Annual Report of Rice and Wheat Processed Food Production edited by the Ministry of Agriculture, Forestry and Fisheries of Japan (http://www.maff.go.jp/j/tokei/index.html).*

The growth period of the spring-sowing cultivation can be set to avoid late spring frost, the rainy season, and be free from typhoon damage in autumn. The spring-sowing cultivation can also increase the chance to couple with other cropping systems. Additionally, there are two periods of peak demand for buckwheat noodles (Soba) in Japan (Fig. 5.10), and new harvest grains are highly appreciated by Japanese consumers. Therefore the spring-sowing cultivation, which can offer new harvest grains in the first peak demand period in summer, is attracting more and more attention. It is noteworthy that the southernmost prefecture, Okinawa, where buckwheat has not been grown traditionally, has recently produced twice a year from Mar. to May and from Oct. to Dec. after extensive research and development (Hara et al., 2008a, 2011, 2014). Note that double cropping is also practiced in other areas of Japan, and triple cropping used to be applicable in some specific areas such as in the Tottori and Kagoshima prefectures and is referred to as "Sando-soba," which means "three-times buckwheat" (Nijima, 2011). Currently, however, seeding of an optimal cultivar and at an optimal time for each environment is strongly recommended.

CONSUMPTION

Eating Quality of Buckwheat Noodle and Flour

Although many types of buckwheat foods, drinks, and consumption styles are available worldwide, the Japanese style of buckwheat consumption as food is almost all noodles, termed "Soba-kiri" or just "Soba," made from common buckwheat flour. Soba are highly appreciated if they are freshly ground, freshly kneaded, and freshly boiled as well as freshly harvested (Ikeda, 2002) to produce good aroma, taste, and texture. Thus palatability is one of the most important characteristics to Japanese consumers. Sensory evaluation of buckwheat noodles can be difficult, especially when the noodles are produced from 100% buckwheat flour, because making the noodles requires artisan skill. Horigane and Yamada (2005) developed a noodle-making system for quality evaluation of 100% buckwheat flour noodles, which consists of a low-temperature mill, a plastic three-rod puddler and pressure device, a rolling pin with two different diameters, and a rotary cutter. They demonstrated with proton nuclear magnetic resonance imaging and a creep test that the process of mixing hydrated buckwheat flour, termed "Mizumawashi," should be completed in 3 min using the puddler, which could minimize drying of the product (Horigane et al., 2004). Kawakami et al. (2008, 2009) analyzed the flavor of buckwheat flour using gas chromatography mass spectroscopy analysis, taste sensors, and a texture analyzer. The results showed that low storage temperatures were suitable for preserving the flavor of buckwheat, and longer storage periods increased the bitterness and astringency and decreased aroma. Additionally, Ikeda et al. (1997) found a negative correlation between total protein content and total starch content, and between total protein content and the hardness, cohesiveness, springiness, and chewiness of heated buckwheat dough using 30 samples derived from 8 countries. Inoue (2004) and his group demonstrated that the amylose content of the tetraploid variety "Shinshu-Osoba" varied from 14.9% to 23.5% with an average of 18.7% when cultivated in 110 different locations from the Iwate to Kumamoto prefectures and was negatively correlated with crude protein content. They also showed that the amylose content correlated positively and negatively with maximum air temperature and relative humidity of each cultivated environment, respectively. Special buckwheat noodles, the ingredient grains of which are harvested on a cool highland area where the daily temperature range becomes wide and makes fog at maturing, are highly appreciated as "Kirishita-soba" or "under-fog soba" (Nijima, 2011). Therefore Inoue (2004) hypothesized that such a "foggy" upland weather condition as well as a high dose of nitrogen fertilization produces buckwheat grain with high protein content, a soft and viscous texture, and the preferred characteristics for making Soba. In any case, the determining factors of the sensory quality of Soba should be minutely elucidated. Note that Ikeda (2002) reviewed the composition, utilization, and processing of buckwheat seed and flour and then proposed "molecular cookery science," in which palatability and acceptability of buckwheat foods are clarified at a molecular basis.

It would be desirable to pay more attention to foreign styles of buckwheat food and to develop a new style of processed food and drink to increase consumption and production of buckwheat in Japan.

Seasonal Patterns and Regional Variation of Domestic Consumption

Buckwheat noodle consumption peaks twice a year in Japan. The first peak is from Jun. to Jul. and the second is in Dec. (Fig. 5.10; Takahashi, 2011). In Jul., it is hot and humid, which influences people to eat a plain, simple, and light food such as Soba. Whereas in Dec. there is a traditional and unique eating custom termed "Tosikosi-soba" or year-crossing noodles, in which people eat Soba for good luck. Therefore buckwheat noodles are not consumed daily, and there are seasonal changes in consumption. Buckwheat raw noodle production, which may also be an indicator of consumption at Soba restaurants because of its short shelf-life, is less varied among prefectures (Fig. 5.11) when compared to yield and cultivation (Figs. 5.4 and 5.5). However, Saitama and Nagano used the highest amount of buckwheat flour for raw noodle production in 2009.

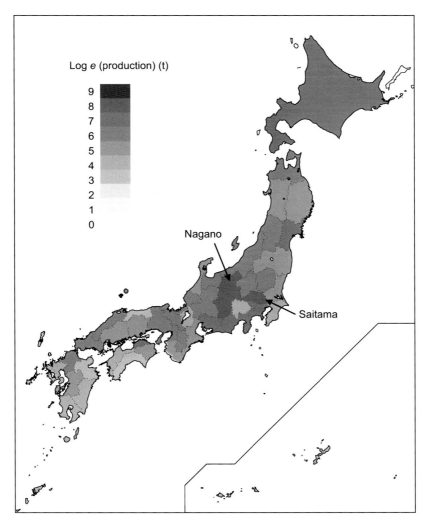

FIGURE 5.11 Regional variation in production of buckwheat raw noodle in 2009. The data (metric ton) were natural log-transformed, classified, and displayed in gray scale by prefecture. *Statistical data were retrieved from the Annual Report of Rice and Wheat Processed Food Production edited by the Ministry of Agriculture, Forestry and Fisheries of Japan (http://www.maff.go.jp/j/tokei/index.html).*

It is believed that grains with green seed coats produce better aroma and taste than white/brown grains. Grains are usually harvested when the blackened husk ratio reaches 80% (ie, when 20% of total grains are still immature). However, some producers harvest 1–2 weeks earlier when the blackened husk ratio is only 40–70% (Wada, 2011). The early harvest results in more green grains with high rutin content and antioxidant properties and good sensory evaluation, although the grains have higher water content (Wada, 2011). Although the color tone of the buckwheat flour does not change much until 180 days at room temperature (Wada, 2011), aroma and taste may be gradually lost, as has been reported in the previous studies (Kawakami et al., 2008, 2009). For these reasons, domestically produced and even locally sourced buckwheat flour is desired.

Development of Novel Cultivars With High Added Value

Recently, in addition to the eating quality, the physiological traits (health-promoting properties) of rutin are attracting more attention. Morishita and Tetsuka (2001, 2002) demonstrated yearly and varietal differences in agronomic traits and seed components of 38–76 common buckwheat varieties grown in Kyusyu. They showed that rutin content was positively and negatively correlated with mean temperature of the grain-filling period and grain yield, respectively. These researchers found several high rutin content varieties originating from Nepal. Morishita et al. (2006) demonstrated that the rutin content and antioxidant activity of Tartary buckwheat grain were 100–250-fold and 3–4-fold higher than that of common buckwheat grain, respectively, in two locational field experiments across 2 years. Morishita et al. (2007) showed that rutin is a major antioxidant source (~90%) in Tartary buckwheat and an unknown compound is a major one in common buckwheat.

Suzuki et al. (2014a) isolated four trace rutinosidase activity lines (originating from eastern Nepal) including "f3g-162" from ~500 lines of Tartary buckwheat. A new variety, "Manten-Kirari," has been developed, in which the activity of two types of rutinosidases that convert rutin to quercetin are extremely reduced, resulting in lowered rutin degradation and a nonbitter taste (Suzuki et al., 2014b).

Additionally, food allergies have become a more and more serious social problem in Japan. Buckwheat-containing food requires a mandatory label because it is one of seven major or intensely acting allergenic foods, which include eggs, milk, wheat, peanuts, shrimp, crab, and buckwheat. In buckwheat grains, the two major allergens are Fag e 1 (β-polypeptides of 13S globulin) and Fag e 2 (16-kDa 2S albumin) (Nair and Adachi, 1999; Yoshimasu et al., 2000). 2S albumin is likely involved in an immediate hypersensitivity reaction that includes anaphylaxis (Tanaka et al., 2002). 2S albumin is encoded by one authentic and three homologous genes that all exist within a ~108 kbp region of buckwheat genome and by one pseudo gene (unpublished data). 13S globulin, which reacts with IgE from almost all patients, is a multigenic protein containing a 0–6 time repeated insertion sequence (Khan et al., 2012). The complicated structure and function of this protein have been gradually elucidated at a protein and DNA level (Fujino et al., 2001; Morita et al., 2006; Sano et al., 2014). It is therefore desirable to develop a hypoallergenic buckwheat plant using such molecular information.

Consequently, improved grain quality in taste, physiological properties, functionality, allergenicity, and widening consumption styles are desirable, as such efforts would further stimulate buckwheat consumption and production in Japan.

ACKNOWLEDGMENT

This work was partly supported by the Sapporo Bioscience Foundation.

REFERENCES

Allotey, D.F.K., Horiuchi, T., Miyagawa, S., 1997a. Influence of split ammonium sulfate on nitrogen availability from ^{15}N-labeled matured soybean as green manure for buckwheat (*Fagopyrum esculentum* Moench). Jpn. J. Crop Sci. 66, 76–84.

Allotey, D.F.K., Horiuchi, T., Miyagawa, S., 1997b. Growth and nutrient dynamics of buckwheat (*Fagopyrum esculentum* Moench) as influenced by different applications of green soybean manure and bio-decomposer. Jpn. J. Crop Sci. 66, 407–417.

Fujimura, Y., Oba, S., Horiuchi, T., 2001. Effects of fertilization and poliploidy on grain shedding habit of cultivated buckwheats (*Fagopyrum* spp.). Jpn. J. Crop Sci. 70, 221–225 [Japanese with English abstract].

Fujino, K., Funatsuki, H., Inada, M., Shimono, Y., Kikuta, Y., 2001. Expression, cloning, and immunological analysis of buckwheat (*Fagopyrum esculentum* Moench) seed storage proteins. J. Agric. Food Chem. 49, 1825–1829.

Funatsuki, H., Suvorova, G.N., Sekimura, K., 1996. Determinate type variants in Japanese buckwheat lines. Breed. Sci. 46, 275–277.

Hara, T., Ohsawa, R., 2013. Accurate evaluation of photoperiodic sensitivity and genetic diversity in common buckwheat under a controlled environment. Plant Prod. Sci. 16, 247–254.

Hara, T., Matsui, K., Noda, T., Tetsuka, T., 2007. Effects of preharvest sprouting on flour pasting viscosity in common buckwheat (*Fagopyrum esculentum* Moench). Plant Prod. Sci. 10, 361–366.

Hara, T., Teruya, H., Shiono, T., Ikoma, H., Tetsuka, T., Matsui, K., Michiyama, H., 2008a. Grain yield and morphological character of buckwheat (*Fagopyrum esculentum* Moench) grown in winter in subtropical Japan. Jpn. J. Crop Sci. 77, 151–158 [Japanese with English abstract].

Hara, T., Tetsuka, T., Matsui, K., Ikoma, H., Sugimoto, A., 2008b. Evaluation of cultivar differences in preharvest sprouting of common buckwheat (*Fagopyrum esculentum* Moench). Plant Prod. Sci. 11, 82–87.

Hara, T., Matsui, K., Ikoma, H., Tetsuka, T., 2009a. Cultivar difference in grain yield and preharvest sprouting in buckwheat (*Fagopyrum esculentum* Moench). Jpn. J. Crop Sci. 78, 189–195 [Japanese with English abstract].

Hara, T., Sasaki, T., Tetsuka, T., Ikoma, H., Kohyama, K., 2009b. Effects of sprouting on texture of cooked buckwheat (*Fagopyrum esculentum* Moench) noodles. Plant Prod. Sci. 12, 492–496.

Hara, T., Arakawa, Y., Takeuchi, M., Sumi, H., Shiono, T., Takamine-Yamaguchi, N., Teruya, H., Ikoma, H., 2011. Effects of composted manure on growth and grain yield of buckwheat (*Fagopyrum esculentum* Moench) in acid soil in subtropical Japan. Jpn. J. Crop Sci. 80, 35–42 [Japanese with English abstract].

Hara, T., Arakawa, Y., Nagahama, R., Yamaguchi, N., Sumi, H., Tanaka, A., Ikoma, H., 2014. Growing buckwheat (*Fagopyrum esculentum* Moench) with composted manure and reduced chemical fertilizer on acid soil in subtropical Japan. Jpn. J. Crop Sci. 83, 118–125 [Japanese with English abstract].

Hayashi, H., 2011. Perspective of Japanese buckwheat viewed from the world. Tokusansyubyo 10, 2–5 [Japanese].

Honda, Y., Mukasa, Y., Suzuki, T., Maruyama-Funatsuki, W., Funatsuki, H., Sekimura, K., Kato, S., Wagatsuma, M., 2009. The breeding and characteristics of a common buckwheat cultivar "Kitanomashu". Res. Bull. Natl. Agric. Res. Cent. Hokkaido Reg. 191, 41–52 [Japanese with English abstract].

Horigane, A., Yamada, S., 2005. Evaluation of buckwheat noodle. Refrigeration 80, 399–401 [Japanese with English abstract].

Horigane, A., Yamada, S., Hikichi, Y., Ohba, S., Matsukura, U., Imai, T., 2004. Analysis of mixing process of hydrated buckwheat flour: development of quality evaluation method of buckwheat noodle: part I. Nippon Shokuhin Kogyo Gakkaishi 51, 346–351 [Japanese with English abstract].

Hosokawa, H., 2011. Mitigation of excess soil moisture damage with simultaneously operating machine for tillage-list-seeding in converted fields from a paddy. Tokusansyubyo 10, 40–44 [Japanese].

Ikeda, K., 2002. Buckwheat: composition, chemistry, and processing. Taylor, S.L. (Ed.), Advances in Food and Nutrition Research, vol. 44, Academic Press, San Diego, CA, pp. 395–434.

Ikeda, K., Kishida, M., Kreft, I., Yasumoto, K., 1997. Endogenous factors responsible for the textural characteristics of buckwheat products. J. Nutr. Sci. Vitaminol. 43, 101–111.

Inoue, N., 2004. Variation of chemical composition in buckwheat seeds as influenced by cultivation site. Hokuriku Crop Sci. 40, 139–145 [Japanese].

Kawakami, I., Murayama, N., Kawasaki, S., Igasaki, T., Hayashida, Y., 2008. Effects of storage temperature on flavor of stone-milled buckwheat flour. Nippon Shokuhin Kagaku Kogaku Kaishi 55, 559–565 [Japanese with English abstract].

Kawakami, I., Murayama, N., Kawasaki, S., Igasaki, T., Hayashida, Y., 2009. Influence of storage period on characteristics of stone-milled buckwheat flour. Nippon Shokuhin Kagaku Kogaku Kaishi 56, 513–519 [Japanese with English abstract].

Khan, N., Takahashi, Y., Katsube-Tanaka, T., 2012. Tandem repeat inserts in 13S globulin subunits, the major allergenic storage protein of common buckwheat (*Fagopyrum esculentum* Moench) seeds. Food Chem. 133, 29–37.

Matano, T., 1990. Recent researches on buckwheat: trends in the world around. Jpn. J. Crop Sci. 59, 582–589 [Japanese].

Matsui, K., Kiryu, Y., Komatsuda, T., Kurauchi, N., Ohtani, T., Tetsuka, T., 2004. Identification of AFLP makers linked to non-seed shattering locus (*sht1*) in buckwheat and conversion to STS markers for marker-assisted selection. Genome 47, 469–474.

Matsui, K., Tetsuka, T., Hara, T., Morishita, T., 2008. Breeding and characterization of a new self-compatible common buckwheat parental line "Buckwheat Norin-PL1". Res. Bull. Natl. Agric. Res. Cent. Kyushu Okinawa Reg. 49, 11–17 [Japanese with English abstract].

Matsuura, A., Inanaga, S., Tetsuka, T., Murata, K., 2005. Differences in vegetative growth response to soil flooding between common and Tartary buckwheat. Plant Prod. Sci. 8, 525–532.

Michiyama, H., Fukui, A., Hayashi, H., 1998. Differences in the progression of successive flowering between summer and autumn ecotype cultivars in common buckwheat (*Fagopyrum esculentum* Moench). Jpn. J. Crop Sci. 67, 498–504 [Japanese with English abstract].

Michiyama, H., Arikuni, M., Hirano, T., Hayashi, H., 2003. Influence of day length before and after the start of anthesis on the growth, flowering and seed-setting in common buckwheat (*Fagopyrum esculentum* Moench). Plant Prod. Sci. 6, 235–242.

Morishita, T., 2011. Buckwheat breeding in NARO Hokkaido Agricultural Research Center. Tokusansyubyo 10, 18–21 [Japanese].

Morishita, T., Tetsuka, T., 2001. Year-to-year variation and varietal difference of agronomic characters of common buckwheat in the Kyushu area. Jpn. J. Crop Sci. 70, 379–386 [Japanese with English abstract].

Morishita, T., Tetsuka, T., 2002. Varietal differences of rutin protein and oil content of common buckwheat (*Fagopyrum esculentum*) grains in Kyushu area. Jpn. J. Crop Sci. 71, 192–197 [Japanese with English abstract].

Morishita, T., Yamaguchi, H., Degi, K., Tetsuka, T., 2006. Agronomic characters and chemical component of grains of Tartary buckwheat. Jpn. J. Crop Sci. 75, 335–344 [Japanese with English abstract].

Morishita, T., Yamaguchi, H., Degi, K., 2007. The contribution of polyphenols to antioxidative activity in common buckwheat and Tartary buckwheat grain. Plant Prod. Sci. 10, 99–104.

Morita, N., Maeda, T., Saia, R., Miyake, K., Yoshioka, H., Urisu, A., Adachi, T., 2006. Studies on distribution of protein and allergen in graded flours prepared from whole buckwheat grains. Food Res. Int. 39, 782–790.

Mukasa, Y., 2011. Studies on new breeding methodologies and variety developments of two buckwheat species (*Fagopyrum esculentum* Moench and *F. tataricum* Gaertn). Res. Bull. NARO Hokkaido Agric. Res. Cent. 195, 57–114 [Japanese with English abstract].

Nagatomo, T., 1984. The Science of Buckwheat. Shinchosha, Tokyo [Japanese].

Nair, A., Adachi, T., 1999. Immunodetection and characterization of allergenic proteins in common buckwheat (*Fagopyrum esculentum*). Plant Biotechnol. 16, 219–224.

Nijima, S., 2011. Dictionary of Buckwheat. Kodansha, Tokyo [Japanese].

Oba, S., Suzuki, Y., Fujimoto, F., 1998. Breaking strength of pedicel and grain shattering habit in two species of buckwheat (*Fagopyrum* spp.). Plant Prod. Sci. 1, 62–66.

Oba, S., Ohta, A., Fujimoto, F., 1999. Breaking strength of pedicel as an index of grain-shattering habit in autotetraploid and diploid buckwheat (*Fagopyrum esculentum* Moench) cultivars. Plant Prod. Sci. 2, 190–195.

Oba, S., Fujimura, Y., Horiuchi, T., 2004. Association of grain shedding habit with polyploidy in Tartary buckwheat (*Fagopyrum tataricum*) strains. Plant Prod. Sci. 7, 212–216.

Ohsako, T., Ohnishi, O., 1998. New *Fagopyrum* species revealed by morphological and molecular analyses. Genes Genet. Syst. 73, 85–94.

Ohsako, T., Ohnishi, O., 2000. Intra- and interspecific phylogeny of wild *Fagopyrum* (Polygonaceae) species based on nucleotide sequences of noncoding regions in chloroplast DNA. Am. J. Bot. 87, 573–582.

Ohsawa, R., 2011. Current status of buckwheat breeding in Japan. Tokusansyubyo 10, 12–17 [Japanese].

Sakata, K., Ohsawa, R., 2005. Effect of flooding stress on the seeding emergence and the growth of common buckwheat. Jpn. J. Crop Sci. 74, 23–29 [Japanese with English abstract].

Sakata, K., Ohsawa, R., 2006. Varietal differences of flood tolerance during germination and selection of the tolerant lines in common buckwheat. Plant Prod. Sci. 9, 395–400.

Sano, M., Nakagawa, M., Oishi, A., Yasui, Y., Katsube-Tanaka, T., 2014. Diversification of 13S globulins, allergenic seed storage proteins, of common buckwheat. Food Chem. 155, 192–198.

Sugimoto, H., 2004a. Effects of nitrogen application on the growth and yield of summer buckwheat cultivated in western Japan with special reference to dry matter production and nitrogen absorption. Jpn. J. Crop Sci. 73, 181–188 [Japanese with English abstract].

Sugimoto, H., 2004b. Effects of incorporation timing and the amount of Chinese milk vetch as green manure on the flowering and ripening of summer buckwheat. Jpn. J. Crop Sci. 73, 424–430 [Japanese with English abstract].

Sugimoto, H., Sato, T., 1999. Summer buckwheat cultivation in the warm southwestern region of Japan: effects of sowing time on growth and seed yield. Jpn. J. Crop Sci. 68, 39–44 [Japanese with English abstract].

Sugimoto, H., Sato, T., 2000. Effects of excessive soil moisture at different growth stages on seed yield of summer buckwheat. Jpn. J. Crop Sci. 69, 189–193 [Japanese with English abstract].

Sugimoto, H., Kurono, M., Takano, K., Khono, Y., Sato, T., 2000. Is Chinese milk vetch useful as green manure for summer buckwheat cultivation? Jpn. J. Crop Sci. 69, 24–30 [Japanese with English abstract].

Sugimoto, H., Sugiyama, H., Morimoto, M., Yamamoto, A., 2007. Effects of seeding rate sowing time and incorporation timing of Chinese milk vetch (*Astragalus sinicus* L.) as green manure on the growth and yield of summer buckwheat (*Fagopyrum esculentum* Moench). Jpn. J. Crop Sci. 76, 45–51 [Japanese with English abstract].

Suzuki, T., Mukasa, Y., Morishita, T., Takigawa, S., Noda, T., 2012. Traits of shattering resistant buckwheat "W/SK86GF". Breed. Sci. 62, 360–364.

Suzuki, T., Morishita, T., Mukasa, Y., Takigawa, S., Yokota, S., Ishiguro, K., Noda, T., 2014a. Discovery and genetic analysis of non-bitter Tartary buckwheat (*Fagopyrum tataricum* Gaertn.) with trace-rutinosidase activity. Breed. Sci. 64, 339–343.

Suzuki, T., Morishita, T., Mukasa, Y., Takigawa, S., Yokota, S., Ishiguro, K., Noda, T., 2014b. Breeding of "Manten-Kirari", a non-bitter and trace-rutinosidase variety of Tartary buckwheat (*Fagopyrum tataricum* Gaertn.). Breed. Sci. 64, 344–350.

Takahashi, K., 2011. Situation concerning buckwheat. Tokusansyubyo 10, 6–11 [Japanese].

Tanaka, K., Matsumoto, K., Akasawa, A., Nakajima, T., Nagasu, T., Iikura, Y., Saito, H., 2002. Pepsin-resistant 16-kD buckwheat protein is associated with immediate hypersensitivity reaction in patients with buckwheat allergy. Int. Arch. Allergy Immunol. 129, 49–56.

Tetsuka, T., Uchino, A., 2002. Variation of plant types in Japanese native cultivars of common buckwheat (*Fagopyrum esculentum* Moench). Jpn. J. Crop Sci. 71, 493–499 [Japanese with English abstract].

Tetsuka, T., Uchino, A., 2005. Variation in seed shape and husk color in Japanese native cultivars of common buckwheat (*Fagopyrum esculentum* Moench). Plant Prod. Sci. 8, 60–64.

Ujihara, A., Matano, T., Yamamoto, Y., 1977. Some properties of autotetraploid buckwheat. Bull. Fac. Agric. Shinshu Univ. 14, 127–136 [Japanese with English abstract].

Wada, Y., 2011. Production and quality improvement technology of early harvested buckwheat. Tokusansyubyo 10, 52–55 [Japanese].

Wagatsuma, T., 2004. Selection of determinate type plants from common buckwheat cultivar "Kitawasesoba" and development of a new line, Horokei 3. J. Rakuno Gakuen Univ. 29, 1–7 [Japanese with English abstract].

Yasui, Y., Mori, M., Aii, J., Abe, T., Matsumoto, D., Sato, S., Hayashi, Y., Ohnishi, O., Ota, T., 2012. *S-Locus Early Flowering 3* is exclusively present in the genomes of short-styled buckwheat plants that exhibit heteromorphic self-incompatibility. PLoS One 7 (2), e31264.

Yoshimasu, M.A., Zhang, J.W., Hayakawa, S., Mine, Y., 2000. Electrophoretic and immunochemical characterization of allergenic proteins in buckwheat. Int. Arch. Allergy Immunol. 123, 130–136.

Chapter six

The Unique Value of Buckwheat as a Most Important Traditional Cereal Crop in Ukraine

L.K. Taranenko*, O.L. Yatsyshen**,
P.P. Taranenko†, T.P. Taranenko‡

**Scientific-Production Enterprise Antaria, Kiev, Ukraine;
**National Scientific Center "Institute of Agriculture" of the National
Academy of Agricultural Sciences, Department of breeding of groat crops,
Kiev, Ukraine; †Scientific-Production Enterprise Antaria, Foreign Relations
Department, Kiev, Ukraine; ‡Scientific-Production Enterprise Antaria,
Marketing Department, Kiev, Ukraine*

THE UNIQUE VALUE OF BUCKWHEAT IN UKRAINE

During 4000 years of cultivation, buckwheat crops have spread from Southeast Asia to the countries of Western Europe, Africa, and South and North America.

Although the interest in growing buckwheat has grown recently, its cultivation area is only 2.5 million hectares and is mainly located in moderate temperate climate zones of the Northern Hemisphere.

The largest planted area of buckwheat is in China (1,584,000 ha or 75–80%), which is followed by smaller areas in Russia, Ukraine, Japan, Canada, the United States, Brazil, and countries of former Yugoslavia. In Western Europe the leading buckwheat grower is France, where in spite of a relatively small planted area (31,900 ha) the country produces the biggest yield (3.5 tons/ha). In other European countries, buckwheat planted areas are of a modest size and are frequently not taken into account in statistics (Parahin, 2010).

Buckwheat is the most important form of traditional cereal crops in Ukraine.

High demand for this product is caused by its unique nutritional and dietary properties, which can improve the activity of digestive and blood vascular systems and possibly reduce negative effects of increased nuclear radiation exposure.

Molecular Breeding and Nutritional Aspects of Buckwheat. http://dx.doi.org/10.1016/B978-0-12-803692-1.00006-7

Buckwheat grains contain 12.6% of proteins. This is much more than other cereals and bread (rice 7.0%, millet 12.0%, oats 11.3%, pearl barley 9.8%, rye bread 5.0%, and white bread 8.6%) (Savytskyi, 1963).

Most buckwheat proteins (up to 80%) are soluble and are well assimilated by the human digestive system. Their high dietary value is caused by the diversity of amino acids (17) and high prevalence of essential amino acids (tryptophan, lysine, methionine), which are insufficient in quantities in other cereals and breads. Buckwheat contains considerable amounts of histidine, which stimulates growth and development in children. Because of the high content of cysteine (300 mg/100 g of buckwheat) the plant has radioprotective properties. The biological value of buckwheat proteins approximates to dry milk (92.3%) and chicken eggs (81.4–99.8%). The high assimilation rates of buckwheat proteins are explained by the presence of small amounts of organic acids (citric acid, oxalic acid, malic acid) in the grains, which improves activity of the human digestive system (Zotikov et al., 2010).

Boiled buckwheat is an important source of valuable minerals such as phosphorus, calcium, iron, copper, manganese, and zinc. Zinc content in buckwheat is 2.6 times higher than in other cereals. Levels of vitamins PP (nicotinic acid), B_1 (thiamine), B_2 (riboflavin), and E (tocopherol) are also higher. All these facts make buckwheat an important dietary product and explain the existing interest in the introduction of biomolecular methods in the breeding process of buckwheat (Zotikov et al., 2010).

Recently, an interest in buckwheat as a plant for use in medicine has appeared. Because of its high rutin (vitamin P) content it is used for maintaining the functional properties of the vascular system. Buckwheat has a positive influence on dysfunctions in blood circulation, cardiovascular collapse, vasospasms, and edemas. Also a buckwheat diet is used for the prevention and treatment of arterial sclerosis with lipidemia (Zotikov et al., 2010).

The proteins in buckwheat in contrast to the proteins of other cereal crops have no gluten. This explains why buckwheat flour is used in the production of different types of thin cookies. In combination with soya bean flour it is used in the production of certain types of chocolate. In China, Korea, and Japan, buckwheat flour is used in the production of vermicelli.

The calorie content of buckwheat is determined by the high content of carbohydrates (85%) and low content of gluten (1%), which does not assimilate in the human digestive system. Buckwheat grains contain 3.8% of fats. The low value of the iodine number (94.2–96%) in buckwheat indicates a high content of nondrying oils (linoleic and lanoline acids), and its low oxidizing number shows its high resistance to oxidizing. Resistance to oxidizing indicates the presence of substantial quantities of vitamin E, which has antioxidant properties. Because of all these facts, buckwheat can be stored for a very long time without loss of nutritional and dietary properties (Zotikov et al., 2010).

Products from buckwheat are very valuable because of their unique composition of vitamins and enzymes. They contain rutin (vitamin P), nicotinic acid (vitamin PP), thiamine (B_1), riboflavin (B_2), folic acid (B_9), cyanocobalamin (B_{12}), and the enzymes proteinase, aminase, maltase, glucosidase, and phytase (Zotikov et al., 2010).

Buckwheat is the only grain crop that contains rutin. The highest content of rutin in buckwheat is in the flowers and leaves; it is lower in the stems and is lowest in grains. Rutin reduces penetrability and fragility of blood vessels, improves the ability of the vascular system to restore, and prevents the occurrence of dying-off tissues after frostbite. Rutin is used for the treatment and prophylaxis of pleurisy, endocarditis, peritonitis, glaucoma, hypertension, pancreatic diabetes, measles, and epidemic typhus. It is also recommended for those who work with X-ray equipment and radioactive materials.

Nicotinic acid (vitamin PP) presents in all organs of the buckwheat plant, but the highest concentration is observed in the seed coat. Concentration of nicotinic acid in the grains is measured as 4.4 mg per 100 g. Vitamin PP is known to take part in redox reactions of animal organisms. In the absence or insufficiency of these vitamin levels, the synthesis of pigments can sustain damage, and damage to the liver and vasoconstriction can occur. Nicotinic acid reduces cholesterol and optimizes the ratio of its fraction. This is why it is recommended for the treatment of arterial sclerosis with lipidemia, ulcer diseases, etc.

From a medical point of view the most important feature of the dietary properties of buckwheat is the high content of the vitamin B group. Pollen and grains of buckwheat are rich in vitamin B_1, which takes part in the regulation of synaptic stimulation. This is why buckwheat is useful for prophylactics and the treatment of neuritis, neuralgia, radiculitis, paralysis, and syncopes. It also used for the treatment of dermatoses and intoxication by mercury, methyl alcohol, and arsenic.

Vitamin B_6 takes part in the metabolism of carbohydrates and prevents vitamin insufficiency in pregnant woman and in children. Being a natural source of the vitamin, buckwheat food is recommended also for patients with vascular and radiation diseases, children's hepatitis and dermatitis, intoxications caused by drugs against tuberculosis, and other diseases.

Folic acid (B_7) and cyanocobalamin (B_{12}) take part in the synthesis of amino acids. They play an important role in the treatment of anemia, gastric diseases, and leukemia caused by nuclear irradiation.

Riboflavin (vitamin B_2) has influence on the metabolism of proteins and carbohydrates, hemoglobin synthesis, and improves eyesight.

Buckwheat contains optimal amounts of copper that improve efficiency of iron consumption for hemoglobin synthesis in red blood cells. Insufficiency of copper in an organism can lead to anemia. Copper content in 100 g of buckwheat is 21.8 mg, whereas in fine ground barley it is 11.1 mg, oatmeal 8.68 mg, semolina 7.24 mg, and millet 18.8 mg.

Buckwheat has unique properties in reducing the temperature of the human body, in treating serious disorders of full processes, the mucous and lymphatic system, and regulating the metabolism of lipids, proteins, and carbohydrates in conditions of ionizing irradiation.

The unique combination of biologically valuable compounds ensures buckwheat's usefulness as an extraordinary food, and a strategic, medical, and pharmaceutical material.

In addition, buckwheat is a valuable feed crop. Defective grains and waste products from processing are used as feed in pig farming and poultry farming. One kilogram of these products contains 57 g of proteins. Its feed value is 0.5 feed units.

Buckwheat grows rapidly and during 50–60 days can form up to 2 tons of biomass per hectare. In this way it can be used as green fodder. However, flowers and fruit coats of the crop contain a fagopyrin pigment, which can cause fagopyrism of animals of a white color. This should be taken into account (Zotikov et al., 2010).

In addition, a sufficiently large biomass of buckwheat with a high content of potassium, phosphorus, magnesium, and other chemical elements, which improves soil fertility, can be used as green manure.

Empty glumes of buckwheat are used in Japan as a filling for pillows. In Western Europe they are used for packaging containers in the food-canning industry. In Russia, buckwheat empty glumes are used as a source for potash. And recently the glumes have been used for the production of natural colors in the food industry.

Buckwheat is one of the best plants for bees. The average yield of buckwheat honey is 70–100 kg/ha. The greatest yield of buckwheat honey was 259.8 kg/ha. In addition, it produces a lot of pollen (up to 225 kg/ha) (Zotikov et al., 2010).

Because of late sowing time and a short vegetation period, buckwheat is a traditional assurance crop in the case of emergency resowing of winter and early spring crops (Taranenko and Yatsyshen, 2014).

Buckwheat, which is grown under good technological conditions, leads to the elimination of harmful weeds such as wheatgrass and sow thistle. Its roots and after-harvesting remains contain a lot of phosphorus and potassium. This is why buckwheat is a good forecrop for most winter and spring crops. In modern crop rotations, which have high saturation by cereal crops, the importance of buckwheat as a forerunner has dramatically improved (Taranenko and Yatsyshen, 2014).

Buckwheat is one of the most promising crops for growing in repeated sowing. Fallow changed and after-harvesting sowing can be attempted in areas where the sum of active temperatures ($>10°C$) after harvesting of winter wheat is $100°C$ (90–100 warm days), which makes possible the growing of early ripening and mid-ripening varieties.

It is especially important to grow buckwheat as a soil-reclamation crop in rice crop rotation. It helps to purify fields from weeds and diseases and improves the yield of rice by 0.5–0.7 tons/ha (Savytskyi et al., 1968).

The unique value of products from buckwheat and the irreplaceability of its dietary and medicinal properties by other sources can permanently increase demand for this crop.

REFERENCES

Parahin, N.V., 2010. Buckwheat: biological capabilities and ways to implement them. Vestnik OrelGAU 4, 4–8.

Savytskyi, K.A., 1963. Culture of buckwheat in Ukraine. State Publishing House of Agricultural Literature of Ukrainian SSR, 204 p.

Savytskyi, K.A., Yashovskyi, I.V., Laktionov, B.I., Bahnenko, V.K., 1968. Groat Crops. Urozhai, Kyiv., 260 p.

Taranenko, L.K., Yatsyshen, O.L., 2014. Principles, Methods and Achievements of Buckwheat Selection (*Fagopyrum esculentum* Moench). Nilan Ltd., Vinnytsia, 224 p.

Zotikov, V.I., Naumkina, T.S., Sidorenko, V.S., 2010. State of the art and perspectives of development of buckwheat production in Russia. Vestnik OrelGAU 4, 18–22.

Chapter seven

Interspecific Crosses in Buckwheat Breeding

G. Suvorova

All-Russia Research Institute of Legumes and Groat Crops,
Laboratory of Genetics and Biotechnology, Orel, Russia

INTRODUCTION

At least 18 known *Fagopyrum* species are divided into two major groups. The *cymosum* group includes species with a large achene not completely covered with remaining perianths and the *urophyllum* group includes species with small achenes completely covered with remaining perianths (Ohnishi, 2013). In spite of a continuing discussion about the systematic status of one or the other species, the subdivision of the genus *Fagopyrum* into two groups first proposed by Ohnishi and Matsuoka (1996) is unquestionable.

For practical utility the *cymosum* group has two cultivated species, common buckwheat *Fagopyrum esculentum* Moench. and Tartary buckwheat *Fagopyrum tataricum* Gaertn., and two wild species, *Fagopyrum homotropicum* Ohnishi and *Fagopyrum cymosum* Meissn. The wild buckwheat species in the *urophyllum* group have grains that are too small to collect in the wild, hence no species of the *urophyllum* group are collected for utilizing them by local people (Ohnishi, 2013).

The first interspecific hybrid in the genus *Fagopyrum* was an artificial allopolyploid *Fagopyrum giganteum* Krotov obtained as a result of conventional crossing between *F. tataricum* and *F. cymosum* (Krotov and Golubeva, 1973). But above all, numerous attempts were made to obtain hybrids of the common buckwheat *F. esculentum* and other species. The development of a biotechnological method of the embryo rescue technique, on the one hand, and a discovery of a new species *F. homotropicum* (Ohnishi, 1995), on the other hand, extended the possibilities of the interspecific hybridization of buckwheat.

The interspecific crosses are aimed at improvement of the existing varieties and acquiring new characters, which are attributed to the wild relatives but are lacking in cultivated species. Here we briefly describe the results and perspectives of interspecific hybridization in the large achene *cymosum* group of *Fagopyrum*.

87

Molecular Breeding and Nutritional Aspects of Buckwheat. http://dx.doi.org/10.1016/B978-0-12-803692-1.00007-9

The Crossing of *F. esculentum* × *F. cymosum*

F. cymosum is a wild perennial buckwheat with white heterostyle flowers, which grows from the southern part of China to southwest Asia (Ujihara et al., 1990). It is utilized in some areas as a green vegetable or as cattle forage (Campbell, 2003). Only *F. cymosum* (from wild species) is used as a medicinal plant, mainly in China (Ohnishi, 2013).

The species exists both in diploid and tetraploid forms, and it is considered that the tetraploid populations of *F. cymosum* arose at least twice (Yasui and Ohnishi, 1998; Yamane et al., 2003). Ohnishi (2010) proposed that all the *cymosum* forms, such as diploid as well as tetraploid, should be classified into one species.

The cross combination of *F. esculentum* × *F. cymosum* is characterized by strong postgamic incompatibility. The pollen tubes of *F. esculentum* and *F. cymosum* grew satisfactorily in their reciprocal pollination (Golyshkin and Fesenko, 1988; Hirose et al., 1994). However, the fertilized embryo could grow to the rod or early globular stage (2–3 days after pollination) and thereafter be gradually degraded (Shaikh et al., 2002b). The postfertilization barrier can be overcome only by using the ovule rescue technique.

For the first time the interspecific hybrids between common buckwheat *F. esculentum* and perennial species *F. cymosum* were obtained by Japanese and Russian researchers independently (Ujihara et al., 1990; Suvorova et al., 1994).

Fourteen interspecific hybrids were recovered by Ujihara et al. (1990) through ovule culture between cultivated tetraploid common buckwheat *F. esculentum* and the wild tetraploid species *F. cymosum*. Six hybrids were successfully grown to the flowering stage. All of the hybrid plants had perennial growth habits and their stems or leaves appeared morphologically intermediate of the parent species. While fertilized ovules could be developed through backcrossing hybrids with common buckwheat, all of the ovaries aborted before maturation. But the authors succeeded in obtaining two whole BC_1F_1 plants at the flowering stage. Later this combination was repeated at the diploid level, where 21 hybrids were obtained, which showed a heterostyle and perennial nature (Hirose et al., 1995).

The first two viable hybrid clones in the crossing of *F. esculentum* and *F. cymosum* were obtained by Suvorova et al. (1994) using the ovule rescue technique. Hybrid plants were periodically regenerated from the in vitro tissue culture characterized by intensive branching and resembled more the paternal species *F. cymosum*. Later, more than 30 embryos were recovered and hybrid plants were produced for 8 years by crossing *F. esculentum* and *F. cymosum* at the tetraploid level (Suvorova, 2001). The plants set ovaries as spontaneously as in artificial pollination, but mature seeds were not formed. By means of repeated use of the ovule rescue technique, 14 clones BC_1F_1 and two clones BC_2F_1 were raised. The plants BC_1 (F_1 × *F. esculentum*) were morphologically similar to the recurrent parent *F. esculentum*. They all had a low level of branching, large thrum flowers, and all were vigorous. The plants BC_1 (F_1 × *F. cymosum*) were less vigorous, but their branching level was more expressed. The clones BC_2F_1 developed weakly both in vitro and in vivo. No mature seeds were obtained from the backcrossed progenies.

The third species, *F. homotropicum*, was used in crosses with F_1 hybrids (Suvorova, 2001, 2010). As a result of multiple crossing and ovule rescue, complex hybrids [(*F. esculentum* × *F. cymosum*) × *F. homotropicum*] × *F. homotropicum* and [(*F. esculentum* × *F. cymosum*) × *F. homotropicum*] × *F. esculentum* were obtained. The presence of a genome of *F. homotropicum* was indicated by random amplified polymorphic DNA (RAPD) analysis, which confirmed the obtaining of a trispecies hybrid between *F. esculentum*, *F. cymosum*, and *F. homotropicum*. The hybrid plants were vigorous in their natural habitat but not able to form mature seeds on their own and needed ovule culture to receive progeny. Various hybrid plants are shown in Fig. 7.1.

FIGURE 7.1 Interspecific buckwheat hybrids and their parents. (A) *F. cymosum* plant, (B) F_1 (*F. esculentum* × *F. cymosum*) plant, (C) F_1 (*F. esculentum* × *F. cymosum*) × *F. esculentum* plant, (D) trispecies hybrid in vitro, (E) plantlet of trispecies hybrid, (F) [(*F. esculentum* × *F. cymosum*) × *F. homotropicum*] × *F. homotropicum* plant in vivo (Suvorova, 2010).

As a result of an application of the embryo culture method, the plants from 10 embryos of *F. esculentum* × *F. cymosum* (4×) were obtained by Rumyantseva et al. (1995). The hybrids were able to produce seeds with crosses with *F. esculentum* or as a result of self-fertilization. However, the embryos in these seeds were gradually degenerated and irremediably damaged if the seeds stayed on plants more than 10–15 days after pollination.

Seven plants in the cross of *F. esculentum* × *F. cymosum* produced by Woo et al. (1999) through embryo rescue at the diploid level reached the flowering stage. All the hybrids were vigorous in their growth; the flowers were normal but were self-sterile.

The plants from three hybrid clones of *F. esculentum* × *F. cymosum* crossed at the diploid level by Yui et al. (2004) grew to flower. All clones were small and not vigorous; some ovaries enlarged to some extent but no seed was produced. Some flowers were backcrossed with *F. esculentum*; however, no seed was produced.

Thus the interspecific hybrids between *F. esculentum* and *F. cymosum* were obtained by many researches independently. In spite of differences in the embryo rescue procedure, the parent accessions, and the geographical location, this cross is characterized by high recurrence of the results. The hybrid plants raised at different times and in different countries were very similar morphologically, and all of them were unable to form mature seeds and produce progenies without recycling of ovule rescue. Hence sterility of interspecific hybrids F$_1$ (*F. esculentum* × *F. cymosum*) is not a particular case but characterizes the hybrid combination as a whole. The raising of backcrossed progenies could not overcome the sterility.

The chromosome number of somatic cells of the F$_1$ plants of *F. esculentum* × *F. cymosum* hybrids was found to be tetraploid ($2n = 32$) or aneuploid ($2n = 30$, $2n = 24$) (Ujihara et al., 1990; Rumyantseva et al., 1995). So we may suppose that mitotic cell division at the tetraploid level in most cases took place without abnormalities and therefore the hybrid plants were vigorous. But meiotic analysis of hybrids revealed some alterations in tetrad formation and meiotic instability seems to be the reason for hybrid sterility (Rumyantseva et al., 1995).

The problem of overcoming of interspecific incompatibility barrier in the crossing of *F. esculentum* with the wild perennial species *F. cymosum* has only been solved partially but only when the ovule culture technique recovered the hybrid embryo. However, the sterility of the interspecific hybrids *F. esculentum* × *F. cymosum* has not yet been overcome.

The Crossing of *F. tataricum* × *F. cymosum*

As previously mentioned the hybrid between *F. tataricum* and *F. cymosum* was the first interspecific hybrid in the genus *Fagopyrum*. The accession of *F. tataricum* from China transferred by colchicine treatment at the tetraploid level was taken as a female parent, and a tetraploid accession of *F. cymosum* from India served as a male parent. The 10 hybrid seeds, obtained by crossing, gave seedlings among which one plant was distinguished with white pin flowers, which was more vigorous and productive in comparison to parental species (Krotov and Golubeva, 1973). The hybrid nature of the plant was confirmed cytologically. The hybrid appeared to be amphidiploid and was determined by the author as a separate species, *Fagopyrum giganteum* Krotov (Krotov, 1975).

The artificial species *F. giganteum* was sometimes used as a male parent in crossing with *F. esculentum*, but all hybrid plants obtained through embryo rescue were female sterile and did not cross back to the parental species (Rumyantseva et al., 1995; Hirose et al., 1995; Woo et al., 1995).

Since *F. giganteum* has been used as a bridge species for introgression of traits of *F. cymosum* into cultivated Tartary buckwheat (Fesenko et al., 1998). In the progenies of the late generations (F_{10}–F_{15}) of the *F. tataricum* × *F. giganteum* hybrids the plants with the different growth habitus later or earlier maturing comparatively to the initial lines of *F. tataricum* ($4x = 2$) were selected (Fesenko and Fesenko, 2010). The most productive progenies of the F_{10} and later generations have been united by the authors into the new species *Fagopyrum hybridum*. *F. hybridum* is similar to the tetraploid *F. tataricum* in taste and characteristics of grain and has a yield at the level of cultivated buckwheat.

The Crossing of *F. tataricum* × *F. esculentum*

Tartary buckwheat *F. tataricum* is a self-pollinated annual species with greenish homostyle flowers. The species is presented both in wild and cultivated forms. Since 1989 Chinese scientists have revised their understanding of Tartary buckwheat because of its unique nutritional value (Lin, 2004). Tartary buckwheat contains a higher level of rutin compared to that of common buckwheat. At the present the consumption of Tartary buckwheat is increasing in China; it is utilized in some regions of Europe and has been introduced into Japan as one of the beneficial foods for human health (Ikeda et al., 2012). *F. tataricum* has long been considered as a parent in interspecific crosses because it has many desirable traits such as higher seed yield, self-pollination ability, frost resistance, and overall plant vigor (Campbell, 2003).

Observation of pollen tube growth and embryo development in the interspecific crossing of *F. esculentum* with *F. tataricum* demonstrated that the pollen tube of common buckwheat entered the embryo sac of Tartary buckwheat, the ovaries began to grow, but the ovules from those ovaries failed to produce any embryo in the culture medium (Samimy, 1991). The interspecific incompatibility was attributed to the failure of fertilization or to the embryo abortion associated with retarded zygotic development and early degeneration of endosperm (Shaikh et al., 2002a). Comparatively fewer abnormalities occurred in the hybrid embryo of the cross between *F. cymosum* and *F. esculentum* than in that between *F. tataricum* and *F. esculentum* (Shaikh et al., 2002b).

Despite the difficulties of crossing, the researchers continued the search for ways to overcome the incompatibility between *F. tataricum* and *F. esculentum* species.

Tartary buckwheat was successfully hybridized with common buckwheat at the diploid level by Samimy et al. (1996), using as the male parent the artificially synthesized unique genotype of common buckwheat with the isozyme alleles similar to that of Tartary buckwheat. Ovule culture was used to rescue the 7–10-day old embryos. Four out of 263 cultured ovules continued to grow as callus and one of these differentiated and formed a callus with buds and shoots from which cloned plants were produced. Flowers produced by the hybrid plants were of the same type (homomorphic) and size as those of Tartary, but with white sepals like common buckwheat. So far the hybrid buckwheat has not produced any seeds.

Hirose et al. (1995) could produce the interspecific hybrid between *F. tataricum* and *F. esculentum* at the tetraploid level. One hybrid plant was obtained out of 367 ovules planted in vitro. The hybrid plant had white flowers with a long pistil and short stamen.

Attempts to improve the success of interspecific hybridization between the two cultivated buckwheat species *F. tataricum* and *F. esculentum* were made by Wang et al. (2002a). Three sterile hybrids were produced from the interspecific cross of the two species from a total of 111 ovules grown with tissue culture. The hybrids had the homomorphic flowers of *F. tataricum* with a flower size intermediate between the two parents. One of the three hybrids was found to be a triploid derived from an unreduced female gamete of *F. tataricum*, and two hybrids were found to be diploid obtained as a result of the fusion of haploid gametes. Chromosome doubling and a bridge cross with the third species *F. homotropicum* were attempted in an effort to restore the hybrid fertility.

Eight seedlings regenerated from 7-day-old embryos resulting from crosses between the Nepalese species *F. tataricum* and *F. esculentum* were produced by Niroula et al. (2006). However, all the regenerated hybrid plants died after field plantation in the soil.

Azaduzzaman et al. (2009) obtained 17 hybrids of *F. tataricum* × *F. esculentum* from 334 ovules (diploid and tetraploid) by using *F. tataricum* as the female parent following ovule culture methods. The hybridity was confirmed using plant morphological characters, cytological observation, and RAPD analysis. The researchers were successful in obtaining F_2 generation. The flower color of the hybrids appeared to be intermediate like pink, but F_2 flower colors were pink, white, and whitish green. The diploid and tetraploid plants F_1 and F_2 in this experiment were more fertile than their parents.

The excellent data presented by Azaduzzaman et al. (2009) disagree with the results described earlier particularly concerning the fertility of *F. tataricum* × *F. esculentum* hybrids. The explanation of this contradiction could be the possible misclassification of the plant material used in the crossing, or it could be that in this case the compatible forms of two species were found.

Researchers over the years have not abandoned their efforts to produce the interspecific hybrids between Tartary and common buckwheat. Different approaches such as ovule culture, artificial creation of a compatible plant, and bridge crosses have been applied to overcome incompatibility and obtain the seed progenies of the hybrids. The first successful results allow the consideration that hybridization between *F. esculentum* and *F. tataricum* will be one of the directions of buckwheat breeding.

The Crossing of *F. esculentum* × *F. homotropicum*

F. homotropicum is a self-pollinated wild species, which was discovered by O. Ohnishi (1995) in Southwest China. Since its discovery, *F. homotropicum* has become a favorite object in interspecific crosses of *Fagopyrum* species especially because of its self-compatible nature and crossability with common buckwheat. Moreover, the *F. homotropicum* is characterized by high levels of seed set, tolerance to environmental stresses, and increased rutin content (Wang et al., 2005).

Campbell (1995) was the first to successfully hybridize *F. esculentum* and *F. homotropicum*, both at the diploid level, using embryo culture to rescue the 18-to-20-day-old embryos. The embryos grew into complete plants and were found to be fully fertile. Flowers produced by hybrid plants were homomorphic with pink sepals and pink pericarp similar to those of *F. homotropicum*. The F_2 progeny segregated for many of the characteristics found in either of the parents. The F_2 were backcrossed to common buckwheat and the second backcrossed progeny produced seeds that were close to common buckwheat in seed size and shape. The cross combination of *F. esculentum* × *F. homotropicum* was successfully used in the breeding process. Several thousand self-pollinated lines were evaluated and thousands of plants selected every year (Wang and Campbell, 1998).

Later on, many researchers crossed common buckwheat with *F. homotropicum*, obtaining hybrid plants and their self- and backcrossed progenies with different levels of fertility. In some cases the embryos were rescued by means of in vitro culture (Campbell, 1995; Woo et al., 1999; Matsui et al., 2003). But sometimes the hybrids between *F. homotropicum* and *F. esculentum* have been produced without the ovule culture (Hirose et al., 1995; Fesenko et al., 1998, 2001; Kim et al., 2002).

In the meantime, new populations *of F. homotropicum* have been found by Ohnishi in Yunnan and Sichuan provinces in China. By counting of chromosome number Ohnishi and Asano (1999) found that *F. homotropicum* consists of diploid ($2n = 16$) and tetraploid ($2n = 32$) populations. Furthermore, the allozyme analyses revealed that the tetraploid is allotetraploid and its diploid ancestors are most probably a diploid *F. homotropicum* and a wild common buckwheat *F. esculentum* ssp. *ancestrale*.

The tetraploid nature of some accessions of *F. homotropicum* was proposed by Wang and Campbell (1998) based on the cross behavior in their intraspecific crosses. This was confirmed as well as the allotetraploid origin of *F. homotropicum* $2n = 4x = 32$ by a wider series of crossings and by cytological characteristics (Wang et al., 2002b). The tetraploid form of *F. homotropicum* was successfully hybridized with diploid *F. esculentum*; the sterile triploid ($2n = 3x = 24$) F_1 hybrids were obtained followed by colchicine treatment, which resulted in the restoration of fertility at the hexaploid as well at the diploid level (Wang et al., 2005).

The evaluation of the genetic diversity of the diploid and tetraploid forms of *F. homotropicum* based on the nucleotide sequences of a nuclear gene AGAMOUS(AG) clustered the populations into three large groups. This revealed that the tetraploid form of *F. homotropicum* has emerged at least three times including the hybridization events between deeply differentiated diploid forms of *F. homotropicum* and the tetraploidization between *F. homotropicum* and *F. esculentum* ssp. *ancestrale* (Tomiyoshi et al., 2012). It is expected that the diploid form of *F. homotropicum* in phylogenetic Group III will have high cross-compatibility not only with *F. esculentum* ssp. *ancestrale* but also with common buckwheat, which would be of great use in buckwheat breeding programs.

Studying the inheritance of the gene SHT (seed shattering habit), Fesenko and Fesenko (2015) have found that a tetraploid form of *F. homotropicum* (accession H#4x) carries four copies of the dominant gene SHT. Two copies were tightly linked to the locus determining homostyly (as in diploid *F. homotropicum*); in contrast, two other copies were not linked with the homostyly gene (as in wild buckwheat *F. esculentum* ssp. *ancestrale*). The results support the hypothesis, proposed by Ohnishi and Asano (1999), that tetraploid *F. homotropicum* has arisen through hybridization between diploid *F. homotropicum* and *F. esculentum* ssp. *ancestrale* followed by genome doubling.

Attempts to use *F. homotropicum* in buckwheat breeding resulted in the creation of self-pollinated breeding lines in most cases. Thus the introgressive autogamous strains from the original cultivar Miyazaki-zarai have been established through recurrent backcrossing for nine generations (Adachi, 2004). The long homostyle buckwheat Norin-PL1 was developed by backcrossing *F. esculentum* cv. Botansoba and an F_1 plant that was produced by crossing Botansoba and *F. homotropicum* from Yunnan (Matsui et al., 2008). The seed size of buckwheat Norin-PL1 was similar to that of Botansoba, but its growth was less vigorous. The author recommended it as a parental line for producing self-compatible buckwheat.

As a result of selection for nonshattering plants among the backcrossed populations of the interspecific hybrid between common buckwheat and *F. homotropicum*, the self-compatible lines were created (Mukasa, 2011). But the yield of self-compatible lines was inferior to that of an open pollinated standard variety and the disadvantage of inbreeding depression was much bigger than the advantage of self-compatibility in pure line breeding.

Fesenko and Fesenko (2013) believe that although the species *F. homotropicum* can serve as a donor of features, which are absent in cultivated buckwheat (low remontant, self-fertility), hybrids with a high proportion of germplasm of wild species in the genome need serious exploration, first and foremost in the selection for earliness and accelerated rhythm of development. An effective method for accelerating breeding may serve the selection in the light of tough competition from more adapted plants of common buckwheat.

Thereby the crossing of common buckwheat and wild species *F. homotropicum* has been the most prospective in regard to the fertility of hybrids and their progenies. The various degrees of compatibility or incompatibility, which is attributed to this combination and sometimes requires ovule rescue to recover the hybrid embryos, are explained by the high genetic diversity of the *F. homotropicum* species. As a result of interspecific hybridization, advanced self-fertile breeding lines have been produced, which are expected to be useful in the breeding process not only by creating self-pollinated varieties but also by improving the cross-pollination of common buckwheat.

We may conclude that the interspecific crosses have occurred several times in the history of the genus *Fagopyrum*. Both natural and artificial allopolyploid species are available in the genus. The environmentally friendly method of interspecific hybridization followed in some cases by the embryo rescue technique may serve as a powerful tool to increase the genetic diversity of cultivated species and improve buckwheat in an evolutionary way. We agree with Kreft (2013) that the methods that keep buckwheat as a health-preserving and healing food should be developed further.

REFERENCES

Adachi, T., 2004. Recent advances in overcoming breeding barriers in buckwheat. In: Proceedings of the 9th International Symposium on Buckwheat. Prague. pp. 22–25.

Azaduzzaman, M., Minami, M., Matsushima, K., Nemoto, K., 2009. Characterization of interspecific hybrid between *F. tataricum* and *F. esculentum*. J. Biol. Sci. 9 (2), 137–144.

Campbell, C., 1995. Inter-specific hybridization in the genus *Fagopyrum*. Proceedings of the 6th International Symposium on Buckwheat, Shinshu, Japan, pp. 255–263.

Campbell, C., 2003. Buckwheat crop improvement. Fagopyrum 20, 1–6.

Fesenko, I.N., Fesenko, N.N., 2010. New species form of buckwheat—*Fagopyrum hybridum*. Vestnik OrelGAU 4, 78–81, (in Russian).

Fesenko, A.N., Fesenko, I.N., 2013. Elements of genetic testing differences between *Fagopyrum esculentum* and *F. homotropicum* and some results of interspecific hybridization in selection of common buckwheat. Vestnik OrelGAU 2 (41), 2–5.

Fesenko, N.N., Fesenko, A.N., Ohnishi, O., 1998. Some genetic peculiarities of reproductive system of wild relatives of common buckwheat *Fagopyrum esculentum*. In: Proceedings of the 7th International Symposium on Buckwheat, Part 6. Canada. pp. 32–35.

Fesenko, N.N., Fesenko, I.N., 2015. Inheritance analysis of theS4/SHT fragment of linkage group #4 supports allopolyploid origin and reveals genome composition of a tetraploid (2n = 32) lineage of *Fagopyrum homotropicum* Ohnishi. Plant Systemat. Evol. 301 (8), 2141–2146.

Fesenko, I.N., Fesenko, N.N., Onishi, O., 2001. Compatibility and congruity of interspecific crosses in *Fagopyrum*. Proceedings of the 8th International Symposium on Buckwheat. Korea, pp. 404–410.

Golyshkin, L., Fesenko, N., 1988. Investigation of process of fertilization in the intra- and interspecific hybridization by the luminescent microscopy. Flower Genetics and Problem of Compatibility in Buckwheat. Nauka, Moscow, pp. 79–92, (in Russian).

Hirose, T., Ujihara, A., Kitabayashi, H., Minami, M., 1994. Interspecific cross-compatibility in *Fagopyrum* according to pollen tube growth. Jpn. J. Breed. 44 (3), 307–314.

Hirose, T., Lee, B.S., Okuno, J., Konishi, A., Minami, M., Ujihara, A., 1995. Interspecific pollen–pistil interaction and hybridization in genus *Fagopyrum*. In: Proceedings of the 6th International Symposium on Buckwheat. Japan. pp. 239–245.

Ikeda, K., Ikeda, S., Kreft, I., Lin, R., 2012. Utilization of Tartary buckwheat. Fagopyrum 29, 27–30.

Kim, Y., Kim, S., Lee, K., Chang, K., Kim, N., Shin, Y., Park, C., 2002. Interspecific hybridization between Korean buckwheat landraces (*Fagopyrum esculentum* Moench) and self-fertilizing buckwheat species (*F. homotropicum* Ohnishi). Fagopyrum 19, 37–42.

Kreft, I., 2013. Buckwheat research from genetic to nutrition. Fagopyrum 30, 3–7.

Krotov, A.S., 1975. Buckwheat—*Fagopyrum* Mill. Cultivated flora of USSR. Kolos, Leningrad, pp. 7–118, (in Russian).

Krotov, A.S., Golubeva, E.A., 1973. Cytological studies on an interspecific hybrid *Fagopyrum tataricum* × *F.cymosum*. Bull. Appl. Bot. Genet. Plant Breed. 51 (1), 256–260, (in Russian).

Lin, R., 2004. The development and utilization of Tartary buckwheat resources. In: Proceedings of the 9th International Symposium on Buckwheat. Prague. pp. 252–258.

Matsui, K., Tetsuka, T., Nishio, T., Hara, T., 2003. Heteromorphic incompatibility retained in self-compatible plants produced by a cross between common and wild buckwheat. New Phytol. 159, 701–708.

Matsui, K., Tetsuka, T., Hara, T., Morishita, T., 2008. Breeding and characterization of a new self-compatible common buckwheat (*Fagopyrum esculentum*) parental line, "Buckwheat Norin-PL1". Bull. Natl. Agric. Res. Cent. Kyushu Okinawa Region 49, 11–17.

Mukasa, Y., 2011. Studies on new breeding methodologies and variety developments of two buckwheat species (*Fagopyrum esculentum* Moench. and *F. tataricum* Gaertn). Res. Bull. NARO Hokkaido Agric. Res. Cent. 195, 57–114.

Niroula, R.K., Bimb, H.P., Sah, B.P., 2006. Interspecific hybrids of buckwheat (*Fagopyrum* spp.) regenerated through embryo rescue. Sci. World 4, 74–77.

Ohnishi, O., 1995. Discovery of new *Fagopyrum* species and its implication for the studies of evolution of *Fagopyrum* and of the origin of cultivated buckwheat. In: Proceedings of the 6th International Symposium on Buckwheat. Japan. pp. 175–190.

Ohnishi, O., 2010. Distribution and classification of wild buckwheat species 1. Cymosum group. Review. Fagopyrum 27, 1–8.

Ohnishi, O., 2013. Distribution of wild species and perspective for their utilization. Fagopyrum 30, 9–14.

Ohnishi, O., Asano, N., 1999. Genetic diversity of *Fagopyrum homotropicum*, a wild species related to common buckwheat. Genet. Resour. Crop Evol. 46, 389–398.

Ohnishi, O., Matsuoka, Y., 1996. Search for the wild ancestor of buckwheat II. Taxonomy of *Fagopyrum* (Polygonaceae) species based on morphology, isozymes and cpDNA variability. Genes Genet. Syst. 72, 383–390.

Rumyantseva, N., Fedoseeva, N., Abdrakhmanova, G., Nikolskaya, V., Lopato, S., 1995. Interspecific hybridization in the genus *Fagopyrum* using *in vitro* embryo culture. In: Proceedings of the 6th International Symposium on Buckwheat. Japan. pp. 211–220.

Samimy, C., 1991. Barrier to interspecific crossing of *Fagopyrum esculentum* with *Fagopyrum tataricum*: I. Site of pollen-tube arrest. II. Organogenesis from immature embryos of *F. tataricum*. Euphytica 54, 215–219.

Samimy, C., Bjorkman, T., Siritunga, D., Blanchard, L., 1996. Overcoming the barrier to interspecific hybridization of *Fagopyrum esculentum* with wild *Fagopyrum tataricum*. Euphytica 91, 323–330.

Shaikh, N.Y., Guan, L.-M., Adachi, T., 2002a. Failure of fertilization associated with absence of zygote development in the interspecific cross of *Fagopyrum tataricum* × *F. esculentum*. Breed. Sci. 52, 9–13.

Shaikh, N.Y., Guan, L.-M., Adachi, T., 2002b. Ultrastructural aspects on degeneration of embryo, endosperm and suspensor cells following interspecific crosses in genus *Fagopyrum*. Breed. Sci. 52, 171–176.

Suvorova, G.N., 2001. The problem of interspecific cross of *Fagopyrum esculentum* Moench. × *Fagopyrum cymosum* Meissn. In: Proceedings of the 8th International Symposium on Buckwheat. Korea. pp. 311–318.

Suvorova, G.N., 2010. Perspectives of interspecific buckwheat hybridization. In: Proceedings of the 11th International Symposium on Buckwheat. Russia. pp. 295–299.

Suvorova, G.N., Fesenko, N.N., Kostrubin, M.M., 1994. Obtaining of interspecific buckwheat hybrid (*Fagopyrum esculentum* Moench. × *F.cymosum* Meissn.). Fagopyrum 14, 13–16.

Tomiyoshi, M., Yasui, Y., Ohsako, T., Li, C., Ohnishi, O., 2012. Phylogenetic analysis of AGAMOUS sequences reveals the origin of the diploid and tetraploid forms of self-pollinating wild buckwheat, *Fagopyrum homotropicum* Ohnishi. Breed. Sci. 62, 241–247.

Ujihara, A., Nakamura, Y., Minami, M., 1990. Interspecific hybridization in genus *Fagopyrum*—properties of hybrids (*F. esculentum* Moench. × *F. cymosum* Meissner) through ovule culture. Gamma Field Symposia 29, 45–53.

Wang, Y., Campbell, C., 1998. Interspecific hybridization in buckwheat among *Fagopyrum esculentum*, *F. homotropicum* and *F. tataricum*. In: Proceedings of the 7th International Symposium on Buckwheat, Part 6. Canada. pp. 1–12.

Wang, Y., Scarth, R., Campbell, C., 2002a. Interspecific hybridization between *Fagopyrum tataricum* (L.) Gartn. and *F. esculentum* Moench. Fagopyrum 19, 31–35.

Wang, Y., Scarth, R., Campbell, C., 2002b. Comparison between diploid and tetraploid forms of *Fagopyrum homotropicum* in intraspecific and interspecific crossability and cytological characteristics. Fagopyrum 19, 23–29.

Wang, Y., Scarth, R., Campbell, C., 2005. Interspecific hybridization between diploid *Fagopyrum esculentum* and tetraploid *F. homotropicum*. Can. J. Plant Sci. 85, 41–48.

Woo, S.H., Tsai, Q.S., Adachi, T., 1995. Possibility of interspecific hybridization by embryo rescue in genus *Fagopyrum*. In: Proceedings of the 6th International Symposium on Buckwheat. Japan. pp. 225–237.

Woo, S.H., Wang, Y.J., Campbell, C.G., 1999. Interspecific hybrids with *Fagopyrum cymosum* in the genus *Fagopyrum*. Fagopyrum 16, 13–18.

Yamane, K., Yasui, Y., Ohnishi, O., 2003. Intraspecific cpDNA variations of diploid and tetraploid perennial buckwheat (Polygonaceae). Amer. J. Bot. 90 (3), 339–346.

Yasui, Y., Ohnishi, O., 1998. Phylogenetic relationships among *Fagopyrum* species revealed by nucleotide sequences of the ITS region of the nuclear rRNA gene. Genes Genet. Syst. 73, 201–210.

Yui, M., Hayashi, T., Yamamori, M., Kato, M., 2004. Inter-specific hybridization between Japanese common buckwheat "ybakawa-zarai" and *Fagopyrum cymosum* Meissn. In: Proceedings of the 9th International Symposium on Buckwheat. Prague. pp. 190–194.

Chapter eight

Crop Evolution of Buckwheat in Eastern Europe: Microevolutionary Trends in the Secondary Center of Buckwheat Genetic Diversity

A.N. Fesenko*, N.N. Fesenko**, O.I. Romanova†, I.N. Fesenko**

*Laboratory of Groats Crops Breeding, All-Russia Research Institute of Legumes and Groats Crops, Orel, Streletskoe, Russia; **Laboratory of Genetics and Biotechnology, All-Russia Research Institute of Legumes and Groats Crops, Orel, Streletskoe, Russia; †Department of Small Grains, N.I. Vavilov's Institute of Plant Industry, Saint-Petersburg, Bolshaya Morskaya, Russia

INTRODUCTION

Geographical distribution and genetic structure of any crop are always influenced by different factors. Having played a role, many of them are simply just history; others have yet to appear. But the main factors are productivity and market demand, particularly in comparison with other crops, which often act as competitors. At the present time, buckwheat looks to be an outsider, with sowing areas decreasing. In this regard, Eastern Europe was an encouraging exception: this region became the secondary center of genetic diversity of the species *Fagopyrum esculentum*, and here the diversity was widely applied in the breeding of buckwheat cultivars. In this chapter we will try to review this phenomenon.

Molecular Breeding and Nutritional Aspects of Buckwheat. http://dx.doi.org/10.1016/B978-0-12-803692-1.00008-0

Buckwheat in Different Periods (Historical Speculation)

Europe is "heated" by the Gulf Stream. This has encouraged the distribution of agriculture toward the northern regions of Eastern Europe, with very long days in the early summer, where a foothold for photoperiodic neutral genotypes and crops was established.

Agriculture had come to Europe from the south. However, in Eastern Europe, the forest connecting the south with the steppe was controlled by different successive nomadic tribes. Some of them founded cities surrounded with agricultural enclaves. Others, however, were against the "mutilation" of beautiful natural pastures by the plow. Tending nomadic livestock was one of the most complicated human activities, requiring knowledge of human and veterinary medicine, the ability to respond to changes, including seasonal (summer to winter), and the ability to hunt and protect their herds from predators and neighbors. These natural warriors often perceived the strange farmers as the legitimate spoils: the slave trade flourished until the inclusion of the northern Black Sea coast in the Russian Empire in the 18th century, after which it became possible for farmers to settle en masse and plow the steppe black soil; as a result Russia would become a leading exporter of agricultural production in the world during the next century and a half.

More than 200 years after the settlement of the nomadic tribes in the 13th century there was a huge bonding of manageable space between the Pacific and Atlantic oceans, which also included Moscow as a subordinated territory. After disintegration of the nomad's empire (the Golden Horde), Moscow gradually reunited the territories in the Russian Empire where half of the noblemen had nomadic ancestry (Gumilev, 2006).

About 1000 years ago, agriculture was established on the border of forest and steppe and penetrated deeply into the forest zone (in the northeast of the Dnieper River, toward contemporary Moscow). Not all crops had stood the test, but rye and buckwheat were the most suitable. In this "northern boiler" the most early-ripening and photoperiodic neutral populations of buckwheat were gradually formed, which had overcome the "Little Ice Age," which had a peak at the turn of 16th to 17th centuries, when there were 7 consecutive years of crop failure after which came the "troubled times", that was the collapse of the state and foreign intervention. Restoration of the state was made possible through the efforts of both the people that organized the Home Guard, which knocked out the invaders from Moscow, and warming of the climate (especially of summer weather). However, significant climate warming began only in the 20th century. The prolonged frost-free period affected the morphotype of new buckwheat varieties.

During the last few centuries, the general trend was a gradual decrease of the share of buckwheat among grain crops. This was facilitated by the spread of the potato, which is also an alkaline food, in addition to sauerkraut and sour rye bread. Because of the comparison with rye, Russian authors have considered buckwheat as the heat-loving crop and well-digestible food (porridge), whereas eastern authors (in comparison with rice) considered it to be a cold-resistant crop and hard-to-digest food.

General Regularities in Common Buckwheat Adaptation

Adaptation of common buckwheat to conditions of the East European area was mainly based on the shortening of the vegetation period. So, in our field experiments, the buckwheat populations from East Asia vegetate until the first autumn frosts (119 days, on average) and populations from the southern part of the East European area of common buckwheat distribution (Ukraine) reached harvesting ripeness within 93 days on average (Table 8.1). The length of the vegetation period of populations from the north of Eastern Europe is about 66 days (Table 8.1).

The length of the vegetation period of buckwheat populations is correlated with the number of vegetative nodes on the main stem (stem branching zone [SBZ]) ($r = 0.741$, $P_0 < 0.001$); therefore it can be described in terms of SBZ structure. The plants of earlier-ripening populations form fewer vegetative nodes on the stem and branches (for details see Fesenko et al., 1998).

Additionally, the tendency towards a certain reduction of the branching potential of common buckwheat morphotypes in the south-to-north direction is linked with the increase of the share of plants with limited secondary branching (Table 8.1). The number of vegetative nodes on the upper branches of these plants is reduced from two to three (on the "common-type" plants) down to zero to one (Fesenko et al., 2006).

The number of vegetative nodes on the stem and on the entire plant had a reliable positive correlation ($r = 0.913$ and $r = 0.792$, respectively) with the

TABLE 8.1 Characteristics of Buckwheat Populations from Asia and Eastern Europe

| Region of Origin | Number of Nodes in SBZ | | | Average Length (Days) | |
	Average	Range	LB[a]	Vegetative Growth	Growing Season
East Asia	10.1	9.0–11.5	0.0	38	119
Ukraine	5.6	3.9–6.8	5.2	33	93
Central region of Russia	4.7	4.2–4.9	6.8	31	79
Belarus	4.2	3.2–4.3	14.1	30	68
Volga region of Russia	4.1	3.3–4.2	21.9	30	69
Northwestern region of Russia	3.8	2.8–3.6	15.1	28	66

[a]Plants with a complete or partial reduction of the branching zone on the upper branch (%)

yielding ability of a buckwheat population (Fesenko et al., 2010). Therefore a reduction in the number of vegetative nodes leads to a lower productivity potential of the earlier-ripening populations.

Selective Forces for Maintenance of Intrapopulation Polymorphism for the Number of Vegetative Nodes

Since common buckwheat lacks special adaptations to ecological stresses (Fesenko et al., 2006), the strategy of its ecological adaptation is based on the "catching" of favorable weather conditions. It is achieved through maintenance in any population of biotypes with a different number of vegetative nodes per plant. Such a polymorphic population is more adapted to contradictive requirements: a particular proportion of morphotypes in the population provides optimal productivity under typical conditions and retains the ability to adapt when the conditions deviate.

Depending on the weather fluctuations, the favorable periods for seed formation coincide with the developmental rhythm of a plant with a peculiar SBZ morphotype, which determines the advantage of this morphotype in terms of its contribution to the yield of the population. Over many years, because of the random variation of weather factors in a locality, a local population accumulates the SBZ morphotypes in peculiar proportions. In any local population the modal SBZ morphotype (or two to three modal morphotypes) is most productive under most typical weather conditions; nonmodal SBZ morphotypes are maintained in the population because of sporadic climate deviations when such morphotypes become more adaptive, at least for one season. Since such a structure is a result of natural selection, the population parameters characterize both local conditions (climate and soil) and the amplitude of the predominant weather fluctuations in a locality. This is clearly reflected in structural peculiarities of local populations from different regions.

In our experiments the highest degree of diversity in terms of SBZ morphotypes was displayed by East Asian populations, which evolved in the most favorable conditions: the range from 5 to 15 nodes in SBZ (Table 8.2).

The landraces from Eastern Europe include fewer SBZ morphotypes within a population. Populations from various regions include different combinations of particular SBZ morphotypes. So most Ukrainian populations were formed under the long vegetation period together with favorable temperatures and high soil fertility. For this reason a characteristic of the structure of local Ukrainian populations is a wide range of modal SBZ morphotypes together with a large portion of relatively late ripening ones: SBZ-4 (20.4%), SBZ-5 (34.2%), and SBZ-6 (31.3%) (Table 8.2).

Climatic and soil conditions of the central region of Russia are favorable for vegetation of varieties of the mid-season type: the morphotypes SBZ-4 (30.8%) and SBZ-5 (32.0%) being the modal ones.

The role of the modal SBZ fraction is most distinct in populations from the northern region of Russia, where heat deficiency is the main limiting factor, and

TABLE 8.2 Stem Branching Zone (SBZ) Polymorphism in Buckwheat Landraces From Different Regions of Eastern Europe and East Asia

Region	Average Share (%) of Plants With SBZ Morphotype												
	3	4	5	6	7	8	9	10	11	12	13	14	15
East Asia			0.4	2.6	5.8	19.6	21.6	8.1	12.6	12.2	8.0	5.6	3.5
Ukraine	4.7	20.4	34.2	31.3	9.4								
Central region of Russia	10.0	30.8	32.0	19.2	8.0								
Belarus	30.5	32.3	26.5	8.3	2.4								
Northwestern region of Russia	44.5	34.0	15.7	5.8									

even insignificant deviation from the mean annual values of temperature represents a considerable stress for plants. In this relation, the modal morphotype here is the earliest SBZ-3 (44.5%). In second place is the SBZ-4 morphotype, whose frequency in the population is 34.0%. Such a structure of population is guaranteed to produce the crop yield during a short and relatively cool summer.

A similar structure is characteristic of the Belarus landraces (Table 8.2). However, it was established under the influence of other factors. In Belarus, local common buckwheat populations grow on low-fertile sandy loams with a low water-retaining capacity. The temperature regime here is similar to that of Central Russia and is sufficient for the growth of mid-season varieties. However, the deficiency of water and mineral nutrients is the main limiting factor, which determines the formation of early populations with SBZ-3 and SBZ-4 as the leading morphotypes.

Preadaptations for Accumulation of Recessive Allele *det* in Landraces

Determinate growth habit, ie, limiting of reproductive system morphogenesis by formation of three to four inflorescences per shoot (IPS), is one of the key tools for breeding high-yielding varieties of common buckwheat. All latter varieties, which were bred in the Institute of Grain Legumes and Groats Crops, have a determinate growth habit. The average yield of buckwheat in Russia was doubled after increasing the share of such varieties in the crop's production up to 56.7% during 2000–2014.

For the first time the determinate buckwheat plants were isolated by Stoletova E.A. (1958) (see Fesenko et al., 2006). Later, such plants were observed in different populations (Fesenko, 1968; Shakhov, 1977; Kreft, 1989): in some samples of the Vavilov Institute's collection the plants' frequency reached 4.7–58.5% (Shakhov, 1977). Apparently, the accumulation of determinate plants in populations from mountainous areas of southern Europe is because of their increased resistance to heat and drought (Fesenko et al., 2006).

Determinate growth is a recessive Mendelian trait (Fesenko, 1968); however, there is also a variability of the number of IPS: at the present time, determinate plants with five or more IPS have been found. It has been shown that the difference between one or two and four or five IPS is controlled by two genes, whose recessive alleles determine large numbers of inflorescences (Fesenko et al., 2009). Determinate plants with only one IPS were found only among hybrids with wild species *Fagopyrum homotropicum*, which apparently carries only dominant alleles of genes that influence the number of inflorescences in the determinants. Recessive plus-alleles of these modifiers are found in populations of determinate varieties from Europe. Allele *det* was a rare admixture in the European populations (Fesenko et al., 2006). Therefore direct natural selection of recessive plus-alleles of the modifier genes to increase the number of IPS of determinate plants looks unlikely. At the same time, determinants were included in breeding programs as soon as they had been found (Fesenko et al., 2006).

Therefore it may be concluded that the determinate plants were developing at least two to three IPS. Besides, plants with only one IPS have been found only among interspecific F_2 hybrids *F. esculentum* × *F. homotropicum*, and, apparently, cultivated populations do not carry a full set of dominant minus-alleles at all the modifiers revealed.

Thus recessive plus-alleles of the modifiers affecting the IPS of plants with determinate growth habit have accumulated randomly in marginal buckwheat populations. Apparently, it promoted accumulation of the recessive allele *det* in these populations. Wide use of the trait "determinate growth habit" in buckwheat breeding programs has increased the concentration of plus-alleles of these modifiers.

A Population Heterogeneity as a Basis for Breeding and Introduction of Buckwheat (the Example of the Phenomenon of cv. Bogatyr)

All populations of common buckwheat are heterogeneous in many ways, including seed size and plant maturation time. This allowed *F. esculentum* to gain a foothold as a grain crop in many agricultural areas, including Northern regions. Early-maturing populations in favorable conditions also demonstrate the ability of gradually prolonging the growing season because of the greater competitiveness of later morphotypes. This phenomenon was well illustrated by the experience of breeding the first commercial variety of buckwheat on Shatilov's station. The work started in 1896. Early-ripening buckwheat was being sowed by local farmers closer to the middle of June. Since moldboardless plowing caused large numbers of quickly growing rhizomatous weeds on the field, the soil was plowed after cattle grazing: cows ate and trampled the sprouting weeds.

On Shatilov's station, moldboard plowing was implemented, and sowing was carried out 3 weeks earlier. In addition, machine separation was used to select the heaviest grain for sowing. As a result, after 12–15 generations, a middle-ripening cultivar Bogatyr was obtained, which was a more powerful because it ripened 2 weeks later than the local population and had the heavier grain on 25% (m_{1000} was about 25 g).

Following Buckwheat Breeding in the USSR and Russia

In the USSR, buckwheat breeding had a very wide geography. So in Siberia (Zamyatkin, 1971; Shumny et al., 1978), Belarus (Anokhina, 1980), and the Moscow region (Zakharov, 1980) breeding work was carried out with self-pollinating (homostylous) buckwheat: these attempts were not very successful. On the contrary, the attempts to increase heterosis (Kiev) or to use mutagenesis (Kamenets-Podolskiy) in buckwheat breeding gave some practical results; also the breeding of tetraploid varieties was successful (Belarus), ultraearly varieties (Moscow region), large-grained varieties (Kazan), and varieties with unusual morphotypes (Ufa) (see Fesenko et al., 2006).

In Orel region (Shatilovskaya station and Institute of Grain Legumes and Groats Crops) the first varieties with large grains were developed: the first, Shatilovskaya 5, was bred with participation of large-grained samples from East Asia. Later, N.N. Petelina in Orel and Kazan bred more large-grained varieties (Mayskaya, Krasnostreletskaya, Karakityanka, etc.); the alleles determining a large size of grain are encountered in the genotypes of many contemporary varieties, but their origin remains debatable. Varieties carrying such alleles are characterized by physiologically determinate growth habit because of the high ability of developing seeds to attract photosynthesis products. This leads also to a relatively fast and early ripening.

In Shatilovo the practice of tetraploid isolation between diploid plots was first developed and used: diploid samples ($2x = 16$) were isolated by a tetraploid ($4x = 32$) variety with winged seeds, which were easily separable from diploid impurities on the sieve. N.V. Fesenko used this approach for wide multivariant breeding work in Shatilovo and later in the Institute of Grain Legumes and Groats Crops. He paid tribute to the selection on heterosis (synthetic variety Orlovchanka), but his main efforts were concentrated on restricting vegetative growth of buckwheat and redirecting the photosynthesis products to developing grain. This was achieved through the use of different growth-limiting mutations. Ultimately, most success has been achieved using the recessive gene *d* (or *det*): the determinate growth habit is a characteristic of all latter varieties bred in the Institute of Grain Legumes and Groats Crops. Some other successful directions were based on using semidwarf, small-foliar, and large-grained morphotypes; also it is the breeding of ultraearly varieties that first stimulates the morphotype selection, ie, selection of earliest-flowering plants with only three vegetative nodes on the stem (SBZ = 3). All these characteristics are usually combined with determinate growth habit. So the first commercial variety with determinate growth habit was Sumchanka. This variety had the usual size of both grain and leaves. Later determinate cultivars, such as Dozhdik, Devyatka, and Dialog, produced larger grains. An exception was the cultivar Dikul, which had smaller grains and smaller leaves: the cultivar was the leader in grain productivity for a long time.

Recently, the mutation *gc* (green corolla) was successfully applied for increasing the nonshattering ability of buckwheat cultivars: the commercial variety Design with determinate growth habit and green flowers has been bred. Also tetraploid varieties show greater resistance to shattering.

The trait limited secondary branching (LSB) was used to develop the high-yielding varieties Ballada and Molva. The general principle of breeding of LSB populations was the increase of SBZ to compensate for the decrease of the number of vegetative nodes on the upper branches. So for breeding of the "usual" ultraearly variety Skorospelaya 81 the morphotype 3 + 2 + 2 has been selected, but LSB variety Skorospelaya 86 has been bred by selection of the morphotype 4 + 0 + 1. The last variety was found to be too genetically aligned: this peculiarity probably impeded its distribution on acid soils in the northern regions (see Budagovskaya, 1998).

REFERENCES

Anokhina, T.A., 1980. The expression of self-incompatibility in monomorphic and dimorphic buckwheat populations. Rus. J. Genet. 16, 136–142.

Budagovskaya, N., 1998. Changes in the state of photoautotrophic and heterotrophic organs of buckwheat plants at iron deficiency and low pH. Fagopyrum 15, 1–7.

Fesenko, I.N., Fesenko, A.N., Biryukova, O.V., Shipulin, O.A., 2009. Genes regulating inflorescences number in buckwheat with a determinate growth habit (homozygote at the recessive allele *det*). Fagopyrum 26, 21–24.

Fesenko, N.N., Martynenko, G.E., Funatsuki, H., Romanova, O.I., 1998. Express evaluation of Russian and Japanese varieties of buckwheat based on characteristics-indicator of duration of vegetative period. Proceedings of the Seventh International Symposium on Buckwheat in Winnipeg, Canada, part 1, pp. 185–192.

Fesenko, N.V., Fesenko, N.N., Romanova, O.I., Alekseeva, E.C., Suvorova, G.N., 2006. Theoretical Basis of Plant Breeding, vol. V. The Gene Bank and Breeding of Groat Crops: Buckwheat. VIR, St. Petersburg, (in Russian).

Fesenko, N.V., 1968. A genetic factor responsible for the determinant type of plants in buckwheat. Rus. J. Genet. 4, 165–166, (in Russian).

Fesenko, N.V., Fesenko, A.N., Shipulin, O.A., Savkin, V.I., Kolomeichenko, V.V., Martynenko, G.E., Mazalov, V.I., 2010. Metameric architecture of plants vegetative sphere as system criterion of adaptive and productive properties of plants and varieties of buckwheat. Proceedings of the 11th International Symposium on Buckwheat at Orel, Russia, pp. 425–428.

Gumilev, L.N., 2006. Drevnyaya Rus' I Velikaya Step' (Ancient Russia and Great Nomads Empire). EKSMO, Moscow, (in Russian).

Kreft, I., 1989. Ideotype breeding of buckwheat. Proceedings of the Fourth International Symposium of Buckwheat Orel, USSR. pp. 3–6.

Shakhov, N.F., 1977. New sources of determinateness and multicorymbness in buckwheat. Bull. VNIIZBK (Orel) 18, 67–69, (in Russian).

Shumny, V.K., Kovalenko, V.I., Kvasova, E.V., Kolosova, L.D., 1978. Some genetic and breeding aspects of reproduction systems in plants. Rus. J. Genet. 14, 25–35.

Zakharov, N.V., 1980. New homostylous form of buckwheat and its evaluation as self-compatibility donor. Bull. VNIIZBK 26, 38–42.

Zamyatkin F.E., 1971. Self-pollinated buckwheat. In: Alekseeva, E.S., Demidenko, G.B., Elagin, I.N. et al. (Eds.), Breeding, Genetics and Biology of Buckwheat. Orel, pp. 103–111.

Chapter nine

Genetic Resources of Buckwheat in India

J.C. Rana*, Mohar Singh*, R.S. Chauhan, R.K. Chahota†,
T.R. Sharma†, R. Yadav*, S. Archak***

**National Bureau of Plant Genetic Resources Regional Station, Shimla, India;
**Department of Biotechnology & Bioinformatics, Jaypee University of Information
Technology, Solan, India; †National Bureau of Plant Genetic Resources,
Pusa Campus, New Delhi, India*

INTRODUCTION

The genus *Fagopyrum* belongs to the family Polygonaceae and around 23 species occur in the highlands of the Euro-Asia region (Arora and Engels, 1992; Ohnishi, 1995; Chen et al., 2001; Rana et al., 2010; Tang et al., 2014). A wide range of ecogeographical conditions in the Indian Himalaya has generated tremendous genetic diversity in buckwheat species, namely, Tartary buckwheat (*Fagopyrum tataricum* Gaertn.) and common buckwheat (*Fagopyrum esculentum* L. Moench), and several wild and weedy forms (Ohnishi, 1988; Joshi and Rana, 1995; Rana, 2004). In India around 1050 accessions of buckwheat have been assembled through collection and introduction. The entire germplasm has been characterized and conserved in medium-term storage facilities at various regional stations of the National Bureau of Plant Genetic Resources (NBPGR) and partly where sufficient seed quantities are available is under long-term storage in the national gene bank at NBPGR, New Delhi. The germplasm has been evaluated for various traits of interest, and trait-specific reference sets have been developed for enhanced utilization of germplasm.

In view of the advancement in molecular biology, the emphasis has also been shifted to develop genomic resources of buckwheat. Expressed sequence tags (ESTs) available at http://compbio.dfci.harvard.edu/tgi/tgipage.html from related families such as Aizoaceae, Amaranthaceae, Plumbaginaceae, and Tamaricaceae, which have 27,191, 26,807, 6387, and 21,709 ETSs, respectively, have been screened. Out of 141 simple sequence repeats (SSRs) identified, based on repeat length, only 13 amplified in buckwheat, indicating low cross-genera transferability of SSRs in buckwheat. The development of eighteen gene-specific

Molecular Breeding and Nutritional Aspects of Buckwheat. http://dx.doi.org/10.1016/B978-0-12-803692-1.00009-2

sequence tagged site (STS) markers and their possible association with phenotypic traits in buckwheat have also been worked out. Four markers—BW10, BW12, BW22, and BW27—showed considerable polymorphism (PIC > 0.5) and overall STS markers produced 2.76 alleles per locus and showed moderate estimates of polymorphic information content (PIC) (0.268) in 91 accessions of buckwheat (unpublished). Gupta et al. (2012) have reported that in addition to structural genes, other classes of genes such as regulators, modifiers, and transporters are also important in biosynthesis and accumulation of flavonoid content in plants, and used cDNA-amplified fragment length polymorphism (AFLP) technology successfully to capture genes that are contributing to differences in rutin content in seed maturing stages of *Fagopyrum* species.

The buckwheat in India is known by various vernacular names such as ogal, phaphar, bresha, mittahe dyat, dro, and brotitae. In the most northern and western states of India, the flour is known as kuttu ka atta and is consumed on fasting days, especially during *Navaratri (religious days according to Hindu mythology)*. Buckwheat has multiple uses and is consumed both as grain and green and valued for its excellent nutritional profile based on the favorable composition of protein complex with high contents of lysine, fibrous material, mineral compounds, vitamins, bioflavonoid rutin, and quercetin. It is also emerging as a healthy alternative to gluten-containing grains in a gluten-free diet because buckwheat seeds are naturally gluten free (Campbell, 1997; Kupper, 2005; Ahmed et al., 2014; Mann et al., 2012; Huang et al., 2014). The seed of Tartary buckwheat contains higher amounts of rutin [about 0.8–1.7% dry weight (dw)] than that of common buckwheat (0.01% dw) and is rich in vitamins, especially vitamin B, and an important source of macroelements (eg, K, Na, Ca, Mg) as well as microelements (eg, Zn, Mn, Cu, Se). Its six major flavonoids are rutin, quercetin, orientin, homoorientin, vitexin, and isovitexin (Fabjan et al., 2003; Stibilj et al., 2004; Morishita et al., 2007; Krahl et al., 2008; Bystricka et al., 2011; Zielinska et al., 2012; Raina and Gupta, 2015). Buckwheat also contains selenium, which ranges from 0.0099 to 0.1208 mg/g with an average of 0.0406 mg/g (Zheng et al., 2011). Selenium is an essential trace element for humans as it enhances human resistance to cancer, cardiovascular and cerebrovascular diseases, AIDS, and diabetes. Because of its nutritional and medicinal profile (Table 9.1) the demand for rutin and other flavonoids derived from buckwheat is growing in the food, pharmaceutical, and cosmetic industries.

The diploid nature ($2n = 16$), short life cycle ($\sim$70 days), and small genome size ($\sim$450 Mbp) make it an ideal species for genetic investigation of the biosynthesis and accumulation of flavonoids. Identification of candidate genes for various traits of economic importance including those involved in the biosynthesis of rutin and other secondary metabolites would be useful for enhancing its medicinal and nutritional properties and will also help in checking the fast depleting genetic resources of buckwheat. This chapter discusses the current status and various aspects of management of genetic resources in India.

TABLE 9.1 Potential Medicinal and Nutritional Uses of Buckwheat

Medicinal and Nutritional Uses	References
Strengthens capillaries and so helps in arteriosclerosis or high blood pressure	Campbell (1997)
Identified as a strong antioxidant, has antiangiogenesis and anticancer properties	Jackson and Venema (2006)
Rutin is a natural flavonoid with antihyperglycemic, antihypertensive, and antioxidative properties	Lee et al. (2007)
Buckwheat seeds are naturally gluten free and thus are currently emerging as healthy alternatives to gluten-containing grains in a gluten-free diet	Kupper (2005)
Antagonizes the increase of capillary fragility associated with hemorrhagic disease and hypertension	Przybylski and Gruczynska (2009)
Contains UV-B radiation-absorbing compounds	Kreft et al. (2002)
Resistant to cancer, cardiovascular and cerebrovascular diseases, AIDS, diabetes	Zheng et al. (2011)
Reduces the risk of arteriosclerosis and has antioxidant activity	Guo et al. (2011); Vojtiskova et al. (2012)
Used as prebiotic food because it increases lactic acid bacteria in the intestine	Prestamo et al. (2003)
Ameliorates spatial memory impairment	Pu et al. (2004)
Toxic to plant pathogenic fungi, Gram-positive and -negative bacteria	Fujimura et al. (2003)
Protects humans from oxidative stress	Schramm et al. (2003)
Has chemopreventive activity and may have a therapeutic role for human leukemias	Ren et al. (2001)
Combats diabetes, obesity, hypertension, and constipation	Li and Zhang (2001)
Suppresses gallstone formation and cholesterol level by enhancing bile acid synthesis	Tomotake et al. (2000)
Reduces serum glucose level	Kawa et al. (2003)
Delays consenescence and prevents body fat accumulation	Du et al. (2004)
Possesses antioxidant properties	Quettier-Deleu et al. (2000)
Treats colon carcinogenesis	Liu et al. (2001)
Ameliorates renal injury	Yokozawa et al. (2001)
Reduces serum triglycerides and total cholesterol level	Wang et al. (2009)

AREAS AND DISTRIBUTION OF GENETIC DIVERSITY IN INDIA

Buckwheat, native to temperate East Asia (Robinson, 1980), exhibits variation in its distribution and diversity including those of wild species throughout the Himalayan region. In India the occurrence of buckwheat ranged from Jammu and Kashmir in the north to Arunachal Pradesh in the east and Tamilnadu in the south (Fig. 9.1). The important areas where buckwheat is largely grown and are rich in genetic diversity are: Kargil and Drass sectors, Gurez valley of Jammu, and Kashmir; Bharmaur, Pangi, outer Saraj, Chopal, Dodra Kuar, Neshang, Pooh subdivision, Sangla, and interiors of Lahaul valley in Himachal Pradesh; Pindari valley, Darma valley, Jolwan, Jaunpur, and Kapkote in Uttrakhand in the north; Siliguri, Darjeeling, and Koch Bihar in West Bengal; Lachan and Lachoong in Sikkim; Tawang, Bomdilla, and Dirang in Arunachal Pradesh and other higher elevations of Meghalaya, Manipur, and Nagaland in the eastern and northeastern regions; and in Nilgiris and Palani hills in the south. Because of its rich nutritional value, a few farmers and pharmaceutical houses attempt its cultivation in the Indian plains in Rabi season but because of its susceptibility to frost the results are not encouraging.

Out of the 23 species, *F. esculentum* and *F. tataricum* are the two cultivated species while others such as *Fagopyrum homotropicum*, *Fagopyrum caudabum*,

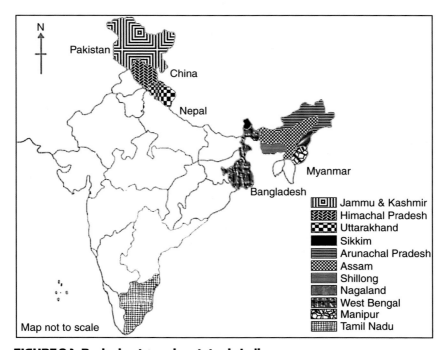

FIGURE 9.1 Buckwheat growing states in India.

Fagopyrum sagittatum, Fagopyrum cymosum, Fagopyrum megacarpum, Fagopyrum gracilipes, Fagopyrum urophyllum, Fagopyrum leptopodum, Fagopyrum lineare, Fagopyrum callianthum, Fagopyrum pleioramosum, Fagopyrum capillatum, Fagopyrum statice, Fagopyrum megaspartnium, Fagopyrum pilus, Fagopyrum crispatofolium, Fagopyrum densovillosum, Fagopyrum hailuogouense, Fagopyrum luojishanense, Fagopyrum wenchuanense, and *Fagopyrum pugense* are occurring mainly in the high lands of the Euro-Asia region (Farooq and Tahir, 1987; Ohnishi, 1995; Chen et al., 2001; Rana, 2004; Liu et al., 2001; Shao et al., 2011; Kalinova and Dadáková, 2013; Tang et al., 2014; Zhou et al., 2015; Hou et al., 2015). Further, *F. tataricum* ssp. *annum* occurs in the Eastern Himalayas and *F. tataricum* ssp. *potanini* in Tibet, Kashmir Himalayas, and northern Pakistan (Ohnishi 1989, 1991, 1992), while *F. tataricum* ssp. *himalianum* and *F. esculentum* ssp. *emerginatum* are occurring in the cold arid region of the Western Himalayas (Rana, 2004; Rana et al., 2012). Munshi (1982) described *Fagopyrum kashmirianum* as separate taxa but morphologically akin to *F. tataricum*, hence it is treated as the same species. The genetic diversity occurs both at the varietal level and species level and the majority of cultivated types are farmers' own selections. Ujihara and Matano (1982) have also reported variations in diploid and tetraploid types in Nepalese buckwheat.

The estimates for area and production under buckwheat in India are not available separately, since all underutilized crops have been dealt with under the term "coarse cereals." The area and production of coarse cereals was 25.67 million ha and 43.05 million tons in 2013–2014 as compared to 43.80 million ha and 30.41 million tons in 1975–1976, respectively as per the report of the Directorate of Economics and Statistics, Government of India, for 2013–2014. This shows that the area under coarse cereals, which also includes buckwheat, has decreased by 41.39% while production has increased by 41.56%. However, the case studies conducted on the trends in the area under buckwheat at different places of Indian Himalayan Region have shown a decline ranging from 60% to 92% (Rana and Sharma, 2000; Rao and Pant, 2001; Rana et al., 2010; Ahmad and Raj, 2012). The reports of other countries also depict the same trends. According to Suzuki, Japan had a total buckwheat acreage of approximately 172,000 ha producing 114,000 MIT in 1897, and it went on declining from 139,000 in 1907 to as low as 40,000 ha producing 25,000–30,000 MIT in 2003. Popović et al. (2014) while analyzing the buckwheat growing trends worldwide reported that the most significant producers of buckwheat in the world are: China (34.25%), Russia (32.43%), and Ukraine (11.46%) and have a trend of increasing areas at the rate of 6.85%, 47.90%, and 43.86%, respectively, while other major producers such as Poland, the United States, and Japan have a trend of decreasing buckwheat production areas. Several factors have been found responsible for the erosion of genetic diversity in buckwheat and important among them are changing cropping patterns, low productivity, changing food habits and life styles, fewer alternative uses/products, and lack of awareness about its food value (Rana et al., 2010).

Nevertheless, increasing awareness of the food value of buckwheat and other coarse cereals among populations and the demand for coarse grains have increased, and therefore we are witnessing the reintroduction and cultivation of buckwheat in some areas of Western Himalaya. A case study of bringing buckwheat back into cultivation in Sangla valley of Kinnaur district in Himachal Pradesh has been depicted in Box 9.1. It is not only because of socioclimatic compulsion, but also because of its medicinal and nutritional properties of which local people are aware.

BOX 9.1 Revived Cultivation of Buckwheat

Buckwheat was a staple food crop in the Sangla valley of Kinnaur district in the state of Himachal Pradesh, India. In the past two decades, planting of apple and offseason vegetables almost wiped out the cultivation of buckwheat from the valley. We conducted a baseline survey and found that people, particularly elders, still love to grow and eat buckwheat. Over the past 3–4 years the NBPGR jointly with nongovernmental organizations, state tourism, and health departments organized food fairs and awareness programs on buckwheat's food value. In this process, many farmers and local political bodies became convinced and stood up for its cultivation by adopting a resolution (compulsory order) through Panchayat (a local political body), which said that at least one buckwheat recipe will be served by every household during any religious or family functions (eg, marriage). This made every household grow at least some buckwheat crops in its field and those who were unable to grow still purchased the seed from others. This has brought back buckwheat cultivation to Sangla valley (Fig. 9.2). Now buckwheat is one of the cash crops in this valley. We have also planned community seed banks in Sangla in Kinaur and Losar in Spiti to ensure a regular supply of seed of these crops.

FIGURE 9.2 Buckwheat cultivation in Sangla Valley of Kinnaur district of Himachal Pradesh in India.

MANAGEMENT OF GENETIC RESOURCES

The NBPGR is a nodal organization in India for the management of plant genetic resources for all the crops important for food and agriculture. The germ-plasm of buckwheat has been augmented both through collections made within India and by introducing it from abroad, and being managed and conserved at the NBPGR. The important activities include exploration and collection, introduction, characterization and evaluation, genetic enhancement and utilization, conservation, documentation, and supply to researchers and other end users.

Exploration and Collection

In the past four decades, the NBPGR has built up 857 germplasm accessions from more than 70 buckwheat growing sites (Fig. 9.3) ranging in elevation from 1000 to 3500 m above sea level. The collections have been made primarily from farmers' fields and occasionally from seed stores and local markets whenever new variants are spotted. The collections represent *F. esculentum* (288), *F. tataricum* (470), *F. sagittatum* (12), *F. cymosum* (18), *F. tataricum* ssp. *himalianium* (34), *F. esculentum* ssp. *emerginatum* (23), and others (12). While collecting germplasms, a wide range of genetic diversity has been recorded for traits like plant types, maturity period, seed shape, seed color, seed size, leaf

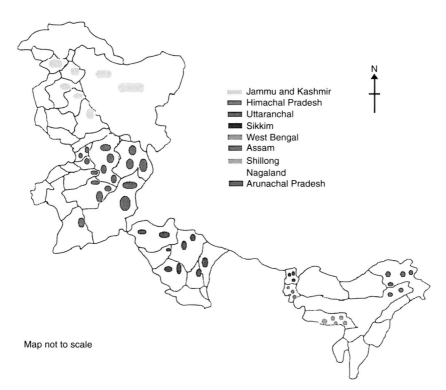

FIGURE 9.3 Genetic diversity rich areas of Buckwheat in India.

shape, leaf size, flower color, stem color, resistance to diseases like leaf spot and powdery mildew, tolerance to lodging, etc.

The physiographic distribution of buckwheat exhibits more genetic variation including those of wild species throughout the Indian Himalayan region with more preponderance in the Western Himalayan region than in northeastern region. This may be because of the common border between the Western Himalayan region and Tibet and migration could have taken place from Southern China and Tibet along several trade routes particularly the silk route and the Hindustan Tibet route. The range of genetic diversity is greater in common buckwheat than Tartary buckwheat, which may be because of its cross-pollinated nature.

INTRODUCTION

Buckwheat is primarily introduced to augment the gaps present in the existing indigenous variability. We have introduced 198 germplasm accessions from different countries representing the United States (55), Canada (48), Nepal (24), Japan (31), Italy (13), Russia (14), Poland (7), and Germany (5). Although sizable germplasm has been introduced from abroad there are still gaps and more germplasm is required to be introduced from other countries that are rich in buckwheat genetic resources such as China, Russia, France, Poland, Slovenia, and Ukraine. Some of the germplasm having ultraearly maturity and bold seeds have been selected from the introduced germplasm. The gap analysis of the entire germplasm (indigenous and exotic) showed that existing collections are still lacking many desirable attributes. Therefore the emphasis now has been shifted from general collection surveys to traits and area-specific surveys. For instance, we are looking for a self-fertile and frost-resistant germplasm in *F. esculentum*. Germplasm for other traits as discussed subsequently is also being sought from other countries under germplasm exchange programs.

- *Ultraearly maturing types*: The majority of buckwheat-growing areas fall into the monocropping season, that is, Apr.–Oct., and remains under snow for the rest of the year. There have been recent changes in the cropping patterns because of cash crops (offseason vegetables), which are sown in early Apr. and harvested in Jul.–Aug. This leaves hardly any time for the next crop and leads to abandonment of buckwheat crops as most of the varieties are long duration and mature in around 4–5 months. We found that only early-maturing varieties can fit into the present crop rotations, that is, varieties sown after potato or green pea and mature before snowfall.
- *Synchronous maturity*: Buckwheat varieties and germplasm have asynchronous maturity, that is, the whole plant does not mature uniformly. This sometimes leads either to multiple picking or seed shattering if not harvested at the same time. It also causes high-yield losses and poor-quality seed harvest.
- *Easier dehulling*: Buckwheat in general and *Fagopyrum tataricum* in particular is difficult to dehull, thus produce black flour. Although few easy dehulling types (called rice buckwheat) have been collected their yield is very low. Further such germplasms need to be introduced and involved in the breeding programs.

- *Lodging resistance*: Lodging is more of a major problem in high rainfall areas than cold dry areas and causes major yield and seed quality losses.
- *Frost tolerance*: Common buckwheat has more potential as a winter crop in the lower elevations and plains of India, but most of it suffers from the problem of frost, thus germplasms having frost tolerance will be useful.
- *Bold seeded and with more groat percentage*: The seed size of the Indian germplasm is low while germplasms that occur in Russia, Ukraine, and France are bold seeded and have increased groat percentage.

Characterization and Evaluation

Characterization and evaluation add value to the collected and introduced germplasm and increase its utilization in crop improvement. We at NBPGR grow the germplasm for 2 years for characterization and record the data as per the descriptors developed by NBPGR/IPGRI (International Plant Genetic Resources Institute). Germplasm found promising during these years is further grown for advanced evaluation for 2 more years either at one location or multilocations. Until now, >950 germplasm accessions have been characterized at different times for various descriptors (IPGRI, 1994; NBPGR, 2000) and a wide range of variability has been noticed for traits such as plant type, maturity period, seed shape, seed color and seed size, leaf shape and leaf size, flower color, stem color, resistance to diseases like leaf spot and powdery mildew, and tolerance to lodging, and promising accessions have been identified (Joshi and Paroda, 1991; Rana et al., 2010). The range of variability and mean values observed for yield and yield-contributing characters is given in Table 9.2. The coefficient of variation (CV) was recorded high for seed yield/plant followed by number of branches, leaf length, leaf width, flower cluster/cyme, and 1000-seed weight. A wide range of differences among genotypes and characters has also been reported by Ujihara and Matano (1977). High heritability coupled with high genetic advance was also observed for seed yield/plant, 1000-seed weight, leaf length, leaf width, number of leaves, number of branches, and days to maturity. This indicates that most of the yield-contributing traits are under the influence of additive gene actions, and use of simple selection methods may give better results for identifying high-yielding genotypes from germplasm and also from the segregating progenies.

The parameters of genetic variability such as CV were found moderate to high for all the traits such as seed yield/plant followed by number of branches, leaf length, leaf width, days to flowering, petiole length, and 1000-seed weight. Similarly, many traits showed high heritability coupled with high genetic advance indicating the presence of an additive type of gene action in the germplasm. Therefore the use of simple selection methods would be effective to make selection of desirable lines from the available germplasm. The germplasms maintained in other countries have also shown a wide range of variability for yield and yield-contributing characters (Ujihara and Matano, 1977; Ujihara, 1983; Choi et al., 1992; Baniya et al., 1995).

TABLE 9.2 Genetic Variability Parameters of Buckwheat

Characters	Range Min.	Max.	Mean	Coefficient of Variation	Heritability	Genetic Advance
Plant height (cm)	36.40	206.05	120.52	38.44	62.58	17.63
Number of internodes	5.00	30.00	14.20	24.56	59.39	16.21
Number of branches	2.00	14.56	8.44	33.66	79.42	33.89
Number of leaves	19.00	147.00	61.47	44.10	51.66	18.59
Petiole length (cm)	1.40	9.25	5.09	18.57	80.32	28.17
Leaf length (cm)	2.70	12.90	7.54	39.65	76.91	37.46
Leaf width (cm)	2.50	14.80	8.72	29.87	72.85	29.68
Days to flower	33.00	82.00	55.75	20.96	80.29	24.37
Cyme length (cm)	1.50	15.00	4.84	17.30	74.08	21.22
Flowers cluster/cyme	4.00	36.00	18.50	31.29	63.24	21.60
Days to mature	65.00	164.00	124.70	40.54	59.12	20.94
1000-seed weight (g)	10.65	32.00	20.56	24.30	68.90	32.44

Frequency distribution constructed for all the traits measured qualitatively showed a wide range of variation among 322 accessions evaluated. Early plant vigor, which determines the subsequent growth of the plant, was good to very good for 98% of accessions while erect growth habit was observed for 84% of accessions. The seed shape was triangular to ovate for the majority of the accessions while seed color was brown for 65% of accessions. Frequency distribution graphs for quantitative traits showed that 96% of accessions fall into the early medium group of flowering while only 68% come under early to mid-maturity indicating a very long reproductive phase of some accessions. Leaf length and leaf width for the majority of the accessions ranged from 5 to 15 cm. The number of leaves/plants varied from 10 to 20 for 92% of accessions. A good range of variation was observed for other traits in a set of 322 accessions.

The matrix developed for correlation coefficients (Table 9.3) showed significant positive correlation of leaf length with leaf width, petiole length, primary branches, cyme length, and seed weight. It is worth mentioning here that important traits such as days to flowering, days to maturity, and seed weight showed negative correlation with most of the yield-contributing traits. This shows that to develop and select early-maturing varieties we should select either a hybridization or recurrent selection approach because simple selection may not give good

TABLE 9.3 Matrix of Correlation Coefficients Among 13 Traits for 322 Accessions

Traits	DF	LL	LW	NOL	NOI	PL	PRB	IPP	CL	PLH	DM	SPI
LL	-0.42*											
LW	-0.16	0.77*										
NOL	-0.17	0.10	0.11									
NOI	-0.15	0.07	0.11	0.87*								
PL	-0.22*	0.51*	0.55*	0.06	0.04							
PRB	-0.49*	0.34*	0.21*	0.56*	0.46*	0.22*						
IPP	-0.45*	0.18	0.02	0.54*	0.47*	-0.01	0.63*					
CL	-0.54*	0.48*	0.28*	0.26*	0.17	0.31*	0.47*	0.41*				
PLH	-0.65*	0.55*	0.29*	0.52*	0.38*	0.30*	0.66*	0.64*	0.65*			
DM	0.67*	-0.27*	-0.04	-0.32*	-0.31	-0.04	-0.61*	-0.58*	-0.45	-0.62*		
SPI	-0.32*	0.21*	0.04	0.54*	0.40*	-0.07	0.55*	0.54*	0.33*	0.58*	-0.53*	
SWT	-0.39*	0.56*	0.30*	-0.06	-0.05	0.23*	0.24*	0.15	0.33*	0.43*	-0.20	0.08

DF, days to flowering; *LL*, leaf length; *LW*, leaf width; *NOL*, number of leaves; *NOI*, number of internodes; *PL*, petiole length; *PRB*, primary branches; *IPP*, inflorescence per plant; *CL*, cyme length; *PLH*, plant height; *DM*, days to maturity; *SPI*, seeds per inflorescence; *SWT*, seed weight.
*P value significant at 5% level of significance.

combinations. Other combinations of traits, which have strong positive correlations among themselves, were number of leaves and number of internodes, number of inflorescences and number of primary branches, days to flowering and days to maturity, plant height and cyme length, and plant height and number of inflorescences/plant. Interestingly, cyme length showed no association with number of inflorescences/plant.

Genetic diversity and principal component analysis (PCA) performed on 322 accessions showed that all the accessions were grouped into three clusters and each cluster was found to have a varied number of accessions. The number of accession falls in each cluster was highest (158) in cluster 2 followed by cluster 1 (282) and cluster 3 (40). The PCA used to eliminate redundancy in the data set revealed that all the 13 quantitatively measured traits were loaded on first the five components; however, the major portion of variance (60.86%) was explained by the first two components (Fig. 9.4). The first component (PC1) accounted for 42.06% of variation, through leaf length, number of leaves, primary branches, plant height, internodes/plants, and number of seeds/inflorescence while PC2 accounted for 18.81% of variation loaded on leaf width, seed weight, and petiole length.

Genetic divergence analysis done on 96 germplasm accessions of Tartary buckwheat categorized the material into 13 different clusters and showed internodes/plant, branches/plant, flower clusters/cyme, 1000-seed weight, days to maturity as most divergent, and potential traits for selecting superior genotypes in buckwheat (Rana, 1998; Rana and Sharma, 2000) (Table 9.4). Stability analysis performed for 13 promising lines for 5 years (1988–1992) showed that IC13374, IC13411, Kullu gangri, and VL7 as high yielding most consistently over the years (Joshi and Rana, 1995). Germplasm lines have also been selected

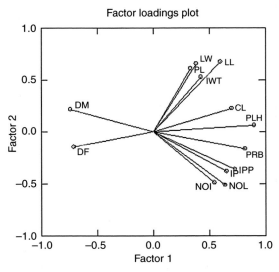

FIGURE 9.4 Biplot of different variables loaded on PC1 and PC2 in buckwheat germplasm.

TABLE 9.4 Trait-Specific Reference Sets of Buckwheat Germplasm

Characters	Germplasm Accessions	Value
Days to flowering	IC16555, IC274429, IC412762, IC521305, IC42415, IC108511, IC16559, IC37273, IC37285, IC42414	<39
Leaf length	IC341081, IC274444, IC310104, IC278957, IC274439	>10 cm
Leaf width	IC341681, IC311004, IC310046, IC318859, IC310047	>10 cm
Number of leaves/plant	IC18889, EC125357, EC58322, IC329568, IC341656, IC26589, IC18757, IC18751, IC42427	>100
Number of internodes	IC258244, IC258230, IC341680, IC313468, IC381077, IC26589, IC18757, IC26587, IC18751, IC42427	>20
Number of primary branches	EC188664, IC274423, IC318859, IC329194, IC318859, IC18890, IC37300, IC107991, IC204079, IC107310	>8
Cyme length	IC361635, IC341631, IC318859, EC125357, EC58322	>7 cm
Days to maturity	IC310104, EC323731, IC341671, EC323729, IC329568	<80 days
Seed yield/plant	IC18869, IC18889, IC318859, IC329401, IC329404	>100 g
1000-seed weight	IC381077, IC381098, IC381049, EC216685, IC58322, IC363973, IC340325, IC125357, IC188669, IC363948, IC412849	>25 g
Protein content	IC108499, IC108500, IC107989, IC291963, IC382287	>14.0%
Total phenols	IC.310045, IC274439, IC341674, IC266947, IC274439	>1.60%
Lysine	IC341674, IC310045, IC274439, IC108499, IC274438	>4.5%
Rutin content	IC042421, IC107962, IC310045, IC014889, IC014253	>17.0 µg/mg

for easy dehulling in Tartary buckwheat popularly known as rice buckwheat and has also been reported in China, Nepal, and Bhutan (Campbell, 1997). Two lines, namely, IC329457 and IC341679, have been identified as easy to dehull (dehulled just by rubbing with hands) from the Indian germplasm (Rana et al., 2004). This germplasm will be a useful genetic resource to overcome the problem of black flour in Tartary buckwheat, which is considered a major bottleneck in its value addition. Genetic enhancement of largely available non-dehulled types will be possible using easily dehulled types as one of the parents in breeding programs. Campbell (1997), however, through his personal data has reported that crosses between the two types are extremely difficult. The genetics of the easy dehulling characteristic of *F. tataricum* has shown that the trait is controlled by a single recessive gene (Wang and Campbell, 2007). We are attempting to genetically map the easy dehulling trait for its eventual incorporation into genetically superior genotypes through molecular breeding.

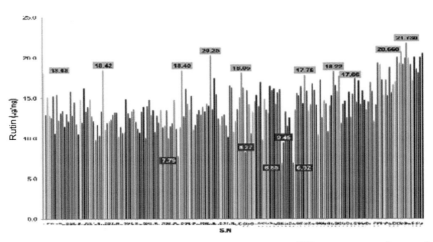

FIGURE 9.5 Variation for rutin content among different accessions of *F. tataricum*.

Germplasms have also been evaluated for quality characters such as protein content, total phenol, free phenol, lysine content, amino acid, and vitamins E and P contents (Keli, 1992; Suzuki et al., 2005a; Lin et al., 2008; Anonymous, 2008). We evaluated 60 promising accessions and found that protein content varied from 8.20% to 15.10%, total phenols from 1.4% to 1.70%, free phenols from 0.27% to 0.94%, and available lysine from 3.89% to 5.60%. Rutin contents analyzed in mature seeds of 200 *F. tataricum* accessions showed relatively large variation, ranging from 0.6% to 2.0% (dw) compared to only 0.07% (dw) in selected accessions of *F. esculentum* (Fig. 9.5). These results agree with previous reports on variation in rutin content in cultivated buckwheat (Eggum et al., 1981; Bonafaccia and Fabjan, 2003). Park et al. (2004) compared rutin content variation in different parts of *Fagopyrum* sp. (*F. tataricum*, *F. cymosum*, and *F. esculentum*), which revealed the highest content in flowers and lowest in roots. There are reports on rutin content variation in different *Fagopyrum* sp. coupled with antioxidant activity, which decreased in the order: *F. tataricum* > *F. homotropicum* > *F. esculentum* (Jiang et al., 2006). Morishita et al. (2007) reported 3–4 times higher antioxidant activity in Tartary than common buckwheat grains. They showed that rutin contributed 2% of the total antioxidant activity in common buckwheat while 11–13% was contributed by (–)-epicatechin in contrast to Tartary buckwheat where rutin appeared to be the major antioxidant (85–90%) (Morishita et al., 2007). The variation in rutin content has been estimated from ≤10 µg/mg to ≥16 µg/mg through reverse phase high-performance liquid chromatography and further research on rutin biosynthesis involving accessions with high and low rutin content lines is in progress.

Studies have also been made and are being carried out on phytochemical screening and evaluation of nutritional and antioxidant potential, amino acid profiling, starch properties, protease inhibitors, and antifungal protein, isolation and mapping of genes, and also on the flavonoid biosynthesis genes (Senthilkumaran et al., 2007; Chrungoo et al., 2010; Gupta et al., 2012). The Smetanska and Sytar (2015) reported that the presence of anthocyanins in the vegetative organs of buckwheat can be a reliable genetic marker for screening plants with high content of rutin. The highest rutin content was measured in inflorescences of *F. esculentum* Moench (cultivars Lileya, Bilshovik, Rubra) than in *F. tataricum* (ssp. *rotundatum* Krot. and ssp. *tuberculatum* Krot.), *F. cymosum* Meissn, and *F. giganteum* Krot. The comparative analysis of total phenolics and phenolic acid composition together with antioxidant activities in inflorescences of *F. esculentum*, *F. tataricum rotundatum*, and *F. esculentum* (green flowers) shows that *F. esculentum* inflorescences (green flowers) have been characterized by the highest content of salicylic acid (115 mg/100 g^{-1} dw) and methoxycinnamic acid (74 mg/100 g^{-1} dw).

Genome Resources

A significant amount of research has been conducted on the functionalities and properties of buckwheat proteins, flavonoids, flavones, phytosteroles, thiamin-binding proteins, and other rare compounds (Li and Zhang, 2001; Tomotake et al., 2002; Kreft et al., 2006; Zielinski et al., 2009). However, the availability of genome resources such as a good linkage map, different classes of molecular markers, EST libraries, large-insert DNA libraries, etc. is limited. There are fragmentary reports on understanding the species relationship by using molecular markers such as polymer chain reaction (PCR)-based DNA fingerprinting, which has been used to demonstrate species relationships in Indian *Fagopyrum*. Of the 75 random 10-mer primers tested on 14 accessions and two subspecies of *Fagopyrum*, only 19 generated reproducible bands (Sharma and Jana, 2002a). A total of 364 bands was observed with an average of 19.15 bands per primer, of which 99.45% were polymorphic, which helped to elucidate interspecies relationship in *Fagopyrum* (Sharma and Jana, 2002b). We also characterized 51 accessions of *F. esculentum* (29), *F. tataricum* (20), and *F. cymosum* (2) using random amplified polymorphic DNA (RAPD) (Sethilkumarn et al., 2007). The species-wise population data indicated that *F. tataricum* was relatively more polymorphic than *F. esculentum* accessions. Expected heterozygosity was more for *F. esculentum* because of its outcrossing nature. The estimated fixation index (FST) value indicated low differentiation among populations of a species zone-wise. Species-wise population structure indicated more diversity between species than diversity between zones. Differentiation between species is strong as indicated by the calculated FST value that falls above the upper 95% limit. The RAPD analysis also revealed that *F. cymosum* was relatively more close to *F. esculentum* than *F. tataricum*. A genetic map for *F. esculentum* and *F. homotropicum* on the basis of 223 and 211 AFLP markers, respectively, has been developed (Yasui et al., 2004). The map of *F. homotropicum* has eight linkage groups with 211 AFLP markers covering 548.9 cM. Microsatellite markers have been developed in common buckwheat by sequencing 2785 clones from the libraries and it was shown that 1483 clones contained microsatellites that were enriched for $(CT)_n$ and $(GT)_n$ repeats. Primer pairs were designed for 237 of the microsatellite loci, of which 180 primer pairs were amplified. Out of this, 44 primer pairs were evaluated for their ability to detect variations in common buckwheat populations and utilized in seven related *Fagopyrum* species including *F. tataricum* (Konishi et al., 2006). One bacterial artificial chromosomal (BAC) library has been constructed from a wild buckwheat species, *F. homotropicum* (Nagano et al., 2001). The applicability of 17 EST primers developed from common buckwheat was tested in other wild and cultivated *Fagopyrum* species (Joshi et al., 2006). The amplification products were different in band intensity. The results indicated that the transferability of EST markers developed for common buckwheat decreased with an increase in genetic distance between species.

The absence of a well-developed linkage map and availability of a limited number of molecular markers in buckwheat prompted us to look for in silico alternatives for rapid identification of additional molecular markers. We utilized ESTs available in other plant species belonging to a taxonomically common order of *Fagopyrum* species. Buckwheat belongs to the family Polygonaceae and order Caryophyllales, therefore we chose those plant species that come under the same order for identification of molecular markers such as SSRs. All ESTs available in a particular plant species (Table 9.5) were downloaded from the database TIGR (http://compbio.dfci.harvard.edu/tgi/tgipage.html).

The SSRs were identified in ESTs using PGG Bioinformatics at (http://hornbill.cspp.latrobe.edu.au/cgi-binpub/autosnip/index_autosnip.pl) and primers were designed for their amplification. Primer pairs were designed and synthesized for 141 SSRs based on repeat length out of which 13 SSRs were

TABLE 9.5 Status of ESTs in Plant Species Related to *Fagopyrum*

Family	Plant Species	Number of EST
Aizoaceae (ice plant family)	*Mesembryanthemum crystallinum*	27,191
Amaranthaceae (goosefoot family)	*Beta vulgaris*	25,834
	Suaeda salsa	973
Plumbaginaceae (leadwort family)	*Limonium bicolor*	4,686
	Plumbago zeylanica	1701
Tamaricaceae (tamarix family)	*Tamarix androssowii,*	4,627
	Tamarix hispida	17,082

EST, Expressed sequence tag.

successfully amplified on *F. tataricum* genotypes, indicating poor transferability of SSRs. Fifty-four SSRs, which were identified by Konishi et al. (2006) in *F. esculentum*, were also tested on selected accessions of *F. tataricum*, but no polymorphism was found on the selected accessions.

We also tried to study the comparative genomics of genes involved in rutin biosynthesis. The rutin biosynthetic pathway has been elucidated in various plant species. Nine genes known to be involved in the rutin biosynthesis pathway are: phenylalanine ammonia-lyase, cinnamate-4-hydroxylase (C4H), 4-coumaryl CoA ligase (4CL), chalcone synthase (CHS), chalcone isomerase, flavonol synthase, flavanone-3-hydroxylase (F3H), flavanone-3'-hydroxylase, and glucosyl/rhamnosyl transferase. Out of these, two genes, CHS and glucosyl transferase, have been identified in *F. esculentum* and *F. tataricum*, respectively (Hrazdina et al., 1986; Suzuki et al., 2005b). We used comparative genomics to identify and clone the remaining rutin biosynthesis genes in Tartary buckwheat. As most of the genes are present in multiple copies in the genomes of plants, we used *Arabidopsis* genome information to identify the most significant copy of each gene.

The nucleotide and protein sequences of genes involved in rutin biosynthesis were retrieved from different plant species and the multiple sequence alignments were done to find out the extent of sequence similarity. Primer pairs were designed from conserved regions of gene sequences retrieved from dicot plants and then tested on *Fagopyrum* species (common buckwheat, Tartary buckwheat, and rice-Tartary buckwheat). All genes were amplified in *Fagopyrum*. We attained single-band amplification in CHS, 4CL, and glucosyl/rhamnosyl transferases, whereas in the case of F3H and C4H, multiple copies of genes were amplified. The *F. tataricum* genotypes showing contrasting variation for rutin content are being used in the identification of DNA sequence variations in genes involved in rutin biosynthesis. The identification of single nucleotide polymorphisms in high versus low rutin content genotypes will be of great significance in molecular breeding of buckwheat for high rutin content. The high rutin content genotypes are also being used in identifying regulatory genes controlling rutin biosynthesis through differential display analysis. We are also constructing a BAC library from a high rutin content, easy to dehull genotype with the ultimate goal of cloning useful genes. The BAC library would be very useful in pursuing genomics of *F. tataricum*.

Gupta et al. (2012) studied differential transcript profiling through cDNA-AFLP in seed maturing stages (inflorescence to seed maturation) with 32 primer combinations generating a total of 509 transcript fragments (TDFs). One hundred and sixty-seven TDFs were then eluted, cloned, and sequenced from *F. tataricum* and *F. esculentum*. The TDFs represented genes controlling different biological processes such as basic and secondary metabolism (33%), regulation (18%), signal transduction (14%), transportation (13%), cellular organization (10%), and photosynthesis and energy (4%) and most of the TDFs except those belonging to cellular metabolism showed relatively higher transcript abundance in *F. tataricum* over *F. esculentum*. They concluded that in addition to structural genes, other classes of genes such as regulators, modifiers, and transporters are also important in biosynthesis and accumulation of flavonoid content in plants. cDNA-AFLP technology was successfully utilized to capture genes that are contributing to differences in rutin content in seed maturing stages of *Fagopyrum* species. Increased transcript abundance of TDFs during transition from flowers to seed maturation suggests their involvement not only in the higher rutin content of *F. tataricum* over *F. esculentum* but also in nutritional superiority of the former.

We also tested gene-specific STS markers against 91 buckwheat accessions to elucidate allelic diversity at these loci. Out of the 27 STS loci screened, only 18 returned assayable amplifications. The remaining nine primers either amplified null allele (less likely) or needed selection of a different region within the gene for a detectable PCR product. BW16 amplified only one allele whereas BW10 amplified as many as five alleles. Among the rest, eight markers amplified two alleles; six markers amplified three; and two markers amplified four alleles. On average these STS primers amplified 2.7 bands per locus. Four of the STS markers, BW10 (*Fe2SA1*, 8 kD allergen protein), BW12 (major allergenic storage protein, *FAGAG1*), BW22 (declined protein during seed development), and BW27 (proteinase inhibitor, *BTIw1*), showed substantial polymorphism (PIC > 0.5) among the 91 buckwheat accessions. However, overall the markers showed moderate estimates of polymorphic information content (0.268), observed heterozygosity (0.259), and expected heterozygosity (0.318). One possible reason for moderate estimates of polymorphism could be that the STS loci were located in genes responsible for major functions and therefore the sequences could be relatively conserved.

Marker phenotype association showed that out of 24 morphological traits, 18 traits exhibited absolutely no linkage with marker profiles. Test weight, days to maturity, leaf length, number of primary branches, plant height, and seed shape showed varied degrees of marker–trait association. The STS marker-hosting genes exhibiting association were BW10 (8 kD allergen protein), BW18 (aspartic proteinase 9), BW13(legumin-like protein), BW17 (chalcone synthase), BW22 (declined protein during seed development), BW09 (13S globulin), BW25 (cysperoxiredoxin), and BW24 (fagopyritol synthase 1). The results encourage the screening of more germplasm lines and deploy the markers identified to be putatively associated with specific traits to screen-mapping populations.

USES OF BUCKWHEAT

Buckwheat is a crop that holds tremendous agronomic and nutritional benefits. Buckwheat flour has numerous uses. It is used in pancake mixes as well as in various breads. It is often blended with wheat flour for use in bread, pasta products, and some breakfast cereals (Robinson, 1980). Buckwheat flour, popularly known as kuttu ka atta, in India, is eaten on brata or fast days, being one of the lawful foods for such occasions. Occasionally, the flour is made into a paste, with vegetables and salt added, and made into small balls, fried, and served hot, locally called pakoras; other similar salted preparations are called chillare and jalebi in India or sil and fulaura in Nepal, or, when prepared with sugar, puwa in the Eastern and halwa in the Western Himalayas. It is also pounded and boiled like rice and consumed as a substitute for rice. Buckwheat is quite complementary to cereal flours, and can be used to improve their nutritional quality, since it is high in essential amino acids. Studies have shown that up to 60% of buckwheat flour mixed with wheat flour can produce an acceptable bread (Pomeranz, 1983). Buckwheat groats are served as part of soldiers' rations and cooked with butter, tallow, or hemp seed oil in Russia.

Recently, buckwheat has also been introduced as a nutraceutical food. A nutraceutical is defined as any substance that is a food or part of a food and provides medical or health benefits, including the prevention and treatment of disease (DeFelice, 1994). Buckwheat contains vitamin P, which contains the flavonoid rutin. Rutin is known for its effectiveness in reducing the cholesterol count in the blood. In addition, buckwheat is an effective preventive measure against high blood pressure. Rutin is known to keep capillaries and arteries strong and flexible. The effectiveness of rutin in buckwheat is strengthened with the addition of vitamin C. Regular consumption of 30 g of buckwheat has been shown to lower blood pressure regardless of other factors such as age and weight. In a study conducted in cooperation with the Johns Hopkins Medical Institute, Jiang et al. (1995) reported that subjects who consumed the greatest amount of buckwheat had the lowest blood pressures. Buckwheat fields in bloom can serve as a valuable source of nectar for bees. Honey produced from buckwheat is typically dark and has a stronger flavor than honey produced from clover, and is preferred by some consumers.

Germplasm Utilization

The need to carry out systematic multidisciplinary research on underutilized crops by using a wide array of promising germplasm was started in 1982 under the All India Coordinated Project. Presently, 27 crops are being researched under the project and buckwheat is one of them, included in 1984. To date, more than 250 promising germplasm accession/breeding lines have been tested for yield and adaptability at various locations, namely, Shimla, Bhowali, Shillong, Almora, Ranichuari, Kukumseri, Palampur, and Sangla. As a result, five varieties, namely, Himpriya and Himgiri by NBPGR Regional Station Shimla, VL7 by VPKAS, Almora and PRB1 by GBPAUT Hill campus, and Ranichauri and Sangla B1 by CSKHPKV, Palampur, have been released. Another variety, namely, "IC109728," has been identified by NBPGR Shimla and is being proposed for release at national level. In addition, there are a large number of germplasm accessions/breeding lines that performed better across the locations but could not be released as commercial varieties and have also been marked as desirable parents for multiple traits for breeding purposes and have been supplied to plant breeders.

Germplasm Conservation

The technological advancement in agriculture has caused large-scale changes in cropping patterns vis-à-vis people's food habits. This has greatly led to the neglect and fast genetic erosion of traditional varieties and crops like buckwheat. Under the circumstances, ex situ conservation of genetic resources, particularly of underutilized crops like buckwheat, has greater significance and is an alternative to protect genetic resources from further erosion. In India, NBPGR is the nodal institute for ex situ conservation of plant genetic resources and it maintains buckwheat germplasm at the following facilities:

1. 994 germplasm collections are stored in the National Gene Bank for long-term storage at −20°C and 4–5% relative humidity at New Delhi as base collections.
2. 1055 germplasm accessions are stored in medium-term storage at 7–8°C and 35% relative humidity at Shimla as working collections.

The working collections are maintained through periodic regeneration and used for the exchange of germplasm. About 350 working collections are also maintained at other institutes such as by G.B. Pant Agricultural University, Hill Campus, Ranichauri, Vivekananda Parvatiya Krishi Anusandhan Shala, Almora, regional stations of the Agricultural University, Palampur, at Sangla and Kukumseri, the NBPGR Regional Station, Shillong, and the ICAR Research Complex for NEH Region. On-farm conservation for some of the crops and landraces including those of buckwheat has also been initiated at some selected traditional farming systems in the Western Himalayan region of India. Traditional farming systems are of particular importance to maintain local genetic diversity on-farm and play an important role in the food and livelihood security of local communities. However, it must be mentioned that under the present boom of commercial farming in India, on-farm conservation on the whole and underutilized crops like buckwheat in particular are difficult tasks to be achieved. Nonetheless, organizing awareness camps and brainstorming sessions at grassroots level involving farmers and extension agencies and highlighting its food and medicinal value have generated a lot of interest among farming communities for growing and conserving buckwheat genetic resources.

1. *Documentation*: Data generated from explorations, characterization, and evaluation have been documented at the NBPGR. At present, passport data of 900 germplasm accessions and characterization data of 775 germplasm accessions are maintained as a database in MS Access and about 1000 accessions in MS Excel. A catalog of 408 germplasm accessions has also been published by the NBPGR (Joshi and Paroda, 1991). The information has also been published in various forms such as research papers, bulletins, popular articles, etc.
2. *Genetic erosion*: The Himalayan region of India is predominantly rural and the day-to-day livelihoods of its people are mainly supported by the agrihorticultural sector. The stability and sustainability of its agriculture

is therefore of paramount importance. The loss of biological diversity and related traditions needs urgent attention. We can no longer afford to ignore this as a concern of only rich people or nations; rather this loss hits the poor the hardest, and makes farmers, particularly the hill farmers where high-input agricultural management systems do not work, generally adopt subsistence farming systems. Genetic erosion is apparent in this region. The area under crops such amaranth, buckwheat, chenopod, and minor millets as well as local landraces have declined substantially and ranges from 60% to 92% (Rana and Sharma, 2000). Wild progenitor and several other wild species depict the same story. Several factors have been found responsible for erosion of genetic diversity in buckwheat:

a. *Changing cropping patterns*: In earlier days, buckwheat was an integral part of the cropping systems of the hills but modernization and developmental activities, particularly the road network, meant that traditional buckwheat growing was ousted in favor of new cash crops like apple, green pea, cabbage, hops, potato, etc. These high-return-oriented crops displaced buckwheat from many traditional growing areas.

b. *Low productivity*: Until now, the major share of buckwheat produce comes only from farmers' own varieties, which are generally poor yielders.

c. *Changing food habits*: The food base has become narrow and totally restricted to wheat, rice, and maize. The younger generation is found to be less interested in the consumption of buckwheat than older people. A survey was conducted in both a major buckwheat growing area (Kinnaur) and a partially buckwheat growing area (Kangra) to discover people's preference for eating buckwheat. The data were generated in terms of number of days in a year for which people were eating buckwheat in 1997–1998 and in 1977–1978. It was observed that there was a reduction of 76% in Kangra and 37% in Kinnaur over the 20 years (Rana and Sharma, 2000).

d. *Less alternative uses/products*: The crop is mainly used only for grain and green purposes. There is a lack of value addition for making alternate products for edible and medicinal uses as has been done in Japan, China, and Russia. In Japan 90% of the flour is used for making popular noodles called "Soba."

e. *Lack of awareness about its food value*: People growing buckwheat in general and younger generations in particular are not aware of its medicinal and nutritive value. Most people consuming buckwheat think that it is a poor man's food and has low food value compared to rice and wheat.

CONCLUSION AND FUTURE RESEARCH NEEDS

Cultivated buckwheat, despite its consumption and use in many parts of the world, lacks breeding research. Future collections should be made from areas where genetic erosion is taking place, largely because of changing cropping patterns and the replacement of buckwheat by other crops, and moreover where high diversity exists but has never been explored. There is a need for a regional and eventually global database on buckwheat ex situ conservation so as to record the basic passport data and a minimum of characterization data for each accession held in ex situ collections worldwide. Systematic and careful germplasm characterization efforts are required to identify the trait-specific germplasms such as ultraearly maturing types, synchronous maturity, resistance to shattering, ease of dehulling especially in Tartary buckwheat, and tolerance to lodging and frost; in addition, increased groat percentages need to be identified. Wild species diversity needs further taxonomic study, a biosystematic approach to classification, and authentication of diversity at species and subspecies levels. Breeding must be encouraged to increase buckwheat quality, productivity, and special characteristics such as nutrition and health. More focus is required on the utilization of material for crop improvement, particularly a wild gene pool for incorporating diverse traits for cold hardiness, disease resistance, etc. Genomic resources in buckwheat are limited. Efforts need to be directed toward developing codominant markers and generating linkage maps with deeper coverage that can assist breeders in quantitative trait locus dissection and marker-assisted breeding. Optimization of in vitro protocols for large-scale production of secondary metabolites of industrial importance like rutin can be very useful.

The nutraceutical potential of buckwheat has not been fully recognized by the government and the public, thus funding and policy support have not been sufficient for its sustainable development and the level of research into buckwheat has remained very low. Industry back-up is also lacking as most buckwheat products are raw products or initial processing products. A lack of advanced processing technology has limited value-adding opportunities for meeting market demand. Therefore more efforts should be focused on developing value-adding opportunities for farmers and enterprises such as advanced processing technology applications for analyzing nutrient and medical composition of buckwheat to develop more diversified products.

REFERENCES

Ahmad, F., Raj, A., 2012. Buckwheat: a legacy on the verge of extinction in Ladakh. Curr. Sci. 103 (1), 13.

Ahmed, A., Khalid, N., Ahmad, A., Abbasi, N.A., Latif, M.S.Z., Randhawa, M.A., 2014. Phytochemicals and biofunctional properties of buckwheat: a review. J. Agric. Sci. 152 (3), 349–369.

Anonymous, 2008. Annual report. All India Coordinated Network on Under-utilized Crops. NBPGR Pusa Campus, New Delhi–110 012, India.

Arora, R.K., Engels, J.M.M., 1992. Buckwheat genetic resources in the Himalayan region: present status and future thrust. International Board for Plant Genetic Resources, Rome, pp. 87–91.

Baniya, B.K., Dongol, D.M.S., Dhungel, N.R., 1995. Further characterization and evaluation of Nepalese buckwheat (*Fagopyrum* spp.) landraces. Shinshu. Proceedings of the Sixth International Symposium on Buckwheat. pp. 295–304.

Bonafaccia, G., Fabjan, N., 2003. Nutritional Comparison of Tartary Buckwheat with Common Buckwheat and Minor Cereals. Zbornik Biotehniške fakultete Univerze v Ljubljani, Kmetijstvo, 81, pp. 349–355.

Bystricka, J., Vollmannova, A., Kupecesek, A., Musilova, J., Polakova, Z., Cicova, I., Bojnanska, T., 2011. Bioactive compounds in different plant parts of various buckwheat (*Fagopyrum esculentum* Moench.) cultivars. Cereal Res. Comm. 39, 436–444.

Campbell, C.G., 1997. Buckwheat. *Fagopyrum esculentum* Moench. Promoting the conservation and use of underutilized and neglected crops.19, IPK, Germany and IPGRI, Rome, Italy.

Chen, Y., Kenaschuk, E., Dribnenki, B., 2001. Inheritance of rust resistance genes and molecular markers in microspore derived populations of flax. Plant Breed. 120, 82–84.

Choi, B.H., Park, K.Y., Park, R.K., 1992. Buckwheat genetic resources in Korea. Proceedings of the Buckwheat Genetic Resources in East Asia. Papers of an IBPGR Workshop, Ibaraki, Japan, December 18–20, 1991. International Crop Network Series No. 6. International Board for Plant Genetic Resources, Rome, pp. 45–52.

Chrungoo, N.K., Devdasan, N., Kreft, I., Licen, M., 2010. Identification and molecular characterization of granule bound starch synthase (GBSS-I) from Buckwheat (*Fagopyrum* spp.). Proceedings of the Eleventh International Symposium on Buckwheat, Orel, Russia, pp. 455–463.

DeFelice, S.L., 1994. Food companies must pursue nutraceutical R & D–now! Food Eng. Dec. 1994.

Du, D.K., Li, Z.X., Yu, X.Z., 2004. Research progress on buckwheat protein. Food Sci. Chin. 25 (910), 409–414.

Eggum, B.O., Kreft, I., Javornik, B., 1981. Chemical composition and protein quality of buckwheat (*Fagopyrum esculentum* Moench). Plant Foods Hum. Nutr. 30, 175–179.

Fabjan, N., Rode, J., Kosir, I.J., Wang, Z., Zhang, Z., Kreft, I., 2003. Tartary buckwheat (*Fagopyrum tataricum* Gaertn.) as a source of dietary rutin and quercitrin. J. Agric. Food Chem. 51, 6452–6455.

Farooq, S., Tahir, S., 1987. Comparative study of some growth attributes in buckwheat. Fagopyrum 7, 9–12.

Fujimura, M., Minami, Y., Watanabe, K., Tadera, K., 2003. Purification, characterization, and sequencing of a novel type of antimicrobial peptides, Fa-AMP1 and Fa-AMP2, from seeds of buckwheat (*Fagopyrum esculentum* Moench.). Biosci. Biotechnol. Biochem. 67, 1636–1642.

Guo, X.D., Ma, Y.J., Parry, J., Gao, J.M., Yu, L.L., Wang, M., 2011. Phenolics content and antioxidant activity of Tartary buckwheat from different locations. Molecules 16, 9850–9867.

Gupta, N., Sharma, S.K., Rana, J.C., Chauhan, R.S., 2012. AFLP fingerprinting of Tartary buckwheat accessions (*Fagopyrum tataricum*) displaying rutin content variation. Fitoterapia 83, 1131–1137.

Hou, L.-L., Zhou, M.-L., Zhang, Q., Qi, L.-P., Yang, X.-B., Tang, Y., Zhu, X.-M., Shao, J.-R., 2015. *Fagopyrum luojishanense*, a new species of Polygonaceae from Sichuan, China. Novon: J. Bot. Nomen. 24 (1), 22–26.

Hrazdina, G., Lifson, E., Weeden, N.F., 1986. Isolation and characterization of buckwheat (*Fagopyrum esculentum* M.) chalcone synthase and its polyclonal antibodies. Arch. Biochem. Biophys. 247, 414–419.

Huang, X.Y., Zeller, F.J., Hung, K.F., Shi, T.X., Chen, Q.F., 2014. Variation of major minerals and trace elements in seeds of Tartary buckwheat (*Fagopyrum tataricum* Gaertn.). Genet. Resour. Crop Evol. 61, 567–577.

IPGRI, 1994. Buckwheat genetic resources in Nepal: a status report (submitted to IPGRI-APO, Singapore). NARC, Nepal, pp. 51.

Jackson, S.J., Venema, R.C., 2006. Quercetin inhibits eNOS, microtubule polymerization, and mitotic progression in bovine aortic endothelial cells. J. Nutr. 36, 1178–1184.

Jiang, H., Klag, M.J., Whelton, P.K., Mo, J.P., Chen, J.Y., Qian, M.C., Mo, P.S., He, G.Q., 1995. Oats and buckwheat intakes and cardiovascular disease risk factors in an ethnic minority in China. Am. J. Clin. Nutr. 61, 366–372.

Jiang, P., Burczynski, F., Campbell, C., Pierce, G., Austria, J.A., Briggs, C.J., 2006. Rutin and flavonoid contents in three buckwheat species *Fagopyrum esculentum*, *F. tataricum*, and *F. homotropicum* and their protective effects against lipid peroxidation. Food Res. Int. 40, 356–364.

Joshi, B.D., Paroda, R.S., 1991. Buckwheat in India. NBPGR, New Delhi, p. 117.

Joshi, B.D., Rana, J.C., 1995. Stability analysis in buckwheat. Indian J. Agric. Sci. 65, 32–34.

Joshi, B.K., Okuno, K., Ohsawa, R., Hara, T., 2006. Common buckwheat-based EST primers in the genome of other species of *Fagopyrum*. Nepal Agric. Res. J. 7, 27–36.

Kalinova, D., Dadáková, E., 2013. Influence of sowing date and stand density on rutin level in buckwheat. Cereal Res. Comm. 41, 348–358.

Kawa, J.M., Taylor, C.G., Przybylski, R., 2003. Buckwheat concentrate reduces serum glucose in streptozotocin-diabetic rats. J. Agric. Food Chem. 51, 7287–7291.

Keli, Y., 1992. Genetic resources of buckwheat (Fagopyrum) in China. In: Buckwheat Genetic Resources in East Asia. Papers of an IBPGR Workshop, Ibaraki, Japan, Dec.18–20, 1991, IBPGR 18–20.

Konishi, T., Iwata, H., Yashiro, K., Tsumura, Y., Ohsawa, R., Yasui, Y., Ohnishi, O., 2006. Development and characterization of microsatellite markers for common buckwheat. Breed. Sci. 56, 277–285.

Krahl, M., Back, W., Zarnkow, M., Kreisz, M., 2008. Determination of optimized malting conditions for the enrichment of rutin, vitexin and orientin in common buckwheat (*Fagopyrum esculentum* Moench.). J. Inst. Brew. 114, 294–299.

Kreft, S., Strukelj, B., Gaberscik, A., Kreft, I., 2002. Rutin in buckwheat herbs grown at different UV-B radiation levels: comparison of two UV spectrophotometric and an HPLC method. J. Exp. Bot. 53, 1801–1804.

Kreft, I., Fabjan, N., Yasumoto, K., 2006. Rutin content in buckwheat (*Fagopyrum esculentum* Moench) food materials and products. Food Chem. 98, 508–512.

Kupper, C., 2005. Dietary guidelines and implementation for celiac disease. Gastroenterology 128, 121–127.

Lee, S.Y., Cho, S.I., Park, M.H., Kim, Y.K., Choi, J.E., Park, S.U., 2007. Growth and rutin production in hairy root cultures of buckwheat (*Fagopyrum esculentum* Moench.). Prep. Biochem. Biotechnol. 37, 239–246.

Li, S.Q., Zhang, Q.H., 2001. Advances in the development of functional foods from buckwheat. Crit. Rev. Food Sci. Nutr. 41, 451–464.

Lin, L.-Y., Peng, C.-C., Yang, Y.-L., Peng, R.-Y., 2008. Optimization of bioactive compounds in buckwheat sprouts and their effect on blood cholesterol in hamsters. J. Agric. Food Chem. 56 (4), 1216–1223.

Liu, Z., Ishikawa, W., Huang, X., Tomotake, H., Kayashita, J., Watanabe, H., Kato, N., 2001. A buckwheat protein product suppresses 1,2-dimethylhydrazine-induced colon carcinogenesis in rats by reducing cell proliferation. J. Nutr. 131, 1850–1853.

Mann, S., Gupta, D., Gupta, R.K., 2012. Evaluation of nutritional and antioxidant potential of Indian buckwheat grains. Indian J. Trad. Knowl. 11 (1), 40–44.

Morishita, T., Yamaguchi, H., Degi, K., 2007. Contribution of polyphenols to antioxidant activity in common buckwheat and Tartary buckwheat grain. Plant Prod. Sci. 10, 99–104.

Munshi, H., 1982. A new species of *Fagopyrum* from Kashmir Himalaya. J. Econ. Taxon. Bot. 3, 627–630.

Nagano, M., Jotaro, A., Campbell, C.G., Kawaskl, S., Adachi, T., 2001. Construction of a BAC library of the genus *Fagopyrum*. Proceedings of the Eighth ISB International Symposium on Buckwheat, pp. 292–297.

NBPGR, 2000. Minimal Descriptors (Part – I). Agro biodiversity (PGR)-9. National Bureau of Plant Genetic Resources, New Delhi.

Ohnishi, O., 1989. Cultivated buckwheat and their wild relatives in the Himalayas and Southern China. Proceedings of the Fourth International Symposium on Buckwheat, pp. 562–571.

Ohnishi, O., 1991. Discovery of wild ancestor of common buckwheat. Fagopyrum 11, 5–10.

Ohnishi, O., 1992. Buckwheat in Bhutan. Fagopyrum 12, 5–13.

Ohnishi, O., 1995. Discovery of new *Fagopyrum* species and its implication for the studies of evolution of *Fagopyrum* and of the origin of cultivated buckwheat. In: Proceedings of the Sixth International Symposium on Buckwheat, pp. 26–29.

Park, B. J., Park, J. I., Chang, K.J., Park, C.H., 2004. Comparison of rutin content in seed and plant of Tartary buckwheat (*Fagopyrum tataricum*). Proceedings of the Ninth International Symposium on Buckwheat, pp. 626–629.

Pomeranz, Y., 1983. Buckwheat: structure, composition, and utilization. Crit. Rev. Food Sci. Nutr. 19 (3), 213–250.

Popović, V., Sikora, V., Berenji, J., Filipović, V., Dolijanović, Ž., Ikanović, J., Dončić D., 2014. Analysis of buckwheat production in the world and Serbia. Economics of Agriculture 1, UDC: 633.2:631.559.

Prestamo, G., Pedrazuela, A., Penas, E., Lasuncion, M.A., Arroyo, G., 2003. Role of buckwheat diet on rats as prebiotic and healthy food. Nutr. Res. 23, 803–814.

Przybylski, R., Gruczynska, E., 2009. A review of nutritional and nutraceutical components of buckwheat. Eur. J. Plant Sci. Biotechnol. 3, 10–22.

Pu, F., Mishima, K., Egashira, N., Iwasaki, K., Kaneko, T., Uchida, T., Irie, K., Ishibashi, D., Fujii, H., Kosuna, K., Fujiwara, M., 2004. Protective effect of buckwheat polyphenols against long-lasting impairment of spatial memory associated with hippocampal neuronal damage in rats subjected to repeated cerebral ischemia. J. Pharmacol. Sci. 94, 393–402.

Quettier-Deleu, C., Gressier, B., Vasseur, J., Dine, T., Brunet, C., Luyckx, M., et al., 2000. Phenolic compounds and antioxidant activities of buckwheat (*Fagopyrum esculentum* Moench) hulls and flour. J. Ethnopharmacol. 72, 35–42.

Raina, A., Gupta, V., 2015. Evaluation of buckwheat (*Fagopyrum* species) germplasm for rutin content in seeds. Indian J. Plant Physiol. 20 (2), 167–171.

Rana, J.C., 1998. Genetic diversity and correlation analysis in Tartary buckwheat (*Fagopyrum tatricum*) gene pool. Proceedings of International Symposium on Buckwheat in Canada, pp. 220–232.

Rana, J.C., 2004. Buckwheat genetic resources management in India. Proceedings of the Ninth International Symposium on Buckwheat, pp. 271–282.

Rana, J.C., Chauhan, R.C., Sharma, T.R., Gupta, N., 2012. Analyzing problems and prospects of buckwheat cultivation in India. Eur. J. Plant Sci. Biotechnol. 6 (2), 50–56.

Rana, J.C., Sharma, B.D., 2000. Variation, genetic divergence and interrelationship analysis in buckwheat. Fagopyrum 17, 9–14.

Rana, J.C., Sharma, T.R., Verma, V.D., Yadav, S.K., Pradheep, K., 2004. An easy de-hulling type buckwheat (*Fagopyrum tataricum* Gaertn.). Indian J. Plant Genet. Resour. 17 (3), 356.

Rana, J.C., Singh, A., Sharma, Y., Pradheep, K., Mendiratta, N., 2010. Dynamics of plant bioresources in Western Himalayan region of India—watershed based case study. Curr. Sci. 98 (2), 192–203.

Rao, K.S., Pant, R., 2001. Land use dynamics and landscape change pattern in a typical micro watershed in the mid elevation zone of central Himalaya, India. Agric. Ecosyst. Environ. 86, 113–123.

Ren, W., Qiao, Z., Wang, H., Zhu, L., Zhang, L., Lu, Y., Cui, Y., Zhang, Z., Wang, Z., 2001. Tartary buckwheat flavonoid activates caspase 3 and induces HL-60 cell apoptosis. Meth. Find. Exp. Clin. Pharmacol. 23, 427–432.

Robinson, R.G., 1980. The buckwheat crop in Minnesota. Agr. Exp. Sta. Bul. 539, Univ. Minnesota, St. Paul.

Schramm, D.D., Karim, M., Schrader, H.R., Holt, R.R., Cardetti, M., Keen, C.L., 2003. Honey with high levels of antioxidants can provide protection to healthy human subjects. J. Agric. Food Chem. 51, 732–735.

Senthilkumaran, R., Bisht, I.S., Bhat, K.V., Rana, J.C., 2007. Diversity in buckwheat (*Fagopyrum* spp.) landrace populations from north-western Indian Himalayas. Genet. Resour. Crop Evol. 55 (2), 287–302.

Shao, J.R., Zhou, M.L., Zhu, X.M., Wang, D.Z., Bai, D.Q., 2011. *Fagopyrum wenchuanense* and *Fagopyrum qiangcai*, two new species of Polygonaceae from Sichuan, China. Novon 21, 256–261.

Sharma, T.R., Jana, S., 2002a. Random amplified polymorphic DNA (RAPD) variation in *Fagopyrum tataricum* Gaertn. accessions from China and Himalayan region. Euphytica 127, 327–333.

Sharma, T.R., Jana, S., 2002b. Species relationships in *Fagopyrum* revealed by PCR-based DNA fingerprinting. Theor. Appl. Genet. 105, 306–312.

Smetanska, I., Sytar, O., 2015. The contribution of buckwheat genetic resources to health and dietary diversity. 8th Euro Biotechnology Congress, August 18–20, 2015, Frankfurt, Germany.

Stibilj, V., Kreft, I., Smrkolj, P., Osvald, J., 2004. Enhanced selenium content in buckwheat (*Fagopyrum esculentum* Moench.) and pumpkin (*Cucurbita pepo* L.) seeds by foliar fertilization. Eur. Food Res. Technol. 219, 142–144.

Suzuki, T., Kim, S., Yamauchi, H., Takigawa, S., Honda, Y., Mukasa, Y., 2005a. Characterization of a flavonoid 3-O-glucosyltransferase and its activity during cotyledon growth in buckwheat (*Fagopyrum esculentum*). Plant Sci. 169, 943–948.

Suzuki, T., Honda, Y., Mukasa, Y., Kim, S.J., 2005b. Effects of lipase, lipoxygenase, peroxidase, and rutin on quality deteriorations in buckwheat flour. J. Agric. Food Chem. 53, 8400–8405.

Tang, Z., Huang, L., Gou, J., Chen, H., Han, X., 2014. Genetic relationships among buckwheat (*Fagopyrum*) species from southwest China based on chloroplast and nuclear SSR markers. J. Genet. 93, 849–853.

Tomotake, H., Shimaoka, I., Kayashita, J., Yokoyama, F., Nakajoh, M., Kato, N., 2000. A buckwheat protein product suppresses gallstone formation and plasma cholesterol more strongly than soy protein isolate in hamsters. J. Nutr. 130, 1670–1674.

Tomotake, H., Shimaoka, I., Kayashita, J., Nakajoh, M., Kato, N., 2002. Physicochemical and functional properties of buckwheat protein product. J. Agric. Food Chem. 50, 2125–2129.

Ujihara, A., 1983. Studies on the ecological features and the potentials as breeding materials of Asian common buckwheat varieties (*Fagopyrum esculentum* M.). PhD thesis, Faculty of Agriculture, Kyoto University, Japan.

Ujihara, A., Matano, T., 1977. On the cultivation and analysis of some agronomical characters of Tartary buckwheat in Nepal. Jpn. J. Breed. 27, 80–81.

Ujihara, A., Matano, T., 1982. Distribution and morphological variation of wild buckwheat *Fagopyrum cymosum* in Nepal. Department of Crop Sciences & Plant Breeding, Faculty of Agriculture, Shinshu University, Japan.

Vojtiskova, P., Kristyna, K., Kuban, V., Kracman, S., 2012. Chemical composition of buckwheat plant (*F. esculentum*) and selected buckwheat products. J. Microbiol. Biotechnol. Food Sci. 1, 1011–1019.

Wang, Y., Campbell, C.G., 2007. Tartary buckwheat breeding (*Fagopyrum tataricum* L. Gaertn.) through hybridization with its rice-Tartary type. Euphytica 156, 399–405.

Wang, M., Liu, J.R., Gao, J.M., Parry, J.W., Wei, Y.M., 2009. Antioxidant activity of Tartary buckwheat bran extract and its effect on lipid profile of hyperlipidemic rats. J. Agric. Food Chem. 57, 5106–5112.

Yasui, Y., Wang, Y., Ohnishi, O., Campbell, C.G., 2004. Amplified fragment length polymorphism linkage analysis of common buckwheat (*Fagopyrum esculentum*) and its wild self-pollinated relative *Fagopyrum homotropicum*. Genome 47, 345–351.

Yokozawa, T., Fujii, H., Kosuna, K., Nonaka, G., 2001. Effects of buckwheat in a renal ischemia-reperfusion model. Biosci. Biotechnol. Biochem. 65, 396–400.

Zheng, S., Cheng-hua, H.A.N., Huang, K.-F., 2011. Research on Se content of different Tartary buckwheat genotypes. Agric. Sci. Technol. 12 (1), 102–104.

Zhou, M.-L., Zhang, Q., Zheng, Y.-D., Tang, Y., Li, F.-L., Zhu, X.-M., Shao, J.-R., 2015. *Fagopyrum hailuogouense* (Polygonaceae), one new species from Sichuan, China. Novon: J. Bot. Nomen. 24 (2), 222.

Zielinska, D., Turemko, M., Kwiatkowski, J., Zielinski, H., 2012. Evaluation of flavonoid contents and antioxidant capacity of the aerial parts of common and Tartary buckwheat plant. Molecules 17, 9668–9682.

Zielinski, H., Michalska, A., Benavent, M.A., Castillo, M., Piskula, M., 2009. Changes in protein quality and antioxidant properties of buckwheat seeds and groats induced by roasting. J. Agric. Food Chem. 57, 4771–4776.

Chapter ten

Phenotypic Plasticity in Buckwheat

N.K. Chrungoo*, L. Dohtdong**, U. Chettry**

*Department of Botany, North Eastern Hill University, Shillong, India;
**Plant Molecular Biology Laboratory, UGC-Centre for Advanced Studies in Botany,
North-Eastern Hill University, Shillong, India

INTRODUCTION

Plasticity is an important attribute that enables plants to maintain fitness in heterogeneous environments. Although plasticity has been a subject of extensive study, most studies ignore the impact of factors other than the environment on plant growth and development. With an initial emphasis on the relationship between phenotypic plasticity and abiotic stress in plants, the scope of work has enlarged to climate change and invasive capacities of plants as well as different biotic stresses (Schlichting, 2002; Sultan, 2000; Valladares et al., 2006). While phenotypic plasticity, the ability of a plant genotype to respond to different environmental conditions by producing different phenotypes, might play an important role in biological invasion, more empirical studies are needed before a general pattern emerges that can be used in predicting the invasive potential of a species. Logically for organisms living in a heterogeneous environment, natural selection should favor an individual with the ability to change its phenotype so that a trait always expresses optimal values. If a single genotype can express the optimal phenotype in each environment, one might expect it to supplant alternatives, leading to phenotypic diversity underlain by genetic uniformity. Yet in most instances, instead of such ubiquitous phenotypic plasticity, most species differentiate into individuals with a range of fixed phenotypes (Hereford, 2009).

Besides genetic variation for a canalized phenotype, phenotypic plasticity also plays an important role in diversification (West-Eberhard, 1989). Even though phenotypic plasticity was earlier considered to be a consequence of developmental inconsistencies, evidence suggests that most of such variations are selectively advantageous (Bronmark and Miner, 1992; Robinson et al., 1996). Thus just like any other phenotypic character phenotypic plasticity is also

137

Molecular Breeding and Nutritional Aspects of Buckwheat. http://dx.doi.org/10.1016/B978-0-12-803692-1.00010-9

considered as a trait subject to selection (Schlichting and Levin, 1986; Scheiner, 1993; Schlichting and Pigliucci, 1998). The relationship between the value of a trait in one environment and its plasticity has been the subject of considerable debate (Via and Lande, 1985; Schlichting and Levin, 1986; Via et al., 1995; Pigliucci, 2005). Via and Lande (1985) proposed that plasticity is a function of the differential expression of the same genes in different environments. Under this model the plasticity of a trait can have a strongly correlated response to direct selection on that trait (as long as the cross-environment genetic correlation does not equal 1). Alternatively, the plasticity of a trait might be under separate regulatory control from the trait itself (Schlichting and Pigliucci, 1995). Under the second model, direct selection on a trait should have little effect on the plasticity of that trait. Although these two mechanisms are not mutually exclusive, it is important to determine empirically whether correlated response to directional selection involves change in plasticity.

Plastic responses of plants have received attention in two interlinked contexts. First, developmental plasticity has profound implications for plant evolution. If a single genotype can result in multiple phenotypes, this weakens the link between selection at the level of the phenotype and changes in allele frequency at the level of the genotype. However, plastic responses of plants also make it possible to investigate the genetic control mechanisms underlying the phenomenon. Although adaptive plasticity is generally considered to be of most evolutionary and ecological interest, plasticity in plant responses to varying conditions of growth has tremendous bearing on the survival and colonization potential of a plant. Nonadaptive plasticity reduces the likelihood of persistence in a new environment that is stressful and thus increases the strength of selection. Grether (2005) has suggested that whenever a species encounters stiff environmental conditions that trigger phenotypic changes with reduced fitness, selection will favor appropriate genetic changes that counteract the phenotypic changes so as to restore the phenotype to its ancestral state. Such an adaptive evolution, referred to as "genetic compensation," leads to genetic differences among populations so that the mean trait values of populations in different environments may appear to be more similar when measured in their native habitat than when grown under common environmental conditions. Thus while genetic compensation would tend to reduce phenotypic variability it would promote genetic divergence in a species. However, stressful environments could also increase phenotypic variation via the expression of cryptic genetic variations. Thus while the phenotypic diversity under a set of environmental conditions may represent only a fraction of the diversity that is actually available, cryptic genetic variation could be revealed under stressful environments.

Although plant functional traits appear to be important indicators of species' responses to land use changes, there is no clear understanding of how the variations in traits and their plasticity determine variations in species performance. Silveira et al. (2010) have shown that variations in productivity of a species in response to change in N supply was mainly explained by a group of traits (a set of predictor variables), rather than by one individual trait, which included average flowering date, tiller density per unit land area and leaf dry matter content. Strategies adopted by a species toward nutrient availability have also been shown to be based on traits linked to leaf morphology (Wright et al., 2005). Thus species with quick returns on investments of nutrients exhibit high leaf nutrient concentrations, high rates of photosynthesis and respiration, short leaf lifespan, and low dry mass investment per leaf area. Thus plasticity of plant traits could influence the survival of a species in an ecological niche and increase ecosystem functioning by reducing niche overlap along resource axes. However, what is not clear is how covariations between traits and their plasticity would determine survival of the species in a particular ecological niche.

Even though the diversity of genetic resources is fundamental for ecosystem functioning, sustainable agricultural production, and attainment of food and nutritional security, only a few crop species represent the world food basket. Currently only 30 plant species are known to provide 95% of the world's food energy needs. A consequence of increased reliance on major food crops is the shrinking of the food basket (Prescott-Allen and Prescott-Allen, 1990). This "nutritional paradox," as it is called (Ogle and Grivetti, 1995), has its roots in "agricultural simplification," a process that favored some crops over others on the basis of their comparative advantages for growing in a wider range of habitats, simple cultivation requirements, easier processing and storability, nutritional properties, taste, etc. While the 20th century witnessed a systematic approach toward rescuing the genetic resources of staple crops (Pistorius, 1997), the 21st century has started with awareness on the necessity of rescuing and improvement of crops left aside by research, technology, marketing systems, as well as conservation efforts. Among these the International Plant Genetic Resources Institute and Consultative Group on International Agriculture have identified buckwheat (*Fagopyrum* spp.), grain amaranth (*Amaranthus* spp.), and (*Chenopodium* spp.) as crops of potential for use in crop improvement programs.

Adaptive Plasticity as a Selective Advantage

Despite the long history of both theoretical and empirical research on plasticity, the mechanisms that regulate plasticity of plant responses are still poorly understood. While natural selection is expected to favor locally adapted plants when the environment is constant, phenotypic plasticity would be more beneficial in spatially or temporally heterogeneous environments. Pigliucci et al. (2006) have defined phenotypic plasticity as a "property of a genotype to produce different phenotypes" in response to a changing environment. While the notion of "phenotypic plasticity as an evolving trait" has received much focus (Scheiner, 1993, 2013; Alpert and Simms, 2002; DeWitt and Scheiner, 2004), the fact that phenotypic plasticity can be a useful paradigm to understand the interactions of genetics, development, ecology, and evolution cannot be underestimated.

Although plastic responses are quite often considered as "adaptive," the response can be considered as just one of several factors that may cause variation in amounts and patterns of plastic response between taxa. In addition, the amount and pattern of plastic response can evolve independently of the character mean (Schlichting, 1984). For example, in the genus *Portulaca*, the character shoot/root ratio has a very similar mean value for *Portulaca grandiflora* (20.4) and

Portulaca oleracea (18.5), but their amounts and directions of plastic response differ markedly (Zimmerman, 1976). Environmental factors responsible for evolutionary changes in phenotypic plasticity can be assigned to three general categories, namely, selection, drift, and disruption of the genetic system. While the phenotypic plasticity caused by selection would include situations where advantageous plastic responses are related to specific environmental regimes, plasticity may also evolve through random changes not driven by selection, that is, inbreeding, hybridization, and polyploidy (Gama and Hallauer, 1980; Pooni and Jinks, 1980; Garbutt and Bazzaz, 1983; Levin, 1983; Schlichting and Levin, 1986). Thus plastic responses can be conceived as "typical" developmental sequences because of the interaction of the organism's genotype with the environment. While seed selection by farmers over seasons exerts selection pressure on populations of genotypes, thereby reducing the plasticity at population level, introduction of new varieties or introgression of genes from hybridization with wild species or varieties would lead to enhanced levels of plasticity and thereby a higher capability to withstand environmental changes. In this context the neglected and underutilized crop genetic resources assume significance for sustainable agriculture (Eyzaguirre et al., 1999; Mal, 2007). Many underutilized species occupy important niches adapted to risky and fragile conditions of rural communities and have a comparative advantage in marginal lands as they can withstand stress. Most of the these crops do not require high inputs and can be successfully grown in marginal, degraded, and wastelands with minimal inputs and at the same time can contribute to increased agricultural production, enhanced crop diversification, and improved environment and have the potential to contribute useful genes to breed better varieties capable of withstanding and sustaining the climate change scenario (Mal and Joshi, 1991; Mal, 2007; Padulosi et al., 2009).

Cultivation, Utilization, and Diversity of Buckwheat

Even though buckwheat is considered to be a minor crop, it is an indispensable food in the temperate and hill regions of East Asia and Europe. Buckwheat is a multipurpose crop used for food, feed, medicine, and manure (Li and Zhang, 2001; Zeller, 2001). Common buckwheat (*Fagopyrum esculentum* Moench), a diploid ($2n = 16$) annual, is widely cultivated in Asia, Europe, and America. Because of its short growth span, capability to grow at high altitudes, and the high-quality protein of its grains it is an important crop in mountainous regions of India, China, Russia, Ukraine, Kazakhstan, parts of Eastern Europe, Canada, Japan, Korea, and Nepal. Buckwheat grains are a rich source of dietary proteins for gluten-sensitive individuals (Wei et al., 2003; Stibilj et al., 2004). Buckwheat leaves and flowers are a rich source of rutin, catechins, and other polyphenols that are potential antioxidants (Luthar, 1992; Oomah and Mazza, 1996; Watanabe, 1998). Buckwheat proteins have also been reported to have anticancer, hypoglycemic, and antihypertention properties (Kayashita et al., 1999; Park and Ohba, 2004; Kayashita et al., 1995a,b, 1996, 1997;

Tomotake et al., 2000, 2006). As a green manure crop, buckwheat produces only modest biomass but offers rapid growth, improves soil, and makes phosphorus more available. Quick, aggressive growth accounts for its success as a smother crop for suppressing weeds, particularly in late summer. The plant is an erect annual profusely branched herb, which grows in height from 0.5 to 1.5 m. The plant has hollow stems that vary in color from green to red and brown at maturity. Buckwheat has a shallow tap root system, with numerous laterals extending from 0.9 to 1.2 m in depth. With an indeterminate growth habit, the plants begin to flower 5–6 weeks after sowing and mature in 80–110 days. Joshi (1999) has reported a wide range of variation in 11 morphological characters including plant height, number of branches, number of internodes, number of leaves, leaf length, leaf width, days to flower, number of seeds per cyme, seed shape, size and color, seed weight, and yield per plant in 577 collections represented by *F. esculentum*, *Fagopyrum emarginatum*, *Fagopyrum tataricum*, *F. tataricum* var. *himalianum*, *Fagopyrum giganteum*, and *Fagopyrum cymosum* from different agroecological regions of the Himalayas (see Figs. 10.1–10.4). They suggested that leaf width and seed weight in buckwheat were the most potential traits for genetic amelioration. The crop has been reported to exhibit significant plasticity with reference to stem biomass allocation, leaf biomass allocation, leaf root ratio, plant stem length, stem diameter, branch length, and number of branches as a function of planting density and sowing date.

Genome Plasticity in Buckwheat

Genome size variations have long been considered as one of the important indicators of genetic complexity. However, lack of correspondence between genome size and morphological or physiological complexity of an organism has been historically termed the "C-value paradox" (Thomas, 1971). Since the discovery of noncoding DNA and its impact on genome size variation, the term "paradox" has been replaced by "enigma" in an attempt to more appropriately identify the topic as a "perplexing subject" (Gregory, 2005). *Fagopyrum* is no exception to this enigma. In comparison with other taxa, the C-values in *Fagopyrum* are distributed rather widely and range from 0.55 in *F. tataricum* to 1.92 in *Fagopyrum urophyllum* (Nagano et al., 2000). The values show 2.53- and 2.82-fold magnitude of variation among species within the *cymosum* and *urophyllum* groups, respectively. The overlapping of C-values among members of the *cymosum* and *urophyllum* groups of the genus *Fagopyrum* has been suggested to imply a dynamic rearrangement of the genome in the genus within a short span of evolution. In an attempt to correlate genome size with seed size, Knight and Ackerly (2002) have suggested that while small genomes can be associated with either small or large seeds, plants having large genomes may not have small seeds. However, Beaulieu et al. (2007) could not find any correlation between genome size and seed mass across 1222 species from 139 families and 48 orders of seed plants. Their observations did, however, indicate that species with very large genome sizes never had small seeds, while species with small genome

FIGURE 10.1 Variations in plant height in accessions of *Fagopyrum esculentum* (A–C), *F. tataricum* (B–D), and *F. cymosum* (E, F) (A, IC-188669; B, IC-319588b; C, IC-13141; D, IC-412744; E, IC-421597; F, IC-412863; G, IC-9248; H, IC-18642) growing in experimental field of Botany Department, North Eastern Hill University, Shillong (India).

size had a large range of seed size. Interestingly, the genome size of *Fagopyrum leptopodum*, *F. tataricum*, and its subspecies, *F. tataricum* spp. *potanini*, all of which produce small seeds, is much smaller than the mean genome size for the entire genus. While Knight and Beaulieu (2008) have highlighted the existence of a positive correlation between genome size variations and morphological characters like cell size and guard cell diameter, they suggested a triangular relationship between genome size and maximum plant height across 324 species of angiosperms. Their observation was that as genome size increases, maximum plant height decreases within angiosperm. Contrary to this suggestion the tallest

FIGURE 10.2 Variations in the shape of leaf blade among different species of the genus *Fagopyrum*.

FIGURE 10.3 Variations in the color of sepals in flowers of *Fagopyrum esculentum*, *F. tataricum*, and *F. cymosum*.

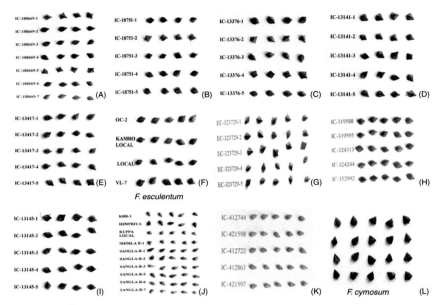

FIGURE 10.4 Variations in grain morphology in different accessions of buck-wheat (*Fagopyrum* spp.). A–I: Grains of accessions/cultivars belonging to *F. esculentum*; (J, K) grains of accessions/cultivars belonging to *F. tataricum*; (L) grains of accessions belonging to *F. cymosum*.

wild species, *F. urophyllum*, has the largest genome size and the smallest species and *Fagopyrum lineare* has the lowest 2C value of 1.08 pg. Contradictions have also been observed in the correlation between reproductive behavior and genome size. While a positive correlation between selfing behavior and low genome size has been reported for many seed plants (Albach and Greilhuber, 2004; Wright et al., 2008), observations on the breeding behavior and genome size in buckwheat do not indicate any correlation between the two. Although *F. tataricum* and *F. tataricum* spp. *potanini*, which have the smallest genome size within the genus *Fagopyrum*, exhibit self-breeding behavior, *Fagopyrum pleioramosum*, which has the second largest genome within the genus, also shows self-breeding habit. On the other hand, *Fagopyrum statice* and *F. leptopodum*, which have comparatively smaller genomes, are outbreeders. *F. cymosum* and *Fagopyrum gracilipes* have exactly same C-value but the former is an outbreeding species, while the latter predominantly displays selfing (Ohnishi, 1995).

F. cymosum had long been suggested to be the wild ancestor of *F. esculentum* and *F. tataricum* until the discovery of *F. esculentum* spp. *ancestrale* and *F. tataricum* spp. *potanini* by Ohnishi (1991). The postulates *F. esculentum* spp. *ancestralis* differentiated from *F. cymosum* much earlier than *F. tataricum* spp. *Potanini*, and *Fagopyrum homotropicum* has probably been derived from *F. esculentum* ssp. *ancestralis* by shifting the breeding system from outcrossing to self-fertilization (Ohnishi and Matsuoka, 1996; Yasui and Ohnishi, 1998), which would tend to indicate that genome evolution in the *cymosum* group could have progressed in the direction of increase as well as decrease in size. Presumably, the processes leading to evolution of *F. esculentum* from *F. cymosum* was accompanied by decrease of genome size while that leading to evolution of *F. tataricum* progressed with increase in size of the genome. On the other hand, evolutionary processes within the *urophyllum* group involved only decrease in size of the genome. Thus DNA loss could have been the predominant factor in evolutionary processes within the *urophyllum* group.

Morphological Plasticity in Buckwheat

A logical corollary for modular organization of plants is the assumption that plasticity of the whole plant would depend on both the components/traits that exhibit plasticity as well as the nature of plastic response. Further, the magnitude of response of different traits may differ in response to the stimulus thereby creating a hierarchy of responses. Most of the studies on phenotypic plasticity have been conducted under controlled environmental conditions, which negate the influence variations caused by combinations of environmental factors on plastic responses. Disentangling the effects of sowing density and timing of germination on phenotypic variations Japhet et al. (2009) have demonstrated true plastic shifts of investment into architectural traits and biomass ratios caused by altered developmental rates in common buckwheat (*F. esculentum*). Allometric analysis of the relationships between branch length and total biomass showed a significant departure from isometry, thereby indicating a linear relationship of branch length as a function of size. However, they did not observe any significant differences in allometric exponents within treatments thereby indicating that the differences observed in branch length were caused by the direct effect of size rather than as a result of true plastic investment in branches. Their observations further revealed that in contrast to plants sown at low density, higher density of sowing resulted in buckwheat plants with significantly longer stems and a significantly greater size to mass ratio (SMR) but thinner stems. This clearly suggests that the most densely grown plants invested a significant proportion of their biomass toward stem length elongation rather than expansion of stem diameter. Dudley and Schmitt (1996) and Pigliucci (2001) have suggested that although such a stem elongation response could improve survival and growth of dense populations, it could be maladaptive when light is not limiting. Another response closely related to that seen with stem elongation was the decrease in length and total number of branches in plants grown at very high density. Since greater branching intensity is generally associated with a greater potential for producing more reproductive meristems in annual species such as *F. esculentum*, which can produce seeds along the length of its axes, a high branching intensity would maximize the number of seeds obtained at the end of the growing season. While investigating the influence of intraspecific competition on meristem plasticity in *F. esculentum*, Zhang et al. (2008) suggested that while allocation of meristems to reproduction could be explained as a function of allometric growth ("apparent plasticity"), variations in meristem allocation patterns under different planting densities appeared to show responses independent of allometric growth. They suggested a higher plasticity in vegetative growth in buckwheat than reproductive growth. Other than the work mentioned previously, not much work has been carried out on phenotypic plasticity in buckwheat. Rout (2007) has reported a larger variation in leaf area ranging in *F. esculentum* grown during summer than that grown during the winter season. On the other hand, it took a much larger number of days to attain 50% flowering and 50% seed set in plants grown during winter than those grown during the summer season. These observations are suggestive of buckwheat as a model system for investigating morphological plasticity including invasiveness, co-evolution with symbionts, and biogeography.

The Way Forward

Even though buckwheat cultivation has been practiced for long time, the crop still has a long way to go in the direction of greater popularity and importance. Important factors responsible for low productivity of buckwheat are self-incompatibility caused by dimorphic heterostylism, seed collapse at the postzygotic stage, and seed shattering caused by indeterminate growth habit (Tahir and Farooq, 1991). The scarcity of photosynthetic products in kernel is also reported to be one of the main factors that cause low yield in buckwheat (Yang et al., 1998). Buckwheat flowering is profuse and the photosynthetic capabilities of the plant do not meet the requirements to properly fill all the seeds. If the yielding ability of common buckwheat were to double, the plants only require 24% of the flowers now being produced. Woo (2006) has suggested that nutrients that are otherwise being expended for production of flowers, which the plant no longer needs, can be redirected towards grain filling, thereby leading to higher productivity. Buckwheat is not only a promising crop in itself but also offers a wide range of desirable qualities for the improvement of other crops. It also holds the possibility of making an important contribution toward the advancement of knowledge linking phenotypic responses with genetic variations under varying environmental conditions.

ACKNOWLEDGMENTS

The authors are thankful to the Department of Biotechnology, Government of India, for supporting the work under its Biotech Hub program, see grant no. BT/04/NE/2009 to NKC.

REFERENCES

Albach, D.C., Greilhuber, J., 2004. Genome size variation and evolution in Veronica. Ann. Bot. 94, 897–911.

Alpert, P., Simms, E., 2002. The relative advantages of plasticity and fixity in different environments: when is it good for a plant to adjust? Evol. Ecol. 16, 285–297.

Beaulieu, J.M., Moles, A.T., Leitch, I.J., Bennett, M.D., Dickie, J.B., Knight, C.A., 2007. Correlated evolution of genome size and seed mass. New Phytol. 173, 422–437.

Bronmark, C., Miner, J.G., 1992. Predator-induced phenotypical change in body morphology in crucian carp. Science 258, 1348–1350.

DeWitt, T.J., Scheiner, S.M., 2004. Phenotypic Plasticity: Functional and Conceptual Approaches. Oxford University Press, New York.

Dudley, S., Schmitt, J., 1996. Testing the adaptive plasticity hypothesis: density-dependent selection on manipulated stem length in *Impatiens capensis*. Am. Nat. 147, 445–465.

Eyzaguirre, P., Padulosi, S., Hodgkin, T., 1999. IPGRI's strategy for neglected and underutilized species and the human dimension of agrobiodiversity. In: Padulosi, S. (Ed.), Priority Setting for Underutilized and Neglected Plant Species of the Mediterranean Region. Report of the IPGRI Conference, Feb. 9-11, 1998. ICARDA, Aleppo, Syria. International Plant Genetic Resources Institute, Rome, Italy.

Gama, E.E.G., Hallauer, A.R., 1980. Stability of hybrids produced from selected and unselected lines of maize. Crop Sci. 20, 623–626.

Garbutt, K., Bazzaz, F.A., 1983. Leaf demography, flower production and biomass of diploid and tetraploid populations of *Phlox drummondii* Hook, on a soil moisture gradient. New Phytol. 93, 129–141.

Gregory, T.R., 2005. The C-value enigma in plants and animals: a review of parallels and an appeal for partnership. Ann. Bot. 95, 133–146.

Grether, G.F., 2005. Environmental change, phenotypic plasticity, and genetic compensation. Am. Nat. 166, 115–123.

Hereford, J., 2009. A quantitative survey of local adaptation and fitness trade-offs. Am. Nat. 173, 579–588.

Japhet, W., Zhou, D., Zhang, H., Zhang, H., Yu, T., 2009. Evidence of phenotypic plasticity in the response of *Fagopyrum esculentum* to population density and sowing date. J. Plant Biol. 52, 303–311.

Joshi, B.D., 1999. Status of buckwheat in India. Fagopyrum 16, 7–11.

Kayashita, J., Shimaoka, I., Nakajoh, M., 1995a. Hypocholesterolemic effect of buckwheat protein extract in rats fed cholesterol enriched diets. Nutr. Res. 15, 691–698.

Kayashita, J., Shimaoka, I., Nakajoh, M., Arachi, Y., Kato, N., 1995. Feeding of buckwheat protein extract reduces body fat content in rats. Proceedings of the 6th International Symposium on Buckwheat, Shinshu, Japan, 935–940.

Kayashita, J., Shimaoka, I., Nakajoh, M., Kato, N., 1996. Feeding of buckwheat protein extract reduces hepatic triglyceride concentration, adipose tissue weight and hepatic lipogenesis in rats. J. Nutri. Biochem. 7, 555–559.

Kayashita, J., Shimaoka, I., Nakajoh, M., Yamazaki, M., Kato, N., 1997. Consumption of buckwheat protein lowers plasma cholesterol and raises fecal neutral sterols in cholesterol-fed rats because of its low digestibility. J. Nutr. 127, 1395–1400.

Kayashita, J., Shimaoka, I., Nakajoh, M., Kishida, N., Kato, N., 1999. Consumption of buckwheat protein extract retards 7,12-dimethylbenz[α]anthracene-induced mammary carcinogenesis in rats. Biosci. Biotechnol. Biochem. 63, 1837–1839.

Knight, C.A., Ackerly, D.D., 2002. Variation in nuclear DNA content across environmental gradients: a quantile regression analysis. Ecol. Lett. 5, 66–76.

Knight, C.A., Beaulieu, J., 2008. Genome size scaling in phenotype space. Ann. Bot. 101, 759–766.

Levin, D.A., 1983. Polyploidy and novelty in flowering plants. Am. Nat. 122, 1–25.

Li, S., Zhang, Q.H., 2001. Advances in the development of functional foods from buckwheat. Crit. Rev. Food Sci. Nutr. 41, 451–464.

Luthar, Z., 1992. Phenol classification and tanin content of buckwheat seeds. Fagopyrum 12, 36–42.

Mal, B., 2007. Neglected and Underutilized crop genetic resources for sustainable agriculture. Indian J. Plant Genet. Resour. 20, 1–14.

Mal, B., Joshi, V., 1991. Underutilized plant resources. Paroda, R.S., Arora, R.K. (Eds.), Plant Genetic Resources—Conservation and ManagementMalhotra Publishing House, New Delhi, India, pp. 211–229.

Nagano, M., All, J., Campbell, C.G., Kawasaki, S., 2000. Genome size analysis of the genus *Fagopyrum*. Fagopyrum 17, 35–39.

Ogle, B.M., Grivetti, L.E., 1995. Legacy of the chameleon: edible wild plants in the Kingdom of Swaziland, Southern Africa. A cultural, ecological, nutritional study.

Part II—Demographics, species availability and dietary use, analyses by ecological zone. Ecol. Food Nutr. 17, 1–30.

Ohnishi, O., 1991. Discovery of wild ancestor of common buckwheat. Fagopyrum 11, 5–10.

Ohnishi, O., 1995. Discovery of new *Fagopyrum* species and its implication for the study of evolution of *Fagopyrum* and the origin of cultivated buckwheat. Proceedings of the 6th International Symposium on Buckwheat, Shinsu, Japan, 175–190.

Ohnishi, O., Matsuoka, Y., 1996. Search for the wild ancestor of buckwheat. II. Taxonomy of *Fagopyrum* (Polygonaceae) species based on morphology, isozymes and cpDNA variability. Genes Genet. Syst. 71, 383–390.

Oomah, B.D., Mazza, C., 1996. Flavonoids and antioxidative activities in buckwheat. J. Agric. Food Chem. 44, 1746–1750.

Padulosi, S., Mal, B., Ravi, S.B., Gowda, J., 2009. Food security and climate change: role of plant genetic resources of minor millets. Indian J. Plant Genet. Resour. 22, 1–16.

Park, S., Ohba, H., 2004. Suppressive activity of protease inhibitors from buckwheat seeds against human T-acute lymphoblastic leukemia cell lines. Appl. Biochem. Biotechnol. 117, 65–74.

Pigliucci, M., 2001. Phenotypic Plasticity: Beyond Nature and Nurture: Syntheses in Ecology and Evolution. Johns Hopkins University Press, Baltimore.

Pigliucci, M., 2005. A strategy on how to deal with anti-science fundamentalism. EMBO Rep. (6), 1106–1109.

Pigliucci, M., Murren, C.J., Schlichting, C.D., 2006. Phenotypic plasticity and evolution by genetic assimilation. J. Exp. Biol. 209, 2362–2367.

Pistorius, R., 1997. Scientists, Plants and Politics—A history of the Plant Genetic Resources Movement. International Plant Genetic Resources Institute, Rome, Italy, 134.

Pooni, H.S., Jinks, J.L., 1980. Nonlinear genotype × environment interactions. II. Statistical models and genetical control. Heredity 45, 389–400.

Prescott-Allen, R., Prescott-Allen, C., 1990. How many plants feed the world? Conserv. Biol. 4, 365–374.

Robinson, B.W., Wilson, D.S., Shea, G.O., 1996. Tradeoffs of ecological specialization: an intraspecific comparison of pumpkinseed sunfish phenotypes. Ecology 77, 170–178.

Rout, A., 2007. Analysis of molecular diversity in accessions of buckwheat (*Fagopyrum esculentum* Moench) from Himalayan ranges. North Eastern Hill University, Shillong, India, PhD thesis.

Scheiner, S.M., 1993. Genetics and evolution of phenotypic plasticity. Annu. Rev. Ecol. Evol. Syst. 24, 35–68.

Scheiner, S.M., 2013. The genetics of phenotypic plasticity. XII. Temporal and spatial heterogeneity. Ecol. Evol. 3, 4596–4609.

Schlichting, C.D., 1984. Studies on phenotypic plasticity in annual *Phlox*. PhD thesis, University of Texas, Austin, p. 114.

Schlichting, C.D., 2002. Phenotypic plasticity in plants. Plant Species Biol. 17, 85–88.

Schlichting, C.D., Levin, D.A., 1986. Effects of inbreeding on phenotypic plasticity in cultivated *Phlox drummondii*. Theor. Appl. Genet. 72, 114–119.

Schlichting, C.D., Pigliucci, M., 1995. Gene regulation, quantitative genetics and the evolution of reaction norms. Evol. Ecol. 9, 154–168.

Schlichting, C.D., Pigliucci, M., 1998. Phenotypic Evolution: A Reaction Norm Perspective. Sinauer Associates, Sunderland, MA.

Silveira, P.L., Louault, F., Carrère, P., Maire, V., Andueza, D., Soussana, J.F., 2010. The role of plant traits and their plasticity in the response of pasture grasses to nutrients and cutting frequency. Ann. Bot. 105, 957–965.

Stibilj, V., Kreft, I., Smrkolj, P., Osvald, J., 2004. Enhanced selenium content in buckwheat (*Fagopyrum esculentum* Moench) and pumpkin (*Cucurbita pepo* L.) seeds by foliar fertilization. Eur. Food Res. Technol. 219, 142–144.

Sultan, S.E., 2000. Phenotypic plasticity for plant development, function and life history. Trends Plant Sci. 5, 537–542.

Tahir, I., Farooq, S., 1991. Growth patterns in buckwheat (*Fagopyrum* spp.) grown in Kashmir. Fagopyrum 11, 63–67.

Thomas, C.A., 1971. The genetic organization of chromosomes. Annu. Rev. Genet. 5, 237–256.

Tomotake, H., Shimaoka, I., Kayashita, J., Yokoyama, F., Nakajoh, M., Kato, N., 2000. A buckwheat protein product suppresses gallstone formation and plasma cholesterol more strongly than soy protein isolate in hamsters. J. Nutr. 130, 1670–1674.

Tomotake, H., Yamamoto, N., Yanaka, N., Ohinata, H., Yamazaki, R., Kayashita, J., Kato, N., 2006. High protein buckwheat flour suppresses hypercholesterolemia in rats and gallstone formation in mice by hypercholesterolemic diet and body fat in rats because of its low protein digestibility. Nutrition 22, 166–173.

Valladares, F., Sanchez-Gomez, D., Miguel, A.Z., 2006. Quantitative estimation of phenotypic plasticity: bridging the gap between the evolutionary concept and its ecological applications. J. Ecol. 94, 1103–1116.

Via, S., Lande, R., 1985. Genotype–environment interaction and the evolution of phenotypic plasticity. Evolution 39, 505–522.

Via, S., Gomulkiewicz, R., De Jong, G., Scheiner, S.M., Schlichting, C.D., Van Tienderen, P.H., 1995. Adaptive phenotypic plasticity: consensus and controversy. Trends Ecol. Evol. 10, 212–217.

Watanabe, M., 1998. Catechins as antioxidants from buckwheat (*Fagopyrum esculentum* Moench) groats. J. Agric. Food Chem. 46, 839–845.

Wei, Y., Hu, X., Zhang, G., Ouyang, S., 2003. Studies on the amino acid and mineral content of buckwheat protein fractions. Nahrung Food 47, 114–116.

West-Eberhard, M.J., 1989. Phenotypic plasticity and the origins of diversity. Annu. Rev. Ecol. Evol. Syst. 20, 249–278.

Woo, S.H., 2006. Breeding improvement of processing buckwheat. J. Agric. Sci. 23, 27–34.

Wright, I.J., Reich, P.B., Cornelissen, J.H.C., Falster, D.S., Groom, P.K., Hikosaka, K., Lee, W., Lusk, C.H., Niinemets, U., Oleksyn, J., Osada, N., Poorter, H., Warton, D.I., Westoby, M., 2005. Modulation of leaf economic traits and trait relationships by climate. Global Ecol. Biogeogr. 14, 411–421.

Wright, S.I., Rob, W.N., Foxe, J.P., Spencer, C.H.B., 2008. Genomic consequences of outcrossing and selfing in plants. Int. J. Plant Sci. 169, 105–118.

Yang, W., Hao, Y., Li, G., Zhou, N., 1998. Relationship between reproductive growth of common buckwheat and light duration. Proceedings of the 7th International Symposium on Buckwheat, Winnipeg, Manitoba, Canada, 44–48.

Yasui, Y., Ohnishi, O., 1998. Interspecific relationships in *Fagopyrum* (Polygonaceae) revealed by the nucleotide sequences of the *rbc*L and *acc*D genes and their intergenic region. Amer. J. Bot. 85, 1134–1142.

Zeller, F., 2001. Buckwheat (*Fagopyrum esculentum* Moench): utilization, genetics, breeding. Bodenkultur 52, 259–276.

Zhang, L., Fletcher, A.G., Cheung, V., Winston, F., Stargell, L.A., 2008. Spn1 regulates the recruitment of Spt6 and the Swi/Snf complex during transcriptional activation by RNA polymerase II. Mol. Cell Biol. 28, 1393–1403.

Zimmerman, C.A., 1976. Growth characteristics of weediness in *Portulaca oleracea* L. Ecology 57, 964–974.

Chapter eleven

Bioactive Compounds in Buckwheat Sprouts

M.-L. Zhou*, G. Wieslander**, Y. Tang[†], Y.-X. Tang*,
J.-R. Shao[‡], Y.-M. Wu*

*Biotechnology Research Institute, Chinese Academy of Agricultural Sciences, Beijing, China; **Department of Occupational and Environmental Medicine, Uppsala University, Uppsala, Sweden; [†]School of Life Sciences, Sichuan Agricultural University, Yaan, Sichuan, China; [‡]Department of Food Science, Sichuan Tourism University, Chengdu, Sichuan, China

INTRODUCTION

Buckwheat is a pseudocereal crop that belongs to the Polygonaceae family and *Fagopyrum* genus, and grows mainly in the northern hemisphere; it is also a commonly eaten food throughout the world (Zhou et al., 2012). The most widely grown buckwheat species include common buckwheat (CB) (*Fagopyrum esculentum*) and Tartary buckwheat (TB) (*F. tataricum*) (Zhou et al., 2012). Buckwheat is rich in antioxidant components, such as rutin, orientin, vitexin, quercetin, isovitexin, isoorientin, and vitamins; most of these have also been proven to be effective as scavengers of active oxygen (reviewed by Zhang et al., 2012). More and more attention is being focused on buckwheat's healing effects in chronic diseases, such as cancer, diabetes, and neurodegenerative diseases, because of its well-balanced amino acid pattern and bioactive compounds (Zhang et al., 2012). Until now, much progress has been made in improving the nutritional factors and healing effects of buckwheat functional food (Zhang et al., 2012). Buckwheat sprouts have been successfully developed as a new vegetable because of their soft and slightly crispy texture and attractive fragrance. In addition, buckwheat sprouts are rich in amino acids such as higher lysine contents, minerals, and crude fiber. Moreover, the rutin contents of buckwheat sprouts were higher than those of buckwheat seeds (Kim et al., 2001, 2004). Hence buckwheat sprouts have excellent nutritional value and can be a popular health product in Asian and European countries. In the present review we summarize buckwheat sprouts' bioactive components and use analytical methods, functional effects, quality control, as well as suggestions to develop buckwheat sprouts for the near future.

151

Molecular Breeding and Nutritional Aspects of Buckwheat. http://dx.doi.org/10.1016/B978-0-12-803692-1.00011-0

BIOACTIVE COMPOUNDS

Compared to the buckwheat seed, buckwheat sprouts have abundant nutrients such as monosaccharides, linoleic acid, and vitamins B1 and B6 as well as vitamin C (Table 11.1) (Kim et al., 2004). As buckwheat seeding days progressed, di-, tri-, and tetrasaccharides (sucrose, maltose, raffinose) were promptly degraded to monosaccharides such as fructose and glucose, to provide the energy and synthetic requirements of buckwheat sprouts (Kim et al., 2004). In addition, unsaturated fatty acids such as linoleic and linolenic acid were increased as germination days progressed (Kim et al., 2004). Linoleic and linolenic acids are also called essential fatty acids because of their necessity in the human body. It has been reported that vitamins B1 and B6 are important for carbohydrate and fat metabolism, while vitamin C plays vital roles as a powerful antioxidant (Zangen et al., 1998; Kim et al., 2004). Interestingly, vitamins B1and B6 and vitamin C were found in buckwheat sprouts; however, they were not observed in buckwheat seeds (Kim et al., 2004). Buckwheat protein quality is very high with biological values above 90%, which can show a strong supplemental effect when combined with vegetable proteins to improve the dietary amino acid balance (Zhang et al., 2012). The amino acids include two nutritional categories: essential and nonessential. Essential amino acids are not synthesized and play important roles in the human body. The latest report showed that the most abundant amino acid in buckwheat sprouts was valine, which accounted for 40% and 62%, in CB and TB, respectively (Woo et al., 2013). Additionally, the content of essential free amino acids in TB sprouts was 53% higher than that in CB, which reveals that TB sprouts are more beneficial to the human body (Woo et al., 2013).

TABLE 11.1 Summary of Bioactive Compounds in Buckwheat Sprouts

Category	Compounds	Detected Methods	References
Monosaccharides	Fructose and glucose	HPLC	Kim et al. (2004)
Unsaturated fatty acids	Linoleic and linolenic acid	Capillary gas chromatography analysis	Kim et al. (2004)
Vitamins	Vitamins B1, B6, and C	HPLC	Kim et al. (2004)
Proteins	Essential amino acids such as valine and tyrosine	High-speed amino acid analyzer	Kim et al. (2004) Woo et al. (2013)
Flavonoids	Rutin, isoorientin, orientin, isovitexin, vitexin, and quercetin-3-O-robinobioside	HPLC, RP-UFLC, UPLC-Q-TOF-MS, CE	Kim et al. (2008) Kim et al. (2011) Lim et al. (2012) Koyama et al. (2013) Nam et al. (2015)
Carotenoids	Lutein and β-carotene	HPLC-UV-HG-AFS	Tuan et al. (2013)

HPLC, High-performance liquid chromatography; *RP-UFLC*, reverse-phase ultrafast liquid chromatography; *UPLC-Q-TOF-MS*, ultra-performance liquid chromatography quadrupole time of flight mass spectrometry; *CE*, capillary electrophoresis; *HPLC-UV-HG-AFS*, high-performance liquid chromatography-UV irradiation-hydride generation-atomic fluorescence spectrometry.

The fact that buckwheat sprouts are also rich in flavonoid compounds was recognized as the most health protective effect because of their antioxidant and antihypertensive activities (Table 11.1) (Zhang et al., 2012). The total contents of phenolic compounds in TB sprouts were similar to those of CB sprouts, while the rutin contents in TB sprouts were fivefold higher than those of CB (Kim et al., 2008). The highest rutin content in buckwheat sprouts was 109.0 mg/100 g fresh weight at 6 days after sowing (Koyama et al., 2013). In addition, buckwheat sprouts contain other kinds of major phenols, including isoorientin, orientin, isovitexin, and vitexin (Koyama et al., 2013). However, quercetin became undetectable on day 7 of germination in both CB and TB sprouts (Ren and Sun, 2014). The estimation of phenylalanine ammonia lyase (PAL) activity in the sprouts revealed the existence of an obvious positive linear relationship between PAL activity and flavonoid accumulation (Ren and Sun, 2014). Carotenoids are the second most abundant pigment in nature and serve as precursors of vitamin A, which is one of the most important micronutrients in the human body. Furthermore, a diet containing carotenoid-rich food can protect against heart disease, cataracts, ultraviolet-induced skin damage, and even some cancers (Mactier and Weaver, 2005). It has been shown that two kinds of carotenoids, lutein and β-carotene, were identified in light-grown TB sprouts (Tuan et al., 2013). Therefore buckwheat sprouts are recommended for their high antioxidative activity, as well as being an excellent source of vegetables, especially TB sprouts, being rich in rutin.

ANALYTICAL METHODS

Phytochemical analyses of buckwheat sprouts have been performed intensively during the last few years. High-performance liquid chromatography (HPLC) is traditionally used to determine the content of flavonoids, water-soluble vitamins, and monosaccharides in buckwheat sprouts (Table 11.1) (Kim et al., 2004, 2007, 2008; Zhang et al., 2012; Ghimeray et al., 2014; Ren and Sun, 2014). In CB sprouts, chlorogenic acid, four C-glycosylflavones (orientin, isoorientin, vitexin, and isovitexin), rutin, and quercetin can be detected by HPLC, whereas only rutin was detected in TB sprouts (Kim et al., 2007; Nam et al., 2015). However, the HPLC technique was unable to separate and quantify individual flavonoids to a satisfactory degree. Reverse-phase ultrafast liquid chromatography (RP-UFLC) is also employed for quantitative and qualitative analysis of isoorientin, orientin, rutin, and vitexin in CB sprouts (Lim et al., 2012). A metabolomic analysis of CB sprouts by ultraperformance liquid chromatography-quadrupole-time-of-flight (UPLC-Q-TOF) mass spectroscopy (MS) and partial least-squares-discriminant analysis was performed and chlorogenic acid, catechin, isoorientin, orientin, rutin, vitexin, and quercitrin were identified (Kim et al., 2011). In addition, a novel HPLC elution gradient method, based on water-acetonitrile, was used to successfully separate rutin and the previously unidentified compound quercetin-3-O-robinobioside in CB sprouts (Nam et al., 2015). The rutin contents in CB sprouts were analyzed by capillary electrophoresis (CE), which showed high efficiency, small sample volume, high speed, and good resolution (Zhang et al., 2012; Koyama et al., 2013).

Therefore to identify newly detected bioactive compounds, we should combine different kinds of methods such as nuclear magnetic resonance spectroscopy, HPLC, and Q-TOF MS.

PHYSIOLOGICAL EFFECTS

Our previous review paper summarized that the buckwheat plants perform many biological values in the human body (Zhang et al., 2012). During the last few years it has been reported that buckwheat sprouts also have many unique physiological functions, such as reducing plasma and blood cholesterol levels as well as antiadipogenic activity with antioxidative properties (Kuwabara et al., 2007; Lin et al., 2008; Nakamura et al., 2013; Lee et al., 2013). A cholesterol-free diet with buckwheat sprouts fed to rats showed that feeding TB sprouts significantly reduced plasma total cholesterol concentrations compared to the control group, whereas there was no significant difference between the CB sprouts and control feeding groups (Kuwabara et al., 2007). Furthermore, fecal bile acid excretion and cecal short-chain fatty acid concentrations as well as hepatic cholesterol 7-alpha-hydroxylase and 3-hydroxy-3-methyl-glutaryl-CoA reductase mRNA expression in the TB sprouts feeding groups were significantly higher than in the control group, which provide a solid explanation that TB sprouts have a lower serum cholesterol concentration (Kuwabara et al., 2007). However, day 8 CB sprout-feeding hamsters performed the most active antioxidative and free-radical scavenging capabilities as well as potent hypocholesterolemic and hypotriglyceridemic activities because of the maximum levels of nutrients and bioactivities in the day 8 sprouts (Lin et al., 2008). In addition, it has been shown that neofermented buckwheat sprouts fed to spontaneously hypertensive rats decreased both systolic and diastolic blood pressure, as well as performed ACE inhibition and vasorelaxation activities (Nakamura et al., 2013). Buckwheat sprouts treated with methyl jasmonate (MeJA) significantly increased the total amount of phenolics and antioxidant activity (Kim et al., 2011). The latest survey report showed that MeJA-treated buckwheat sprouts had antiadipogenesis activity with antioxidative properties in 3T3-L1 cells (Lee et al., 2013). Thus there is still a long way to go to determine the physiological effects of buckwheat sprouts because of their biopharmaceutical activities and high levels of phytochemicals.

FACTORS AFFECT THE QUALITY OF SPROUTS

In recent years, buckwheat sprouts have been under the spotlight in the international market because they contain high levels of bioactive compounds. Buckwheat sprouts have light yellow-colored cotyledons and bright white-colored hypocotyls, which are similar to soybean sprouts (Kim et al., 2004). The pericarp shedding of buckwheat sprouts is one of the most important quality-related characteristics because the pericarp sprouts that remained affect the taste of fresh vegetables. In general, the water-spray culture method was used to accelerate pericarp shedding of buckwheat sprouts, while the pericarp shedding rate is not so effective. Later on, Kim and his colleagues developed a new culture system that could completely remove the pericarps (Kim et al., 2004). In addition, many other factors, such as light, salinity, and phytohormone MeJA, greatly affected the quality of buckwheat sprouts in the cultivation process (Table 11.2).

The effects of various light compositions on the concentration of bioactive compounds and physiological activity in buckwheat sprouts have been evaluated. It has been shown that the light-grown TB sprouts had a higher content of carotenoids, including lutein and β-carotene, compared to the dark-grown sprouts (Tuan et al., 2013). Real-time polymer chain reaction analysis showed that the carotenoid biosynthetic genes' mRNA levels were also induced in light-grown TB sprouts (Tuan et al., 2013), which reveals that light can increase the production of carotenoids by activating the expression of carotenoid biosynthesis-related genes. Flavonoids have been recognized as protective substances and produced in response to UV-B radiation (Zhang et al., 2012). The dark-grown CB sprouts exhibited thin, yellow cotyledons and bright white hypocotyls, whereas sprouts irradiated with different types of light (white, red, green, blue, UV-A, and UV-B > 300 nm) had green cotyledons and pink hypocotyls (Tsurunaga et al., 2013). Interestingly, irradiation with UV-B > 300 nm increased the anthocyanin and rutin concentration, as well as the 1,1-diphenyl-2 picrylhydrazyl radical scavenging activity of

TABLE 11.2 Main Factors to Improve the Bioactive Compounds of Buckwheat Sprouts

Factors	Increased Compounds	Species	References
UV-B > 300 nm	Anthocyanin and rutin	CB	Tsurunaga et al. (2013)
Blue LED light	Rutin and cyanidin 3-*O*-rutinoside	TB	Thwe et al. (2014)
Red LED light	Catechin	TB	Thwe et al. (2014)
LED light + L-Phe	Rutin and chlorogenic acid	TB	Seo et al. (2015)
FIR at 120°C	Quercetin	TB	Ghimeray et al. (2014)
TEW	Cu, Zn, Mn, and Fe	CB	Liu et al. (2007)
TEW	Cu, Zn, and Fe	TB	Hsu et al. (2008)
Se solutions	Se	CB	Cuderman et al. (2010)
50 or 100 mM NaCl	Isoorientin, orientin, rutin, vitexin, and carotenoids	CB	Lim et al. (2012)
MeJA	Chlorogenic acid, catechin, rutin, vitexin, isoorientin, orientin, and quercitrin	CB	Kim et al. (2011)
YPS	Rutin and quercetin	TB	Zhao et al. (2012)

LED, Light-emitting diode; *FIR*, far infrared irradiation; *TEW*, trace element water; *MeJA*, methyl jasmonate; *YPS*, yeast polysaccharide; *L-Phe*, L-phenylalanine; *CB*, common buckwheat; *TB*, Tartary buckwheat.

CB sprouts (Tsurunaga et al., 2013). However, irradiation with UV-B light at wavelengths of 260–300 nm were detrimental to the growth of CB sprouts (Tsurunaga et al., 2013). Light-emitting diodes (LEDs) are artificial light sources that exhibit many advantages for controlling plant growth and development. It has been reported that the TB sprouts irradiated with LED light exhibited higher accumulation of phenylpropanoids and expression of phenylpropanoid biosynthetic genes (Thwe et al., 2014). Their results showed that the sprouts irradiated with blue LED light induced the accumulation of rutin and cyanidin 3-O-rutinoside, whereas exposure to red LED light increased the catechin content (Thwe et al., 2014). However, another research group showed that the TB sprouts irradiated with various LEDs such as red, blue, and red + blue lamps did not affect the levels of phenolic compounds; in contrast, there were significant differences in the CB sprouts (Lee et al., 2014). TB sprouts irradiated with LED light combined with L-phenylalanine (L-Phe) feeding increased the phenolic compounds, such as rutin and chlorogenic acid (Seo et al., 2015). Additionally, TB sprouts treated with far infrared irradiation (FIR) exhibited the maximum production of quercetin at 120°C (Ghimeray et al., 2014). Because of the health benefits and commercial potential, efficient production of buckwheat sprouts with high levels of bioactive compounds is desirable. Therefore an optimal light source is necessary to produce the high biological values of buckwheat sprouts.

Environmental conditions such as water quality and salinity commonly influence the growth and development of plants. Trace elements, such as manganese (Mn), zinc (Zn), and selenium (Se), play important roles in involving the main cellular defenses against reactive oxygen species (ROS), as cofactors of superoxide dismutase (SOD), and glutathione peroxidase or thioredoxin reductase, respectively (Mukhopadhay and Sharma, 1991; McCall et al., 2000; Cuderman et al., 2010). The CB sprouts grown in 300 ppm of trace element water (TEW) exhibited higher Cu, Zn, Mn, and Fe contents than those grown in the control deionized water (DIW) (Liu et al., 2007). Similarly, the TB sprouts grown in TEW also increased Cu, Zn, and Fe contents, but not Se and Mn contents (Hsu et al., 2008). The levels of rutin, quercitrin, and quercetin did not change between both CB and TB sprouts grown in TEW and DIW (Liu et al., 2007; Hsu et al., 2008). Additionally, a physiological test showed that the extract from TEW-treated CB and TB sprouts increased intracellular SOD activity as well as reduced ROS and superoxide anions in the human Hep G2 cell (Liu et al., 2007; Hsu et al., 2008). Therefore TEW could improve the antioxidant activities of buckwheat sprouts. However, both CB and TB sprouts grown in TEW did not increase the amount of Se, thus it is necessary to grow the sprouts in Se solution. Cuderman and colleagues investigated the metabolism of Se in CB sprouts from seeds soaked in various Se solutions, which provided some useful information to improve the Se nutrition of buckwheat sprouts (Cuderman et al., 2010). Besides water quality, salinity stress also affects the quality and nutritional value of buckwheat sprouts. It has been reported that treatment with low NaCl concentrations (10 mM) significantly increased the germination of CB seeds, but did not affect the fresh weight of sprouts compared to the control (0 mM NaCl). However, treatment with high NaCl concentrations (>100 mM) significantly decreased the germination of buckwheat seeds as well as the growth rate of sprouts (Lim et al., 2012). Interestingly, the levels of phenolic compounds (isoorientin, orientin, rutin, and vitexin) and carotenoid, and the antioxidant activity of the sprouts treated with 50 and 100 mM after day 7 germination, were significantly higher than those of the control (0 mM NaCl) (Lim et al., 2012). Therefore treatment with an appropriate concentration of NaCl (between 10 and 50 mM NaCl) and trace elements could improve the nutritional quality of sprouts, including the level of phenolic compounds, carotenoids, and antioxidant activity.

Jasmonic acid and its methyl ester, collectively called jasmonates, have been found to induce the biosynthesis of a variety of secondary metabolites in different plant species, including phenylpropanoids (Zhou et al., 2011). MeJA induced the activity of enzymes involved in the phenylpropanoid pathway, including PAL, thereby resulting in the accumulation of phenolic compounds (Kim et al., 2011). Exogenous MeJA-treated CB sprouts showed higher antioxidant activity and contents of the phenolic compounds, including chlorogenic acid, catechin, isoorientin, orientin, rutin, vitexin, and quercitrin (Kim et al., 2011). Therefore understanding the mechanism of MeJA signaling in buckwheat may be useful to better contribute to improving the functional quality of the sprouts. Additionally, the TB sprouts treated with the yeast polysaccharide (YPS) exhibited higher contents of rutin and quercetin, resulting from the activation of the phenylpropanoid pathway, compared to the control (Zhao et al., 2012). The understanding of buckwheat sprout responses to biotic and abiotic factors has advanced rapidly in recent years. The signal transduction of the phenylpropanoid pathway is activated upon perception of these factors, eventually leading to transcriptional activation of key enzyme genes and eventually to promote the synthesis of various flavonoid compounds (Zhou et al., 2011). This would provide novel opportunities to improve the quality of buckwheat sprouts.

CONCLUDING REMARKS

In the last few years, buckwheat sprouts have received increasing attention because of their potential dietary consumption as a new vegetable and various potential health benefits. In this review we summarized the current progress of buckwheat sprout research with regard to bioactive components, analytical methods, physiological effects, as well as the exogenous factors used to improve the quality of sprouts. How to move ahead? Based on previous research results, we provided a number of perspectives to promote and improve the cultivation quality of buckwheat sprouts. First, building new strategies to develop new buckwheat cultivars with the characteristics of self-removing pericarp during the growth of the sprouts could reduce the cost of buckwheat sprout production and therefore the cost to consumers. Second, optimizing cultivation conditions, such as modified TEW combined with low concentration of NaCl and MeJA, L-Phe feeding and UV-B irradiation, to maximize the production of bioactive compounds in buckwheat sprouts. Third, a more detailed picture should be drawn about the molecular mechanism of exogenous factors and how they regulate bioactive secondary metabolite biosynthesis. Lastly, discovering new key enzyme genes and their functionality is still very important in exploring the biosynthetic mechanism of bioactive metabolites; currently, coexpression analyses of "systems biology" datasets is a powerful tool to elucidate unknown key enzyme genes in the pathway of specialized metabolites, such as flavonoids and carotenoids.

ACKNOWLEDGMENTS

This research was supported by the Key Project of Science and Technology of Sichuan, China (grant no. 04NG001-015, "Protection and exploitation of wild-type buckwheat germplasm resource").

REFERENCES

Cuderman, P., Ožbolt, L., Ivan Kreft, I., Stibilj, V., 2010. Extraction of Se species in buckwheat sprouts grown from seeds soaked in various Se solutions. Food Chem. 123, 941–948.

Ghimeray, A.K., Sharma, P., Phoutaxay, P., Salitxay, T., Woo, S.H., Park, S.U., Park, C.H., 2014. Far infrared irradiation alters total polyphenol, total flavonoid, antioxidant property and quercetin production in Tartary buckwheat sprout powder. J. Cereal Sci. 59, 167–172.

Hsu, C.-K., Chiang, B.-H., Chen, Y.-S., Yang, J.-H., Liu, C.-L. 2008. Improving the ntioxidant activity of buckwheat (*Fagopyrum tataricm* Gaertn) sprout with trace element water. Food Chem. 108, 633–641.

Kim, S.L., Son, Y.K., Hwang, J.J., Kim, S.K., Hur, H.S., Park, C.H., 2001. Development and utilization of buckwheat sprouts as functional vegetables. Fagopyrum 18, 49–54.

Kim, S.L., Kim, S.K., Park, C.H., 2004. Introduction and nutritional evaluation of buckwheat sprouts as a new vegetable. Food Res. Int. 37, 319–327.

Kim, S.J., Zaidul, I.S.M., Maeda, T., Suzuki, T., Hashimoto, N., Takigawa, S., Noda, T., Matsuura-Endo, C., Yamauchi, H., 2007. A time-course study of flavonoids in the sprouts of Tartary (*Fagopyrum tataricum* Gaertn.) buckwheats. Sci. Hort. 115, 13–18.

Kim, S.J., Zaidul, I.S.M., Suzuki, T., Mukasa, Y., Hashimoto, N., Takigawa, S., Noda, T., Matsuura-Endo, C., Yamauchi, H., 2008. Comparison of phenolic compositions between common and Tartary buckwheat (*Fagopyrum*) sprouts. Food Chem. 110, 814–820.

Kim, H.J., Park, K.J., Lim, J.H., 2011. Metabolomic analysis of phenolic compounds in buckwheat (*Fagopyrum esculentum* M.) sprouts treated with methyl jasmonate. J. Agric. Food Chem. 59, 5707–5713.

Koyama, M., Nakamura, C., Nakamura, K., 2013. Changes in phenols contents from buckwheat sprouts during growth stage. J. Food Sci. Technol. 50, 86–93.

Kuwabara, T., Han, K.H., Hashimoto, N., Yamauchi, H., Shimada, K., Sekikawa, M., Fukushima, M., 2007. Tartary buckwheat sprout powder lowers plasma cholesterol level in rats. J. Nutr. Sci. Vitaminol. (Tokyo) 53, 501–507.

Lee, Y.J., Kim, K.J., Park, K.J., Yoon, B.R., Lim, J.H., Lee, O.H., 2013. Buckwheat (*Fagopyrum esculentum* M.) sprout treated with methyl jasmonate (MeJA) improved anti-adipogenic activity associated with the oxidative stress system in 3T3-L1 adipocytes. Int. J. Mol. Sci. 14, 1428–1442.

Lee, S.-W., Seo, J.M., Lee, M.-K., Chun, J.-H., Antonisamy, P., Arasu, M.V., Suzuki, T., Al-Dhabi, N.A., Kim, S.-J., 2014. Influence of different LED lamps on the production of phenolic compounds in common and Tartary buckwheat sprouts. Ind. Crops Prod. 54, 320–326.

Lim, J.H., Park, K.J., Kim, B.K., Jeong, J.W., Kim, H.J., 2012. Effect of salinity stress on phenolic compounds and carotenoids in buckwheat (*Fagopyrum esculentum* M.) sprout. Food Chem. 135, 1065–1070.

Lin, L.Y., Peng, C.C., Yang, Y.L., Peng, R.Y., 2008. Optimization of bioactive compounds in buckwheat sprouts and their effect on blood cholesterol in hamsters. J. Agric. Food Chem. 56, 1216–1223.

Liu, C.-L., Chen, Y.-S., Yang, J.-H., Chiang, B.-H., Hsu, C.-K., 2007. Trace element water improves the antioxidant activity of buckwheat (*Fagopyrum esculentum* Moench) sprouts. J. Agric. Food Chem. 55, 8934–8940.

Mactier, H., Weaver, L.T., 2005. Vitamin A and preterm infants: what we know, what we don't know and what we need to know. Arch. Dis. Child.—Fetal and Neonatal Edition 90, 103–108.

McCall, K.A., Huang, C., Fierke, C.A., 2000. Function and mechanism of zinc metalloenzymes. J. Nutr. 130, 1437S–1446S.

Mukhopadhay, M.J., Sharma, A., 1991. Manganese in cell metabolism of higher plants. Bot. Rev. 57, 117–149.

Nakamura, K., Naramoto, K., Koyama, M., 2013. Blood-pressure-lowering effect of fermented buckwheat sprouts in spontaneously hypertensive rats. J. Funct. Foods 5, 406–415.

Nam, T.G., Lee, S.M., Park, J.H., Kim, D.O., Baek, N.I., Eom, S.H., 2015. Flavonoid analysis of buckwheat sprouts. Food Chem. 170, 97–101.

Ren, S.C., Sun, J.T., 2014. Changes in phenolic content, phenylalanine ammonia-lyase (PAL) activity, and antioxidant capacity of two buckwheat sprouts in relation to germination. J. Funct. Foods 7, 298–304.

Seo, J.-M., Arasu, M.V., Kim, Y.-B., Park, S.U., Kim, S.-J., 2015. Phenylalanine and LED lights enhance phenolic compound production in Tartary buckwheat sprouts. Food Chem. 177, 204–213.

Thwe, A.A., Kim, Y.B., Li, X., Seo, J.M., Kim, S.J., Suzuki, T., Chung, S.O., Park, S.U., 2014. Effects of light-emitting diodes on expression of phenylpropanoid biosynthetic genes and accumulation of phenylpropanoids in *Fagopyrum tataricum* sprouts. J. Agric. Food Chem. 62, 4839–4845.

Tsurunaga, Y., Takahashi, T., Katsube, T., Kudo, A., Kuramitsu, O., Ishiwata, M., Matsumoto, S., 2013. Effects of UV-B irradiation on the levels of anthocyanin, rutin and radical scavenging activity of buckwheat sprouts. Food Chem. 141, 552–556.

Tuan, P.A., Thwe, A.A., Kim, J.K., Kim, Y.B., Lee, S., Park, S.U., 2013. Molecular characterisation and the light–dark regulation of carotenoid biosynthesis in sprouts of Tartary buckwheat (*Fagopyrum tataricum* Gaertn.). Food Chem. 141, 3803–3812.

Woo, S.H., Kamal, A.H.M., Park, S.M., Kwon, S.O., Park, S.U., Roy, S.K., Lee, J.Y., Choi, J.S., 2013. Relative distribution of free amino acids in buckwheat. Food Sci. Biotechnol. 22, 665–669.

Zangen, A., Botzer, D., Zanger, R., Shainberg, A., 1998. Furosemide and digoxin inhibit thiamine uptake in cardiac cells. Eur. J. Pharmacol. 361, 151–155.

Zhang, Z.L., Zhou, M.L., Tang, Y., Li, F.L., Tang, Y.X., Shao, J.R., Xue, W.T., Wu, Y.M., 2012. Bioactive compounds in functional buckwheat food. Food Res. Int. 49, 389–395.

Zhao, G., Zhao, J., Peng, L., Zou, L., Wang, J., Zhong, L., Xiang, D., 2012. Effects of yeast polysaccharide on growth and flavonoid accumulation in *Fagopyrum tataricum* sprout cultures. Molecules 17, 11335–11345.

Zhou, M.L., Zhu, X.M., Shao, J.R., Tang, Y.X., Wu, Y.M., 2011. Production and metabolic engineering of bioactive substances in plant hairy root culture. Appl. Microbiol. Biotechnol. 90, 1229–1239.

Zhou, M.L., Bai, D.Q., Tang, Y., Zhu, X.M., Shao, J.R., 2012. Genetic diversity of four new species related to southwestern Sichuan buckwheats as revealed by karyotype, ISSR and allozyme characterization. Plant Syst. Evol. 298, 751–759.

Chapter | twelve

Bioactive Flavonoids in Buckwheat Grain and Green Parts

I. Kreft*, G. Wieslander**, B. Vombergar†

*Department of Forest Physiology and Genetics, Slovenian Forestry Institute, Ljubljana, Slovenia; **Department of Medical Sciences, Uppsala University, Occupational and Environmental Medicine, Uppsala, Sweden; †Education Centre Piramida, Maribor, Slovenia

INTRODUCTION

Buckwheat is an ancient Chinese noncereal. From China it was brought in the 9th to 12th centuries to Russia and in the 13th to 15th centuries to Central and Northern Europe. Old documents tell that Christian cross-riders brought buckwheat in the 13th century to Nuremberg, and to many European countries. The names attributed to buckwheat were many, for example, boghvede (Denmark) and bohvete (Sweden), probably because of their similarity to the shape of the beech nut and their wheat-like characteristics. The Swedish botanist Carl von Linné (Carolus Linneaus, 1707–1778) classified buckwheat in the Polygonaceae family and *Fagopyrum* genus. In Europe, common buckwheat (*Fagopyrum esculentum*) was the most common species. Linné was familiar as well with Tartary buckwheat (*Fagopyrum tataricum*). Other names in Europe are pohanka (Czech Republic and Slovakia), gryka (Poland), grečka (Ukraine), grečiha (Russia), grano saraceno (Italy), sarrasin, blé noir (France), gwinizh-du (Brittany, France), Buchweizen (Germany), Heidenkorn or Haidl (Austria), Wëllkar (Luxemburg), Plenten (South Tyrol), tattari (Finland), ajda (Slovenia), ajdina (Prekmurje, Slovenia), and heljda (Croatia, Serbia, and Bosnia and Herzegovina). In 1842 buckwheat was an important crop in Denmark, and C.H. Andersen wrote a story of how flowering buckwheat was a proud plant. Buckwheat entered the United States later than Europe; it was brought there in the 17th century. In olden times buckwheat was known to tolerate poor soil and sandy lands. Buckwheat is rich in phytonutrients: rutin, quercetin, and others.

Molecular Breeding and Nutritional Aspects of Buckwheat. http://dx.doi.org/10.1016/B978-0-12-803692-1.00012-2

In this review, we will summarize buckwheat flavonoid bioactive components and some related epidemiological and other research on the health effects of buckwheat.

FLAVONOIDS

Buckwheat is an important food material not only because of its pleasant taste (it is very popular in some countries where it is traditionally grown), but also as a basis for functional foods with antioxidant activity, based on the flavonoid content and other effects, which are important for human health (He et al., 1995; Holasová et al., 2002). The antioxidant activity is based on the content of the flavonoids rutin and quercetin in buckwheat. Tartary buckwheat contains about 100-fold more rutin in comparison to common buckwheat (Fabjan et al., 2003). Rutin is important for protecting buckwheat plants from solar UV radiation and from the cold, desiccation, and pests (Kreft et al., 2002; Germ, 2004; Suzuki et al., 2005). The combination of high levels of rutin content and rutinosidase activity produces a strong bitterness after grazing, which protects buckwheat, especially Tartary buckwheat, from being eaten by animals (Suzuki et al., 2015a). Chitarrini et al. (2014) suggest that rutin-derived quercetin is more efficient in inhibiting aflatoxin biosynthesis by *Aspergillus flavus* in comparison to rutin. This is especially important in climate change conditions and wet weather during the buckwheat harvest to repress aflatoxin biosynthesis in buckwheat grain.

During the germination of buckwheat seeds there is an increase in the flavonoid content, but also in the content of phototoxic fagopyrin, and the presence of both is more intense in conditions of light than in dark (Kreft et al., 2013). The flavonoid composition of common and Tartary buckwheat sprouts was reported by Nam et al. (2015). Buckwheat anthraquinones, because of their molecular structure, are possible precursors of fagopyrins, and were studied by Wu et al. (2015). During the development of young plants of common and Tartary buckwheat there is also a rise in tannin concentration (Gadžo et al., 2010). Buckwheat grain tannins were studied by Luthar and Tišler (1992).

Rutin content in food products processed from groats, leaves, and flowers of buckwheat is reported by Park et al. (2000). Flavonoids are present in bread and cakes made from buckwheat flour (Vogrinčič et al., 2010; Wieslander et al., 2011). Buckwheat grain contains mainly rutin and sometimes a low concentration of quercetin, but during the bread-making process, the levels of these two flavonoids are reversed; that is, there is more quercetin than rutin (Vogrinčič et al., 2010). This is caused by the transformation of rutin into quercetin during the bread-making process by the activity of the rutin-degrading enzymes, which become active in buckwheat samples in the presence of water (Yasuda and Nakagawa, 1994; Suzuki et al., 2002; Vogrinčič et al., 2010).

To follow the transformation of rutin to quercetin and to record possible losses of flavonoids, the quercetin equivalents of rutin should be calculated (rutin concentration divided by 2.02). The transformation of rutin to quercetin is expected mainly because of the rutin-degrading enzyme in the buckwheat seeds (Yasuda and Nakagawa, 1994). According to our unpublished results, during husking of buckwheat groats by a traditional method, rutinase molecules are inactivated after cooking the grain and much of the rutin is preserved. It seems that rutin is heat stable during the baking, boiling, grilling, microwaving, and steaming of foods, but some could be leached out (Drinkwater et al., 2015). Decomposition of rutin could be prevented by using nonbitter buckwheat varieties (Suzuki et al., 2014).

In bread made with Tartary buckwheat flour, while the rutin concentration decreased, the quercetin concentration increased and remained more stable during processing than rutin (Vogrinčič et al., 2010).

The lower concentrations of flavonoids established in buckwheat-processed foods after hydrothermal treatment (dough making, baking) might theoretically also be caused by the integration of phenolic compounds into starch (amylose) structures, which transform their structure during the baking process and can incorporate smaller molecules into their structure during this transformational process (Skrabanja et al., 2000; Goderis et al., 2014; Ryno et al., 2014). But not much is known yet about these processes.

There is a possibility that under conditions of high flavonoid concentrations, as in the case of Tartary buckwheat, the molecules may protect one another from degradation. A similar tendency was reported by Vogrinčič et al. (2010). In dough and bread with low starting concentrations of rutin and quercetin, the rutin was degraded completely, but when there were high initial concentrations of rutin and quercetin, some rutin molecules remained unaffected through the technological process.

Rutin molecules are degraded during dough fermentation (Vogrinčič et al., 2010). We do not yet know whether this is the effect of the rutinase enzyme present in buckwheat flour (Yasuda and Nakagawa, 1994; Suzuki et al., 2002), or if there are enzymes with a similar effect that are delivered by the yeast microorganisms or by wheat in the case of bread made from flour mixtures. If rutin is degraded during dough fermentation, and if it is the effect of buckwheat rutinase, this means that the rutinase molecules are persisting and active during dough fermentation. According to our unpublished results, hydrothermal treatment (treating water-soaked buckwheat grain at a temperature above 90°C) is damaging for the activity of rutinase. Another way to avoid the transformation of rutin to quercetin, and thus to avoid a bitter taste of quercetin in food products, is to use the naturally low rutinase buckwheat variety discovered by Suzuki et al. (2014). Rutin and rutin-degrading enzymes may be allocated in different structures of buckwheat grain (Suzuki et al., 2014). Buckwheat grain morphology in relation to composition was studied by Kreft and Kreft (2000) and by Skrabanja et al. (2004) (Figs. 12.1 and 12.2).

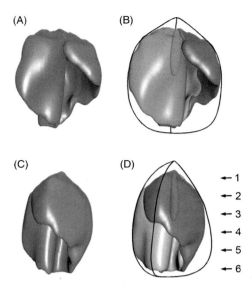

FIGURE 12.1 An equatorial view of the three-dimensional computer generated model of the buckwheat embryo. Top two panels represent one view (from the bottom at Fig. 12.2) and the bottom two panels represent the view perpendicular to the first one (from the left at Fig. 12.2). Black lines on the panels (B and D) represent buckwheat grain edges. Light gray and dark gray are respective cotyledons; the embryo proper is visual in the center at the distal end of cotyledons. On the panels A and C one cotyledon and the embryonic axis has been removed. Arrows in the panel D are pointing at the sections, which are shown on Fig. 12.2. *Kreft and Kreft (2000) from* Fagopyrum *17, by the permission of authors and publisher.*

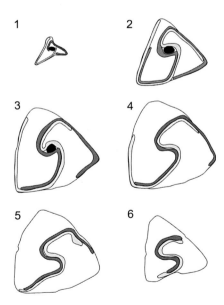

FIGURE 12.2 Cross-sections of the computer generated three-dimensional model of the buckwheat embryo. Black line represents testa. Light gray and dark gray areas are the respective cotyledons; black spot in the center is the embryonic axis. *Kreft and Kreft (2000) from* Fagopyrum *17, by the permission of authors and publisher.*

HEALTH EFFECTS

For the intestinal uptake of dietary quercetin glycosides, sugar moiety is of foremost importance (Arts et al., 2004). It is known that quercetin from the diet can be widely distributed in animal tissues (De Boer et al., 2005). Quercetin has mainly similar bioavailability and health-maintaining effects as rutin (Sikder et al., 2014). Thus from this point of view the transition of rutin to quercetin during food processing is acceptable. But besides the bitter taste, quercetin is less active in comparison to rutin in antiinflammatory activities (preventing ileitis and colitis) since rutin is taken up by the ileal mucosa and slowly released to the lumen (Mascaraque et al., 2015).

Buckwheat products are able to decrease cholesterol levels and improve lung capacity in humans (Sikder et al., 2014; Wieslander et al., 2011, 2012; Yang et al., 2014). Buckwheat extracts can also protect DNA from damage caused by hydroxyl radicals (Cao et al., 2008; Vogrinčič et al., 2013). Tartary buckwheat improves total antioxidant capacity of the bread; chia grain and Tartary buckwheat represent excellent raw materials for the formulation of gluten-free bread with high nutritional value (Costantini et al., 2014). Rutin antagonizes the increase of capillary fragility associated with hemorrhagic disease or hypertension in humans (Griffith et al., 1944).

Buckwheat is an important source of essential trace elements such as Se, Fe, and Zn (Pongrac et al., 2013; Golob et al., 2015). Rutin binds Zn preferentially at the 5-hydroxyl-4-keto site (Wei and Guo, 2014). The flavonoid kelation of other trace elements has not yet been studied in detail.

Possible toxicity of rutin-rich Tartary buckwheat was studied by Suzuki et al. (2015b), but no acute or subacute toxicity was established when rats were fed for 14 or 28 days with 10 and 5 g flour/kg of body weight, respectively, with buckwheat flour containing up to 1.57% rutin. Nor were any other adverse effects detected.

Thus from the standpoint of maintaining health and respecting traditional food habits, common and Tartary buckwheat are a welcome enrichment of the daily diet.

ACKNOWLEDGMENTS

This research was supported by the Key Project of Science and Technology of Sichuan, China (grant no. 04NG001-015, "Protection and exploitation of wild-type buckwheat germplasm resource"); by the Slovenian Research Agency, through program P3-0395 "Nutrition and Public Health," and projects J4-4224 and J4-5524, supported by EUFORINNO 7th FP EU Infrastructure Programme (Reg. Pot No. 315982) and cofinanced by the European Community under project No. 26220220180: Building Research Centre "AgroBioTech." The authors acknowledge Professor Mateja Germ for critical reading of the manuscript and Jean McCollister for English language editing.

REFERENCES

Arts, I.C.W., Sesink, A.L.A., Faassen-Petersa, M., Hollman, P.C.H., 2004. The type of sugar moiety is a major determinant of the small intestinal uptake and subsequent biliary excretion of dietary quercetin glycosides. Br. J. Nutr. 91, 841–847.

Cao, W., Chen, W.J., Suo, Z.R., Yao, Y.P., 2008. Protective effect of ethanolic extracts of buckwheat groats on DNA damage caused by hydroxyl radicals. Food Res. Int. 41, 924–929.

Chitarrini, G., Nobili, C., Pinzari, F., Antonini, A., De Rossi, P., Del Fiore, A., Procacci, S., Tolaini, V., Scala, V., Scarpari, M., Reverberi, M., 2014. Buckwheat achenes antioxidant profile moderates Aspergillus flavus growth and aflatoxin production. Int. J. Food Microbiol. 189, 1–10.

Costantini, L., Lukšič, L., Molinari, R., Kreft, I., Bonafaccia, G., Manzi, L., Merendino, N., 2014. Development of gluten-free bread using Tartary buckwheat and chia flour rich in flavonoids and omega-3 fatty acids as ingredients. Food Chem. 165, 232–240.

De Boer, V.C.J., Dihal, A.A., Van der Woude, H., Arts, I.C.W., Wolffram, S., Alink, G.M., Rietjens, I.M., Keijer, J., Hollman, P.C., 2005. Tissue distribution of quercetin in rats and pigs. J. Nutr. 135, 1718–1725.

Drinkwater, J.M., Tsao, R., Liu, R.H., Defelice, C., Wolyn, D.J., 2015. Effects of cooking on rutin and glutathione concentrations and antioxidant activity of green asparagus (*Asparagus officinalis*) spears. J. Funct. Foods 12, 342–353.

Fabjan, N., Rode, J., Kosir, I.J., Wang, Z.H., Zhang, Z., Kreft, I., 2003. Tartary buckwheat (*Fagopyrum tataricum* Gaertn.) as a source of dietary rutin and quercitrin. J. Agric. Food Chem. 51, 6452–6455.

Gadžo, D., Djikić, M., Gavrić, T., Štrekelj, P., 2010. Comparison of tannin concentration in young plants of common and Tartary buckwheat. Acta Agric. Sloven. 95, 75–78.

Germ, M., 2004. Environmental factors stimulate synthesis of protective substances in buckwheat. In: Faberová, I. (Ed.), Proceedings of the 9th International Symposium on Buckwheat, August 18–22, 2004. Advances in Buckwheat Research. Research Institute of Crop Production, Prague, pp. 55–60.

Goderis, B., Putseys, J.A., Joke, A., Gommes, C.J., Bosmans, G.M., Delcour, J.A., 2014. The structure and thermal stability of amylose–lipid complexes: a case study on amylose–glycerol monostearate. Cryst. Growth Des. 14, 3221–3233.

Golob, A., Stibilj, V., Kreft, I., Germ, M., 2015. The feasibility of using Tartary buckwheat as a Se-containing food material. J. Chem. (Hindawi) (1–4), 246042.

Griffith, J.Q., Couch, J.F., Lindauer, M.A., 1944. Effect of rutin on increased capillary fragility in man. Proceedings of the Society for Experimental Biology and Medicine 55, 228–229.

He, J., Klag, M.J., Whelton, P.K., Mo, J.P., Chen, J.Y., Qian, M.C., Mo, P.S., He, G.Q., 1995. Oats and buckwheat intakes and cardiovascular disease risk factors in an ethnic minority of China. Am. J. Clin. Nutr. 61, 366–372.

Holasová, M., Fiedlerová, V., Smrcinová, H., Orsák, M., Lachman, J., Vavreinova, S., 2002. Buckwheat—the source of antioxidant activity in functional foods. Food Res. Int. 35, 207–211.

Kreft, S., Kreft, M., 2000. Localization and the morphology of buckwheat embryo. Fagopyrum 17, 15–19.

Kreft, S., Štrukelj, B., Gaberščik, A., Kreft, I., 2002. Rutin in buckwheat herbs grown at different UV-B radiation levels: comparison of two UV spectrophotometric and an HPLC method. J. Exp. Bot. 53, 1801–1804.

Kreft, S., Janeš, D., Kreft, I., 2013. The content of fagopyrin and polyphenols in common and Tartary buckwheat sprouts. Acta Pharm. 63, 553–560.

Luthar, Z., Tišler, V., 1992. Tannin–carbohydrate complex in buckwheat seeds (*Fagopyrum esculentum* Moench). Fagopyrum 12, 21–26.

Mascaraque, C., Lopez-Posadas, R., Monte, M.J., Romero-Calvo, I., Daddaoua, A., Gonzalez, M., Martinez-Plata, E., Suarez, M.D., Gonzalez, R., Marin, J.J.G., Zarzuelo, A., Martinez-Augustin, O., de Medina, F.S., 2015. The small intestinal mucosa acts as a rutin reservoir to extend flavonoid anti-inflammatory activity in experimental ileitis and colitis. J. Funct. Foods 13, 117–125.

Nam, T.-G., Lee, S.M., Park, J.H., Kim, D.O., Baek, N.I., Eom, S.H., 2015. Flavonoid analysis of buckwheat sprouts. Food Chem. 170, 97–101.

Park, C.H., Kim, Y.B., Choi, Y.S., Heo, K., Kim, S.L., Lee, K.C., Chang, K., Lee, H.B., 2000. Rutin content in food products processed from groats, leaves, and flowers of buckwheat. Fagopyrum 17, 63–66.

Pongrac, P., Vogel-Mikuš, K., Jeromel, L., Vavpetič, P., Pelicon, P., Kaulich, B., Gianoncelli, A., Eichert, D., Regvar, M., Kreft, I., 2013. Spatially resolved distributions of the mineral elements in the grain of Tartary buckwheat (*Fagopyrum tataricum*). Food Res. Int. 54, 125–131.

Ryno, L.M., Levine, Y., Iovine, P.M., 2014. Synthesis, characterization, and comparative analysis of amylose–guest complexes prepared by microwave irradiation. Carbohyd. Res. 383, 82–88.

Sikder, K., Kesh, S.B., Das, N., Manna, K., Dey, S., 2014. The high antioxidative power of quercetin (aglycone flavonoid) and its glycone (rutin) avert high cholesterol diet induced hepatotoxicity and inflammation in Swiss albino mice. Food Funct. 5, 1294–1303.

Skrabanja, V., Laerke, H.N., Kreft, I., 2000. Protein–polyphenol interactions and in vivo digestibility of buckwheat groat proteins. Pflüg. Arch.—Eur. J. Physiol. 440, 129–131.

Skrabanja, V., Kreft, I., Golob, T., Modic, M., Ikeda, S., Ikeda, K., Kreft, S., Bonafaccia, G., Knapp, M., Košmelj, K., 2004. Nutrient content in buckwheat milling fractions. Cereal Chem. 81, 172–176.

Suzuki, T., Honda, Y., Funatsuki, W., Nakatsuka, K., 2002. Purification and characterization of flavonol 3-glucosidase, and its activity during ripening in Tartary buckwheat seeds. Plant Sci. 163, 417–423.

Suzuki, T., Honda, Y., Mukasa, Y., 2005. Effects of UV-B radiation, cold and desiccation stress on rutin concentration and rutin glucosidase activity in Tartary buckwheat (*Fagopyrum tataricum*) leaves. Plant Sci. 168, 1303–1307.

Suzuki, T., Morishita, T., Mukasa, Y., Takigawa, S., Yokota, S., Ishiguro, K., Noda, T., 2014. Breeding of "Manten-Kirari," a non-bitter and trace-rutinosidase variety of Tartary buckwheat (*Fagopyrum tataricum* Gaertn.). Breed. Sci. 64, 344–350.

Suzuki, T., Morishita, T., Kim, S.-J., Park, S.U., Woo, S.H., Noda, T., Takigawa, S., 2015a. Physiological roles of rutin in the buckwheat plant. JARQ—Japan Agric. Res. Quart. 49, 37–43.

Suzuki, T., Morishita, T., Noda, T., Ishiguro, K., 2015b. Acute and subacute toxicity studies on rutin-rich Tartary buckwheat dough in experimental animals. J. Nutr. Sci. Vitaminol. 61, 175–181.

Vogrinčič, M., Timoracká, M., Melichacova, S., Vollmannová, A., Kreft, I., 2010. Degradation of rutin and polyphenols during the preparation of Tartary buckwheat bread. J. Agric. Food Chem. 58, 4883–4887.

Vogrinčič, M., Kreft, I., Filipič, M., Žegura, B., 2013. Antigenotoxic effect of Tartary (*Fagopyrum tataricum*) and common (*Fagopyrum esculentum*) buckwheat flour. Journal of Medicinal Food 16, 944–952.

Wei, Y., Guo, M., 2014. Zinc-binding sites on selected flavonoids. Biol. Trace Elem. Res. 161, 223–230.

Wieslander, G., Fabjan, N., Vogrinčič, M., Kreft, I., Janson, C., Spetz-Nyström, U., Vombergar, B., Tagesson, C., Leanderson, P., Norbäck, D., 2011. Eating buckwheat cookies is associated with the reduction in serum levels of myeloperoxidase and cholesterol: a double blind crossover study in day-care centre staffs. Tohoku J. Exp. Med. 2011, 123–130.

Wieslander, G., Fabjan, N., Vogrinčič, M., Kreft, I., Vombergar, B., Norbäck, D., 2012. Effects of common and Tartary buckwheat consumption on mucosal symptoms, headache and tiredness: a double-blind crossover intervention study. Int. J. Food Agric. Environ.—JFAE 10, 107–110.

Wu, X.O., Ge, X.S., Liang, S.X., Lv, Y.K., Sun, H.W., 2015. A novel selective accelerated solvent extraction for effective separation and rapid simultaneous determination of six anthraquinones in Tartary buckwheat and its products by UPLC-DAD. Food Anal. Meth. 8, 1124–1132.

Yang, N., Li, Y.M., Zhang, K.S., Jiao, R., Ma, K.Y., Zhang, R., Ren, G.X., Chen, Z.Y., 2014. Hypocholesterolemic activity of buckwheat flour is mediated by increasing sterol excretion and down-regulation of intestinal NPC1L1 and ACAT2. J. Funct. Foods 6, 311–318.

Yasuda, T., Nakagawa, H., 1994. Purification and characterization of rutin-degrading... enzymes in Tartary buckwheat seeds. Phytochemistry 37, 133–136.

Chapter thirteen

Nutritional Value of Buckwheat Proteins and Starch

V. Škrabanja*, I. Kreft**

*Department of Food Science and Technology, Biotechnical Faculty, Ljubljana, Slovenia; **Department of Forest Physiology and Genetics, Slovenian Forestry Institute, Ljubljana, Slovenia

INTRODUCTION

Recently, buckwheat grain has been introduced to many countries because the flour and groats are nutritionally rich, and the foods prepared from buckwheat have beneficial effects on human health. Among the important health effects are a decrease in the serum concentration of cholesterol (Wieslander et al., 2011, 2012; Stokić et al., 2015) and a decrease in glycemic and insulin indexes (Skrabanja et al., 2001) after consumption of buckwheat foods/meals. Much attention is being given to the educational and cultural aspects of growing and eating buckwheat as well as to its utilization and health effects (Ikeda et al., 2013; Škrabanja, 2014; Vombergar et al., 2014).

FUNCTIONAL VALUE OF BUCKWHEAT PROTEINS

Buckwheat is considered a pseudocereal with high nutritional value because of its protein composition. Although the grain has a low protein content (10.6 and 10.3 g/100 g of dry weight in common and Tartary buckwheat, respectively), it has a balanced amino acid composition, with high levels of essential amino acids such as leucine and lysine (6.92, 5.84, and 7.11, 6.18 g/100 g of protein in common and Tartary buckwheat, respectively) (Bonafaccia et al., 2003).

The high content of proteins, flavonoids, and trace elements in certain buckwheat grain milling fractions suggests a potential application of these materials in special dietary products (Skrabanja et al., 2004). Buckwheat seed proteins can contain selenium (Golob et al., 2015), an essential trace element in human nutrition.

169

Molecular Breeding and Nutritional Aspects of Buckwheat. http://dx.doi.org/10.1016/B978-0-12-803692-1.00013-4

Javornik and Kreft (1984) and Wei et al. (2003) reported on the amino acid composition of buckwheat protein solubility fractions. There are a number of differences in the amino acid composition of different solubility fractions (Javornik and Kreft, 1984), but these differences are minor. As lysine content is higher in albumins and globulins, this contributes to the well-balanced amino acid composition of buckwheat.

The nomenclature otherwise used for wheat proteins is often used in reports of buckwheat solubility fractions. According to the reports (Javornik and Kreft, 1984; Wei et al., 2003), among buckwheat proteins 1–7% are "prolamins" and 11–23% are "glutelins," depending on the sample and extraction method. Interestingly, "glutelins" have a reported content of lysine on the same level as "albumins."

In cereals a classification system of proteins is used that is based on solubility fractions: albumins—soluble in water or hypotonic solutions and coagulated by heat; globulins—soluble in salt solutions or in "isotonic" solutions; prolamins—soluble in aqueous ethanol or other alcohols; glutelins—soluble in diluted acids or bases, detergents, or by reducing agents; the rest are insoluble.

The use of the terms albumins, globulins, glutelins, and prolamins is reasonable, but it is at least in the case of buckwheat better to name the fractions according to the extraction method as (1) water soluble, (2) salt soluble, (3) ethanol soluble, (4) alkali soluble, and (5) the rest as insoluble. If, prior to extraction by (2) solution, extraction and/or repeated washing in (1) extraction medium is not performed, then most or all of (1) proteins could also be extracted by the (2) method. It is not necessarily the case that by extraction in ethanol the extract also contains all or most of the proteins extractable by the (1) and/or (2) extraction medium. In any event we could name the first two extraction fractions as albumins and globulins, respectively, but there is no evidence for sufficient similarity between wheat prolamins and glutelins and the respective buckwheat solubility fractions. Thus the use of wheat protein nomenclature for the solubility fractions in the case of buckwheat is not adequate. To label the ethanol soluble fraction as prolamins could be misleading, since buckwheat is a dicot and hence distant from cereals in the botanical system. Not only is there a low amount of prolamins, but there is no evidence to suggest that ethanol-soluble proteins labeled as "prolamins" have any structural or other similarity to cereal prolamins in addition to the solubility in ethanol.

Experimental buckwheat samples were obtained by means of different hydrothermal treatments to study the impact of the levels of polyphenols on protein digestibility (Skrabanja et al., 2000). A rat model system was used and considerable interaction between polyphenols and proteins was observed during the hydrothermal treatment. This interaction reduces the digestion of proteins through the small and large intestine. But microbial processes in the colon enhance the digestibility of protein, otherwise blocked by polyphenols in hydrothermally processed buckwheat (Skrabanja et al., 2000). The authors cited established that polyphenols naturally present in buckwheat husks lower the true digestibility of buckwheat proteins but do not adversely affect the biological value of buckwheat proteins. As reported by Ikeda et al. (1986), tannic acid and catechin exhibited a significant inhibitory effect on the in vitro peptic and pancreatic digestion of buckwheat globulin.

Ikeda et al. (1986) and Ikeda and Kishida (1993) studied the in vitro digestibility of buckwheat protein and the impact of secondary buckwheat metabolites. Luthar and Tišler (1992) and Luthar and Kreft (1996) studied the composition of tannin in buckwheat. According to Riedl and Hagerman (2001), the tannin–protein complex can act as radical scavenger and radical sink.

Evidence in the literature indicates that buckwheat proteins can reduce the concentration of cholesterol in the serum by increasing the fecal excretion of steroids, which is induced by the binding of steroids to undigested proteins (Takahama and Hirota, 2011). According to Ma and Xiong (2009) digestion-resistant peptides are largely responsible for bile acid elimination. Buckwheat proteins have been reported to prevent gallstone formation more strongly than soy protein isolates, and they may slow mammary carcinogenesis by lowering serum estradiol, and suppress colon carcinogenesis by reducing cell proliferation (Kayashita et al., 1999; Tomotake et al., 2000; Liu et al., 2001). These effects are most probably connected with the limited digestibility of buckwheat proteins.

As buckwheat does not contain gluten proteins, it is used as food material for patients with celiac disease (Costantini et al., 2014; Giménez-Bastida et al., 2015). Although buckwheat allergy is not very common, allergic disorders caused by eating buckwheat food or by using pillows filled with buckwheat husks have been reported in some cases (Yamada et al., 1995; Nair and Adachi, 2002; Wieslander et al., 2000; Nagata et al., 2000; Kondo et al., 2001). Buckwheat pillows filled with husk from hydrothermally processed groats are mostly dust free and considered to be safer.

FUNCTIONAL VALUE OF BUCKWHEAT STARCH

Buckwheat flour contains 70–91% (w/w) of starch depending on the milling method (Skrabanja et al., 2004). The amylose content of starch is the basis for the appearance of retrograded starch during the hydrothermal processing of food materials (Skrabanja et al., 2001). Buckwheat is known as having relatively small starch granules and an amylose content of starch higher than those of cereals (Skrabanja and Kreft, 1998).

Starch in Tartary buckwheat has an amylose content close to 39.0% in addition to a high flavonoid content, indicating that Tartary buckwheat has the potential to be exploited for the production of functional foods with a low glycemic index (Gao et al., 2016).

Through fractionating in the milling and sieving process it is possible to obtain starch-rich fractions and fractions rich in high-quality proteins from buckwheat seeds (Bonafaccia et al., 2003). Through this process, fractions low in proteins (4.5–7.5%), which contain up to about 92% starch and make up 32% of the entire milled seeds fractions, are obtained in addition to protein-rich products (up to 30% protein by dry matter), as reported by Skrabanja et al. (2004). Physicochemical and pasting properties of common buckwheat starches have been studied by Li et al. (1997) and Gao et al. (2016). In common buckwheat the apparent amylose concentration ranges from 22% to 26% (Li et al., 1997), and the total amylose in starch ranges from 36z% to 43% (Gao et al., 2016). The amylose content, starch granule diameter, and solubility were found to be different among buckwheat samples. As the varieties have different starch characteristics, it is necessary to take this into account in processing buckwheat (Li et al., 1997; Gao et al., 2016). It is possible that buckwheat growing conditions and thus the assimilate availability during the buckwheat grain-filling process, in addition to the variety characteristics, may have some influence on the size of starch granules and/or amylose content in starch.

The rate and extent of starch digestibility, as influenced by the technological factors, were investigated in buckwheat samples: cooked buckwheat groats, breads with added buckwheat flour or intact groats, and buckwheat pasta (ie, noodles). The resistant starch contents in the buckwheat samples were low; the highest amount was determined in the cooked buckwheat groats (6%, total starch basis) and the lowest in bread with 70% of buckwheat flour (0.9% resistant starch, total starch basis). Skrabanja et al. (1998, 2001), Kreft and Škrabanja (2002). Treated buckwheat samples contained about 4% retrograded starch in dry matter, as compared to untreated and dry-heated buckwheat, which had about 1% retrograded starch. There is a good correlation ($r = 0.91$, $P < 0.01$) between the retrograded starch by the in vitro method and the undigested starch in vivo in rats. Groats prepared by the traditional method of boiling buckwheat grain before dehusking followed by slow drying contain less than 48% (dry matter basis) rapidly available starch, in comparison to white wheat bread, where the corresponding value is close to 59%. Buckwheat groats starch with a reduced rate of digestion could be a possible complement to or a substitute for other more commonly used sources of carbohydrates (white wheat bread, rice, potatoes).

Cooked buckwheat groats with the lowest glycemic index were accepted as the most satiating meal among the buckwheat samples tested (satiety index = 114). Despite the low glycemic index, no such satiating property was found after consumption of bread containing 50% of buckwheat groats. The satiety index of this particular bread was equal to the satiety index of the white wheat bread (ie, 100).

The ultrastructure of common buckwheat grain was started to be studied by Pomeranz and Sachs (1972). Tartary buckwheat grain ultrastructure was studied by Javornik and Kreft (1980), revealing that buckwheat aleurone and protein bodies are of a size less than 1000 nm. Gregori and Kreft (2012) established that in the low amylose mutant, spherical starch granules are seen in the middle of endosperm. In the mutant, on the nanoscale level, pinpricks on the surface of starch granules were visible, and within starch granules empty spaces among the alernate amylopectin layers. Growth rings of alternate layers of buckwheat starch grain were studied as well by atomic force microscopy by Neethirajan et al. (2012), and by scanning electron microscopy combined with confocal laser microscopy by Chrungoo et al. (2013). Progress in the research of common and Tartary starch grain size and shape is reviewed and reported by Gao et al. (2016).

The decrease in glycemic and insulin indexes is attributed to the formation of amylase-resistant starch produced by heating (Skrabanja et al., 2000). Although buckwheat flour starch becomes amylase resistant after heating, according to Takahama and Hirota (2010) it is not yet fully understood how this amylase resistance is obtained. The acquisition of amylase resistance could be caused by the formation of starch–lipid complexes since buckwheat grain contains lipids. Lipids and free fatty acids can inhibit amylase-catalyzed digestion of starch by forming starch–fatty acid complexes (Tufvesson et al., 2001). Buckwheat flour contains rutin and other phenolic substances including flavan-3-ols. These flavonoids could also make amylase-resistant starch by heating, as polyphenols can make noncovalent complexes with starch. Polyphenols including flavonoids have been reported to be inhibitors of alpha-amylase activity (Kim et al., 2000; He et al., 2007; Li et al., 2009; Kim et al., 2010).

During in vivo digestion the bile salts bind to starch, mainly to the helical structures of the amylose, which are not occupied by other molecules such as fatty acids or polyphenols. Such binding results in the inhibition of starch digestion by pancreatin. Thus the bile salts can be discussed from the standpoint of not only lipid digestion but also starch digestion (Takahama and Hirota, 2011).

The compounds with flavonoid structure may associate with starch and other polysaccharides. Isoflavones bind to receptors of the steroid hormone estrogen (Messina and Wood, 2008). The bile salts also bind to starch to inhibit its digestion by amylase. To study the effects of bile and bile salts on starch–iodine complex formation and pancreatin-induced starch digestion it is necessary to compare nonextracted buckwheat starch with organic solvent-extracted buckwheat starch. Methanol and some other solvents are able to extract fatty acids and polyphenols from buckwheat starch. Takahama and Hirota (2011) established that bile salts can form complexes with buckwheat starch, based on the helical structures not occupied by other molecules such as lipids and polyphenols, in the duodenum; the formation of complexes of starch with bile salts makes starch amylase resistant in the small intestine. However, the complexes can be degraded in the colon.

INTERACTIONS

According to Takahama and Hirota (2010), fatty acids from buckwheat flour are bound to amylose and inhibit its digestion by pancreatin. Rutin and epicatechin-dimethylgallate, bound to both amylose and amylopectin, may inhibit their digestion as well. The lower glycemic indexes of bread made from mixtures of wheat flour and buckwheat flour is caused by binding of fatty acids, rutin, and proanthocyanidins, including flavan-3-ols, to buckwheat starch. Such binding results in lower susceptibility of buckwheat starch to amylase. The ability of buckwheat flour to decrease the glycemic and insulin indexes of products made from wheat and buckwheat flour mixtures could be attributed to the binding of fatty acids, rutin, and proanthocyanidins including flavan-3-ols, which are contained in buckwheat flour, as well as to wheat constituents (Skrabanja et al., 2001; Takahama and Hirota, 2010).

During the traditional preparation of buckwheat groats, there is a significant migration of the substances between the pericarp and the groats (Janeš et al., 2010). By processing buckwheat and making products and dishes containing buckwheat, there are possible diverse interactions among constituents in the mixture during the procedure, especially during the hydrothermal treatments.

ACKNOWLEDGMENTS

This study was financed by the Slovenian Research Agency, through program P3-0395 "Nutrition and Public Health," and projects J4-4224 and J4-5524, supported by EUFORINNO 7th FP EU Infrastructure Programme (RegPot No. 315982) and cofinanced by the European Community under project No. 26220220180: Building Research Centre "AgroBioTech."

REFERENCES

Bonafaccia, G., Marocchini, M., Kreft, I., 2003. Composition and technological properties of the flour and bran from common and Tartary buckwheat. Food Chem. 80, 9–15.

Chrungoo, N.K., Devadasan, N., Kreft, I., Gregori, M., 2013. Identification and characterization of granule bound starch synthase (GBSS-I) from common buckwheat (*Fagopyrum esculentum* Moench). J. Plant Biochem. Biotechnol. 22, 269–276.

Costantini, L., Lukšič, L., Molinari, R., Kreft, I., Bonafaccia, G., Manzi, L., Merendino, N., 2014. Development of gluten-free bread using Tartary buckwheat and chia flour rich in flavonoids and omega-3 fatty acids as ingredients. Food Chem. 165, 232–240.

Gao, J., Kreft, I., Chao, G., Wang, Y., Liu, W., Wang, L., Wang, P., Gao, X., Feng, B., 2016. Tartary buckwheat (*Fagopyrum tataricum* Gaertn.) starch, a side product in functional food production, as a potential source of retrograded starch. Food Chem. 190, 552–558.

Giménez-Bastida, J.A., Piskula, M., Zielinski, H., 2015. Recent advances in development of gluten-free buckwheat products. Trends Food Sci. Technol. 44, 58–65.

Golob, A., Stibilj, V., Kreft, I., Germ, M., 2015. The feasibility of using Tartary buckwheat as a Se-containing food material. J. Chem., doi: 10.1155/2015/246042.

Gregori, M., Kreft, I., 2012. Breakable starch granules in a low-amylose buckwheat (*Fagopyrum esculentum* Moench) mutant. Int. J. Food Agric. Environ.—JFAE 10, 258–262.

He, Q., Lv, Y., Yao, K., 2007. Effects of tea polyphenols on the activities of alpha-amylase, pepsin, trypsin and lipase. Food Chem. 101, 1178–1182, I.

Ikeda, K., Kishida, M., 1993. Digestibility of proteins in buckwheat seed. Fagopyrum 13, 21–24.

Ikeda, K., Oku, M., Kusano, T., Yasumoto, K., 1986. Inhibitory potency of plant antinutrients towards the in vitro digestibility of buckwheat protein. J. Food Sci. 51, 1527–1530.

Ikeda, K., Asami, Y., Konishi, T., Nishihana, A., Ikeda, S., Lin, R., Kreft, I., 2013. Molecular-cookery-scientific characterization of buckwheat products. In: Germ, M. (Eds.), The Proceedings of Papers. Pernica: Fagopyrum—slovensko društvo za promocijo ajde, pp. 30–31.

Janeš, D., Prosen, H., Kreft, I., Kreft, S., 2010. Aroma compounds in buckwheat (*Fagopyrum esculentum* Moench) groats, flour, bran, and husk. Cereal Chem. 87, 141–143.

Javornik, B., Kreft, I., 1980. Structure of buckwheat kernel. In: Kreft, I., Javornik, B., Vombergar, B. (Eds.), Buckwheat: Genetics, Plant Breeding, Utilization. Biotehniška fakulteta, Ljubljana, pp. pp. 105–113.

Javornik, B., Kreft, I., 1984. Characterization of buckwheat proteins. Fagopyrum 4, 30–38.

Kayashita, I., Shimaoka, I., Nakajoh, M., Kishida, N., Kato, N., 1999. Consumption of a buckwheat protein extract retards 7,12-dimethylbenz[alpha] anthracene-induced mammary carcinogenesis in rats. Biosci. Biotechnol. Biochem. 63, 1837–1839.

Kim, J.-S., Kwon, C.-S., Son, K.H., 2000. Inhibition of alpha-glucosidase and amylase by luteolin, a flavonoid. Biosci. Biotechnol. Biochem. 64, 2458–3461.

Kim, H., Kim, J.-K., Kang, L., Jeong, K., Jung, S., 2010. Docking and scoring of quercetin and quercetin glycosides against alpha-amylase receptor. Bull. Korean Chem. Soc. 31, 461–463.

Kondo, Y., Urisu, A., Tokuda, R., Ishida, N., Yasuda, T., 2001. Molecular characterization of a 24-kDa buckwheat protein, one of the major allergens of buckwheat seed. Fagopyrum 18, 21–25.

Kreft, I., Škrabanja, V., 2002. Nutritional properties of starch in buckwheat noodles. J. Nutr. Sci. Vitaminol. 48, 47–50.

Li, W.D., Lin, R.F., Corke, H., 1997. Physicochemical properties of common and Tartary buckwheat starch. Cereal Chem. 74, 79–82.

Li, Y., Gao, F., Gao, F., Shan, F., Bian, J., Zhao, C., 2009. Study on the interaction between 3 flavonoid compounds and alpha-amylase by fluorescence spectroscopy and enzyme kinetics. J. Food Sci. 74, C199–C203.

Liu, Z., Ishikawa, W., Huang, X., Tomotake, H., Kayashita, J., Watanabe, H., Kato, N., 2001. A buckwheat protein product suppresses 1,2-dimethylhydrazine-induced colon carcinogenesis in rats by reducing cell proliferation. J. Nutr. 131, 1850–1853.

Luthar, Z., Kreft, I., 1996. Composition of tannin in buckwheat (*Fagopyrum esculentum* Moench) seedsvol. 67Zbornik BF Univerze v Ljubljani, Kmetijstvo, pp. 59–65.

Luthar, Z., Tišler, V., 1992. Tannin–carbohydrate complex in buckwheat seeds (*Fagopyrum esculentum* Moench). Fagopyrum 12, 21–26.

Ma, Y.Y., Xiong, Y.L.L., 2009. Antioxidant and bile acid binding activity of buckwheat protein in vitro digests. J. Agric. Food Chem. 57, 4372–4380.

Messina, M.J., Wood, C.E., 2008. Soy isoflavones, estrogen therapy, and breast cancer risk: analysis and commentary. Nutr. J. 7, 17–27.

Nagata, Y., Fujino, K., Hashiguchi, S., Abe, N., Zaima, Y., Ito, Y., Takahashi, Y., Maeda, K., Sugimura, K., 2000. Molecular characterization of buckwheat major immunoglobulin E-reactive proteins in allergic patients. Allergol. Int. 49, 117–124.

Nair, A., Adachi, T., 2002. Screening and selection of hypoallergenic buckwheat species. Sci. World J. 2, 818–826.

Neethirajan, S., Tsukamoto, K., Kanahara, H., Sugiyama, S., 2012. Ultrastructural analysis of buckwheat starch components using atomic force microscopy. J. Food Sci. 71, N2–N7.

Pomeranz, Y., Sachs, I.B., 1972. Scanning electron microscopy of the buckwheat kernel. Cereal Chem. 49, 23–25.

Riedl, K.M., Hagerman, A.E., 2001. Tannin–protein complex as radical scavengers and radical sinks. J. Agric. Food Chem. 49, 4917–4923.

Škrabanja, V. 2014. Rad bi vedel več - o ajdi = Ich würde gern mehr wissen - über Buchweizen = I'd like to know more about buckwheat = Vorrei saperne di più - sul grano saraceno. Arte4, Novo Mesto.

Skrabanja, V., Kreft, I., 1998. Resistant starch formation following autoclaving of buckwheat (*Fagopyrum esculentum* Moench) groats: an in vitro study. J. Agric. Food Chem. 46, 2020–2023.

Skrabanja, V., Laerke, H.N., Kreft, I., 1998. Effect of hydrothermal processing of buckwheat (*Fagopyrum esculentum* Moench) groats on starch enzymatic availability in vitro and in vivo in rats. J. Cereal Sci. 28, 209–221.

Skrabanja, V., Laerke, H.N., Kreft, I., 2000. Protein–polyphenol interactions and in vivo digestibility of buckwheat groat proteins. Pflüg. Arch.—Eur. J. Physiol. 440, 129–131.

Skrabanja, V., Elmstahl, H.G.M.L., Kreft, I., Bjorck, I.M.E., 2001. Nutritional properties of starch in buckwheat products: studies in vitro and in vivo. J. Agric. Food Chem. 49, 490–496.

Skrabanja, V., Kreft, I., Golob, T., Modic, M., Ikeda, S., Ikeda, K., Kreft, S., Bonafaccia, G., Knapp, M., Kosmelj, K., 2004. Nutrients content in buckwheat milling fractions. Cereal Chem. 81, 172–176.

Stokić, E., Mandić, A., Sakać, M., Misan, A., Pestorić, M., Simurina, O., Jambrec, D., Jovanov, P., Nedeljković, N., Milovanović, I., Sedej, I., 2015. Quality of buckwheat-enriched wheat bread and its antihyperlipidemic effect in statin treated patients. LWT—Food Sci. Technol. 63, 556–561.

Takahama, U., Hirota, S., 2010. Fatty acids, epicatechin-dimethylgallate, and rutin interact with buckwheat starch inhibiting its digestion by amylase: implications for the decrease in glycemic index by buckwheat flour. J. Agric. Food Chem. 58, 12431–12439.

Takahama, U., Hirota, S., 2011. Inhibition of buckwheat starch digestion by the formation of starch/bile salt complexes: possibility of its occurrence in the intestine. J. Agric. Food Chem. 59, 6277–6283.

Tomotake, H., Shimaoka, I., Kayashita, J., Yokoyama, F., Nakajoh, M., Kato, N., 2000. A buckwheat protein product suppresses gallstone formation and plasma cholesterol more strongly than soy protein isolate in hamsters. J. Nutr. 130, 1670–1674.

Tufvesson, F., Škrabanja, V., Björck, I., Liljeberg Elmståhl, H., Eliasson, A.-C., 2001. Digestibility of starch systems containing amylose-glycerol monopalmitin complexes. Lebensm. Wiss. Technol. 34, 131–139.

Vombergar, B., Kreft, I., Horvat, M., Vorih, S., 2014. Ajda = Buckwheat. Kmečki glas, Ljubljana.

Wei, Y., Hu, X., Zhang, G., Ouyang, S., 2003. Studies on the amino acid and mineral content of buckwheat protein fractions. Nahrung/Food 47, 114–116.

Wieslander, G., Norbäck, D., Wang, Z.H., Zhang, Z., Mi, Y.H., Lin, R.F., 2000. Buckwheat allergy and reports on asthma and atopic disorders in Taiyuan city, Northern China. Asian Pacific J. Allerg. Immunol. 18, 147–152.

Wieslander, G., Fabjan, N., Vogrinčič, M., Kreft, I., Janson, C., Spetz-Nystrom, U., Vombergar, B., Tagesson, C., Leanderson, P., Norbäck, D., 2011. Eating cookies is associated with the reduction in serum levels of myeloperoxidase and cholesterol: a double blind crossover study in day-care centre staffs. Tohoku J. Exp. Med. 225, 123–130.

Wieslander, G., Fabjan, N., Vogrinčič, M., Kreft, I., Vombergar, B., Norbäck, D., 2012. Effects of common and Tartary buckwheat consumption on mucosal symptoms, headache and tiredness: a double-blind crossover intervention study. Int. J. Food Agric. Environ.—JFAE 10, 107–110.

Yamada, K., Urisu, A., Morita, Y., Kondo, Y., Wada, E., Komada, H., Yamada, M., Inagaki, Y., Torii, S., 1995. Immediate hypersensitive reactions to buckwheat ingestion and cross allergenicity between buckwheat and rice antigens in subjects with high-levels of IGE antibodies to buckwheat. Ann. Allerg. Asthma Immunol. 75, 56–61.

Chapter fourteen

Nutritional Aspects of Buckwheat in the Czech Republic

D. Janovská, P. Hlásná Čepková

Gene Bank, Crop Research Institute, Prague, Ruzyně, Czech Republic

INTRODUCTION

History of Buckwheat in the Czech Territory

Common buckwheat is originally from China; however, it was evidently brought to Europe by Ottoman Turk invaders. In many European languages common buckwheat is considered a wheat from pagans or brought by pagans (trigo sarraceno, sarracin, grano saraceno, pohanka, tatarka, etc.) (Petr, 1995). In the Czech territory the first remains of common buckwheat pollen were documented from the turn of the 9th century in Prague (Čulíková, 1998). The oldest archeological findings of grains were found in Opava and they came from the 12th century (Petr, 1995). The first written record is dated to 1365. In 1562, Italian botanist Pietro Andrea Mattioli in his Commentaries on the *Materia Medica* of Dioscorides wrote that: "in Bohemia there is plenty of common buckwheat, it is grown for home and ordinary meal, and I have heard that they make bread from it." The biggest boom in common buckwheat growing was during the 16th and 17th centuries in Těšín territory and in Moravian Wallachia. The popularity of common buckwheat had been declining since the 18th century because of changes in diet, preference for pastries from the white wheat flour, and the import of cheap rice from Asia. The only region in the Czech territory where common buckwheat remained as a national meal until the beginning of the 20th century was Moravian Wallachia. Buckwheat cultivation has great popularity and tradition in this region of the Czech Republic. There are records that in the Nový Hrozenkov region, daily family consumption was about 2 kg, mainly as kasha. In 1920, common buckwheat was grown on 3046 ha and the yield was 0.79 t per ha. The area of common buckwheat declined and in 1988 (Figs. 14.1 and 14.2) it was only 2 ha in the former Czechoslovakia (FAOStat, 2015).

177

Molecular Breeding and Nutritional Aspects of Buckwheat. http://dx.doi.org/10.1016/B978-0-12-803692-1.00014-6

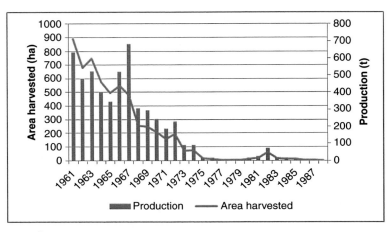

FIGURE 14.1 Area harvested and production of common buckwheat in the former Czechoslovakia.

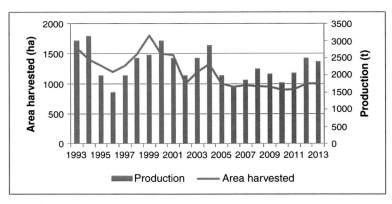

FIGURE 14.2 Area harvested and production of common buckwheat in the Czech Republic.

The renaissance of utilization and growing of common buckwheat started in the 1990s because of its nutritional aspects and relatively low demands on growing conditions. Nowadays, common buckwheat is one of the main cash crops in Czech organic production. In 2013 the amount of common buckwheat exported was 479,000 tons and 547,000 tons were imported, mainly from Ukraine, China, and Poland. The majority of buckwheat processing involves mostly small-scale production and manufacturing.

Tartary buckwheat is from China as well; however, it was not grown in the Czech territory on a wider scale. There are only a few mentions of Tartary buckwheat cultivation from the 18th and 19th centuries from the mountainous border between Slovakia and Moravia (the Czech Republic consists of three parts—Bohemia, Moravia, and Silesia). This region was inhabited by very poor people and buckwheat was important, mainly in the years when there were crop failures. They used the Tartary buckwheat grains for flour production and added it to bread dough (Petr, 1995). Nowadays, Tartary buckwheat is being assessed as a potential new food source in the Czech Republic.

Nutritional Quality of Buckwheat in the Czech Republic

Common buckwheat has important nutritional qualities in the Czech Republic nowadays. Part of the Czech population, as well as in other counties in Europe, prefers a gluten-free diet either because of celiac disease or just because they believe that gluten-free food is healthier. The renaissance of common buckwheat consumption is mainly caused by a desire to return to traditional crops.

The quality of common buckwheat is important as well. In most countries, as well as in the Czech Republic, buckwheat is not eaten purely to satiate people's hunger and for food energy. It is eaten because of its tradition, its taste, its diversity on the menu, and because of knowledge of the importance of buckwheat food products for human health. These demands and reasons for consuming buckwheat may be met solely by buckwheat products of the highest possible quality. They have a typical and clearly pronounced taste, and their nutrients are important for human health and free from artificial chemicals (Kreft, 2001).

The excellent nutritional value of common buckwheat is conditioned by favorable composition of the protein complex with well-balanced amino acid composition, fiber, minerals, fat, vitamins, and bioflavonoids such as rutin (Kalinová et al., 2007; Kalinová and Dadáková, 2006; Kalinova and Vrchotová, 2011; Michalová et al., 1998); these attributes have been the subject of agricultural research and studies in the Czech Republic since the 1990s (Michalová and Hutař, 1998). From the published articles the average crude protein content oscillates around 12%, which is comparable with other cereals (Eggum et al., 1980; Saturni et al., 2010). The average content of crude protein of 20 accessions tested in the Czech gene bank (Crop Research Institute) was 13.64% during a 3-year experiment (Table 14.1). Krkošková and Mrázová (2005) published the average protein content in buckwheat seeds in a range from 8.51% to 18.87%. The protein content is dependent on variety and on yearly conditions (Dvořáček and Čepková, 2004). In the case of Tartary buckwheat, our data (unpublished) showed the level of total protein around 10.4 g/100 g in grains, which is less than in common buckwheat (Campbell, 1997).

Protein fractionation, according to the Osborne classification for buckwheat flour, has been reported by many authors (Aufhammer, 2000; Javornik, 1981; Bonafaccia et al., 1994; Wei et al., 2003). Buckwheat varieties cultivated in the Czech Republic have the following composition of protein fractions: as a major fraction albumins and globulins 47.44% and glutelins 50.80% with the lowest proportions of prolamins 1.78%, respectively (Čepková, 2006). Because of the low content of prolamins, common buckwheat is generally considered as an important gluten-free pseudocereal with a large potential for highly nutritious gluten-free products (Giménez-Bastida et al., 2015). Common buckwheat proteins possess well-balanced amino acid compositions, as shown in Table 14.2. Data

TABLE 14.1 Grain Composition of Common Buckwheat Accessions Cultivated in the Czech Republic

Variety	Crude Protein[a] (%)	Starch[a] (%)	Oil[a] (%)	P[a] (%)	K[a,b] (%)	Ca[a,b] (%)	Mg[a] (%)	Zn[a] (mg/kg)	Fe[a] (mg/kg)	Rutin (%)		Vitamins (%)[c]		
										Dried Leaves and Stem[a]	Seeds[a]	B1	B2	B3
Aelita	13.39 ± 2.34	61.45 ± 4.03	2.33 ± 0.35	0.36 ± 0.03	0.50 ± 0.04	0.09 ± 0.01	0.22 ± 0.01	45.00 ± 4.24	77.00 ± 28.28	2.85 ± 1.18	0.02 ± 0.001	0.27	0.12	0.90
Astoriya	14.13 ± 0.75	60.85 ± 1.91	2.37 ± 0.37	0.37 ± 0.01	0.50 ± 0.04	0.09 ± 0.01	0.22 ± 0.001	43.50 ± 3.54	94.00 ± 2.28	2.92 ± 1.52	0.019 ± 0.004	0.26	0.11	3.19
Chernoplodnaya	13.75 ± 2.17	59.10 ± 1.13	2.44 ± 0.29	0.38 ± 0.02	0.52 ± 0.04	0.09 ± 0.01	0.23 ± 0.01	43.00 ± 2.83	81.50 ± 26.16	2.78 ± 1.41	0.02 ± 0.001	0.22	0.10	1.20
Monori	13.60 ± 0.61	58.65 ± 2.90	2.35 ± 0.28	0.36 ± 0.02	0.51 ± 0.04	0.11 ± 0.03	0.22 ± 0.004	41.50 ± 2.12	77.50 ± 12.02	2.84 ± 1.47	0.02 ± 0.01	0.22	0.10	1.62
Pyra	14.27 ± 0.49	58.70 ± 2.97	2.30 ± 0.17	0.35 ± 0.05	0.52 ± 0.05	0.09 ± 0.01	0.23 ± 0.00	42.5 ± 2.12	91.00 ± 0.01	2.84 ± 1.52	0.02 ± 0.001	0.23	0.11	2.54
0125000024	13.45 ± 1.00	57.75 ± 5.87	2.41 ± 0.45	0.37 ± 0.01	0.50 ± 0.06	0.12 ± 0.03	0.22 ± 0.004	43.5 ± 4.95	87.00 ± 5.66	2.94 ± 1.59	0.02 ± 0.01	0.25	0.11	1.30
Tetraharpe	13.33 ± 0.77	54.50 ± 0.42	2.66 ± 0.15	0.35 ± 0.03	0.57 ± 0.07	0.12 ± 0.02	0.21 ± 0.01	44.00 ± 0.01	88.00 ± 8.49	2.57 ± 1.58	0.02 ± 0.004	0.29	0.12	3.14
Yaita Zairai	13.59 ± 1.38	57.15 ± 2.19	2.52 ± 0.04	0.33 ± 0.02	0.55 ± 0.06	0.11 ± 0.02	0.21 ± 0.01	42.50 ± 0.71	91.00 ± 48.08	2.47 ± 1.70	0.02 ± 0.003	0.30	0.12	1.81
Stoyoama Zairai	14.19 ± 1.02	57.40 ± 1.13	2.59 ± 0.37	0.37 ± 0.02	0.54 ± 0.03	0.11 ± 0.02	0.23 ± 0.003	43.5 ± 0.71	93.50 ± 51.62	2.71 ± 1.65	0.02 ± 0.004	0.25	0.12	1.48

Botansoba	13.45 ± 1.00	57.80 ± 1.70	2.35 ± 0.10	0.36 ± 0.03	0.57 ± 0.04	0.10 ± 0.001	0.22 ± 0.005	45.5 ± 0.71	77.50 ± 12.02	2.15 ± 1.55	0.02 ± 0.004	0.26	0.12	1.89
0125000056	13.97 ± 1.19	58.60 ± 0.42	2.35 ± 0.05	0.35 ± 0.02	0.54 ± 0.05	0.12 ± 0.01	0.23 ± 0.009	45.5 ± 0.71	90.00 ± 9.90	2.94 ± 1.97	0.02 ± 0.004	0.31	0.13	1.17
0125000057	14.54 ± 0.14	56.45 ± 7.42	2.40 ± 0.26	0.39 ± 00.03	0.50 ± 0.04	0.11 ± 0.02	0.24 ± 0.007	44.5 ± 2.12	94.50 ± 20.51	2.83 ± 1.89	0.02 ± 0.001	0.25	0.12	0.96
Balada	13.60 ± 1.31	61.55 ± 2.19	2.29 ± 0.14	0.35 ± 0.05	0.50 ± 0.03	0.10 ± 0.01	0.22 ± 0.02	44.00 ± 2.83	78.50 ± 2.12	3.35 ± 1.97	0.02 ± 0.001	0.25	0.10	1.74
Krupinka	13.22 ± 1.17	60.70 ± 2.83	2.34 ± 0.20	0.36 ± 0.02	0.50 ± 0.06	0.09 ± 0.01	0.22 ± 0.01	42.5 ± 0.71	77.00 ± 19.80	3.10 ± 1.72	0.02 ± 0.002	0.28	0.11	1.15
Sumčanka	13.79 ± 0.79	60.05 ± 0.64	2.13 ± 0.18	0.36 ± 0.02	0.49 ± 0.06	0.09 ± 0.01	0.22 ± 0.01	47.00 ± 1.41	91.00 ± 19.09	2.81 ± 1.48	0.02 ± 0.001	0.22	0.11	1.40
Skorospelaya	13.84 ± 1.72	59.70 ± 1.70	2.37 ± 0.50	0.37 ± 0.02	0.52 ± 0.02	0.11 ± 0.02	0.22 ± 0.01	43.00 ± 1.41	81.50 ± 6.36	2.93 ± 161	0.02 ± 0.001	0.22	0.11	0.94
Bolševik	13.06 ± 2.27	55.25 ± 1.20	2.28 ± 0.14	0.34 ± 0.05	0.50 ± 0.03	0.12 ± 0.003	0.22 ± 0.01	44.50 ± 0.71	95.50 ± 45.96	3.34 ± 1.60	0.02 ± 001	0.30	0.11	1.80
Le Harpe	13.14 ± 1.46	58.50 ± 0.85	2.58 ± 0.34	0.37 ± 0.01	0.51 ± 0.03	0.14 ± 0.01	0.23 ± 0.01	43.50 ± 2.12	70.00 ± 22.63	2.70 ± 1.15	0.03 ± 0.01	0.23	0.11	1.40
Hruzsowska	13.82 ± 1.02	61.30 ± 3.82	2.41 ± 0.38	0.35 ± 0.04	0.48 ± 0.05	0.09 ± 0.002	0.21 ± 0.02	40.50 ± 0.71	99.50 ± 47.38	2.64 ± 1.08	0.02 ± 0.001	0.22	0.09	1.10
Prego	14.35 ± 1.43	60.65 ± 2.05	2.30 ± 2.30	0.38 ± 0.03	0.49 ± 0.05	0.09 ± 0.01	0.22 ± 0.01	41.50 ± 0.71	103.00 ± 8.49	2.59 ± 0.95	0.02 ± 0.003	0.24	0.10	1.32
Mean	13.67 ± 1.18	58.81 ± 2.91	2.39 ± 0.23	0.36 ± 0.03	0.51 ± 0.04	0.11 ± 0.02	0.22 ± 0.01	43.62 ± 2.18	86.25 ± 20.89	2.82 ± 1.31	0.02 ± 0.003	0.25	0.11	1.6

[a] Variability within years.
[b] Variability among varieties.
[c] 1 year experiment.

TABLE 14.2 Amino Acid Composition (in %) of Common Buckwheat Accessions Cultivated in the Czech Republic

Variety	Aspartic Acid[a]	Threonine[a]	Serine[a]	Glutamic Acid[a,b]	Proline[a]	Glycine[a]	Alanine[a]	Cysteine[a]	Valine[a]	Methionine[a,b]	Isoleucine[a]	Leucine[a]	Tyrosine[a]	Phenylalanine[a]	Histidine[a]	Lysine[a]	Arginine[a,b]
Aelita	18.77 ± 5.33	5.68 ± 1.90	6.27 ± 1.28	25.19 ± 8.39	7.55 ± 2.73	9.05 ± 2.22	6.69 ± 1.59	4.29 ± 2.06	6.47 ± 3.27	2.07 ± 0.32	4.82 ± 2.40	9.94 ± 2.92	3.59 ± 1.16	7.62 ± 1.36	3.53 ± 1.04	8.67 ± 2.82	17.64 ± 8.73
Astoriya	18.22 ± 5.26	6.07 ± 2.03	6.39 ± 1.46	25.07 ± 6.99	8.00 ± 3.75	9.07 ± 2.53	6.91 ± 1.95	4.06 ± 1.79	6.88 ± 3.02	2.18 ± 0.60	5.10 ± 2.22	10.30 ± 3.25	3.47 ± 1.47	7.16 ± 2.01	3.65 ± 1.17	8.97 ± 2.96	18.25 ± 8.36
Chernoplodnaya	18.62 ± 2.76	5.80 ± 1.01	6.02 ± 1.12	25.70 ± 5.11	7.57 ± 1.96	8.69 ± 1.34	6.70 ± 0.87	3.48 ± 0.96	6.72 ± 1.79	2.12 ± 0.37	4.98 ± 1.24	9.72 ± 1.55	3.85 ± 1.06	6.67 ± 0.73	3.37 ± 0.46	8.66 ± 1.72	15.89 ± 3.52
Monori	18.46 ± 6.11	5.45 ± 1.88	5.90 ± 1.34	22.42 ± 6.17	7.50 ± 3.46	8.59 ± 2.39	6.52 ± 1.79	3.46 ± 2.53	6.39 ± 3.37	1.87 ± 0.44	4.71 ± 2.42	9.98 ± 3.24	3.78 ± 1.57	6.70 ± 1.80	3.38 ± 1.15	8.24 ± 2.94	17.61 ± 10.37
Pyra	17.68 ± 5.20	5.42 ± 1.72	5.48 ± 1.75	23.08 ± 7.25	6.87 ± 2.39	8.92 ± 1.42	6.31 ± 1.88	3.25 ± 1.95	6.71 ± 2.62	1.77 ± 0.47	4.94 ± 1.84	9.28 ± 2.77	3.39 ± 1.01	6.59 ± 2.16	3.21 ± 1.27	9.38 ± 0.93	16.61 ± 7.32
0125000024	19.02 ± 3.69	5.53 ± 1.55	5.64 ± 1.37	22.99 ± 3.97	7.23 ± 2.71	8.78 ± 1.72	6.70 ± 1.29	3.49 ± 2.67	6.37 ± 2.73	1.73 ± 0.58	4.77 ± 2.05	9.74 ± 2.10	2.98 ± 1.66	7.05 ± 1.43	3.36 ± 0.93	8.38 ± 2.18	16.98 ± 8.51
Tetraharpe	19.42 ± 2.93	5.92 ± 1.91	5.97 ± 1.63	24.08 ± 4.18	8.07 ± 3.68	9.09 ± 1.92	7.80 ± 0.81	3.84 ± 1.60	6.68 ± 2.90	2.72 ± 0.99	4.81 ± 2.33	12.12 ± 0.16	2.74 ± 2.04	7.61 ± 0.84	3.62 ± 1.05	9.25 ± 1.96	17.39 ± 8.50
Yaita Zairai	18.40 ± 3.81	5.37 ± 1.50	5.37 ± 1.53	20.43 ± 10.19	7.22 ± 2.46	8.60 ± 0.77	6.36 ± 1.40	2.65 ± 1.69	6.57 ± 2.29	1.57 ± 0.32	4.87 ± 1.62	9.39 ± 1.90	3.16 ± 1.35	6.28 ± 1.98	2.90 ± 1.04	9.15 ± 0.89	14.65 ± 4.56
Stoyoama Zairai	20.17 ± 5.00	5.57 ± 2.00	5.74 ± 1.73	23.72 ± 6.25	8.03 ± 3.05	9.59 ± 1.15	6.93 ± 1.63	3.38 ± 2.60	6.82 ± 3.17	1.70 ± 0.53	5.14 ± 2.30	10.16 ± 2.75	4.04 ± 1.03	6.64 ± 2.46	3.30 ± 1.38	9.58 ± 1.30	18.23 ± 10.18
Botansoba	19.22 ± 3.43	5.63 ± 1.48	5.81 ± 1.49	24.92 ± 5.64	7.23 ± 2.04	8.90 ± 1.56	6.94 ± 1.28	3.50 ± 1.78	6.70 ± 2.70	2.28 ± 0.31	4.88 ± 2.04	10.07 ± 1.84	3.43 ± 1.75	6.82 ± 1.17	3.48 ± 0.88	8.79 ± 2.35	17.00 ± 7.14

0125000056	19.01 ± 3.48	5.76 ± 1.52	5.98 ± 1.77	24.11 ± 5.73	6.77 ± 1.41	9.13 ± 0.56	6.64 ± 1.37	3.21 ± 1.25	6.76 ± 2.16	1.35 ± 0.53	5.16 ± 1.61	9.87 ± 1.64	3.74 ± 0.56	6.64 ± 1.77	3.23 ± 1.21	9.57 ± 0.59	16.90 ± 6.79
0125000057	16.57 ± 6.64	5.54 ± 2.72	6.25 ± 3.33	22.64 ± 10.34	6.79 ± 3.15	8.53 ± 2.67	6.18 ± 2.70	3.22 ± 2.88	6.44 ± 3.32	1.00 ± 0.54	4.88 ± 2.51	9.68 ± 3.95	3.07 ± 2.47	6.03 ± 2.66	3.05 ± 1.71	8.60 ± 2.85	16.14 ± 9.41
Balada	17.69 ± 6.77	5.52 ± 2.22	6.35 ± 2.62	22.49 ± 7.79	7.32 ± 3.17	8.42 ± 2.88	6.33 ± 2.17	3.68 ± 2.43	6.46 ± 3.25	1.18 ± 0.48	4.77 ± 2.33	9.04 ± 3.51	3.27 ± 1.50	6.61 ± 2.43	3.30 ± 1.41	8.24 ± 3.46	17.69 ± 10.82
Krupinka	21.08 ± 6.49	6.29 ± 1.99	6.58 ± 1.56	25.90 ± 7.10	8.55 ± 3.91	10.22 ± 1.58	7.18 ± 1.98	4.36 ± 2.38	7.36 ± 3.00	1.50 ± 0.66	5.43 ± 2.05	11.12 ± 2.32	3.17 ± 2.50	7.49 ± 2.11	3.12 ± 1.68	10.71 ± 1.75	21.40 ± 10.68
Sumčanka	17.51 ± 5.12	5.75 ± 1.61	6.47 ± 2.13	23.72 ± 6.21	7.67 ± 3.06	9.13 ± 1.26	6.54 ± 1.51	3.44 ± 1.50	6.47 ± 2.48	1.63 ± 0.93	4.63 ± 2.03	10.76 ± 0.82	3.30 ± 1.26	7.59 ± 0.32	3.39 ± 0.77	9.31 ± 0.87	16.96 ± 9.19
Skorospelaya	19.34 ± 5.81	5.78 ± 1.62	6.17 ± 1.43	24.08 ± 5.88	8.12 ± 3.05	9.44 ± 1.22	6.66 ± 1.65	3.69 ± 1.48	6.52 ± 2.37	1.57 ± 0.52	4.85 ± 1.76	9.92 ± 2.46	4.04 ± 0.65	6.53 ± 2.42	3.25 ± 1.36	9.43 ± 1.01	18.49 ± 8.72
Bolševik	18.37 ± 6.93	5.48 ± 2.03	6.20 ± 1.32	22.52 ± 7.29	7.78 ± 3.52	7.95 ± 2.53	6.63 ± 2.34	3.10 ± 2.03	6.41 ± 2.63	1.67 ± 0.45	4.78 ± 2.04	9.73 ± 3.65	3.38 ± 1.13	6.50 ± 2.41	3.12 ± 1.33	7.97 ± 3.01	16.25 ± 8.60
Le Harpe	19.20 ± 3.92	5.60 ± 1.48	6.04 ± 1.58	28.00 ± 8.74	7.79 ± 2.35	8.98 ± 2.12	7.41 ± 1.93	3.22 ± 1.22	7.10 ± 2.69	1.54 ± 0.75	5.27 ± 1.68	9.73 ± 2.45	3.41 ± 0.91	7.09 ± 1.14	3.77 ± 1.02	8.70 ± 2.27	18.70 ± 8.20
Hruzsowska	17.37 ± 2.98	4.99 ± 0.84	5.11 ± 1.27	27.00 ± 5.90	6.59 ± 1.12	8.82 ± 1.62	6.37 ± 1.79	2.92 ± 0.37	6.24 ± 1.42	1.44 ± 0.60	4.66 ± 0.95	9.04 ± 1.44	3.31 ± 0.07	6.01 ± 1.32	3.06 ± 1.06	8.67 ± 1.61	18.24 ± 7.15
Prego	17.07 ± 3.33	4.80 ± 1.03	5.06 ± 0.71	24.90 ± 6.93	6.79 ± 1.72	8.04 ± 1.19	6.32 ± 1.36	3.28 ± 0.76	5.82 ± 2.17	1.64 ± 0.34	4.30 ± 1.63	8.59 ± 1.74	2.89 ± 1.57	5.83 ± 0.92	3.11 ± 0.63	7.41 ± 1.61	17.53 ± 8.62
Mean	18.56 ± 4.21	5.59 ± 1.48	5.94 ± 1.49	24.14 ± 6.02	7.47 ± 2.40	8.90 ± 1.60	6.71 ± 1.47	3.46 ± 1.63	6.59 ± 2.26	1.73 ± 0.61	4.89 ± 1.66	9.91 ± 2.20	3.40 ± 1.25	6.77 ± 1.57	3.31 ± 0.99	8.88 ± 1.89	17.43 ± 7.10

[a]Variability within years.
[b]Variability among varieties.

show a considerable variability in the content of methionine and glutamic acid in tested collections. The amino acid composition is characterized by higher lysine content, amounting to about 8.88% on average, threonine (5.59%), and glutamic acid (24.14%). According to Karlubik et al. (1997) the percentage of essential amino acids of common buckwheat proteins is higher than in cereals (2.8–3.5% lysine) and is almost equal to the protein content in legumes (5.7–5.98% lysine). Comparing data on amino acid composition of Tartary and common buckwheat, our preliminary observations revealed almost similar levels of tested amino acids in both species; in Tartary buckwheat the content of tyrosine and leucine was even higher. This statement is consistent with the data of Bonafaccia et al. (2003b). The content of amino acids is not influenced by the processing of common buckwheat seed (Kreft et al., 1994), but it may be modified by locality conditions (Moudrý et al., 2005). Common buckwheat is recommended for increasing the lysine content in cereal products (Matuz et al., 2000).

The total content of fat in common buckwheat achene ranges from 1.5% to 3.7% with the highest concentration in the embryo (7–14%) and lowest in the seed coat (0.4–0.9%) (Campbell, 1997). Results of common buckwheat seed assessment in conditions in the Czech Republic had an average fat content of around 2.93% with slight oscillations across all tested years and accessions (minimum value 2.13% in variety "Sumčanka" and maximum value 2.66% in variety "Tetraharpe"). This result is lower than published by Steadman et al. (2001), who found 3.8% of fat in whole groats. From a health point of view, the content of unsaturated fatty acid is important. In the case of tested varieties the most abundant was linoleic acid (37.19–48.36%) together with linolenic acid (1.93–2.76%); from saturated acid the presence of palmitic fatty acid (11.26–15.86%) was observed (Table 14.3). Similarly, in the study of Bonafaccia et al. (2003b), unsaturated fatty acids prevailed. The fat content of common buckwheat is slightly higher compared to common wheat, even though it is characterized by a higher content of unsaturated fatty acids, which are considered more important for heart disease prevention (Saturni et al., 2010). The fat content in Tartary buckwheat accessions in the Czech Republic was 3.0% on average. Tartary buckwheat contained higher amounts of saturated fatty acids (about 3%) than common buckwheat.

The primary source of carbohydrates in common buckwheat is starch, and its content generally varies from 55% to 75% of the d.m. (Izydorczyk et al., 2014; Campbell, 1997; Steadman et al., 2001). Common buckwheat starch together with proteins significantly influences the textural characteristics. Starch quality is different in comparison with other cereals. Common buckwheat accessions tested in a 3-year experiment under conditions in the Czech Republic showed a range of starch content (57.75–61.55%), which is consistent with the results of other studies. In Tartary buckwheat flour a higher content of total starch, amylose, and resistant starch was detected than in common buckwheat flour (Qin et al., 2010).

Generally, common buckwheat is also considered as a rich source of dietary fiber (Steadman et al., 2001). The fiber content in common buckwheat is higher

(Continued)

TABLE 14.3 Fatty Acid Composition (in %) of Common Buckwheat Accessions Cultivated in the Czech Republic

Variety	Oleic[a]	Palmitic[a]	Palmitoleic[a]	Stearic	Oleic[a,b]	Linoleic[a,b]	Linolenic	Arachic[a]	Eicoseionic[a]
Aelita	2.64 ± 0.59	15.86 ± 2.74	0.56 ± 0.49	2.06 ± 0.07	32.71 ± 7.71	42.49 ± 6.50	1.93 ± 0.66	1.90 ± 0.70	2.53 ± 0.89
Astoriya	2.77 ± 0.74	14.99 ± 5.51	0.62 ± 0.61	2.80 ± 0.87	27.73 ± 11.77	48.36 ± 8.02	1.97 ± 0.38	1.22 ± 0.57	2.38 ± 0.89
Chernoplodnaya	2.61 ± 0.37	14.67 ± 4.06	0.44 ± 0.24	2.78 ± 1.30	31.62 ± 4.61	44.73 ± 2.57	2.14 ± 0.92	1.35 ± 0.61	2.28 ± 1.80
Monori	2.38 ± 0.20	14.80 ± 4.23	0.49 ± 0.33	2.52 ± 1.24	32.68 ± 6.06	43.22 ± 3.93	2.17 ± 1.03	1.46 ± 0.71	2.68 ± 2.09
Pyra	2.73 ± 075	14.98 ± 3.69	0.36 ± 0.45	2.63 ± 1.46	33.85 ± 9.00	42.04 ± 5.47	2.12 ± 0.77	1.32 ± 0.60	2.71 ± 0.85
0125000024	2.50 ± 0.36	13.37 ± 0.99	0.25 ± 0.03	1.72 ± 0.15	38.33 ± 0.97	39.46 ± 1.17	2.51 ± 0.26	1.54 ± 0.34	3.48 ± 0.63
Tetraharpe	2.66 ± 0.10	13.66 ± 1.05	0.36 ± 0.27	1.71 ± 0.16	38.38 ± 2.22	39.21 ± 1.01	2.76 ± 0.50	1.58 ± 0.37	2.90 ± 1.04
Yaita Zairai	2.51 ± 0.03	15.18 ± 1.17	0.32 ± 0.08	1.60 ± 0.25	39.27 ± 2.18	37.19 ± 2.60	1.94 ± 0.51	1.67 ± 0.09	2.83 ± 1.18
Stoyoama Zairai	3.12 ± 0.97	14.73 ± 0.83	0.55 ± 0.22	1.57 ± 0.22	38.00 ± 1.91	38.22 ± 2.46	2.20 ± 0.12	1.79 ± 0.24	2.97 ± 1.19
Botansoba	2.33 ± 0.08	14.11 ± 1.21	0.35 ± 0.08	1.69 ± 0.21	36.79 ± 1.96	39.42 ± 1.63	2.48 ± 0.04	1.48 ± 0.27	3.67 ± 2.27
0125000056	2.59 ± 0.43	13.95 ± 0.47	0.35 ± 0.23	1.69 ± 0.23	37.71 ± 2.75	39.31 ± 2.00	2.41 ± 0.85	1.39 ± .36	3.19 ± 0.53
0125000057	2.50 ± 0.26	13.49 ± 1.81	0.34 ± 0.22	1.61 ± 0.28	38.65 ± 1.07	39.61 ± 0.48	2.45 ± 0.40	1.06 ± 0.62	2.76 ± 0.73
Balada	2.34 ± 0.14	13.32 ± 1.69	0.36 ± 0.11	1.84 ± 0.39	40.51 ± 3.57	36.53 ± 3.59	2.30 ± 0.45	1.52 ± 0.50	3.61 ± 1.29

TABLE 14.3 Fatty Acid Composition (in %) of Common Buckwheat Accessions Cultivated in the Czech Republic (cont.)

Variety	Oleic[a]	Palmitic[a]	Palmitoleic[a]	Stearic	Oleic[a,b]	Linoleic[a,b]	Linolenic	Arachic[a]	Eicoseionic[a]
Krupinka	3.00 ± 1.15	13.09 ± 1.42	0.29 ± 0.28	1.71 ± 0.18	38.61 ± 2.38	39.15 ± 1.55	2.53 ± 0.22	1.52 ± 0.42	3.11 ± 0.59
Šumčanka	2.44 ± 0.42	13.19 ± 0.14	0.38 ± 0.25	1.67 ± 0.20	38.70 ± 2.04	39.27 ± 1.91	2.60 ± 0.16	1.41 ± 0.54	3.12 ± 0.59
Skorospelaya	2.67 ± 0.62	13.60 ± 0.18	0.44 ± 0.14	1.70 ± 0.14	37.63 ± 0.77	40.03 ± 0.50	2.32 ± 0.40	1.19 ± 0.54	3.10 ± 0.47
Bolševik	2.82 ± 0.93	11.26 ± 0.69	0.27 ± 0.12	1.57 ± 0.21	42.24 ± 0.85	38.03 ± 0.82	2.20 ± 0.19	1.13 ± 0.96	3.30 ± 0.68
Le Harpe	2.54 ± 0.25	14.01 ± 0.66	0.31 ± 0.10	1.67 ± 0.18	34.34 ± 0.85	40.98 ± 2.31	2.70 ± 0.20	1.39 ± 0.36	4.27 ± 1.63
Hruzsowska	2.79 ± 0.71	13.25 ± 1.01	0.24 ± 0.23	1.76 ± 0.15	37.55 ± 1.05	39.93 ± 1.03	2.37 ± 0.33	1.59 ± 0.27	3.32 ± 0.54
Prego	2.33 ± 0.16	12.70 ± 0.92	0.30 ± 0.09	1.75 ± 0.07	38.04 ± 0.75	39.23 ± 0.80	2.34 ± 0.24	1.75 ± 0.37	3.78 ± 0.50
Mean	2.61 ± 0.51	13.91 ± 2.13	0.38 ± 0.25	1.90 ± 0.63	36.67 ± 4.98	40.32 ± 3.78	2.32 ± 0.48	1.46 ± 0.48	3.10 ± 1.08

[a]Variability within years.
[b]Variability among varieties.

than in regular cereals (Saturni et al., 2010). Grubben and Parthohardjono (1996) presented the content of fiber in whole seeds in the range 10–11%; however, the fiber content oscillates depending on the milling products used (Bonafaccia et al., 2003b). Izydorczyk et al. (2014) indicated lower water solubility of dietary fiber (28–36%) with the major proportion as water-insoluble fiber. The dietary fiber was considered to have a positive effect on blood serum, lowered cholesterol content and blood sugars, decreased the risk of heart attack and body weight, and increased stool weight (Slavin, 2013; Kendall et al., 2010). The same trend suggested evaluation of Tartary buckwheat accessions where the content of fiber was higher in comparison with common buckwheat (unpublished data).

Common buckwheat is also seen as a source of vitamins and minerals. In the Czech Republic the vitamin content assessment showed that evaluated common buckwheat groats contained 0.25% vitamin B1 (thiamin), 0.11% vitamin B2 (riboflavin), and 2.95% vitamin B3 on average. The results obtained corresponded with published contents of vitamins by Bonafaccia et al. (2003b). In the case of Tartary buckwheat grains, vitamins B1, B2, and B3 were represented in minor amounts in comparison with common buckwheat (unpublished data).

Common buckwheat is a good source of several minerals. The mineral content varied in the range 2.0–2.5%. Its content is affected by processing as well as by seasonal conditions. Except for calcium, common buckwheat is a richer source of nutritionally important minerals than many cereals such as rice, sorghum, millet, and maize (Saturni et al., 2010; Bonafaccia et al., 2003a). On the other hand, the content of calcium was observed as higher in Tartary buckwheat accessions cultivated in the Czech Republic (unpublished data). Common buckwheat might be an important source of zinc, copper, selenium, manganese, and other trace elements (Table 14.1). Various cereal fractions may have had different mineral contents. Dark buckwheat flour is much more interesting from this point of view (Michalová and Hutař, 1998; Michalová, 1998).

Buckwheat as a Functional Food

Common and Tartary buckwheat are very good sources of antioxidants. Many researchers have highlighted the presence of various phytochemicals present in seeds (Kreft et al., 2006; Jiang et al., 2007; Lee et al., 2016; Kiprovski et al., 2015), flour (Hung and Morita, 2008), leaves, and sprouts (Hsu et al., 2008; Kim et al., 2008). The most important component with antioxidative property is rutin. Rutin has several pharmacological features such as antioxidant, antiinflammatory, antiallergic, antiviral, anticarcinogenic, etc. (Kamalakkannan and Prince, 2006; Bishnoi et al., 2007). It was reported that rutin supplementation from common buckwheat food product sources, such as soba noodles or groats, might improve memory impairment and decrease hippocampal pyramidal neuronal death, such as in Alzheimer's disease. Rutin has the ability to suppress microglial activation and proinflammatory cytokines, it has a neuroprotective effect on different animal models of neurodegeneration, and prevents cognitive impairments by ameliorating oxidative stress and neuroinflammation of sporadic dementia of Alzheimer's type (Koda et al., 2009; Khan et al., 2009; Javed et al., 2012). The rutin content in common and Tartary buckwheat in the conditions of the Czech Republic was assessed in different parts of the plants. A higher rutin content was found in Tartary buckwheat. The differences were determined in the different organs as well. The highest rutin content was found in the leaves and the lowest in the stems in both species. The antioxidant activity of common and Tartary buckwheat leaves was found superior to antioxidant activities of buckwheat seeds, dehulled seeds, and stems and hulls in model samples containing 20% of tested material. The leaves had more than triple the antioxidant activity when compared with seeds, whereas the straws and hulls had lower antioxidant activities than seeds (Holasova et al., 2002; Janovská et al., 2010; Kalinová and Dadáková, 2013).

Utilization of Common Buckwheat as a Food

From the current range of products, common buckwheat is mostly consumed as kasha or dehulled grains (buckwheat groats). Buckwheat noodles are popular in Asia (Ikeda et al., 2003); however, they are becoming popular with Czech consumers as well. In the Czech gene bank, an assessment of noodle preparation with the addition of common buckwheat dried leaves flour (0.5, 1.0, 2.0, and 3.0%) was carried out. The flour was a mixture of common wheat and common buckwheat flour in the ratio 3:7. The results showed that the addition of dried leaves flour improved the antioxidant capacity and rutin content in the noodles. The evaluation of the raw material, raw noodles, and cooked noodles as well as of the water used for boiling noodles showed that the biggest impact on rutin content and antioxidant activity was possessed by dough that was kneaded rather than dough that was boiled. The losses of rutin content to the boiling water were from 10% to 20%. The common buckwheat leaves are also used for the preparation of infusion.

The Milling of Buckwheat

Milling is one of the most common ways of processing buckwheat, which results in various types of products. Dehulling prior to milling generates whole groats (dehulled achenes) and hulls. Traditionally, milling is done with milling stones. Another processing method used is thermic dehulling, but it can influence the sensory and biochemical changes of the product (Moudrý et al., 2005). Common buckwheat seeds contain 26–28% of hulls and the yield of groats should be around 75% in quality varieties. The yield of groats in Tartary buckwheat is less because of hard cutting of the seed coat from the endosperm (Moudrý et al., 2005). Common buckwheat seed milling fractions consist mainly of the endosperm and embryo containing most of the proteins, starch, lipids, and minerals, and are used for preparations of various types of flour (Steadman et al., 2001).

Utilization of Common Buckwheat as Sprouts and Microgreens

Sprouts are sprouted seeds of vegetables or herbs, which are eaten directly or as an addition to salads or other meals. Microgreens became popular in

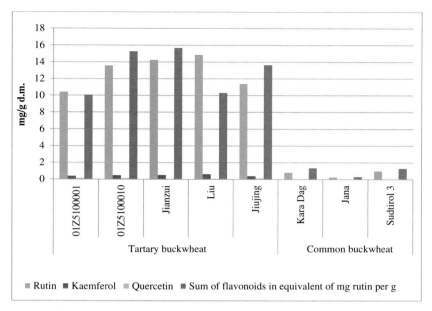

FIGURE 14.3 Flavonoids in microgreens of common and tartary buckwheat.

the mid-1990s in California, USA. Although there is no precise definition of microgreens, many publications show microgreens as a very specific type of vegetable produced from vegetable or herb seeds, very similar to sprouts but grown for longer and, unlike sprouts, grown under light conditions. They are planted 7–10 days before or until the first pair of true leaves appear. They are served as a vegetable and as a "topping" for salads, soups or sandwiches, etc. for enhancing the color and flavor (Brentlinger, 2007; Treadwell et al., 2010; Xiao, 2013). Janovská et al. (2010) referred to the higher content of flavonoids in microgreens of Tartary buckwheat than in common buckwheat and the significant content of phenolic acids present in both species of buckwheat (Fig. 14.3).

CONCLUSIONS

All the information mentioned in this chapter provides a basic overview of current research, production, and potential use of common and Tartary buckwheat, with an emphasis on the dietary benefits for human health in the Czech Republic. Both buckwheat species are gluten free and they have great potential for human nutrition.

ACKNOWLEDGEMENT

This work was supported by the research project RO0415 and by the National Programme on Conservation and Utilization of Plant Genetic Resources and Agro-biodiversity of the Ministry of Agriculture of the Czech Republic.

REFERENCES

Aufhammer, W., 2000. Pseudogetreidearten—Buchweizen, Reismelde und Amarant. Herkunft, Nutzung und Anbau. Stuttgart. Ulmer. p. 262.

Bishnoi, M., Chopra, K., Kulkarni, S.K., 2007. Protective effect of rutin, a polyphenolic flavonoid against haloperidol-induced orofacial dyskinesia and associated behavioural, biochemical and neurochemical ganges. Fundam Clin Pharmacol. 21 (5), 521–552.

Bonafaccia, G., Acquistucci, R., Luthar, Z., 1994. Proximate chemical composition and protein characterization of the buckwheat cultivated in Italy. Fagopyrum 14, 43–48.

Bonafaccia, G., Gambelli, L., Fabjan, N., Kreft, I., 2003a. Trace elements in flour and bran from common and Tartary buckwheat. Food Chem. 83, 1–5.

Bonafaccia, G., Marocchini, M., Kreft, I., 2003b. Composition and technological properties of the flour and bran from common and Tartary buckwheat. Food Chem. 80, 9–15.

Brentlinger, D.J., 2007. New trends in hydroponic crop production in the US. In: Chow, K.K. (Eds.), Proceedings of the International Conference & Exhibition on Soilless Culture Book Series, Singapore, September 5–8, 2005. Acta Hort. pp. 31–33.

Campbell, C.G., 1997. Buckwheat. *Fagopyrum esculentum* Moench. Promoting the conservation and use of underutilized and neglected crops. 19. Institute of Plant Genetics and Crop Plant Research, Gatersleben/International Plant Genetic Institute, Rome, Italy, p. 93.

Čepková, P., 2006. Studium genofondu pohanky seté a pohanky tatarské. Doctoral thesis. Czech University of Life Sciences Prague, p. 137 (in Czech).

Čulíková, V., 1998. Rostlinné makrozbytky z raně středověkých sedimentů na III. nádvoří Pražského hradu. Archaeologica Pragensia 14, 329–341, (in Czech).

Dvořáček, V., Čepková, P., M.A., 2004. Protein content evaluation of several buckwheat varieties. Advances in Buckwheat Research. Proceedings of the IX International Symposium on Buckwheat. Prague, August 18–22, Czech Republic, pp. 734–736.

Eggum, B.O., Kreft, I., Javornik, B., 1980. Chemical composition and protein quality of buckwheat (*Fagopyrum esculentum* Moench). Qual. Plantarum—Plant Foods Hum. Nutr. 30, 175–179.

FAOStat, 2015. Statistical database of the Food and Agricultural Organization of the United Nations. Available from: http://faostat.fao.org.

Giménez-Bastida, J.A., Piskuła, M., Zieliński, H., 2015. Recent advances in development of gluten-free buckwheat products. Trends Food Sci. Technol. 44, 58–65.

Grubben, G.J.H., Parthohardjono, S., 1996. *Fagopyrum esculentum* Moench. In: Plant Resources of South-East Asia, No. 10. Cereals. Backhuys Publisher, pp. 95–99.

Holasova, M., Fiedlerova, V., Smrcinova, H., Orsak, M., Lachman, J., Vavreinova, S., 2002. Buckwheat the source of antioxidant activity in functional foods. Food Res Intern. 35, 207–211.

Hsu, C.K., Chiang, B.H., Chen, Y.S., Yang, J.H., Liu, C.L., 2008. Improving the antioxidant activity of buckwheat (*Fagopyrum tataricm* Gaertn) sprout with trace element water. Food Chem. 108, 633–641.

Hung, P.V., Morita, N., 2008. Distribution of phenolic compounds in the graded flours milled from whole buckwheat grains and their antioxidant capacities. Food Chem. 109, 325–331.

Ikeda, K., Asami, Y., Lin, R., Arai, R., Honda, Y., Suzuki, T., Yasumuto, K., 2003. Comparison of mechanical and chemical characteristics between common and Tartary buckwheat. Fagopyrum 20, 53–58.

Izydorczyk, M.S., McMillan, T., Bazin, S., Kletke, J., Dushnicky, L., Dexter, J., 2014. Canadian buckwheat: a unique, useful and under-utilized crop. Can. J. Plant Sci. 94, 509–524.

Janovská, D., Štočková, L., Stehno, Z., 2010. Evaluation of buckwheat sprouts as microgreens. Acta Agric. Slov. 95, 157–162.

Javed, H., Khan, M.M., Ahmad, A., Vaibhav, K., Ahmad, M.E., Khan, A., Ashafaq, M., Islam, F., Siddiqui, M.S., Safhi, M.M., Islam, F., 2012. Rutin prevents cognitive impairments by ameliorating oxidative stress and neuroinflammation in rat model of sporadic dementia of Alzheimer type. Neuroscience 210, 340–352.

Javornik, B., Eggum, B.O., Kreft, I., 1981. Studies protein fractions and protein quality of buckwheat. Genetika 13 (2), 115–121.

Jiang, P., Burczynski, F., Campbell, C., Pierce, G., Austria, J.A., Briggs, C.J., 2007. Rutin and flavonoid contents in three buckwheat species *Fagopyrum esculentum, F. tataricum,* and *F. homotropicum* and their protective effects against lipid peroxidation. Food Res. Int. 40, 356–364.

Kalinová, J., Dadáková, E., 2006. Varietal and year variation of rutin content in common buckwheat (*Fagopyrum esculentum* Moench). Cereal Res. Commun. 34, 1315–1321.

Kalinová, J., Dadáková, E., 2013. Influence of sowing date and stand density on rutin level in buckwheat. Cereal Res. Commun. 41, 348–358.

Kalinova, J., Vrchotová, N., 2011. The influence of organic and conventional crop management, variety and year on the yield and flavonoid level in common buckwheat groats. Food Chem. 127, 602–608.

Kalinová, J., Tříska, J., Vrchotová, N., 2007. Distribution of vitamin E, squalene, epicatechin, and rutin in common buckwheat plants (*Fagopyrum esculentum* Moench). J. Agric. Food Chem. 54 (15), 5330–5335.

Kamalakkannan, N., Prince, P.S., 2006. Antihyperglycaemic and antioxidant effect of rutin, a polyphenolic flavonoid, in streptozotocin-induced diabetic wistar rats. Basic Clin Pharmacol Toxicol. 98 (1), 97–103.

Karlubik, M., Michalik, I., Urminska, D., 1997. Content of amino acids and the biological value of buckwheat grain proteins in comparison with other crops. Acta Zootech. 53, 97–105. (in Slovak)

Kendall, C.W.C., Esfahani, A., Jenkins, D.J.A., 2010. The link between dietary fibre and human health. Food Hydrocoll. 24, 42–48.

Khan, M.M., Ahmad, A., Ishrat, T., Khuwaja, G., Srivastawa, P., Khan, M.B., Raza, S.S., Javed, H., Vaibhav, K., Khan, A., Islam, F., 2009. Rutin protects the neural damage induced by transient focal ischemia in rats. Brain Res. 1292, 123–135.

Kim, S.J., Zaidul, I.S.M., Suzuki, T., Mukasa, Y., Hashimoto, N., Takigawa, S., Noda, T., Matsuura-Endo, C., Yamauchi, H., 2008. Comparison of phenolic compositions between common and Tartary buckwheat (*Fagopyrum*) sprouts. Food Chem. 110, 814–820.

Kiprovski, B., Mikulic-Petkovsek, M., Slatnar, A., Veberic, R., Stampar, F., Malencic, D., Latkovic, D., 2015. Comparison of phenolic profiles and antioxidant properties of European *Fagopyrum esculentum* cultivars. Food Chem. 185, 41–47.

Koda, T., Kuroda, Y., Imai, H., 2009. Rutin supplementation in the diet has protective effects against toxicant-induced hippocampal injury by suppression of microglial activation and pro-inflammatory cytokines: protective effect of rutin against toxicant-induced hippocampal injury. Cell Mol Neurobiol. 29, 523–531.

Kreft, I., 2001. Origin and present situation in buckwheat growing in Europe. Sborník referátů a posterů z odborné conference: Pěstování opomíjených a netradičních plodin v ČR. Výzkumný ústav rostlinné výroby Praha Ruzyně, pp. 19–23.

Kreft, I., Bonafaccia, G., Zigo, A., 1994. Secondary metabolites of buckwheat and their importance in human nutrition. Prehrambeno-tehnol. Biotehnol. Rev. 32, 195–197.

Kreft, I., Fabjan, N., Yasumoto, K., 2006. Rutin content in buckwheat (*Fagopyrum esculentum* Moench) food materials and products. Food Chem. 98, 508–512.

Krkošková, B., Mrázová, Z., 2005. Prophylactic components of buckwheat. Food Res. Int. 38, 561–568.

Lee, L.-S., Choi, E.-J., Kim, C.-H., Sung, J.-M., Kim, Y.-B., Seo, D.-H., Choi, H.-W., Choi, Y.-S., Kum, J.-S., Park, J.-D., 2016. Contribution of flavonoids to the antioxidant properties of common and Tartary buckwheat. J. Cereal Sci. 68, 181–186.

Matuz, J., Bartok, T., Morocz-Salamon, K., Bona, L., 2000. Structure and potential allergenic character of cereal proteins—I. Protein content and amino acid composition. Cereal Res. Commun. 28, 263–270.

Michalová, A., 1998. Study of relationships between yield and quality characters of common buckwheat (*Fagopyrum esculentum* Moench). Advances in Buckwheat Research. Proceedings of the VII International Symposium on Buckwheat. Winnipeg, August 12–14, Canada, pp. I88–I96.

Michalová, A., Hutař, M., 1998. Pohanka setá (*Fagopyrum esculentum*). Výživa a Potravin 53, 138–140, (in Czech).

Michalová, A., Dotlačil, L., Čejka, L., 1998. Evaluation of common buckwheat cultivars. Advances in Buckwheat Research. Proceedings of the VII International Symposium on Buckwheat. Winnipeg, August 12–14, Canada, pp. I97–I108.

Moudrý, J., Kalinová, J., Petr, J., Michalová, A., 2005. Pohanka a proso. Ústav zemědělských a potravinářských informací, Praha (in Czech).

Petr, J., 1995. Pěstování pohanky a prosa. Metodiky pro zavádění výsledků výzkumu do praxe. ÚZPI, Praha, 7, pp. 7 (in Czech).

Qin, P., Wang, Q., Shan, F., Hou, Z., Ren, G., 2010. Nutritional composition and flavonoids content of flour from different buckwheat cultivars. Int. J. Food Sci. Technol. 45, 951–958.

Saturni, L., Ferretti, G., Bacchetti, T., 2010. The gluten-free diet: safety and nutritional quality. Nutrients 2, 16–34.

Slavin, J., 2013. Fiber and prebiotics: mechanisms and health benefits. Nutrients 5, 1417–1435.

Steadman, K.J.J., Burgoon, M.S.S., Lewis, B.A.A., Edwardson, S.E.E., Obendorf, R.L.L., 2001. Buckwheat seed milling fractions: description, macronutrient composition and dietary fibre. J. Cereal Sci. 33, 271–278.

Treadwell, D., Hochmuth, R., Landrum, L., Laughlin, W., 2010. Microgreens: a new specialty crop. University of Florida IFAS Extension HS1164.

Wei, Y.-M., Hu, X.-Z., Zhang, G.-Q., Ouyang, S.-H., 2003. Studies on the amino acid and mineral content of buckwheat protein fractions. Food/Nahrung 47, 114–116.

Xiao, Z., 2013. Nutrition, sensory, quality and safety evaluation of a new specialty produce: microgreens. Doctoral Thesis. Faculty of the Graduate School of the University of Maryland, College Park. Available from: http://drum.lib.umd.edu/bitstream/handle/1903/14900/Xiao_umd_0117E_14806.pdf.

Chapter fifteen

Factors Important for Structural Properties and Quality of Buckwheat Products

K. Ikeda, S. Ikeda

Kobe Gakuin University, Faculty of Nutrition, Kobe, Japan

INTRODUCTION

Buckwheat (*Fagopyrum esculentum* Moench) is an important crop in some areas of the world (Ikeda, 2002a; Kreft et al., 2003). Buckwheat flour contains beneficial components for human health such as protein, dietary fiber, vitamins, polyphenolics (Ikeda, 2002a), and minerals (Ikeda and Yamashita, 1994) at high levels. Thus buckwheat flour is an important dietary source of such beneficial components. Buckwheat is widely utilized in various forms throughout the world. There is a variety of buckwheat foods produced on a global basis, such as noodles in Asian countries, cake, bread, and pasta in Europe, crepe and galette in France, kasha in Europe and Russia, zlevanka in Slovenia, and blini in Russia (Kreft et al., 2007; Ikeda, 2002a). In view of their processing and cooking methods, increasing attention has been paid to the palatability and acceptability of various buckwheat foods. In this chapter, we will discuss factors important for structural properties and quality of buckwheat products. Table 15.1 summarizes the quality characteristics of buckwheat (Ikeda, 2002b). There are various factors responsible for the quality characteristics of buckwheat (Table 15.1).

PALATABILITY AND ACCEPTABILITY OF FOODS: PROPOSAL OF THE MOLECULAR COOKERY SCIENCE

There is globally a large variety of traditional foods including buckwheat foods. In general, how do humans obtain the palatability and acceptability of foods? It is well known that there are various factors involved in the palatability and

193

Molecular Breeding and Nutritional Aspects of Buckwheat. http://dx.doi.org/10.1016/B978-0-12-803692-1.00015-8

TABLE 15.1 Quality Characteristics of Buckwheat (Ikeda, 2002b)

1. Yield
2. Uniformity of buckwheat grain
3. Milling characteristics
4. Color of grain
5. Genetic trait of buckwheat grain
6. Water content of buckwheat grain
7. Nutritional characteristics
   a. Nutritional characteristics of components present: quality and quantity
   b. Presence of antinutrients as protease inhibitors, amylase inhibitors, tannin, etc.)
   c. Presence of allergens
   d. Safety
8. Characteristics concerning palatability and acceptability
   a. Components concerning palatability
   b. Components concerning antipalatability
   c. Mechanical characteristics (characteristics on masticating)
9. Processing characteristics (characteristics responsible for various uses)
10. Cooking characteristics (characteristics responsible for various cooking forms)

acceptability of foods. Table 15.1 summarizes various factors involved in the palatability and acceptability of foods (Kawabata and Tamura, 1997). There are various factors involved in palatability with acceptability, that is, taste, flavor, color, mechanical characteristics, etc. (Table 15.2) (Kawabata and Tamura, 1997). In these factors responsible for the palatability of foods, especially of cereal foods, mechanical characteristics are an important quality attribute, which affects consumers' acceptance of and palatability of foods (Table 15.1).

Much attention has been paid recently to the food science clarifying the palatability and acceptability of foods. As described previously there is a large variety of buckwheat products globally. Through his buckwheat research, K. Ikeda, who is one of the present authors, proposed a new science for the 21st century, that is, "molecular cookery science" (Ikeda, 1997). The palatability and acceptability of foods, including buckwheat foods, should be clarified from the standpoint of molecular cookery science.

In the case of buckwheat products, mechanical characteristics are an important factor responsible for its palatability with acceptability. Therefore clarifying the mechanical characteristics of buckwheat products is a subject of great interest. Furthermore, molecular characterization of the palatability and acceptability of foods including buckwheat foods is an interesting subject. The present authors have attempted to clarify a molecular basis of the mechanical characteristics of buckwheat foods. They have shown that the major endogenous components, such as protein and starch, are important factors responsible for mechanical characteristics, such as hardness, springiness, and chewiness of buckwheat products (Ikeda et al., 1997).

TABLE 15.2 Factors Responsible for the Palatability and Acceptability of Foods (Kawabata and Tamura, 1997, some modified by K. Ikeda)

Chemical factors

1. Taste factors
    a. Sweet taste
    b. Sour taste
    c. Salt taste
    d. Bitter taste
    e. Umami taste
    f. Hot taste
    g. Pungent taste
    h. Other high grade taste
2. Flavor

Physical factors

1. Appearance
2. Mechanical factors
3. Textural factors
4. Temperature
5. Sound on masticating

Inherent factors: nationality, sex distinction, age, physical constitution

Acquired factors: climate, region, dietary habit, education, life

Psychological factors

Physiological factors

Environmental factors

SCIENTIFIC ANALYSIS OF TRADITIONAL PREPARING METHODS OF BUCKWHEAT NOODLES

Noodles made from buckwheat flour water dough are popular in some regions of the world such as Japan, China, and Korea (Ikeda and Ikeda, 1999). In Japan, buckwheat noodles are a popular, traditional food. Traditional buckwheat noodle preparation methods have been reported in Japanese history for about 400 years or more. In particular, traditional buckwheat noodle preparation methods have been developed in Edo, which is the former name for Tokyo. Therefore these methods are called Edo-style buckwheat noodle preparation methods. Traditional buckwheat noodle preparation generally consist of five successive processes (Fig. 15.1, cited from Asami et al., 2008): first, mixing with water and often with wheat flour in a wooden ball; second, kneading the dough; third, making the dough into a round shape; fourth, rolling out; and lastly, cutting. Although excellent techniques, if any, might lie behind in each process, the scientific principles behind the techniques of preparing buckwheat noodles remain to be clarified.

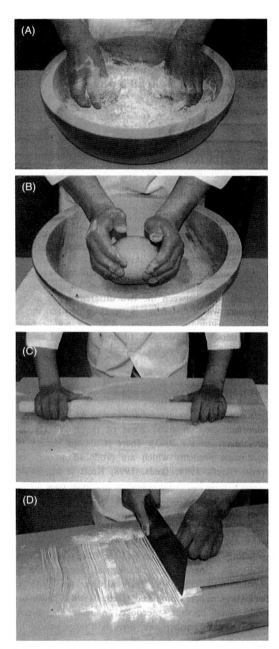

FIGURE 15.1 Scheme of traditional preparation of buckwheat noodles.
(A) Buckwheat flour was put with wheat flour in a mixing bowl (*kone-bachi*) (mixing step, *mizu-mawash*); (B) buckwheat flour-water dough was well kneaded by hand in the mixing bowl and then rolled into balls (packing-together step, *kukuri*); (C) buckwheat dough was extended by the rolling pin (*men-bou*) (extending step, *nobashi*); and (D) buckwheat dough was cut off with a noodle knife (cutting step, *houchou*).

Clarifying the scientific principles of the excellent techniques in buckwheat noodle preparation processes is a subject of great interest. There are various unanswered questions concerning traditional buckwheat noodle preparation methods.

There is a common proverbial saying concerning the palatability and acceptability of buckwheat noodles in Japan, that is, buckwheat noodles prepared with all parts of the following four conditions are believed to be more palatable and acceptable: first, noodles made from just-harvested and dried buckwheat seed; second, noodles made from just-ground buckwheat flour; third, noodles made from just-prepared buckwheat noodles; and lastly, just-cooked buckwheat noodles (Ikeda, 2002a).

Regarding this proverbial saying about milling, there are two milling methods: one is traditional milling with a stone mill; the other is modern milling with a roll milling machine. Buckwheat grain is traditionally milled using a stone mill. On the other hand, in the modern buckwheat industry, buckwheat is usually milled with an industrial scale roller milling machine. However, the mechanical characteristics of such various buckwheat flours have not yet been fully clarified. Our analysis (Asami et al., 2009) is that the distribution of both stone-milled buckwheat flour and roller-milled flour fitted the Rosin–Rammler distribution (Rosin and Rammler, 1933) very well. In addition, the distribution of stone-milled buckwheat flour was significantly ($p < 0.001$) wider than that of roller-milled flour (Asami et al., 2009). In Japan there is a common, proverbial saying concerning stone milling of buckwheat among buckwheat flour-making experts, that is, both a coarse type of flour and a fine type of flour may be produced from traditional stone milling, the coarse type of flour may be mainly responsible for producing acceptable flavor, whereas the fine type of flour may be mainly responsible for binding particles to each other that are present in the buckwheat flour. Therefore the coexistence of both types of flour in stone-milled buckwheat flour may be important for preparing buckwheat noodles with high palatability and acceptability. The present analysis supports this common proverbial saying scientifically.

Concerning the previous proverbial saying about storage, we have analyzed mechanical characteristics and components of noodles from buckwheat grain stored under various storage conditions. A decrease in mechanical characteristics, which means a decrease in palatability, results from noodles prepared from prolonged stored buckwheat grain. Our statistical analysis shows that temperature, relative humidity, and the length of storage of buckwheat grain may be important factors affecting the mechanical characteristics of resultant noodles (Asami and Ikeda, 2005).

MECHANICAL VARIETY OF ASIAN NOODLES INCLUDING BUCKWHEAT NOODLES AND CLASSIFICATION OF THE NOODLES IN VIEW OF MECHANICAL CHARACTERISTICS

Besides buckwheat foods there is a variety of noodles, the so-called Asian noodles, made mainly from wheat flour in Asian countries (Nagao, 1996; Kruger et al., 1996). Various kinds of noodles, called men in Japanese, such as udon noodles, kishi-men noodles, so-men noodles, and ra-men noodles, are very popular and widely available in Japan. Udon noodles are made by kneading medium wheat flour with salt as a gluten strength enhancer (Nagao, 1995; Kruger et al., 1996). Kishi-men noodles, flat-shaped noodles, are made by kneading medium wheat flour with a large amount of salt. So-men noodles are prepared by kneading medium wheat flour, then spreading the dough with edible oil and aging (Nagao, 1995; Kruger et al., 1996). Ra-men noodles are made by kneading mellower strong wheat flour with alkaline salts (kansui) such as potassium carbonate, sodium carbonate, potassium phosphate, etc. (Nagao, 1995; Kruger et al., 1996). In addition to wheat noodles, bifun noodles made from rice flour, and harusame noodles made from starch isolated from certain plant foodstuffs such as potato, are also popular and widely available in Japan. In view of such a background, the characterization of buckwheat foods, especially noodles, in comparison with other cereal foods with respect to palatability and acceptability appears to be an especially interesting subject. The mechanical characteristics of buckwheat foods may be an important quality attribute affecting their palatability and acceptability. In comparison with various kinds of oriental noodles, clarifying the question of how buckwheat noodles may exhibit mechanical characteristics is an interesting subject. Our studies showed that various noodles including buckwheat noodles were classified into some groups mainly in view of their brittleness characteristics (Asami et al., 2011).

COMPARISON BETWEEN COMMON AND TARTARY BUCKWHEAT PRODUCTS IN VIEW OF MECHANICAL CHARACTERISTICS

There are two different species of cultivated buckwheat, that is,. common buckwheat and Tartary buckwheat. The major difference between the two cultivated species of buckwheat, that is, common buckwheat (*Fagopyrum esculentum* Moench) and Tartary buckwheat (*Fagopyrum tataricum* Gaertner), is that common buckwheat is utilized worldwide, whereas Tartary buckwheat is utilized as a traditional food in relatively limited regions such as the south region of China (Zhang et al., 2003), Bhutan (Norbu and Roder, 2003), the region of the Himalayan hills from northern Pakistan to eastern Tibet (Ohnishi, 2003), and in Islek in Europe (Kreft et al., 2007). Recent studies have suggested that Tartary buckwheat may exhibit beneficial effects on human health (Lin et al., 1998). Thus much attention has been currently paid to Tartary buckwheat, and the development of new products made from Tartary buckwheat is currently the subject of great interest.

We have undertaken to clarify mechanical characteristics of common and Tartary buckwheat products. Textural and principal component analyses showed that the textural characteristics of various cereal doughs including common and Tatary buckwheat were classified into six groups: common buckwheat group, Tartary buckwheat group, wheat group, other cereal (OC) group I, OC group II, and OC group III (Asami et al., 2006). This finding indicates that mechanical characteristics of doughs made from common and Tartary buckwheat may be clearly different from those made from other cereals. Another analysis showed that the mechanical characteristics of doughs made from common buckwheat are clearly different from those of doughs made from Tartary buckwheat (Ikeda et al., 2003). Dough made from Tartary buckwheat flour exhibited higher hardness and lower cohesiveness than that made from common buckwheat flour (Asami et al., 2007). Higher hardness may lead to low palatability, whereas low cohesiveness may lead to poorly preparing resultant products. Our chemical and mechanical analysis (Asami et al., 2007) concluded that higher hardness and lower cohesiveness of dough made from Tartary buckwheat flour may be because of a high level of endogenous rutin present in the flour. Our analysis showed the hardness and cohesiveness of dough made from Tartary buckwheat flour significantly correlated the rutin content of the flour (Asami et al., 2007). In this connection, treatment of freeze-dried soybean curd (kori-tofu) with gaseous ammonia during its processing has been performed to soften the kori-tofu by permission of the Food Sanitation Law, Japan. Another study has been undertaken to evaluate the effects of some improvers for the mechanical defects, that is, high hardness and low cohesiveness, of dough made from Tartary buckwheat flour: our analysis that ammonia gas, which is known to improve mechanical characteristics of some foods such as soybean tofu, exhibited marked effects on the mechanical characteristics of dough made from Tartary buckwheat flour (Asami et al., 2010).

Increasing attention is presently being paid to the palatability and acceptability of buckwheat foods, including noodles, from the perspective of their cooking and processing characteristics (Ikeda, 2002a). Mechanical characteristics of buckwheat foods may be an important quality attribute affecting their masticatory characteristics in relation to palatability and acceptability. Improvement of the masticatory characteristics of noodles including buckwheat and wheat noodles was currently attempted by adding certain dough improvers. Preparation of wheat noodles, by incorporation of cassava starch into the noodles as a dough improver, was reported to improve the masticatory characteristics of the wheat noodles (Yokoyama, 2002).

DEVELOPMENT OF BUCKWHEAT PRODUCTS USED IN MASS FOOD SERVING

Mass food serving in various institutions provides mass-cooked meals for people who are specified as a mass group. These include various institutions such as hospitals, nursing insurance institutions for the elderly, schools, offices, factories, etc. Because meals provided by these institutions may have a key role to play in maintaining customers' health, producing nutritionally well-balanced meals is an important function. At the same time the meals provided should have high palatability and acceptability. In the case of certain food-serviced institutions such as offices, factories, etc., meals are often provided using a so-called self-service style. Customers can select their favorite dishes from the various prepared dishes. Therefore meals provided by these institutions should combine high nutritional quality together with high palatability and acceptability. In Japan a large variety of buckwheat noodles are available, for example, hand-made noodles (te-uti-men), dried noodles (kan-men), and precooked noodles (yude-men). Especially, hand-made buckwheat noodles are a most acceptable food. However, it is well known that masticatory characteristics remarkably decrease as buckwheat noodles stand for a long period after cooking. In this connection, dishes provided in food-serviced institutions usually are preserved for some periods after cooking. It is difficult or noticeably impossible to serve hand-made buckwheat noodles just after cooking. For utilization of buckwheat noodles in mass food serving, preparing buckwheat noodles with stable masticatory characteristics after cooking and subsequent standing is an essential requirement.

The present authors have developed methods to prepare buckwheat noodles with stable masticatory characteristics after cooking and subsequent standing, and to characterize the buckwheat noodles in view of food-serving science. Mechanical analysis of the noodles showed that incorporation of cassava starch as a dough improver in buckwheat noodles stabilized the mechanical characteristics of the resultant buckwheat noodles. Sensory evaluation with human panels showed that buckwheat noodles with cassava starch were preferred when compared with noodles without cassava starch. We concluded that buckwheat noodles made with cassava starch may be acceptable for mass food serving.

MOLECULAR COOKERY SCIENTIFIC CHARACTERIZATION OF MECHANICAL CHANGES ARISING FROM INTERACTIONS BETWEEN BUCKWHEAT COMPONENTS AND METAL IONS

Clarifying the effects of metal ions on the mechanical characteristics of buckwheat products is an important subject in view of molecular cookery science. There was a variation in mechanical characteristics and solubility of soluble proteins among various buckwheat doughs into which various salts had been added. The observed hardness and gumminess of buckwheat doughs prepared in the presence of various salts significantly correlated to the observed alterations in solubility of the soluble protein present in the buckwheat doughs. The present finding suggests that possible structural changes in the soluble proteins of buckwheat dough may be an important factor affecting the mechanical characteristics of buckwheat dough.

CONCLUSIONS

Globally, there is a large variety of buckwheat products including noodles. Increasing attention has been paid to the characteristics of buckwheat in view of nutrition and palatability. In this chapter, we proposed a number of suggestions concerning factors important for structural properties and quality of buckwheat products. On the other hand, there are a number of unanswered topics concerning characteristics of buckwheat. The present data may hopefully stimulate further investigation to full y answer them.

REFERENCES

Asami, Y., Ikeda, K., 2005. Mechanical characteristics of noodles prepared from buckwheat grain stored under different conditions. Fagopyrum 22, 57–62.

Asami, Y., Arai, R., Lin, R., Honda, Y., Suzuki, T., Ikeda, K., 2006. Comparison of textural characteristics of buckwheat doughs with cereal doughs. Fagopyrum 23, 53–59.

Asami, Y., Arai, R., Lin, R., Honda, Y., Suzuki, T., Ikeda, K., 2007. Analysis of components and textural characteristics of various buckwheat cultivars. Fagopyrum 24, 41–48.

Asami, Y.T., Konishi, N., Mochida, S., Ikeda, S., Ikeda, K., 2008. Comparison of mechanical characteristics between buckwheat noodles prepared by two different traditional techniques. Fagopyrum 25, 45–48.

Asami, Y.K., Fujimura, K., Ishii, T., Konishi, N., Mochida, S., Ikeda, S., Ikeda, K., 2009. Mechanical characteristics of buckwheat noodles made by traditional preparing methods. Fagopyrum 26, 77–83.

Asami, Y.N., Mochida, S., Ikeda, S., Azuma, S., Ikeda, K., 2010. Effect of pretreatment with some improvers on mechanical characteristics of Tartary buckwheat. Fagopyrym 27, 47–50.

Asami, Y., Mochida, N., Ikeda, S., Ikeda, K., 2011. Comparison of brittleness characteristics of buckwheat noodles with other cereal noodles. Fagopyrum 28, 57–63.

Ikeda, K., 1997. Molecular cooery science. In: Kawabata, A., Tamura, S. (Eds.), Cookry Science for the 21th Century, vol. 4. Culinary Functions of Foods. Kenpaku-sha Publishing Company, Tokyo, pp. 113–143.

Ikeda, K., 2002a. Buckwheat: composition, chemistry and processing. In: Taylor, S.L. (Ed.), Advances in Food and Nutrition Research. Academic Press, Nebraska, USA, pp. 395–434.

Ikeda, K., 2002b. Buckwheat as a traditional food and probing its beneficial effects on human health. J. Jpn. Soc. Nutr. Food Sci. 55, 295–297.

Ikeda, K., Ikeda, S., 1999. Dietary-cultural comparison among Japan, China and Europe in buckwheat utilization. Jpn. Rep. Diet.-Cult. Stud. 16, 1–42.

Ikeda, S., Yamashita, Y., 1994. Buckwheat as a dietary source of zinc, copper and manganese. Fagopyrum 14, 29–34.

Ikeda, K., Kishida, M., Kreft, I., Yasumoto, K., 1997. Endogenous factors responsible for the textural characteristics of buckwheat products. J. Nutr. Sci. Vitaminol. 43, 101–111.

Ikeda, K., Asami, Y., Lin, R., Honda, Y., Suzuki, T., Yasumoto, K., 2003. Comparison of mechanical and chemical characteristics between common and Tartary buckwheat. Fagopyrum 20, 53–58.

Kawabata, A., Tamura, S., 1997. Culinary functions of foods. Cookery Science for the 21st Centuryvol. 4Kenpaku-sha Press, Tokyo.

Kreft, I., Chang, K.J., Choi, Y.S., Park, C.H. (Eds.), 2003. Ethnobotany of Buckwheat. Jinsol Publishing Co., Seoul.

Kreft, I., Ries, C., Zewen, C. (Eds.), 2007. Das Buchweizen Buch mit Rezepten aus aller Welt, Islek ohne Grenzen EWIV.

Kruger, J.E., Matsuo, R.B., Dick, J.W. (Eds.), 1996. Pasta and Noodle Technology. Am. Assoc. Cereal Chem., Minnesota, USA.

Lin, R., Jia, W., Ren, J., 1998. Research and utilization of Tartary buckwheat. Buckwheat Trend 28, 1–7.

Nagao, S., 1995. The Science of Wheat (in Japanese). Asakura-shoten Press, Tokyo.

Nagao, S., 1996. The science of Wheat. Asakura-shoten Publishing Company, Tokyo.

Norbu, S., Roder, W., 2003. Traditional uses of buckwheat in Bhutan. In: Kreft, I., Chang, K.J., Choi, Y.S., Park, C.H. (Eds.), Ethnobotany of Buckwheat. Jinsol Publishing Co., Seoul, pp. 34–38.

Ohnishi, O., 2003. Buckwheat in the Himalayan hills. In: Kreft, I., Chang, K.J., Choi, Y.S., Park, C.H. (Eds.), Ethnobotany of Buckwheat. Jinsol Publishing Co., Seoul, pp. 21–33.

Rosin, P., Rammler, E., 1933. Laws governing the fitness of powdered coal. J. Inst. Fuel 7, 29–36.

Yokoyama, M., 2002. New texture development of noodles with processed cassava starch. Technical J. Food Chem. Chem. 41, 33–36.

Zhang, Z., Wang, Z., Zhao, Z., 2003. Traditional buckwheat growing and utilization in China. In: Kreft, I., Chang, K.J., Choi, Y.S., Park, C.H. (Eds.), Ethnobotany of Buckwheat. Jinsol Publishing Co., Seoul, pp. 9–20.

Chapter sixteen

Genetic Diversity Among Buckwheat Samples in Regards to Gluten-Free Diets and Coeliac Disease

D. Urminska, M. Chnapek

Department of Biochemistry and Biotechnology, Slovak University of Agriculture in Nitra, Faculty of Biotechnology and Food Sciences, Nitra, Slovakia

INTRODUCTION

Contemporary and state-of-the-art diagnostic methods used in human medicine provide evidence that coeliac disease is a widespread food intolerance around the globe. This condition is caused by a genetically predisposed sensitivity to gluten proteins present in wheat as well as similar proteins found in barley and rye. A potential reactivity to oat proteins has not been confirmed; however, because of various technological processes related to the processing of this crop, as well as a high risk of contamination by other cereals, oat belongs to cereals not suitable for a coeliac disease diet.

The only known effective therapy for coeliac disease patients is a strict gluten-free diet. The patient has to exclude all cereals and cereal-related products, as well as foods containing cereal starch and brans. Eventually, the starting point nutrition of a coeliac disease patient is based on rice, soy, and corn. Pseudo-cereals are an additional option, providing higher amounts of nutritionally important substances as opposed to traditional cereals; however, their disadvantages include poorer technological and sensory properties. Subsequently, buckwheat may have great potential with respect to the production of gluten-free products.

The most simple and straightforward classification of plant proteins was developed by T.B. Osborne (2009), being based on their solubility in a series of solvents. Despite each of the fractions representing a complex mixture of

203

peptides and polypeptides, and the protein solubility is not strict, absolute, and may be variable (Gianibelli et al., 2001), the method is still currently being used (Matta et al., 2009).

Proteins found in cereal grains may be divided into five groups, based on their solubility in different solvents, specifically: proteins soluble in water (albumins), proteins soluble in saline solutions (globulins), proteins soluble in 70–80% aqueous alcohol (prolamins), and proteins soluble in dilute acids or bases (glutelins). A different classification widely used is connected to the differentiation of cereal proteins according to their function, such as cytoplasmic and storage proteins (Prugar and Hraška, 1986; Černý and Šašek, 1996; Ciccocioppo et al., 2005; Yalçin, 2010).

Cytoplasmic proteins (albumins and globulins) may be found primarily in the sprouts and aleurone layer, and their function is generally related to their structural or catalytic properties. Storage proteins (prolamins and glutelins) account for a substantial part of the actual cereal grain, being essential with respect to the technological and nutritional quality of the grain (Kučerová, 2004).

Cytoplasmic proteins add up to around 20–30% of the total protein amount in most cereals, and up to 60% in oat (Alais and Linden, 1991; Eliasson and Larsson, 1993; Belitz et al., 2009). On the other hand, wheat, barley, and rye contain large amounts of storage proteins (Southgate, 2000). The main group of proteins found in buckwheat are proteins soluble in saline solutions (Hiller et al., 1975; Pomeranz, 1983).

With respect to the coeliac disease storage proteins found in cereals, prolamins (wheat gliadins, barley hordeins, rye secalins, and oat avenins) exhibit specific properties that may be the determining cause of this food intolerance. A high concentration of prolin and associated peptide bonds may cause a relatively high resistance toward the proteases in the gastrointestinal tract. High amounts of glutamine enable the deamination of the peptides by an intestinal enzyme called tissue transglutaminase, having a serious impact on their binding activity to the HLA complex of the immune system (Catassi and Fasano, 2008; Gregorini et al., 2009). Both processes have been shown to play critical roles in the pathogenesis of coeliac disease.

Storage proteins of cereal grains may be divided and distinguished into subgroups, primarily based on their relative molecular weight, using polyacrylamide gel electrophoresis (PAGE), high-performance liquid chromatography, and amino acid analysis. Currently, the most studied and characterized proteins are the whey gliadins and glutenins, which may be divided into high molecular weight (HMW) and low molecular weight (LMW) proteins. LMW proteins include monomeric α/β-proteins and γ-gliadins (wheat), γ-secalins (rye), γ-hordeins (barley), and avenins (oat) as well as LMW-GS protein aggregates (wheat), γ-secalins (rye), and B-hordeins (barley). LMW-GS as well as α-, β-, and γ-gliadins are rich in amino acids containing sulfur (Wieser and Koehler, 2008). Moreover, an additional medium molecular weight group of proteins includes whey ω-gliadins, rye ω-secalins, and barley C-hordeins (Shewry, 2004). The lack of cysteine is the key feature of these proteins.

Wheat prolamins and gliadins are defined as active agents in coeliac disease, and the electrophoretic separation enables us to divide them into α-, β-, γ-, and ω-gliadins, with a decreasing mobility from α- to ω-gliadins (Wieser, 2007). Nevertheless, this classification is not strict, and nowadays a more accurate categorization may be used, according to which the gliadins are classified into four different types: ω5-, ω1,2-, α/β-, and γ-gliadins, where ω-gliadins typically have high amounts of glutamine, proline, and phenylalanine, ω5-gliadins have a higher molecular weight (approx. 50 kDa) in comparison to ω1,2-gliadins (approx. 40 kDa), while α/β- and γ-gliadins have a molecular weight of around 28–35 kDa (Vaccino et al., 2009).

Coeliac disease is widely associated with gluten, which is a complex of storage proteins able to form a sticky, viscous, and lattice structure during technological processing (Petr et al., 2003). Technologically speaking, the term gluten is primarily associated with whey dough, because its amount and quality have an important impact on the quality of baked goods, primarily bread (Sciarini et al., 2010; Xie et al., 2010). Nevertheless, it is important to note that gluten contains components other than proteins, for example, water, starch, lipids, and fiber. Gliadin fractions of proteins are responsible for the viscous and elastic properties of the dough. Glutenins, displaying the ability to form polymers with a molecular weight of 10^5–10^7 Da with their disulfide bonds, are responsible for dough extensibility and elasticity.

Based on the molecular weight, glutenins may be divided into several fractions: A (relative molecular weight of 95–140 kDa), B (40–51 kDa), C (31–36.5 kDa), and D (55–80 kDa). A proteins belong to the HMW-GS, while proteins B, C, and D are associated with LMW-GS (Gianibelli et al., 2001; Figueroa et al., 2009).

Originally, it was just the fraction of whey α-gliadins (together with corresponding barley and rye proteins) with a relative molecular weight of approximately 30 kDa, containing peptides with specific amino acid sequences QQPFP, QQQFP, LQPFP, and QLPFP (Vaccino et al., 2009; Xie et al., 2010), that was associated with coeliac disease. Later the activity of β-, γ-, and ω-gliadins was confirmed as well, with reactivity decreasing from α-gliadins to ω-gliadins (Ensari et al., 1998; Stern et al., 2001).

To detect the proteins reactive in coeliac disease the enzyme-linked immunosorbent assay (ELISA) method is the most used technique, where the R5 antibody will specifically interact with the QQPFP, QQQFP, LQPFP, and QLPFP rye peptides (Osman et al., 2001; Valdés et al., 2003), which may be found in α-, γ-, and ω-gliadins (Kahlenberg et al., 2006; van Eckert et al., 2010). Thus ELISA and the R5 antibody are a very suitable choice to detect whey gliadins, barley hordeins, and rye secalins, with a detection limit of approximately 1.5 ppm gliadin in so-called gluten-free products (Méndez et al., 2005). Although the QQQ/PFP sequence is thought to be the immunodominant structure (Kahlenberg et al., 2006), not to be found in the oat avenins, the method is widely used to detect a possible contamination of oat grains, primarily caused by barley proteins (grains, malt) as well as to confirm the absence of gluten in food products containing 100% oat. The structure of gliadins contains different peptides, which have the ability to stimulate the T cells and to subsequently activate the immune response. The most common and known protein is a 33-meric peptide (LQLQPFPQPQLPYPQPQLPYPQPQLPYPQPQPF), which is a suitable substrate for the tissue transglutaminase, leading to a deamination and creation of a specific agent affecting the T cells (Catassi and Fasano, 2008) leading to inflammation processes within the mucous membrane of the small intestine of coeliac patients. With respect to the α-gliadins the immune system is able to detect the following sequences rich in prolin: PFPQPQLPY, PQPQLPYPQ, and PYPQPQLPY (Mowat, 2003; Morón et al., 2008). Besides the primary structure, the secondary structure of proteins active in coeliac disease plays a crucial role. In this sense, a reverse β-helix has a dominant position (Cornell and Johnson, 2001).

The genetic diversity is a direct consequence of the biological evolution, a long-term and spontaneous process. The living systems, distinguished by polymorphisms, may be considered to be both the subject as well as the object of biological evolution. Polymorphisms are phenomena where individuals from the same species, sex, and age differ in numerous qualitative and quantitative traits. Polymorphisms may be additionally defined as the ability of organisms from the same species to adapt to the environment, ultimately ensuring the survival of the species (Flegr, 2009).

Polymorphisms of specific traits may be of a nonhereditary nature. These polymorphisms result as a response of an individual toward specific environmental conditions, to which the individuals or their closest ancestors were exposed during their life or ontogenesis. In the meantime, most of the polymorphic traits are genetically predisposed. Genetic polymorphisms are caused by the presence of at least two alleles within a specific gene (Flegr, 2009). Khlestkina et al. (2004) emphasize that diverse anthropogenic activities such as urbanization, modernization of agricultural production, as well as the introduction of new highly productive cultivars may eventually lead to a decreased biodiversity. Their research focused on the genetic diversity of wheat and showed that the allele migration between specific cultivars occurred during the adaptation of traditional agricultural systems toward artificial modernization, particularly with respect to intensive breeding. Nevertheless, genetic diversity was not particularly affected.

Evaluation of genetic diversity may be realized on different levels. Analysis of DNA polymorphisms is the basic and principal evaluation method (Fletcher, 2004). Fletcher (2004) adds that the genetic diversity may be studied by means of transcriptomics, metabolomics, as well as phenotyping. Protein polymorphisms may emerge as a result of changes in genetic information, as well as environmental conditions, and their mutual interactions may have a deep impact on the synthesis of individual proteins, subsequently affecting the general cellular metabolism (Skylas, 2004). To evaluate the genetic diversity by the means of protein polymorphisms the storage proteins found in plant seeds are the most suitable material to work with (Javaid et al., 2004).

Buckwheat (*Fagopyrum esculentum* Moench and *Fagopyrum tataricum* L.) has been widely used in the food industry mainly because of its nutritional and health benefits (Petr et al., 2003; Yu-Xia et al., 2008). Buckwheat grains contain large amounts of carbohydrates (73.3%), primarily in the form of starch (55%), approximately 6.5% lipids, and 8.5–18.9% proteins (Skrabanja et al., 2001; Steadman et al., 2001; Wijngaard and Arendt, 2006), B vitamins, microelements (Zn, Cu, Mn, Se) and macroelements (K, Na, Ca, Mg), flavonoids, favones, phenolic acids, condensed tannins, phytosterols, and fagopyrins (Fabjan et al., 2003; Bojňanská et al., 2009).

During the processing of the cereal, particularly grinding, most of the proteins (21.6% with respect to *F. esculentum* Moench and 25.3% in relation to *F. tataricum* L.) will be retained in the fraction defined as brans, while the resulting flour will carry approximately 10% proteins (Bonafaccia et al., 2003).

The proteins have a very high biological value, defined by means of the content and ratio of specific essential amino acids. If ovalbumin is the standard (100), buckwheat proteins may reach values up to 93, as opposed to soy (68) or wheat proteins (93) (Cai et al., 2004). The digestibility of buckwheat proteins is, however, relatively low (Wronkowska and Soral-Śmietana, 2008) and because of the content of toxic compounds in the husk, buckwheat may be consumed in the form of peeled grains exclusively (Aubrecht and Biacs, 1999).

Albumins are the dominant fraction of buckwheat (28–42%), followed by glutelins (11–21%) and globulins (14–20%) (Cai et al., 2004). With respect to coeliac disease the prolamins or proteins soluble in alcohols are crucial. Buckwheat comprises up to 2.4–4.2% (Aubrecht and Biacs, 1999), 1.7–2.3% (Cai et al., 2004), or 1–4% (Nałęcz et al., 2009) of these proteins, respectively. Guo and Yao (2006) evaluated the percentage of individual fractions in *F. tataricum* L. concluding that the albumin fraction was the dominant one (43.8%), while glutelins and prolamins constituted 14.6% and 10.5%, respectively. Skerritt (1986) obtained two fractions based on their solubility, specifically a fraction soluble in NaCl (75.2%) and a fraction soluble in ethanol (13.5%).

The analysis of buckwheat proteins using SDS-PAGE under reducing and native conditions carried out by Guo and Yao (2006) revealed that albumins and globulins were constituted by protein subunits with a molecular weight of 38, 41, 57, and 64 kDa. Nonreducing conditions caused the degeneration of the 51–52 kDa subunits in the albumin fraction. Globulins were made up by subunits with a molecular weight of 15–34 kDa. Prolamins displayed four subunits, two of which were present in a very low concentration, while two were detected in high amounts with a molecular weight of 15 and 17 kDa, and being the dominant ones. Glutenins were difficult to be detected because of a low solubility. Reducing conditions caused by the addition of 2-mercaptoethanol led to the detection of two albumin and globulin subunits with an identical molecular weight (38 and 41 kDa). Changes were recorded after the reduction of albumins and globulins in the protein subunits with a molecular weight of 57–64 kDa. After the reduction, the prolamins exhibited two additional subunits with 15 and 17 kDa. These observations are in accordance with Skerritt et al. (1994), who reported that the buckwheat proteins soluble in alcohol (prolamins) have a relatively low molecular weight (10–28 kDa). Skerritt et al. (1994) compared the electrophoretic spectra of wheat and buckwheat proteins revealing that the molecular weight of individual wheat proteins ranged between 15 and 150 kDa, while the buckwheat protein fractions were made up by proteins of 10–80 kDa. The most significant differences were detected with respect to the prolamin fraction, as the gliadins were defined as proteins with a molecular weight of 30–70 kDa, while buckwheat prolamins had a significantly lower molecular weight (10–28 kDa).

Low concentrations of proteins soluble in alcohol do not automatically mean the suitability of buckwheat for the preparation of gluten-free products, as it is necessary to examine their reactivity with antibodies. Codex Alimentarius defines two food categories suitable for coeliac patients—gluten-free foods with gluten concentrations below 20 mg/kg (20 ppm) and foods with decreased gluten concentrations or foods with very low gluten concentrations up to 100 mg/kg.

MATERIAL AND METHODS

Buckwheat grains collected from the cultivars Buckwheat Východ, Darja Semenarna, Orehovica, Kostanjevica na Krki 2009, Mihovo, Šentjernejsko Polje, Bilje SE (Tartary buckwheat), Bilje Brez (Tartary buckwheat), Ajoke Čemažar Kleče, Bela Krajina Slovenia, Dolenje Vrhpolje, and Gorenje Vrhpolje (Slovenian origin) were homogenized in a laboratory mill. The Kjeldahl method (Velp Scientifica, Italy) was used to determine the total nitrogen content as well as the concentration of specific protein fractions ($N \times 6.25$; Steadman et al., 2001). Proteins soluble in water and saline solutions were extracted using 10% (w/v) NaCl, while the proteins soluble in alcohols were extracted by means of 70% (w/v) ethanol. Proteins soluble in basic solutions were extracted using 0.2% NaOH. Each extraction was performed for 1 h at 20°C, repeated three times, and the supernatants obtained by centrifugation (6000 g) were condensed by evaporation and finally lyophilized (Lyovac GT 2, Amsco/Finn-Aqua, Germany).

Polyacrylamide gel electrophoresis in the presence of sodium dodecyl sulfate (SDS-PAGE, Bio-Rad Laboratories Inc., USA) was performed according to a standard ISTA methodology (Wrigley, 2004). The electropherograms were evaluated with the help of a CCD camera UVP and GelWorks 1D software.

To determine the reactivity of coeliacally active buckwheat proteins a sandwich ELISA using the R5 antibody was performed. The RIDASCREEN® Gliadin test (R-Biopharm, Germany) applied gliadin standards with specific concentrations of 0, 5, 10, 20, 40, and 80 ppm, while the R5 antibody was bound to a complex with the horseradish peroxidase. Carbamide peroxide served as a substrate for the enzyme, tetramethylbenzidine acted as a chromogen, and the photometric evaluation of the resulting immunocomplexes was performed by measuring the absorbance at a wavelength of 450 nm (ELISA-reader BioTek, USA). The analysis was carried out using a standard protocol designed by the RIDASCREEN Gliadin test.

All chemicals used in the analyses were purchased from Sigma (USA), Fluka (Switzerland), and Applichem (Germany). Selected markers used to determine the specific molecular weights using SDS-PAGE were purchased from Fermentas International Inc. (Canada).

RESULTS AND DISCUSSIONS

The analysis of the protein content and individual protein fractions suggests that buckwheat contains an average of 10.98% proteins (crude protein; Table 16.1). Significant differences were found between individual cultivars. Bilje Brez had the lowest concentration of proteins, while protein concentrations up to 1.4 times higher were recorded in Kostanjevica na Krki and Bela Krajina Slovenia. Aubrecht and Biacs (2001) indicate that buckwheat flour may contain up to 8.51–18.87% proteins, depending on the cultivar used.

TABLE 16.1 Determination of the Protein Content and Individual Protein Fractions in Buckwheat Grains ($n = 6$)

Cultivar	N_{total} (%)	Crude Protein ($N \times 6.25$) (g/100 g)	Proteins Soluble in NaCl (%)	Proteins Soluble in Ethanol (%)	Proteins Soluble in NaOH (%)
Buckwheat Východ	1.45 ± 0.10	9.06	48.05	7.68	19.26
Darja Semenarna	1.75 ± 0.11	10.93	50.40	3.99	19.21
Orebovica	1.76 ± 0.12	11.00	51.58	3.17	12.67
Kostanjevica na Krki 2009	1.96 ± 0.09	12.25	47.15	2.85	15.73
Mihovo	1.68 ± 0.10	10.50	53.36	2.50	16.70
Šentjernejsko Polje	1.79 ± 0.11	10.93	51.56	3.90	15.65
Bilje SE (T)	1.82 ± 0.10	11.37	40.73	3.84	17.71
Bilje Brez (T)	1.40 ± 0.09	8.75	38.99	2.99	24.02
Ajoke Čemažar Kleče	1.90 ± 0.12	11.87	44.13	4.40	20.60
Bela Krajina Slovenia	1.96 ± 0.10	12.25	46.44	3.56	16.45
Polenje Vrhopolje	1.74 ± 0.12	10.87	40.29	4.02	16.15
Gorenje Vrhopolje	1.93 ± 0.09	12.06	44.94	2.89	18.13

The dominant protein fractions in the buckwheat grain were albumins and globulins, constituting up to 38.99–53.36% of the total protein content. On the contrary, the least representative fraction was soluble in alcohol (prolamins), with a percentage of 2.50–7.68. Proteins soluble in solutions of acids or bases (glutelins) constituted about 12.67–24.02% of buckwheat. Our results are in accordance with Guo and Yao (2006) who suggest that albumins and globulins are the main storage proteins in buckwheat, while prolamins and glutelins represent only a small percentage. Steadman et al. (2001) hypothesize that the globulin content may rise up to 70%. 13S globulins are considered to be the main storage proteins in buckwheat (Aubrecht and Biacs, 1999; Li and Zhang, 2001).

The examined cultivars exhibited significant differences in the composition of individual protein fractions. With respect to coeliac disease, the most interesting fraction of proteins soluble in ethanol represented up to 7.68% in the Buckwheat Východ cultivar, as opposed to the Mihovo cultivar containing only 2.50% of these proteins. It is generally acknowledged that from a nutritional point of view the proteins containing the most favorable amino acid composition are related to albumins and globulins. The highest concentration of the aforementioned proteins was detected in the Mihovo cultivar, building up more than 50% of the total protein content in the buckwheat grain.

The Bilje Brez cultivar diverged from the collection of cultivars used for our analysis, as it carried the lowest amount of proteins (8.41%), out of which the albumin and globulin fraction had the lowest concentration out of the whole buckwheat collection; however, the fraction of proteins soluble in basic solutions (glutelins) was the highest (24.02%).

The technique of protein fractionation does not have the ability to provide absolute and unambiguous results, as the solubility of specific fractions is not strict and well defined. More precise results may be obtained by means of electrophoretic methods.

Numerous markers may be used to examine the diversity of plants and animals. A common requirement for a suitable marker to evaluate diversity is sufficient polymorphism able to divide the population into clusters containing individuals similar to a specific tracking trait. Markers used to evaluate populations of organisms from different buckwheat collections used in this study may be basically divided into morphological, genetic, and molecular markers, including protein and isoenzymatic polymorphisms.

Currently no studies are available on the use of genetic markers in the identification of genetic diversity of buckwheat proteins in relation to the coeliac disease. Analysis of the buckwheat proteome is therefore studied primarily at the level of the fractional composition, while the identification of proteins responsible for the symptoms of coeliac disease stems from immunochemical reactions.

Proteins from the buckwheat cultivars were studied using SDS-PAGE, and individual fractions were identified based on the comparison with the molecular weight marker and wheat standard. We proved the presence on three electrophoretic profiles, which were characterized by identical composition of the dominant protein subunits with minimal deviations in the presence or absence of minor protein elements (Fig. 16.1).

Buckwheat proteins with high molecular weight, similar to macromolecular glutenin subunits, were located in the areas with a relative molecular weight ranging from 65.30 to 93.20 kDa, and represented only 4.15% from the total protein content. We identified one to two dominant and two to four minority proteins, and all buckwheat cultivars contained a dominant protein of 80 kDa. Mihovo and Šentjernejsko Polje cultivars exhibited the presence of a protein of 93 kDa. Additionally, these two cultivars comprised two proteins of 101 and 110 kDa. The rest of the cultivars had up to four minority proteins with a molecular weight of 86, 94, 108, and 114 kDa.

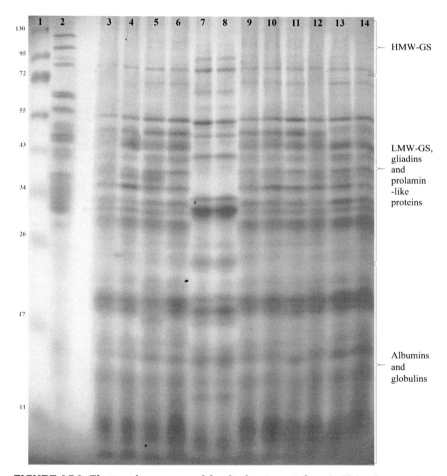

FIGURE 16.1 Electropherogram of buckwheat proteins. 1. Molecular standard (kDa). 2. Bread wheat standard. 3. Buckwheat Východ. 4. Darja Semenarna. 5. Orebovica. 6. Kostanjevica na Krki 2009. 7. Mihovo. 8. Šentjernejsko Polje. 9. Bilje SE. 10. Bilje Brez. 11. Ajoke Čemažar Kleče. 12. Bela Krajina Slovenia. 13. Dolenje Vrhpolje. 14. Gorenje Vrhpolje.

The dominant protein fraction consisted of proteins with a molecular weight of 23–80 kDa, representing 49.9% of the total buckwheat protein content (Table 16.2). Overall, 22 proteins were identified in this area (29–53 kDa), while their quantity ranged between 11 and 13 depending on individual genotypes. All buckwheat cultivars, with the exception of Mihovo and Šentjernejsko Polje (exhibiting a different electrophoretic profile), contained proteins with a molecular weight of 70 and 53 kDa. Mihovo and Šentjernejsko Polje revealed different proteins of 58, 62, and 52 kDa. This fraction exhibited the highest degree of polymorphism: 23, 25, and 31 kDa proteins were found in all cultivars, while 33, 42, 46, 48, 50, 52, 58, and 65 kDa were typical for the Mihovo and Šentjernejsko Polje cultivars. Proteins of 23, 31, 33, 42, 48, 52, and 60 kDa may be considered to be dominant. The dominant protein found in the Mihovo and Šentjernejsko Polje cultivars had a molecular weight of 31 kDa. The 37-kDa protein was absent in Buckwheat Východ, Konstanjevica na Krki 2009, and Bilje SE cultivars.

TABLE 16.2 The Composition of Individual Protein Fractions

Cultivar	Proteins Soluble in 0.2% NaOH (%)	Proteins Soluble in 70% ethanol (%)	Proteins Soluble in 10% NaCl (%)
Standard wheat Chinese spring	11.30	51.53	37.17
Buckwheat Východ	5.25	35.78	58.98
Darja Semenarna	3.88	47.07	49.05
Orehovica	3.67	50.01	46.32
Kostanjevica na Krki 2009	3.67	52.21	44.12
Mihovo	5.24	50.78	43.98
Šentjernejsko Polje	5.49	46.32	48.19
Bilje SE (Tartary)	4.38	48.88	46.74
Bilje Brez (Tartary)	4.21	42.31	53.48
Ajoke Čemažar Kleče	3.72	48.40	47.88
Bela Krajina Slovenia	3.91	45.43	50.66
Dolenje Vrhpolje	2.89	51.28	45.83
Gorenje Vrhpolje	3.58	49.90	46.53

The SDS-PAGE results indicate that the dominant protein fraction found in buckwheat grains consisted of albumins and globulins (46.53%). Furthermore, this protein fraction revealed polymorphisms enabling us to differentiate the Mihovo and Šentjernejsko Polje genotypes from the rest. Excluding the Orebovica genotype (10 proteins), the remaining genotypes exhibited 9 proteins in the albumin and globulin fraction, and all cultivars revealed proteins with molecular weights of 13, 14, 18, 19, and 21 kDa. On the other hand, Mihovo and Šentjernejsko Polje contained proteins of 9, 12, 15, and 17 kDa. Similarly, Orebovica cultivar displayed a 17-kDa protein. These results confirm that the genetic diversity of buckwheat proteins is satisfactory large and suitable for genotypic comparison. At the same time we may suggest that the identification of plant proteins based on their solubility is not sufficient and accurate enough. Indeed, the electrophoretic separation showed that the dominant protein fraction found in buckwheat grains were proteins with a relative molecular weight of 23–80 kDa, corresponding to the molecular weight of wheat gliadins. Guo and Yao (2006) indicate that four buckwheat protein fractions contain disulfide bonds, which aggregate to create more complex units. Using nonreducing conditions, the authors identified albumins of 64, 57, 41, and 38 kDa, as well as globulins of 57, 28, 23, 19, and 15 kDa. Reduced albumins and globulins contained two joined peptides of 41 and 38 kDa. Reduced prolamins were built up by proteins of 29, 26, 17, and 15 kDa.

With respect to coeliac disease it is crucial to perform an immunochemical reaction. The reaction was performed with the help of the R5 antibody, which is the only commercially available antibody enabling the formation of immunocomplexes with wheat, rye, barley, as well as oat prolamins in gluten-free products, with a very low detection limit. It is suitable to detect both native and denaturated prolamins, as the QQPFP motif is resistant toward proteolysis during thermal food processing (Sorell et al., 1998; Thompson and Méndez, 2008).

All buckwheat cultivars subjected to the analysis tested below the detection limit (1.5 ppm), which means that buckwheat does not contain the target amino acid sequences QQPFP, QQQFP, LQPFP, and QLPFP commonly found in wheat gliadins, barley hordeins, and rye secalins, and presumably being the cause of abnormal reactions related to the intolerance of the aforementioned amino acids (van Eckert et al., 2010).

As seen in earlier experiments, a correct protein extraction procedure is essential for the ELISA test. Méndez et al. (2005) suggest that the extraction buffer containing 2-mercaptoethanol and guanidine chloride is satisfactory, as the reduction of disulfide bonds enables the release of the aggregatory structure of prolamins, without a secondary negative impact on the ELISA method using the R5 antibody. Therefore we may hypothesize that all examined buckwheat cultivars may be a suitable raw material for a gluten-free diet.

At the same time it is important to note that the immunogenic properties of coeliacally active proteins depend upon the presence of specific amino acid sequences as well as on the secondary structure of peptides, which may be changed during the technological processing, for example, deamination, which may lead to alterations of the ELISA reactivity.

Rout and Chrungoo (1996, 1999) found that buckwheat proteins create oligomeric complexes with a molecular weight of 280 kDa and containing clusters of 55–60, 32–44, and 16–29 kDa. The basic subunit represents a polypeptide of 26 kDa, whose structure derived from the circular dichroism spectra is composed of alpha-helices (22%), beta-sheets (36%), beta-helices (12%), and 30% have a random shape.

A different source of possible discrepancies in the analysis are the standards, as these are wheat gliadins, although it has been previously shown that certain glutenin fractions may show a reactivity toward coeliac disease (Lester, 2008).

Amino acid sequences of prolamins from various cereals show specific differences, translated into a different solubility, because, for instance, hordeins and secalins are more soluble in diluted solutions of alcohols, as opposed to gliadins, which may have an impact of the accuracy of ELISA. As oat avenins do not have reactive epitopes, the ELISA method is often used to detect the contamination of oat products by other cereals.

Kanerva et al. (2006) analyzed a mixture of avenins and hordeins using a reaction with the R5 antibody, followed by an antibody to ω-gliadins. The authors hypothesize that overly positive results would eliminate the possibility of using hordeins as standards. In terms of a practical application this statement indicates that only a known source of prolamins enables a correct ELISA quantification against a suitable standard, followed by a definitive conclusion with respect to the suitability of a specific cereal for the preparation of gluten-free foods.

ACKNOWLEDGMENTS

This contribution is the result of the project implementation: Centre of excellence for white-green biotechnology, ITMS 26220120054, supported by the Research & Development Operational Programme funded by the ERDF.

REFERENCES

Alais, Ch., Linden, G., 1991. Food Biochemistry. Ellis Horwood, New York.
Aubrecht, R., Biacs, P.Á., 1999. Immunochemical analysis of buckwheat proteins, prolamins and their allergenic character. Acta Aliment. 28 (3), 261–268.
Aubrecht, E., Biacs, P.Á., 2001. Characterization of buckwheat grain proteins and its products. Acta Aliment. 30 (1), 71–80.
Belitz, H.D., Grosch, W., Schieberle, P., 2009. Cereals and cereal products. In: Belitz, H.D., Grosch, W., Schieberle, P. (Eds.), Food Chemistry. Springer, Berlin, pp. 670–745.
Bojňanská, T., Frančáková, H., Chlebo, P., Vollmannová, A., 2009. Rutin content in buckwheat enriched bread and influence of its consumption on plasma total antioxidant status. Czech J. Food Sci. 27 (1), 236–241.
Bonafaccia, G., Gambelli, L., Fabjan, N., Kreft, I., 2003. Trace elements in flour and bran from common and Tartary buckwheat. Food Chem. 83 (1), 1–5.
Cai, Y.Z., Corke, H., Li, W.D., 2004. Buckwheat. In: Wrigley, C., Corke, H., Walker, Ch. (Eds.), Encyclopedia of Grain Science, Elsevier, Oxford, pp. 120–128.
Catassi, C., Fasano, A., 2008. Celiac disease. In: Arendt, E.K., Dal Bello, F. (Eds.), Gluten-Free Cereal Products and Beverages. Academic Press, San Diego, pp. 1–27.

Černý, J., Šašek, A., 1996. Bílkovinné signální geny pšenice obecné, Ústav zemědělských a potravinářskych informací, Praha, p. 62.

Ciccocioppo, R., Di Sabatino, A., Corazza, G.R., 2005. The immune recognition of gluten in celiac disease. Clin. Exp. Immunol. 140 (3), 408–416.

Cornell, H.J., Johnson, G.W., 2001. Structure–activity relationship in coeliac–toxic gliadin peptides. Amino Acids 21 (3), 243–253.

Eliasson, A.C., Larsson, K., 1993. Cereals in breadmaking: a molecular colloidal approach. M. Dekker, New York, 376 pp.

Ensari, A., Marsh, M.N., Moriarty, K.J., Moore, C.M., Fido, R.J., Tatham, A.S., 1998. Studies in vivo of ω-gliadins in gluten sensitivity (coeliac sprue disease). Clin. Sci. 95 (4), 419–424.

Fabjan, N., Rode, J., Koir Iztok, J., Wang, Z., Zhang, Z., Kreft, I., 2003. Tartary buckwheat (*Fagopyrum tataricum* Gaertn) as a source of dietary rutin and quercitrin. J. Agric. Food Chem. 51 (22), 6452–6455.

Figueroa, J.D.C., Maucher, T., Reule, W., Peña, R.J., 2009. Influence of high molecular weight glutenins on viscoelastic properties of intact wheat kernel and relation to functional properties of wheat dough. Cereal Chem. 86 (2), 139–144.

Flegr, J., 2009. Evoluční Biologie. Academia, Praha, Vyd. 1, p. 559.

Fletcher, R.J., 2004. Pseudocereals—overview. In: Wrigley, C., Corke, H., Walker, Ch. (Eds.), Encyclopedia of Grain Science, Elsevier, Oxford, pp. 488–493.

Gianibelli, M.C., Larroque, O.R., Macritchie, F., Wrigley, C.W., 2001. Biochemical, genetics and molecular characterization of wheat endosperm proteins. Cereal Chem. 78 (6), 635–646, <http://www.aaccnet.org/cerealchemistry/freearticle/gianibelli.pdf>.

Gregorini, A., Colomba, M., Ellis, H.J., Ciclitira, P.J., 2009. Immunogenicity characterization of two ancient wheat α-gliadin peptides related to coeliac disease. Nutrients 1 (2), 276–290.

Guo, X.N., Yao, H.Y., 2006. Fractionation and characterization of Tartary buckwheat flour proteins. Food Chem. 98, 90–94.

Hiller, A., Mlodecki, H., Tomczyk, M., 1975. Preliminary characterization of proteins of buckwheat grain by extraction with selected solvents and molecular filtration on Sephadex G-75 gel. Bromatol. Chem. Toksykol. 8, 205.

Javaid, A., Ghafoor, A., Anwar, R., 2004. Seed storage protein electrophoresis in groundnut for evaluating genetic diversity. Pakistan J. Bot. 36 (1), 25–29.

Kahlenberg, F., Sanchez, D., Lachmann, I., Tuckova, I., Tlaskalova, H., Méndez, E., Mothes, T., 2006. Monoclonal antibody R5 for detection of putatively coeliac–toxic gliadin peptides. Eur. Food Res. Technol. 222 (1), 78–82.

Kanerva, P.M., Sontag-Strohm, T.S., Ryöppy, P.H., Alho-Lehto, P., Salovaara, H.O., 2006. Analysis of barley contamination in oats using R5 and ω-gliadin antibodies. J. Cereal Sci. 44 (3), 347–352.

Khlestkina, E.K., Huang, X.Q., Quenum, F.J.-B., Chebotar, S., Röder, M.S., Börner, A., 2004. Genetic diversity in cultivated plants—loss or stability? Theoret. Appl. Genet. 108 (8), 1466–1472.

Kučerová, J., 2004. Technologie Cereálií. MZLU, Brno, p. 141.

Lester, D.R., 2008. Gluten measurement and its relationship to food toxicity for celiac disease patients. Plant Meth. 4 (26), 1–5.

Li, S., Zhang, Q.H., 2001. Advances in the development of functional foods from buckwheat. Food Sci. Nutr. 41 (6), 451–464.

Matta, N.K., Singh, A., Kumar, Y., 2009. Manipulating seed storage proteins for enhanced grain quality in cereals. Afr. J. Food Sci. 3 (13), 439–446.

Méndez, E., Vela, C., Immer, U., Janssen, F.W., 2005. Report of a collaborative trial to investigate the performance of the R5 enzyme linked immunoassay to determine gliadin in gluten-free food. Eur. J. Gastroenterol. Hepatol. 17 (10), 1053–1063.

Morón, B., Cebolla, Á., Manyani, H., Álvarez-Maqueda, M., Megías, M., Del Carmen Thomas, M., López, M.C., Sousa, C., 2008. Sensitive detection of cereal fractions that are toxic to celiac disease patients by using monoclonal antibodies to a main immunogenic wheat peptide. Am. J. Clin. Nutr. 87 (2), 405–414.

Mowat, A.M., 2003. Coeliac disease—a meeting point for genetics, immunology, and protein chemistry. Lancet 361 (9365), 1290–1292.

Nałęcz, D., Dziuba, J., Minkiewicz, P., Dziuba, M., Szerszunowicz, I., 2009. Identification of oat (*Avena sativa*) and buckwheat (*Fagopyrum esculentum*) proteins and their prolamin fractions using two-dimensional polyacrylamide gel electrophoresis. Eur. Food Res. Technol. 230 (1), 71–78.

Osborne, T.B., 2009. In: Matta, N.K., Singh, A., Kumar, Y. Manipulating seed storage proteins for enhanced grain quality in cereals. Afr. J. Food Sci. 3(13), pp. 439–446.

Osman, A.A., Uhlig, H.H., Valdes, I., Amin, M., Méndez, E., Mothes, T., 2001. A monoclonal antibody that recognizes a potential celiac-toxic repetitive pentapeptide epitope in gliadins. Eur. J. Gastroenterol. Hepatol. 13 (10), 1189–1193.

Petr, J., Michalík, I., Tlaskalová, H., Capouchová, I., Faměra, O., Urminská, D., Tučková, L., Knoblochová, H., 2003. Extension of the spectra of plant products for the diet in coeliac disease. Czech J. Food Sci. 21 (2), 59–70.

Pomeranz, Y., 1983. Grain endosperm structure and end-use properties. Progress in Cereal Chemistry and Technology, Developments in Food Science, vol. 5. Czechoslovak Medical Press, Praha, p. 694.

Prugar, J, Hraška, Š., 1986. Kvalita Pšenice. Príroda, Bratislava, p. 224.

Rout, M.K., Chrungoo, N.K., 1996. Partial characterization of the lysine rich 280 kD globulin from common buckwheat *Fagopyrum esculentum* Moench: its antigenic homology with seed proteins of some other crops. Biochem. Biol. Int. 40 (3), 587–595.

Rout, M.K., Chrungoo, N.K., 1999. The lysine and methionine rich basic subunit of buckwheat grain legumin: some results of astructural study. Biochem. Biol. Int. 47 (6), 921–926.

Sciarini, L.S., Ribotta, P.D., León, A.E., Pérez, G.T., 2010. Influence of gluten-free flours and their mixtures on batter properties and bread quality. Food Bioprocess Technol. 3 (4), 577–585.

Shewry, P.R., 2004. Improving the protein content and quality of temperate cereals: wheat, barley and rye. Çakmak, I., Welch, R.M. (Eds.), Impacts of Agriculture on Human Health and Nutrition, vol. 1, EOLSS Publishers Co Ltd., Exford.

Skerritt, J.H., 1986. Molecular comparison of alcohol-soluble wheat and buckwheat proteins. Cereal Chem. 63 (4), 365–369.

Skerritt, J.H., Andrews, J.L., Blundell, M., Beasley, H.L., Bekes, F., 1994. Applications and limitations of immunochemical analysis of biopolymer quality in cereals. Food Agric. Immunol. 6 (2), 173–184.

Skrabanja, V., Elmståhl Lijberg, H.G.M., Kreft, I., Björck Inger, M.E., 2001. Nutritional properties of starch in buckwheat product: studies in vitro and in vivo. J. Agric. Food Chem. 49, 490–496.

Skylas, D.J., 2004. Proteomics of grains. In: Wrigley, C., Corke, H., Walker, Ch. (Eds.), Encyclopedia of Grain Science, Elsevier, Oxford, pp. 480–488.

Sorell, L., López, J.A., Valdés, I., Alfonso, P., Camafeita, E., Acevedo, B., Chirdo, F., Gavilondo, J., Méndez, E., 1998. An innovative sandwich ELISA system based on an antibody cocktail for gluten analysis. FEBS Lett. 439 (1), 46–50.

Southgate, D.A.T., 2000. Cereals and cereal products. In: Garrow, J.S., James, W.P.T., Ralph, A. (Eds.), Human Nutrition and Dietetics. Elsevier, Netherlands, pp. 333–347.

Steadman, K.J., Burgoon, M.S., Lewis, B.A., Edwardson, S.E., Obendorf, R.L., 2001. Buckwheat seed milling fractions: description, macronutrient composition and dietary fibre. J Cereal Sci. 33, 271–278.

Stern, M., Ciclitira, P.J., van Eckert, R., Feighery, C., Janssen, F.W., Méndez, E., Mothes, T., Troncone, R., Wieser, H., 2001. Analysis and clinical effects of gluten in coeliac disease. Eur. J. Gastroenterol. Hepatol. 13 (6), 741–747.

Thompson, T., Méndez, E., 2008. Commercial assays to assess gluten content of gluten-free foods: why they are not created equal. J. Am. Dietet. Assoc. 108 (10), 1682–1687.

Vaccino, P., Becker, H.A., Brandolini, A., Salamini, F., Kilian, B., 2009. A catalogue of *Triticum monococcum* genes encoding toxic and immunogenic peptides for celiac disease patients. Mol. Genet. Genom. 281 (3), 289–300.

Valdés, I., García, E., Llorente, M., Méndez, E., 2003. Innovative approach to low-level gluten determination in foods using a novel sandwich enzyme-linked immunosorbent assay
protocol. Eur. J. Gastroenterol. Hepatol. 15 (5), 465–474.

van Eckert, R., Bond, J., Rawson, P., Klein, Ch.L., Stern, M., Jordan, T.W., 2010. Reactivity
of gluten detecting monoclonal antibodies to a gliadin reference material. J. Cereal Sci.
51 (2), 198–204.

Wieser, H., 2007. Chemistry of gluten proteins. Food Microbiol. 24 (2), 115–119.

Wieser, H., Koehler, P., 2008. The biochemical basis of celiac disease. Cereal Chem. 85 (1),
1–14.

Wijngaard, H.H., Arendt, E.K., 2006. Buckwheat. Cereal Chem. 83 (4), 391–401.

Wrigley, C., 2004. Cereals—overview. In: Wrigley, C., Corke, H., Walker, Ch. (Eds.), Encyclopedia of Grain Science, Elsevier, Oxford, pp. 187–201.

Wronkowska, M., Soral-Śmietana, M., 2008. Buckwheat flour—a valuable component of
gluten-free formulations. Polish J. Food Nutr. Sci. 58 (1), 59–63.

Xie, Z., Wang, C., Wang, K., Wang, S., Li, X., Zhang, Z., Ma, W., Yan, Y., 2010. Molecular
characterization of the celiac disease epitope domains in a-gliadin genes in *Aegilops
tauschii* and hexaploid wheats (*Triticum aestivum* L.). Theoret. Appl. Genet. 121 (7),
1239–1251.

Yalçin, E., 2010. Effect of partial removal of prolamins on some chemical and functional
properties of barley flours. Food Sci. Biotechnol. 19 (3), 735–742.

Yu-Xia, Y., Wei, W., You-Liang, Z., Wen-Ting, Y., Qian-Rong, C., 2008. Genetic diversity
of storage proteins in cultivated buckwheat. Pakistan J. Biol. Sci. 11 (13), 1662–1668.

Chapter seventeen

Toward the Use of Buckwheat as an Ingredient for the Preparation of Functional Food

A. Brunori*, C. Nobili**, S. Procacci*

*ENEA, SSPT-BIOAG Department, Laboratory of BioProducts and BioProcesses, Rome, Italy; **ENEA, SSPT-BIOAG Department, Laboratory of Sustainable Development and Innovation of Agro-industrial System, Rome, Italy

MODERN DIET AND HEALTH SCENARIO

The concept that diet may have a significant impact on human health and well-being is a fact widely accepted by the scientific community. Also consumers are gaining awareness about the strong relationship between food and health and would be positively attracted by functional food products, because of the limited access to these kinds of aliments and higher costs their choice is forcedly directed to more easily available and affordable food of compromised nutritive value, which is heavily marketed, readily accessible, and rich in calories and contributes to foster several health problems such as obesity, diabetes, and cardiovascular disorders.

In this context the improvement of the population's diet has recently been identified by the World Health Organization (WHO), within the European Food and Nutrition Action Plan 2015–20 (EUR/RC64/14), as the most efficient strategy to counteract the insurgence of diet-related noncommunicable chronic diseases. Together with regular physical activity, dietary habits can significantly contribute to maintain a good state of health.

For the aforementioned reasons, the common use of dietary supplements and functional foods is suggested as a must in today's lifestyle (Burdock et al., 2006;

Molecular Breeding and Nutritional Aspects of Buckwheat. http://dx.doi.org/10.1016/B978-0-12-803692-1.00017-1

Choi et al., 2006; Olmedilla-Alonso et al., 2006; Sieber, 2007). The terminology "physiologically functional foods" or simply "functional foods" was conceived over three decades ago in the context of nutrition during space travel (Dymsza, 1975). In the early 1980s the Japanese Academic Society promoted the broadening of the consumption of functional food among the common people as a mean toward the prevention of diet-related diseases. Successively, in the early 1990s still in Japan, a first classification of aliments that possess health beneficial properties, identified as Foods for Specified Health Use, was attempted (Ashwell, 2002). Similarly, in Europe during at the same time, the Functional Food Science in Europe Concerted Action had the purpose of reaching consensus on scientific concepts of functional foods, their identification, development, and the scientific substantiation of their effects. This action proposed as a conclusion "a working definition" of functional foods: foods can be regarded as functional if they can be satisfactorily demonstrated to affect beneficially one or more target functions in the body, beyond adequate nutritional effects, in a way relevant to an improved state of health and well-being and/or reduction of risk of disease. The better functionality of foods can be promoted in different ways (Ashwell, 2002). A better understanding of the body's physiology helps in establishing the relationships between nutrition, lifestyle, and genetic predisposition including its effects on health and quality of life (VV.AA., 2014).

Functional foods must remain foods and they must be able to attain their effects in amounts normally consumed in a diet (Contor, 2001). These types of aliments are comprised of foods containing naturally healthy molecules as well as foods enriched with health-promoting additives. The search for functional foods, or functional food ingredients, is one of the leading trends in today's food industry.

REGULATORY HINTS ABOUT FUNCTIONAL FOOD

To propose an ingredient/product as a functional food with substantiated health claims, which may be labeled, the applicant has to submit an authorization request to the European Food Safety Authority (EFSA) that in turn will express a scientific opinion to its approval. Within this procedure, the key element is the requirement of scientific evidences that demonstrate cause-and-effect relationships between the intake of a compound and a health benefit.

Food health claims must be supported by appropriate and acceptable research results and clinical studies, and their approval is as rigorous as in the case of drugs. This may be difficult to achieve as foods are mixtures of known and unknown chemicals difficult to be described completely in their wholeness.

All foods credited with nutritional properties or in possession of substantiated health claims have been subject to specific legislation since 2006: Regulation 1924/2006. The system put in place is based on premarketing approval of all claims. In addition, the scope of the regulation is broad, since it not only covers the labeling of foodstuffs but also the presentation and advertising of the food.

The regulation covers all possible claims relative to health. It covers nutritional claims (claims that state, suggest, or imply that a food has particular nutritional properties because of the energy it provides, at a reduced or increased rate, or does not provide, and/or the nutrients or other substances it contains, in reduced or increased proportions, or does not contain), health claims (claims that state, suggest, or imply that a relationship exists between a food category, a food, or one of its constituents and health), reduction of disease risk claims (claims that state, suggest, or imply that the consumption of a food category, a food, or one of its constituents significantly reduces a risk factor in the development of a human disease), and claims referring to the development and health of children (VV.AA., 2014).

The Regulation on Food Information to Consumers [Regulation (EU) 1169/2011], containing a long list of labeling requirements, including some specified warnings, all of which must be included on the product label, was integrated by the Commission Regulation (EU) No. 432/2012 of May 16, 2012, establishing a list of permitted health claims made on foods, other than those referring to the reduction of disease risk and to children's development and health.

BUCKWHEAT: AN OPPORTUNITY

Growing concern about the negative health effects of modern diets rich in refined wheat flour, sugar, fat, and protein of animal origin (Wadden et al., 2002), recognized as the main causes of obesity, diabetes, cardiovascular disorders, and degenerative diseases like cancer, has prompted renewed interest in underutilized minor crops rich in health beneficial bioactive compounds (Brunori et al., 2012). Minor crops (buckwheat, oat, millet, rye, etc.) are grown in limited areas or produced in small quantities, and have the main limitation to wider utilization in their reduced grain yield potential. The development of improved cultural practices, together with the creation of more productive cultivars, can afford to increase the yield and the quality of these crops, allowing full exploitation and use, thus offering to consumers the tools to take care of their health in a simple and cost-effective way.

Among minor crops, buckwheat (*Fagopyrum* spp.), a pseudocereal belonging to the Polygonaceae family (Fig. 17.1), has received increasing interest thanks to the growing evidence of beneficial properties of some grain components on health, including reduced starch digestibility (Skrabanja and Kreft, 1998) and consequently lowering of glycemic indices (Skrabanja et al., 2001), anticholesterolemic properties of the protein fraction (Kayashita et al., 1995, 1997; Tomotake et al., 2000, 2001), the well-balanced amino acid composition (Pomeranz and Robbins, 1972), and good source of dietary fiber (Steadman et al., 2001a) and minerals (Ikeda et al., 1995, 2006; Steadman et al., 2001b). Even more impressive are rutin (quercetin-3-rutinoside), a polyphenol contained in buckwheat achenes (Fig. 17.2A,B), health benefits. Some of these are: improved capillary fragility (Griffith et al., 1944), retarded development of diabetes (Odetti et al., 1990), antilipoperoxidant activities (Negre-Salvayre et al., 1991), anticancer activity (Deschner et al., 1991), antihyperglycemic effect (Wang et al., 1992), protective effects against hemoglobin

FIGURE 17.1 Buckwheat crop.

oxidation (Grinberg et al., 1994), a mitigation effect on cardiovascular diseases (He et al., 1995), antioxidative property (Oomah and Mazza, 1996), antimutagenic activity (Aheme and O'Brien, 1999), antiinflammatory activity (Guardia et al., 2001), mitigation of diabetes consequences (Je et al., 2002), suppression of protein glycation (Nagasawa et al., 2003), antiplatelet formation property (Sheu et al., 2004), antiangiogenic effect (Guruvayoorappan and Kuttan, 2007), and neuroprotective effect (Pu et al., 2007).

Thanks to the richness of bioactive compounds, buckwheat grain can be proposed as a functional ingredient and, moreover, being gluten free can be recommended for celiac disease-affected people (Alvarez-Jubete et al., 2010).

FIGURE 17.2 (A) *F. esculentum*, and (B) *F. tataricum* achenes.

USE OF BUCKWHEAT AS A POTENTIAL FUNCTIONAL INGREDIENT

Two buckwheat species are mainly used for human consumption. Common buckwheat (*Fagopyrum esculentum* Moench; Fig. 17.3A) originates from Southwest China and is cultivated in Russia, Japan, Canada, and Europe. This species was historically cultivated in Italy in limited areas of the Alps and the Central and Northern Apennines. Gradually replaced by wheat and other cereal with higher yields, recently this crop has aroused new interest with respect to using it for the preservation of biodiversity and the recovery of marginal areas. Tartary

FIGURE 17.3 (A) *F. esculentum* and, (B) *F. tataricum* plants.

FIGURE 17.4 Bakery products prepared with buckwheat flour.

buckwheat (*Fagopyrum tataricum* Gaertn.; Fig. 17.3B) grows and is used in the mountainous areas of Southwest China, Northern India, Bhutan, and Nepal, and is also cultivated in limited regions of Slovenia, Italy, and northern Europe. This crop, compared to common buckwheat, is less widespread because of its bitter taste although it possesses a more interesting nutritional profile. Tartary buckwheat grain is richer in rutin and quercetin than common buckwheat (Fabjan et al., 2003), so the introduction of its flour in the recipes of traditional foods products (eg, bakery products and pasta; Fig. 17.4) has led to innovative and potential functional foods (De Rossi et al., 2013; Wieslander et al., 2011). The regular intake of buckwheat cookies may favor the reduction of myeloperoxidase levels, an indicator of inflammation, and may lower serum cholesterol levels (Wieslander et al., 2011).

The antioxidant properties of polyphenols have been extensively studied, and it has become clear that the mechanisms of action of polyphenols go beyond the modulation of oxidative stress (Scalbert et al., 2005). Prooxidant, antioxidant, or any of the many other biological effects potentially exerted by polyphenols accounting for or contributing to the health benefits of diets rich in plant-derived foods and beverages remains to be definitively proven (Halliwell, 2007). In fact, up to now, there are no approved health claims about buckwheat or molecules contained in its grain, although some attempts have been made. Proposal 1482, suggesting the immune-system support of buckwheat extract containing a flavonoid–mineral complex, was not approved [EFSA opinion reference/Journal reference 2011; 9(6): 2228]; in addition, the beneficial effect of rutin and quercetin on human health has yet to be demonstrated (proposal ID 1884, 1845, 1846, 1847, 1783, 1784).

Nevertheless, scientists will not forsake the opportunity to propose additional health claims to demonstrate the beneficial effects of buckwheat grain.

As it is well known, in nature, phytochemicals are present as a mixture that renders it to be functionally active, which is in contrast to the "one active principle" hypothesis. In other words, the functional activity of the phytochemicals is not linked to the component alone but is triggered by the presence of other selected components that constitute the mixture. Although dietary phytochemicals play a major role in regulating human health and disease (Jatoi et al., 2007), further attempts could be made concentrating research activities on a pool of biomolecules contained in buckwheat grains (digestible starch, protein, vitamins, minerals, and amino acids).

CONCLUSIONS

There is an increasing societal awareness of the opportunities to improve life quality through healthy eating and of the contribution that sustainable production can make to overall improvement of the environment. The preference of consumers for quality, convenience, diversity, and health, and their justifiable expectations of safety, ethics, and sustainable food production, serve to highlight the opportunities for innovation (The Milan Charter, 2015). To provide an answer to the foregoing issues, an opportunity is offered by the use of underutilized minor crops. The promotion of minor crop cultivation is in accordance with the commitments of sustainable agriculture emphasized in EASAC Policy Report (2011), thus ensuring eco-efficient production, more nutritious and quality foods, and minimizing land use and inputs.

In this context, research activities were carried out, and results obtained by chemical characterization of prototype foods from Tartary buckwheat flour showed an interesting nutritional profile. From this could stem the identification of a pool of biomolecules within the same matrix, thus allowing the opportunity to obtain a more effective natural "multifunctional" ingredient/food without the need to go through a process of functionalization (ie, external addition of biomolecules).

REFERENCES

Aheme, S.A., O'Brien, N.M., 1999. Protection by the flavonoids myricetin, quercetin, and rutin against hydrogen peroxide-induced DNA damage in Caco-2 HepG2 cells. Nutr. Cancer 34, 160–166.

Alvarez-Jubete, L., Wijngaard, H., Arendt, E.K., Gallaghe, R.E., 2010. Polyphenol composition and in vitro antioxidant capacity of amaranth, quinoa, buckwheat and wheat as affected by sprouting and bread baking. Food Chem. 119 (2), 770–778.

Ashwell, M., 2002. Concepts of Functional Foods. ILSI Europe Concise Monograph Series pp. 1–39.

Brunori, A., Baviello, G., Teixeira da Silva, J., Győri, T., Végvári, G., 2012. Grain yield and rutin content of common and Tartary buckwheat varieties grown in north-western Hungary. Eur. J. Plant Sci. Biotechnol. 6 (2), 70–74.

Burdock, G.A., Carabin, I.G., Griffiths, J.C., 2006. The importance of GRAS to the functional food and nutraceutical industries. Toxicology 221, 17–27.

Choi, Y.M., Bae, S.H., Kang, D.H., Suh, H.J., 2006. Hypolipidemic effect of lactobacillus ferment as a functional food supplement. Phytother. Res. 20, 1056–1060.

Contor, L., 2001. Functional food science in Europe. Nutr. Metab. Cardiovasc. Dis. 11 (Suppl. 4), 20–23.

De Rossi, P., Del Fiore, A., Tolaini, V., Presenti, O., Antonini, A., Procacci, S., Nobili, C., Baviello, G., Zannettino, C., Corsini, G., Vitali, F., Brunori, A., 2013. Gli alimenti funzionali: potenzialità di utilizzo del grano saraceno tartarico. Molini d'Italia 9, 30–34.

Deschner, E.E., Ruperto, J., Wong, G., Newmark, H.L., 1991. Quercitin and rutin as inhibitors of azoxymethanol-induced colonic neoplasia. Carcinogenesis 12, 1193–1196.

Dymsza, H.A., 1975. Nutritional application and implication of 1,3-butanediol. Fed. Proc. 34, 2167–2170.

EASAC, 2011. Plant genetic resources for foods and agriculture: roles and research priorities in the European Union. Policy Report, 17.

Fabjan, N., Rode, J., Kosir, I.J., Wang, Z., Kreft, I., 2003. Tartary buckwheat (*Fagopyrum tataricum* Gaertn.) as a source of dietary rutin and quercetin. J. Agric. Food Chem. 51, 6452–6455.

Griffith, J.Q., Couch, J.F., Lindauer, M.A., 1944. Effect of rutin on increased capillary fragility in man. Proc. Soc. Exp. Biol. Med. 55, 228–229.

Grinberg, L.N., Rachmilewitz, E.A., Newmark, H., 1994. Protective effects of rutin against hemoglobin oxidation. Biochem. Pharmacol. 48, 643–649.

Guardia, T., Rotelli, A.E., Juárez, A.O., Pelzer, L.E., 2001. Anti-inflammatory properties of rutin, quercetin and hesperidin on adjuvant arthritis in rat. Farmaco 56, 683–687.

Guruvayoorappan, C., Kuttan, G., 2007. Antiangiogenic effect of rutin and its regulatory effect on the production of VEGF, IL-1β and TNF-α in turnover associated macrophages. J. Biol. Sci. 7, 1511–1519.

Halliwell, B., 2007. Dietary polyphenols: good, bad, or indifferent for your health? Cardiovasc. Res. 73, 341–347.

He, J., Klag, M.J., Whelton, P.K., Mo, J.P., Chen, J.Y., Qian, M.G., Mo, P.S., He, G.Q., 1995. Oats and buckwheat intake and cardiovascular disease risk factors in an ethnic minority of China. Am. J. Clin. Nutr. 61, 366–372.

Jatoi, S.A., Kikuchi, A., Gilani, S.A., Watanabe, K.N., 2007. Phytochemical, pharmacological and ethnobotanical studies in mango ginger (*Curcuma amada* Roxb; Zingiberaceae). Phytother. Res. 21 (6), 507–516.

Ikeda, S., Yamashita, Y., Murakami, T., 1995. Minerals in buckwheat. In: Matano, T., Ujihara, A. (Eds.),Current advances in buckwheat research. Proceedings of the 6th International Symposium on Buckwheat, Shinshu University Press, Shinshu, Japan, 789–792.

Ikeda, S., Yamashita, Y., Tomura, K., Kreft, I., 2006. Nutritional comparison in mineral characteristics between buckwheat and cereals. Fagopyrum 23, 61–65.

Je, H.D., Shin, C.Y., Park, S.Y., Yim, S.H., Kum, C., Huh, I.H., Kim, J.H., Sohn, U.D., 2002. Combination of vitamin C and rutin on neuropathy and lung damage of diabetes mellitus rats. Arch. Pharm. Res. 25 (2), 184–190.

Kayashita, J., Shimaoka, I., Nakajoh, M., 1995. Hypocholesterolemic effect of buckwheat protein extract in rat fed cholesterol enriched diets. Nutr. Res. 15, 691–698.

Kayashita, J., Shimaoka, I., Nakajoh, M., Yamazaki, M., Kato, N., 1997. Consumption of buckwheat protein lowers plasma cholesterol and raises fecal neutral sterols in cholesterol-fed rats because of its low digestibility. J. Nutr. 127, 1395–1400.

Nagasawa, T., Tabata, N., Ito, Y., Aiba, Y., Nishizawa, N., Kitts, D.D., 2003. Dietary G-rutin suppresses glycation in tissue proteins of streptozoticin-induced diabetic rats. Mol. Cell. Biochem. 252, 141–147.

Negre-Salvayre, A., Affany, A., Hariton, C., Salvayre, R., 1991. Additional antilipoperoxidant activities of alpha-tocopherol and ascorbic acid on membranelike systems are potentiated by rutin. Pharmacology 42 (5), 262–272.

Odetti, P.R., Borgoglio, A., De Pascale, A., Rolandi, R., Adezati, L., 1990. Prevention of diabetes-increased aging effect on rat collagen-linked fluorescence by aminoguanidine and rutin. Diabetes 39 (7), 796–801.

Olmedilla-Alonso, B., Granado-Lorencio, F., Herrero-Barbudo, C., Blanco-Navarro, I., 2006. Nutritional approach for designing meat-based functional food products with nuts. Crit. Rev. Food Sci. Nutr. 46, 537–542.

Oomah, B.D., Mazza, G., 1996. Flavonoids and antioxidative activities in buckwheat. J. Agric. Food Chem. 44, 1746–1750.

Pomeranz, Y., Robbins, G.S., 1972. Amino acid composition of buckwheat. J. Agric. Food Chem. 20, 270–274.

Pu, F., Mishima, K., Irie, K., Motohashi, K., Tanaka, Y., Orito, K., Egawa, T., Kitamura, Y., Egashira, N., Iwasaki, K., Fujiwara, M., 2007. Neuroprotective effects of quercetin and rutin on spatial memory impairment in an 8-arm radial maze task and neuronal death induced by repeated cerebral ischemia in rats. J. Pharm. Sci. 104, 329–334.

Scalbert, A., Johnson, I.T., Saltmarsh, M., 2005. Polyphenols: antioxidants and beyond. Am. J. Clin. Nutr. 81, 215S–217S.

Sheu, J.R., Hsiao, G., Chou, P.H., Shen, M.Y., Chou, D.S., 2004. Mechanisms involved in the antiplatelet activity of rutin, a glycoside of the flavonoid quercetin, in human platelets. J. Agric. Food Chem. 52, 4414–4418.

Sieber, C.C., 2007. Functional food in elderly persons. Ther. Umsch. 64, 141–146.

Skrabanja, V., Kreft, I., 1998. Resistant starch formation following autoclaving of buckwheat (*Fagopyrum esculentum* Moench) groats. An in vitro study. J. Agric. Food Chem. 46, 2020–2023.

Skrabanja, V., Liljeberg Elmsttahl, E.H.G.M., Kreft, I., Bjorck, I.M.E., 2001. Nutritional properties of starch in buckwheat products: Studies in vitro and in vivo. J. Agric. Food Chem. 49, 490–496.

Steadman, K.J., Burgoon, M.S., Lewis, B.A., Edwardson, S.E., Obendorf, R.L., 2001a. Buckwheat seed milling fractions: description, macronutrient composition and dietary fibre. J. Cereal Sci. 33 (3), 271–278.

Steadman, K.J., Burgoon, M.S., Lewis, B.A., Edwardson, S.E., Obendorf, R.L., 2001b. Minerals, phytic acid, tannin and rutin in buckwheat seed milling fractions. J. Sci. Food Agric. 81, 1094–1100.

The Milan Charter, 2015. Available from: http://carta.milano.it/en/.

Tomotake, H., Shimoaka, I., Katashita, J., Yokoyama, F., Nakajoh, M., Kato, M., 2000. A buckwheat protein product suppresses gallstone formation and plasma cholesterol more strongly than soy protein isolate in hamster. J. Nutr. 130, 1670–1674.

Tomotake, H., Shimoaka, I., Kayashita, J., Nakajoh, M., Kato, M., 2001. Buckwheat protein suppresses plasma cholesterol more strongly than soy protein isolate in rats by enhancing fecal excretion of steroids. In: Ham, S.S., Choi, Y.S., Kim, N.S., Park, C.H. (Eds.), Advances in buckwheat research. Proceedings of the Eighth International Symposium on Buckwheat, Organizing Committee of the Eighth International Symposium on Buckwheat, Chunchon, Korea, pp. 595–601.

VV.AA., 2014. Nutraceutical and Functional Food Regulations in the United States and Around the World. Edited by Debasis Bagchi, PhD MACN CNS MAIChE, Department of Pharmacological and Pharmaceutical Sciences, University of Houston College of Pharmacy Houston, TX, USA.

Wadden, T.A., Brownell, K.D., Foster, G.D., 2002. Obesity: responding to the global epidemic. J. Consult. Clin. Psychol. 70 (3), 510–525.

Wang, J., Liu, Z., Fu, X., Run, M., 1992. A clinical observation on the hypoglycemic effect of Xinjiang buckwheat. In: Lin, R., Zhou, M., Tao, Y., Li, J., Zhang, Z. (Eds.), Proceedings of the 5th International Symposium on Buckwheat,Agriculture Publishing House, Beijing, China, 465–467.

Wieslander, G., Fabjan, N., Vogrinčič, M., Kreft, I., Janson, C., Spetz-Nyström, U., Vombergar, B., Tagesson, C., Leanderson, P., Norbäck, D., 2011. Eating buckwheat cookies is associated with the reduction in serum levels of myeloperoxidase and cholesterol: a double blind crossover study in day-care centre staffs. Tohoku J. Exp. Med. 225, 123–130.

Chapter eighteen

Buckwheat in the Nutrition of Livestock and Poultry

F. Leiber

FiBL, Research Institute of Organic Agriculture, Frick, Switzerland

INTRODUCTION: POTENTIAL GLOBAL SIGNIFICANCE OF BUCKWHEAT IN LIVESTOCK AND POULTRY NUTRITION

Common buckwheat (*Fagopyrum esculentum*) and Tartary buckwheat (*Fagopyrum tataricum*) are grown on a broad range of climatic regions across all continents (Benvenuti et al., 2011). It is still very common in Asian and Eastern European human diets; however, it has almost disappeared from Western diets. For centuries, buckwheat was also present in many regions of Western Europe because it was a stable crop also on poor soils or in difficult, dry climates (Zeller and Hsam, 2004). Due to its rapid vegetative development until ripening, buckwheat can be grown in many regions as a second crop after the harvest of main cereal crops like wheat or barley in late summer (Baumgärtner et al., 1998; Gupta et al., 2002). This makes it currently a particularly interesting crop for animal feed: it can be produced as a second crop after the main cereals, which can be used for human nutrition. The utilization of main arable crops for animal feed production is increasingly facing global ecological constraints, and claims for restricting feed to other resources are made frequently (Wilkinson, 2011; Schader et al., 2015). In this situation the utilization of buckwheat as a second crop could be a valuable option for feeding livestock and poultry, thereby avoiding the direct feed–food competition.

Besides, buckwheat cultivation comprises some beneficial ecological effects. Due to deep rooting and fast leaf development, buckwheat improves soil structure and fertility and effectively suppresses weeds (Bjorkman, 2008). A further function of late-summer buckwheat plantations is the provision of insect feed by this intensively flowering culture. This is of particular value for honeybees, but also for wild insects, thus maintaining ecological equilibriums and

229

Molecular Breeding and Nutritional Aspects of Buckwheat. http://dx.doi.org/10.1016/B978-0-12-803692-1.00018-3

supporting biological control agents and other beneficial organisms (Lee and Heimpel, 2005; Pontin et al., 2006). However, in many industrialized countries, buckwheat has largely disappeared from human diets and fields. Reintroduction of this crop as an animal feed could be a means to support its cultivation and thus establish the mentioned ecological side effects.

Briefly, the introduction of buckwheat as animal feed would increase the demand for its cultivation. By this, it could thus promote several effects, which are of particular importance in the context of the contemporary ecological pressure on crop production systems: (1) it could partly contribute to mitigate the feed–food competition regarding arable land areas, (2) it would increase biological diversity in crop rotations and improve soil fertility, and (3) it would provide an important late-summer and fall nutrient source for insects, thus maintaining faunal biodiversity.

Buckwheat grain is mainly a starch source (Wijngaard and Arendt, 2006). Protein concentration in common buckwheat is rather low for poultry and pig nutrition. However, the high concentrations of lysine and methionine (Pomeranz and Robbins, 1972; Jacob, 2007) make it a suitable protein source, in particular for chickens. High concentrations of nutritionally important minerals in buckwheat such as Mg, Zn, K, P, Cu, and Mn (Wijngaard and Arendt, 2006), as well as Se (Li and Zhang, 2001) are further characteristics for a high-value animal feed. Of particular interest are the exceptionally high concentrations of tocopherols (Wijngaard and Arendt, 2006) and of rutin (Kalinova et al., 2006), which are both effective antioxidants, thus potentially contributing to animal health and product quality.

A constraint in using buckwheat as animal feed can be the presence of fagopyrin in the buckwheat herb, which may cause photosensitivity in the skin, and lead to allergic reactions and even death of animals after consumption. However, the reported cases are scarce, old, and partly rather anecdotal (Wender et al., 1943; Van Wyk et al., 1952; Jover, 1968), and a scientific update on this issue appears to be necessary to define dietary thresholds for different animal categories and forms of buckwheat supplementation.

BUCKWHEAT FOR RUMINANT SPECIES

For ruminant feed, buckwheat can be considered as whole grain, shelled kernel, or as total plant, fed freshly, or in conserved forms. Only few data on feeding buckwheat to ruminants are reported; however, it appears that the nutritive value is reasonable and some beneficial side effects could be achieved by integrating this plant into diets. Thus buckwheat forage offers a valuable opportunity to utilize less fertile arable lands (O'Meara, 2013) or late summer gaps in crop rotations (Baumgärtner et al., 1998) to produce a sustainable feed component for ruminants.

Total Buckwheat Plant as Ruminant Feed

For ruminants the whole buckwheat plant can serve as a valuable feed resource. This offers more opportunities because forage must be harvested before the plant reaches full maturity (Kälber et al., 2014), thus a shorter timeframe for feed production is needed and cultivation can be still successful if buckwheat is sown lately. Data ranges for the most relevant nutrients to ruminant livestock are presented in Table 18.1.

Grazing buckwheat stubbles with sheep is common in several regions; however, palatability and nutritive value are rather poor (Mulholland and Coombe, 1979).

Feeding freshly harvested total buckwheat plant to dairy cows (at 70% of the total dietary dry matter) resulted in higher intakes and similar milk yields compared with a ryegrass-based diet (Kälber et al., 2011). In vitro, fresh buckwheat herb improved ruminal protein utilization compared to ryegrass (Amelchanka et al., 2010). However, there is a clear decrease in efficiency with

TABLE 18.1 Ranges of Nutrient Values of Different Forms of Buckwheat for Ruminant Nutrition (per kg dry matter)

	Crude Protein (g)		ADF[a] (g)		NEL[b] (MJ)		References
	Min	Max	Min	Max	Min	Max	
Total Plant Forages							
Fresh	110	202	299	435	4.27[c]	4.27[c]	Amelchanka et al. (2010), Kälber et al. (2011), Kälber et al. (2014), Kling and Wöhlbier (1983), Leiber et al. (2012)
Silage	119	154	427	445	4.91[c]	4.91[c]	Amelchanka et al. (2010), Kälber et al. (2012), Kara and Yüksel (2014)
Hay	68	178	311	443			Alencastro (2014), Kara and Yüksel (2014), O'Meara (2013)
Grains							
Whole grain	10.8	137	154	169	7.51[c]	7.51[c]	Amelchanka et al. (2010), Farrell (1978), Gupta et al. (2002), Kara and Yüksel (2014), Leiber et al. (2009b), Leiber et al. (2012), Mulholland and Preston (1995)
Dehulled grain	119	140	28	28			Gupta et al. (2002), Jacob (2007), Leiber et al. (2009b), Wijngaard and Arendt (2006)

[a]Acid detergent fiber.
[b]Net energy for lactation.
[c]Only determined in Amelchanka et al. (2010).

ongoing phenological development. Buckwheat fed in the flowering stage resulted in clearly lower milk yields than when fed in the vegetative stage (Kälber et al., 2014).

Ensiling buckwheat did not change ruminal degradability (Amelchanka et al., 2010), and buckwheat silage may reach crude protein concentrations of up to 15% (Kara and Yüksel, 2014). Intake and performance of dairy cows being fed buckwheat silage was similar to ryegrass-based control diets (Amelchanka et al., 2010; Kälber et al., 2012). A numerically lower milk yield of cows fed buckwheat silage compared to ryegrass silage (Kälber et al., 2012) was explainable by a lower crude protein concentration compared to the control. However, the protein conversion efficiency from feed to milk was better with buckwheat than with ryegrass silage.

Only few data are achievable regarding buckwheat hay, showing a broad range of nutrient concentrations (Table 18.1). One practice feeding experiment was reported (O'Meara, 2013), where dairy cows received 6.8 kg/day of buckwheat hay, partly replacing high-quality pasture hay. Despite low crude protein concentrations in the buckwheat hay, cows performed equally with the control group over several months.

Buckwheat Grains for Ruminants

Feeding buckwheat grains to ruminants has been rarely evaluated; however, always with favorable results. Hulled grains of Tartary and common buckwheat have been demonstrated to have fair feeding values to sheep (Mulholland and Preston, 1995) and cattle (Nicholson et al., 1976; Amelchanka et al., 2010). Data ranges for important nutrients to ruminant livestock are presented in Table 18.1.

Nicholson et al. (1976) reported that although the nutrient digestibility for ruminants was lower than in barley, steam-rolled Tartary buckwheat could serve as a reasonable replacer for cereals in diets for fattening steers, without impairing their growth rates. However, steaming seemed to be a precondition, since feeding dry-rolled buckwheat failed to reach sufficient palatability and comparable weight gains in steers. This suggests the presence of antinutritive factors in Tartary buckwheat, which were inhibited by the steaming process. Amelchanka et al. (2010) fed crushed grains with hulls of common buckwheat to dairy cows at a rate of 94g/kg dry matter of the total diet, replacing partly a commercial wheat-based energy concentrate. This did not have any effects on dietary nutrient composition, intake rates, as well as milk yield and composition. Comparison of wheat and buckwheat grain in an in vitro fermentation test revealed similar nutrient degradability, though slightly less efficient ruminal N utilization for buckwheat (Amelchanka et al., 2010). Feeding whole buckwheat grain instead of wheat or oats to sheep resulted in similar wool growth performance, although buckwheat digestibility was lower (Mulholland and Preston, 1995).

Based on these studies it appears that whole buckwheat grains can serve at least partially as a reasonable replacer for cereal-based energy concentrates in ruminant nutrition. The fact that intake and digestibility can be reduced in buckwheat compared to wheat (Nicholson et al., 1976; Mulholland and Preston, 1995) might indicate the presence of antinutritive factors, the technological inhibition options of which still have to be identified.

Effects of Buckwheat on Ruminal Methanogenesis

Buckwheat feeding to ruminants may imply several functional effects on the rumen and the endogenous metabolism because of high concentrations of secondary metabolites, in particular rutin, and vitamins (Kalinova et al., 2006; Wijngaard and Arendt, 2006), which may modulate microbiological processes in the gut and the antioxidative potential of the tissues. Potential modulating effects on rumen fermentation include mitigation of methane production. However, only Leiber et al. (2012) found in vitro a significant methane mitigation with buckwheat herb and with isolated rutin incubations. Other in vitro studies found no significant buckwheat or rutin effects on ruminal methanogenesis (Broudiscou et al., 2000; Amelchanka et al., 2010; Berger et al., 2015).

Effects of Buckwheat Feed on Milk Quality

High concentrations of plant secondary compounds in forages may alter ruminal lipid metabolism and prevent essential unsaturated (in particular n–3) fatty acids from biohydrogenation (Jayanegara et al., 2011), finally leading to higher concentrations of these fatty acids in the animals' tissues and in milk (Morales and Ungerfeld, 2015). This effect has been demonstrated for buckwheat as well. In feeding trials with fresh forages and silages, transfer efficiency for α-linolenic acid (n–3) from feed to milk was 100% higher, when dairy cows were fed buckwheat herb compared to ryegrass (Kälber et al., 2011, 2013), associated with higher phenol concentrations in the buckwheat forages. A positive effect on conjugated linoleic acid in milk was only found with buckwheat silage (Kälber et al., 2013).

Buckwheat is reported to comprise particularly high concentrations of tocopherols (Wijngaard and Arendt, 2006). In cows' milk, tocopherol contents were increased after consumption of fresh buckwheat if compared to berseem clover, phacelia, and chicory (Kälber et al., 2011).

Feeding buckwheat silage improved the rennet coagulation properties of milk, which is an indicator for cheese-processing qualities (Kälber et al., 2013). The mechanism behind this is not clear, but it may be associated with the changes in protein metabolism induced by buckwheat forages (Amelchanka et al., 2010; Kälber et al., 2012).

BUCKWHEAT FOR PIGS

Including buckwheat into pigs' diets appears reasonable because of high concentrations of starch and the favorable concentrations of essential amino acids, in particular lysine, in the buckwheat protein (Table 18.2). Also the high concentrations of antioxidative phenols and vitamins (Kalinova et al., 2006; Wijngaard

TABLE 18.2 Ranges of Nutrient Values of Buckwheat for Poultry and Pigs

	Whole Grain		Dehulled Grain		References
	Min	Max	Min	Max	
Starch (g/kg dry matter)	545[a]		750[b]		Wijngaard and Arendt (2006), Zheng et al. (1998)
Ether extract (crude fat) (g/kg DM)	21	30	23	29	Eggum et al. (1981), Gupta et al. (2002), Leiber et al. (2009b), Pomeranz (1983), Zheng et al. (1998)
Soluble carbohydrates (g/kg dry matter)	640	733	678	710	Li and Zhang (2001), Pomeranz (1983)
Crude protein (g/kg dry matter)	108	137	120	168	Amelchanka et al. (2010), Anderson and Bowland (1984), Gupta et al. (2002), Leiber et al. (2009b), Li and Zhang (2001), Wijngaard and Arendt (2006), Zheng et al. (1998)
Lysine (g/100 g crude protein)	6.0[c]	5.0	7.0		Anderson and Bowland (1984), Jacob (2007), Li and Zhang (2001), Pomeranz and Robbins (1972)
Methionine (g/100g crude protein)	2.3[c]	1.6	3.0		Anderson and Bowland (1984), Jacob (2007), Li and Zhang (2001), Pomeranz and Robbins (1972)
Cysteine (g/100g crude protein)	1.6[c]	1.7	2.9		Anderson and Bowland (1984), Jacob (2007), Li and Zhang (2001)

[a]Only once determined by Wijngaard and Arendt (2006).
[b]Only once determined by Zheng et al. (1998).
[c]Only once determined in Li and Zhang (2001).

and Arendt, 2006) could be beneficial by improving the antioxidative stability of pork. However, only very few studies have been published on buckwheat in pigs' diets, yet. Regarding performance, the available data support that buckwheat can be rather efficiently used as pig food. Replacing wheat with buckwheat in diets for growing pigs led to similar nutrient digestibility and growth performance (Farrell, 1978; Anderson and Bowland, 1984). However, at high dietary proportions of buckwheat (above 60%), Van Wyk et al. (1952) observed impaired digestibility.

The hypothesis that buckwheat would positively affect lipid metabolism and antioxidatve stability in pigs could so far not be proven. Flis et al. (2010) found no effect of a phenol-rich buckwheat bran in pigs' diets on fatty acid profiles or oxidative stability in the muscle.

Pigs with access to open sky could be challenged by photosensitization when buckwheat is fed in high dosages (Van Wyk et al., 1952). Systematic studies of this issue are, however, lacking.

BUCKWHEAT FOR POULTRY

Due to its high nutrient density, buckwheat appears particularly suitable as a diet component for poultry. It is mainly a source of carbohydrates, but the comparably high methionine concentration in the protein (Table 18.2) makes also the latter fraction interesting for poultry nutrition. The fiber, as part of the hulls of the whole buckwheat grain, may be even a health and digestion advantage, if properly considered in the formulation of rations. A constraint may be the risk of increased photosensitivity due to fagopyrin in birds having ingested buckwheat. However, this risk is related to buckwheat forage (Jover, 1968), but not to the grain, which contains much lower fagopyrin concentrations compared to the herb (Stojilkovski et al., 2013).

Feeding Buckwheat to Layers

Including whole grain buckwheat or shelled buckwheat at a dietary level of 40% dry matter into a layers' diet resulted in equal laying performance and significantly heavier eggs compared to a wheat-based control diet (Leiber and Messikommer, 2011). Farrell (1978) found an opposite result, however, with a generally less performing genotype. Partially substituting corn and soybean components with buckwheat bran (30% dry matter) in a diet for layers maintained their performance on the same level as the control (Benvenuti et al., 2011). Another form of offering buckwheat to chickens is cultivation on free range areas (Horsted et al., 2006). It was shown that the buckwheat herb was well accepted by the hens without an effect on performance.

Buckwheat diets may have distinct effects on egg quality. Feeding whole buckwheat grains can improve the shell strength of the eggs (Leiber and Messikommer, 2011). The tocopherol concentration in egg yolk may be more than doubled when whole grain buckwheat replaces wheat (Leiber and Messikommer, 2011).

Feeding Buckwheat to Broiler Chickens

Replacing wheat or corn by buckwheat in broiler diets up to 40% of the total dietary dry matter results in similar growth rates, slaughter weights, and feed conversion coefficients (Gupta et al., 2002; Jacob and Carter, 2008; Leiber et al., 2009b). Farrell (1978) even found that buckwheat as a main basis of broiler feed was superior to several cereals, regarding growth rate and feed conversion. However, above 40% buckwheat in dietary dry matter, the feed conversion rate may considerably decrease due to decreasing weight gain (Gupta et al., 2002) or increasing intake (Jacob and Carter, 2008). This might be caused by reaching a critical threshold of fiber concentration in the feed because of the hulls of the whole buckwheat grain. Feeding dehulled buckwheat results in a similar performance to whole grain buckwheat (Leiber et al., 2009b); however, the technical process of shelling increases the feed production costs. Thus a reasonable threshold for the dietary inclusion of whole buckwheat grain into broiler diets is apparently at 40% dry matter.

High tocopherol concentrations in buckwheat (Wijngaard and Arendt, 2006) may lead to significantly increased tocopherol concentrations in broiler meat, when buckwheat replaces wheat in the diet (Leiber et al., 2009a). Thus dietary buckwheat clearly improves the nutritional value of poultry products, including eggs and meat.

CONCLUSIONS

Although buckwheat has so far not been exhaustingly explored as an animal feed component, the existing data indicate that this plant species may be successfully utilized as well for ruminants, pigs, and poultry. In ruminants the feeding of buckwheat forages appears to bear a high potential for the utilization of catch crop cultivations in late summer. Also buckwheat grain is feasible as an energy concentrate for cattle. For pigs and poultry the whole buckwheat grain provides sufficient nutrients to replace partly wheat or corn without reducing growing or laying performances. Due to its high phenolic and tocopherol concentrations, buckwheat positively affects product quality by improving fatty acid profiles in cows' milk and increasing tocopherol concentrations in eggs and poultry meat. Rather unexplored so far are the effects of the high concentrations of essential minerals in buckwheat on animal health and performance. Also the toxicity thresholds with respect to the photosensitization caused by fagopyrin have to be reassessed. Due to the ecological benefits of buckwheat cultivation and the opportunity to avoid feed–food competition, this plant appears to be of high future relevance for the development of sustainable animal feeding systems in agriculture.

REFERENCES

Alencastro, R.B.G., 2014. Produtividade e qualidade da forragem de Trigo Mourisco (Fagopyrum esculentum Moench L.) para a alimentação de ruminantes. Dissertação de Mestrado. Faculdade de Agronomia e Medicina Veterinária, Universidade de Brasília.

Amelchanka, S.L., Kreuzer, M., Leiber, F., 2010. Utility of buckwheat (*Fagopyrum esculentum* Moench) as feed: effects of forage and grain on in vitro ruminal fermentation and performance of dairy cows. Anim. Feed Sci. Technol. 155, 111–121.

Anderson, D.M., Bowland, J.P., 1984. Evaluation of buckwheat (*Fagopyrum esculentum*) in diets of growing pigs. Can. J. Anim. Sci. 64, 985–995.

Baumgärtner, J., Schilperoord, P., Basetti, P., Baiocchi, A., Jermini, M., 1998. The use of a phenology model and of risk analyses for planning buckwheat (*Fagopyrum esculentum*) sowing dates in alpine areas. Agric. Syst. 57, 557–569.

Benvenuti, M.N., Giuliotti, L., Pasqua, C., Gatta, D., Bagliacca, M., 2011. Buckwheat bran (*Fagopyrum esculentum*) as partial replacement of corn and soybean meal in the laying hen diet. Ital. J. Anim. Sci. 11 (e2), 9–12.

Berger, L.M., Blank, R., Zorn, F., Wein, S., Metges, C.C., Wolffram, S., 2015. Ruminal degradation of quercetin and its influence on fermentation in ruminants. J. Dairy Sci. 98, 5688–5698.

Bjorkman, T., 2008. Buckwheat cover crops for vegetable rotations to reduce weeds and improve soil condition. Hortscience 43, 1139.

Broudiscou, L.-P., Papon, Y., Broudiscou, A.F., 2000. Effects of dry extracts on fermentation and methanogenesis in continuous culture of rumen microbes. Anim. Feed Sci. Technol. 87, 263–277.

Eggum, B.O., Kreft, I., Javornik, B., 1981. Chemical composition and protein quality of buckwheat (*Fagopyrum esculentum* Moench). Plant Foods Hum. Nutr. 30, 175–179.

Farrell, D.J., 1978. A nutritional evaluation of buckwheat (*Fagopyrum esculentum*). Anim. Feed Sci. Technol. 3, 95–108.

Flis, M., Sobotka, W., Antoszkiewicz, Z., Lipiński, K., Zduńczyk, Z., 2010. The effect of grain polyphenols and the addition of vitamin E to diets enriched with α-linolenic acid on the antioxidant status of pigs. J. Anim. Feed Sci. 19, 539–553.

Gupta, J.J., Yadav, B.P.S., Hore, D.K., 2002. Production potential of buckwheat grain and its feeding value for poultry in Northeast India. Fagopyrum 19, 101–104.

Horsted, K., Hammershoj, M., Hermansen, J.E., 2006. Short-term effects on productivity and egg quality in nutrient restricted versus non-restricted organic layers with access to different forage crops. Acta Agric. Scand. A 56, 42–54.

Jacob, J.P., 2007. Nutrient content of organically grown feedstuffs. J. Appl. Poult. Res. 16, 642–651.

Jacob, J.P., Carter, C.A., 2008. Inclusion of buckwheat in organic broiler diets. J. Appl. Poult. Res. 17, 522–528.

Jayanegara, A., Kreuzer, M., Wina, E., Leiber, F., 2011. Significance of phenolic compounds in tropical forages for the ruminal bypass of polyunsaturated fatty acids and the appearance of biohydrogenation intermediates as examined in vitro. Anim. Prod. Sci. 51, 1127–1136.

Jover, P., 1968. Photosensibilisation de la poule par l'ingestion de la farine de caroube (*Ceratonia siliqua*) et par le ble de sarrasin (*F. esculentum*). In: Proceedings of the European Poultry Conference, The World's Poultry Science Association, Israel Branch, Jerusalem. pp. 438–442.

Kälber, T., Meier, J.S., Kreuzer, M., Leiber, F., 2011. Flowering catch crops used as forage plants for dairy cows: influence on fatty acids and tocopherols in milk. J. Dairy Sci. 94, 1477–1489.

Kälber, T., Kreuzer, M., Leiber, F., 2012. Silages containing buckwheat and chicory: quality, digestibility and nitrogen utilisation by lactating cows. Arch. Anim. Nutr. 66, 50–65.

Kälber, T., Kreuzer, M., Leiber, F., 2013. Effect of feeding buckwheat and chicory silages on fatty acid profile and cheese-making properties of milk from dairy cows. J. Dairy Res. 80, 81–88.

Kälber, T., Kreuzer, M., Leiber, F., 2014. Milk fatty acid composition of dairy cows fed green whole-plant buckwheat, phacelia or chicory in their vegetative and reproductive stage. Anim. Feed Sci. Technol. 193, 71–83.

Kalinova, J., Triska, J., Vrchotova, N., 2006. Distribution of vitamin E, squalene, epicatechin, and rutin in common buckwheat plants (*Fagopyrum esculentum* Moench). J. Agric. Food Chem. 54, 5330–5335.

Kara, M., Yüksel, O., 2014. Can we use buckwheat as animal feed? Turk. J. Agric. Nat. Sci. 1, 295–300.

Kling, M., Wöhlbier, W., 1983. Handelsfuttermittel 2B. Verlag Eugen Ulmer, Stuttgart, Germany.

Lee, J.C., Heimpel, G.E., 2005. Impact of flowering buckwheat on Lepidopteran cabbage pests and their parasitoids at two spatial scales. Biol. Control 34, 290–301.

Leiber, F., Messikommer, R., 2011. Buckwheat grain as a feed component in layer diets: effects on egg quality. Proc. Soc. Nutr. Physiol. 20, 119.

Leiber, F., Messikommer, R., Salzmann, M., Baumann, A., Wenk, C., 2009a. Einfluss von Futterbuchweizen auf das Fettsäuremuster und die Vitamin E Gehalte im Broilerfleisch. In: Wenk, C., Simon, O., Flachowsky, G., Dupuis, M. (Eds.), Schriftenreihe Institut für Nutztierwissenschaften ETH Zürich, 32, pp. 266–267.

Leiber, F., Messikommer, R., Wenk, C., 2009b. Buckwheat: a feed for broiler chicken? Agrarforschung 16, 448–453.

Leiber, F., Kunz, C., Kreuzer, M., 2012. Influence of different morphological parts of buckwheat (*Fagopyrum esculentum*) and its major secondary metabolite rutin on rumen fermentation in vitro. Czech J. Anim. Sci. 57, 10–18.

Li, S.-Q., Zhang, Q.H., 2001. Advances in the development of functional foods from buckwheat. Crit. Rev. Food Sci. Nutr. 41, 451–464.

Morales, R., Ungerfeld, E.M., 2015. Use of tannins to improve fatty acids profile of meat and milk quality in ruminants: a review. Chilean J. Agric. Res. 75, 239–248.

Mulholland, J.G., Coombe, J.B., 1979. A comparison of the forage value for sheep of buckwheat and sorghum stubbles grown on the Southern Tablelands of New South Wales. Aust. J. Exp. Agric. 19, 297–302.

Mulholland, J.G., Preston, G.K., 1995. A comparison of buckwheat, oats, and wheat for the maintenance of liveweight and wool production in sheep. Aust. J. Exp. Agric. 35, 339–342.

Nicholson, J.W.G., McQueen, R., Grant, E.A., Burgess, P.L., 1976. The feeding value of tartary buckwheat for ruminants. Can. J. Anim. Sci. 56, 803–808.

O'Meara, J., 2013. Buckwheat hay: a valuable crop for dairy farmers. Farming Magazine 2013 (3), 26–27.

Pomeranz, Y., 1983. Buckwheat: structure, composition, and utilization. Crit. Rev. Food Sci. Nutr. 19, 213–258.

Pomeranz, Y., Robbins, G.S., 1972. Amino acid composition of buckwheat. J. Agric. Food Chem. 20, 270–274.

Pontin, D.R., Wade, M.R., Kehrli, P., Wratten, S.D., 2006. Attractiveness of single and multiple species flower patches to beneficial insects in agroecosystems. Ann. Appl. Biol. 148, 39–47.

Schader, C., Muller, A., El-Hage Scialabba, N., Hecht, J., Isensee, A., Erb, K.H., Smith, P., Makkar, H.P.S., Klocke, P., Leiber, F., Schwegler, P., Stolze, M., Niggli, U., 2015. Impacts of feeding less food-competing feedstuffs to livestock on global food system sustainability. J. Roy. Soc. Interf. 12, 20150891.

Stojilkovski, K., Kocevar Glavac, N., Kreft, S., Kreft, I., 2013. Fagopyrin and flavonoid contents in common, Tartary, and cymosum buckwheat. J. Food Compos. Anal. 32, 126–130.

Van Wyk, H.P.D., Verbeek, W.A., Osthuizen, S.A., 1952. Buckwheat in rations for growing pigs. Farming South Africa 27, 399–402.

Wender, S.H., Gortner, R.A., Inman, O.L., 1943. The isolation of photosensitizing agents from buckwheat. J. Am. Chem. Soc. 65, 1733–1735.

Wijngaard, H.H., Arendt, E.K., 2006. Buckwheat. Cereal Chem. 83, 391–401.

Wilkinson, J.M., 2011. Re-defining efficiency of feed use by livestock. Animal 5, 1014–1022.

Zeller, F.J., Hsam, S.L.K., 2004. Funktionelles Lebensmittel Buchweizen—Die Vergessene Kulturpflanze. Biologie unserer Zeit 34, 24–31.

Zheng, G.H., Sosulski, F.W., Tyler, R.T., 1998. Wet-milling, composition and functional properties of starch and protein isolated from buckwheat groats. Food Res. Int. 30, 493–502.

Chapter nineteen

Biochemical Properties of Common and Tartary Buckwheat: Centered with Buckwheat Proteomics

D.-G. Lee*, S.H. Woo**, J.-S. Choi*,†

*Biological Disaster Research Group, Korea Basic Science Institute, Daejeon, Korea;
**Department of Crop Science, Chungbuk National University, Cheongju, Korea;
†Department of Analytical Science and Technology, Graduate School of Analytical Science and Technology, Chungnam National University, Daejeon, Korea

GENERAL PROPERTIES OF COMMON AND TARTARY BUCKWHEAT

Buckwheat has been considered a traditional pseudocereal crop for obtaining nutritional and medicinal materials. Today buckwheat is consumed as a functional food, such as naengmyun (Korean cold noodles), Chinese noodles, and Japanese soba, and as an ingredient of pancake mixes and muffins in North America and Canada. The grain contains 12.6% dry weight of protein, mainly consisting of digestible soluble globulin and albumin. Compared to wheat grains the major proteins in buckwheat grain have a well-balanced amino acid composition. In particular, the essential amino acids lysine and threonine are abundant in buckwheat grains (Cepkova and Dvoracek, 2006). The high content of another essential amino acid, histidine, has a positive influence on infant growth (Guozhu et al., 2007). The leaves, stems, and roots of buckwheat also possess relatively high amounts of free amino acids; valine is the most abundant followed by tyrosine, both being metabolizable to phenolic compounds, and glutamine, which is a major nitrogen source (Woo et al., 2013).

Molecular Breeding and Nutritional Aspects of Buckwheat. http://dx.doi.org/10.1016/B978-0-12-803692-1.00019-5

Common buckwheat   Tartary buckwheat

FIGURE 19.1 Comparison of the general morphology of common and Tartary buckwheat. Common buckwheat is partly cultivated for ornamental purposes, and as a noncereal crop. Tartary buckwheat is cultivated for medicinal uses and as a functional ingredient because of a large amount of rutin. Common buckwheat is a self-incompatible species whereas Tartary buckwheat is capable of self-pollination. The seed of common buckwheat tastes sweet whereas the seed of Tartary buckwheat tastes bitter, because of the high content of rutin.

Wild-type buckwheat is generally classified into two types: common buckwheat (*Fagopyrum esculentum* Moench) and Tartary buckwheat (*Fagopyrum tataricum*) (Fig. 19.1). Common buckwheat is a widely cultivated species, producing either a long pistil and short stamen (pin type) or a short pistil and long stamen (thrum type) flower in almost equal frequency, which leads to self-incompatibility (Woo et al., 2010). The other wild-type species, Tartary buckwheat, is morphologically similar to common buckwheat; however, its homomorphic flowers, fragile premature seeds, and ability to self-fertilize allow discrimination between those species. Tartary buckwheat has rarely been cultivated worldwide because of its bitter taste. A previous study revealed that interspecies crosses between *F. esculentum* and *F. tataricum* are partially compatible, suggesting a possible breeding of a better-tasting buckwheat species. Common buckwheat known as "sweet buckwheat" is taller (total length including main root) and has a thicker stem than Tartary buckwheat. Common buckwheat is widely distributed in Korea, Japan, and China, through central Asia to Europe, North America, South Africa, and even Australia. In contrast, Tartary buckwheat appears mainly in high lands, such as Tibet, Guizhou Province in southwestern China, and the Himalayas. Tartary buckwheat is freeze tolerant, suggesting that it is associated to polyphenolic compounds, which allow Tartary buckwheat to endure harsher environments than common buckwheat.

The reason why buckwheat has been considered a healthy food is related to the strong antioxidant activity of buckwheat grains. This benefit is because of the functional metabolites of buckwheat, mainly flavonoids, which play key roles in antioxidant activities. The contents and type of flavonoids differ between buckwheat species, but these compounds are known to lower the levels of plasma cholesterol (Kayashita et al., 1997) and hypertension (Ma et al., 2006), to reduce inflammation (Ishii et al., 2008), and to regulate diabetes and obesity (Li and Zhang, 2001). In general, Tartary buckwheat is richer in physiologically active compounds than common buckwheat. For instance, the amount of flavonoids is four times higher in Tartary buckwheat than in common buckwheat (Li and Zhang, 2001). Besides the functional metabolites, buckwheat flour contains a high content of proteins and functional components that reduce plasma cholesterol, body fat, and cholesterolemic gallstones (Tomotake et al., 2006). However, little is known about how many buckwheat metabolites are expressed and which buckwheat proteins play a role in synthesizing such metabolites. Various functional buckwheat metabolites are explained in the next section in detail. To understand the phenotypes of buckwheat, protein profiling and comparisons are required, using advanced proteomic technologies. Currently, such proteomic approaches in buckwheat are in the initial stages. Innovative proteomic studies and future strategies for buckwheat will be described next.

BIOCHEMICAL PROPERTIES OF BUCKWHEAT METABOLITES

The studies of buckwheat metabolites have been widely performed on the basis of functional natural product. The functional metabolites in buckwheat include small molecules and proteins to prevent and protect against cancer, metabolic syndromes, and immune diseases. This review aims at highlighting recent studies regarding biochemical properties to improve human health.

Lectin

Lectin is a glycoprotein that is widely distributed in various buckwheat species (Bai et al., 2015a). Lectin shows a variety of beneficial functions in biological systems, such as antiinsect (Hilder et al., 1995), antifungal (Herre et al., 2004), anticancer (Dhuna et al., 2005), antiviral activities (Balzarini et al., 1992), and immunomodulatory activity (Rubinstein et al., 2004). To investigate the biological activities of lectin derived from Tartary buckwheat grain, the purified lectins are required of their characterization. Tartary buckwheat lectin can induce the maturation and proliferation of peripheral blood dendritic cells (DCs). It effectively promotes immune response of T-cell. Lectin in a cell evokes the production of proinflammatory elements: interleukin-10 and interleukin-12. Production of these cytokines for maturing peripheral blood DCs will affect the helper T-cell 1 or T-cell 2 immune response. In addition, the lectin proteins of Tartary buckwheat can induce the apoptosis of human leukemia U937 cells. Thus it is suggested that Tartary buckwheat-derived lectins play good potential dietary supplements as an apoptosis inducer for the treatment and prevention of solid and blood cancers.

Trypsin Inhibitor

There is evidence that the protease inhibitors can induce apoptosis in various tumor cell lines; however, their underling mechanisms are not clear. Trypsin inhibitor from buckwheat is known to significantly suppress the proliferation of two blood cancer cell lines: chronic myelogenous leukemia, K-562, and multiple myeloma, IM-9, in vitro (Wang et al., 2007; Zhang et al., 2007). The antitumor activity of recombinant buckwheat trypsin inhibitor (rBTI) has been reported specifically to be exerted to cell proliferation and inhibition of H22 hepatic carcinoma cells in vivo and in vitro (Bai et al., 2015b). This apoptotic induction is controlled by a dose- and time-dependent manner in the mitochondrial pathway via caspase-9. In addition, it was reported that a similar function of rBTI can induce the apoptosis of cancer cells (Wang et al., 2015). The preventive effect of rBTI on human liver carcinoma cell line (Hep G2) was involved in the mitochondrial dysfunction by increasing intracellular total reactive oxygen species (ROS) to result in mitophagy by inducing subcellular degradation. It is notable that a subunit of the translocase of the mitochondrial outer membrane complex assumes to be the direct target of rBTI. Taken together, it concludes that rBTI plays a potential role in cancer drug development by targeting mitochondria-penetrating peptides for the treatment of cancer.

Rutin

The major flavonoids present in buckwheat include rutin, orientin, isoorientin, vitexin, isovitexin, and quercetin (Liu et al., 2008). The basic structure of major flavonoids in buckwheat such as rutin, isoquercetin, and quercetin are shown in Fig. 19.2. Among the antioxidant compounds, rutin is regarded as the best health-preserving flavonoid, which is also known to be highly preventive against inflammation and carcinogenesis (Nile and Park, 2014). In particular, the amount of rutin is 47 times higher in seeds and seed sprouts of Tartary buckwheat than common buckwheat (Kim et al., 2008). As a flavonoid glycoside, rutin can absorb UV-B light to protect the plants from solar radiation. At 28 days after sowing, rutin glucosidase, the enzyme that degrades rutin, showed the highest enzyme activity in buckwheat leaves (Suzuki et al., 2005). It was found that the treatment of buckwheat rutin inhibits cardiomyocyte hypertrophy (Chu et al., 2014). Angiotensin II-induced hypertrophy can dramatically affect the increase of intracellular Ca^{2+} levels, calcineurin activities, and protein expression level as well as protooncogene *c-fos* mRNA transcripts. The treatment of buckwheat rutin in a hypertrophy model system decreased the calcineurin activity in a concentration-dependent manner. However, the regulatory mechanism related to signal transduction pathways in cardiac myocytes by buckwheat rutin remains to be elucidated.

Besides the biochemical property of rutin, there is also the regulation of rutin content by a rutin-degrading enzyme. Rutinosidase can effectively hydrolyze rutin when water is added. The enzyme is present in at least two isoforms in Tartary buckwheat seed, in which the enzyme can release the disaccharide

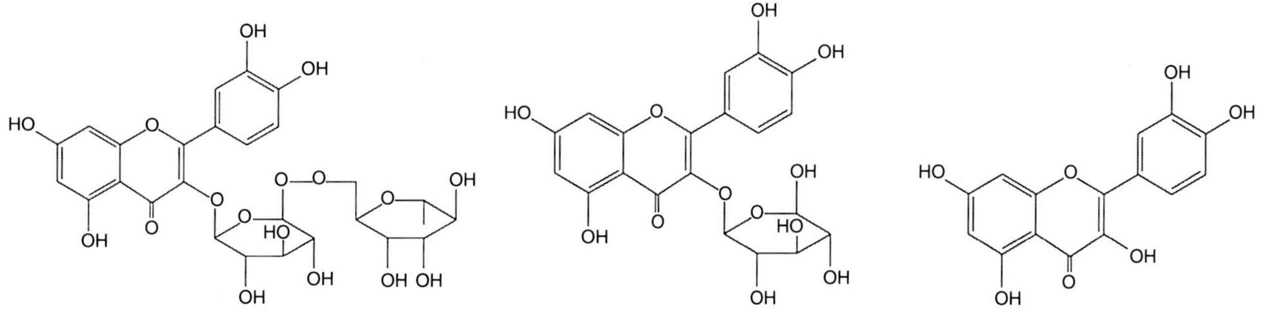

FIGURE 19.2 Molecular structure of rutin, isoquercetin, and quercetin. Rutin is called quercetin-3-O-rutinoside. Rutin has the structure of flavonoid quercetin and disaccharide rutinoside (α-L-rhamnopyranosyl-(1→6)-β-D-glucopyranose).

Rutin   Isoquercetin   Quercetin

rutinose from rutin. The newly developed buckwheat variety "Manten-Kirari" is required to evaluate the toxicity of rutin because of the lower rutinosidase activity than in other breeds. Oral administration of rutin in experimental rats resulted in nontoxicity in acute (14 day) and subacute (27 day) treatment (Suzuki et al., 2015). Rutin may temporarily increase the urine protein and the serum albumin concentration, but these changes are not assumed to be caused by rutin toxicity. Apart from the earlier observation, the oral administration of rutin from buckwheat "Manten-Kirari" did not change the pathological–anatomical phenotypes in major organs such as liver, kidneys, heart, lung bronchi, and pituitary gland in rats. Thus rutin-rich Tartary buckwheat dough might not be linked to toxicity at least in animal experiments.

Quercetin

Apart from rutin, other flavonoid compounds include quercetin, isoquercetin, and fagopyrin in buckwheat seeds and brans, which show a cytotoxic effect against the human hepatoma HepG2 cells (Li et al., 2014). Among the various flavonoids, quercetin shows the most effective antioxidant capacity against the HepG2 cell line. Also quercetin showed the most potent growth inhibition against hepatocellular carcinoma in a time- and dose-dependent manner. Furthermore, quercetin significantly increased intracellular ROS level and apoptotic cell death, leading to the G_2/M phase cell cycle arrest. Thus quercetin derived from buckwheat is expected to function as an effective antitumor agent, and it showed a synergistic effect in the clinical trial using combination therapy with a known anticancer drug.

Anthocyanin

In European countries, buckwheat leaves have been dried to produce herbal tea sold under the brand name "Fagorutin." Buckwheat also produces a water-soluble anthocyanin, the pigment determining the color of flowers, as it is highly concentrated in petals (Suzuki et al., 2007). A Tartary buckwheat cultivar "Hokkai 10" was altered by introducing a chemical mutagen to produce "Hokkai 8," a variant in which the anthocyanins cyanidin-3-*O*-glucoside and cyanidin-3-O-rutinoside were found in high concentrations in the sprout (Kim et al., 2007).

Other Phenolic Compounds

The rise in nutritional quality and functionality happens during the crop seed germination. Various phenolic compounds including phenolic acid, rutin, and C-glycosyl flavones are accumulated in buckwheat seeds. The nutritional effect of buckwheat seed shows strong antioxidant activities as well as various chemical compositions over 72 h of seed germination (Zhang et al., 2015). Besides the representative phenolic compound rutin, other phenolic components such as vitexin, orientin, and isoorientin show antioxidant activity. The antioxidant activities of the phenolic compounds are strongly correlated with a variety of phytochemical contents in buckwheat. The improved antioxidant activities of buckwheat phenolic compounds play the key role of a fundamental prophylactic property for human health, thus germinated buckwheat seed can be a great functional food source for human health promotion.

Tatariside F

Traditional medicines have been known to be an effective disease treatment in northeastern Asian countries. In particular, novel natural product compounds have been in the spotlight as possible new sources for chemotherapy adjuvant or anticancer agents in the case of lessened toxicity. The novel small molecule was isolated from the root of Tartary buckwheat and named tatariside F, which possesses a remarkable antitumor effect against human hepatocellular carcinoma (H22 cell line) (Peng et al., 2015). Tatariside F inhibited the proliferation of H22 cells effectively. In in vivo analysis using a mouse xenograft model system, tatariside F also showed an antitumor effect against hepatic carcinoma. In addition, it was confirmed that it has a synergistic effect in combination with a known anticancer drug, cytophosphamide.

Fagopyrin

Fagopyrin is used for naturally occurring substances in the buckwheat plant. The chemical structure of fagopyrin has a naphthodianthrone skeleton similar to that of hypericin, which causes the phototaxic effect in humans. Fagopyrin is a red-colored anthraquinone derivative and is present in the buckwheat cotyledon (Kreft et al., 2013). If fagopyrin is ingested after exposure to sunlight, the compound causes hyperactivity in humans, similarly to hypericin. Thus an intake of up to 40 g per day is recommended for buckwheat.

Inositol Derivatives

Fagopyritols are one to three galactosyl adduct derivatives of D-chiro-inositol that accumulates in buckwheat seed. Fagopyritol is important for seed germination and is used as a food supplement (Steadman et al., 2000). In particular, fagopyritol B1, O-α-D-galactopyranosyl-(1→2)-D-chiro-inositol is a galactosyl cyclitol in maturing buckwheat seeds associated with desiccation tolerance (Horbowicz et al., 1998). D-Chiro-inositol contained in buckwheat plays the role of a secondary messenger, which is effective in improving type 2 diabetes and treating polycystic ovary syndrome (Yao et al., 2008).

PROTEOMIC APPROACHES: GEL-BASED VERSUS SHOTGUN METHODS

Complete proteomic analysis of buckwheat requires straightforward proteomic strategies. The term "proteome" implies the whole set of proteins expressed in a cell or tissue and it complements the genome. "Proteomics," defined as "the study of the proteome," was coined in 1995 by the Australian biochemist Mark Wilkins (Wilkins et al., 1996). Three technological breakthroughs enabled proteomics to progress: (1) the gathering of genes and proteins in a searchable database, ultimately a complete genome database; (2) the emergence of user-friendly bioinformatics to handle the vast data; and (3) the empowerment of genome-wide expression profiling platforms such as microchip arrays and RNA sequencing. Because of the continuing development of proteomic techniques, these technical approaches have been extended to crop sciences. Crop proteomics allowed the identification of a wide range of storage, allergenic, and abiotic/biotic stress–response proteins, and the elucidation of host–pathogen interactions. Crop proteomics definitely helps improving the productivity and stress-endurable ability of crop species.

Proteomic techniques are broadly classified into two types of strategies: gel-based and gel-free shotgun proteomic methods (Fig. 19.3). Details of both are well described in a previous review (Choi et al., 2006). In brief, the gel-based proteomic approach involves an initial separation of proteins followed by a trypsin digestion to produce the resultant peptides. Protein identification is performed by peptide mass fingerprinting and peptide fragmentation using tandem mass spectrometry (MS/MS) (Choi et al., 2008). Individual proteins are separated from the biological proteome complex by two-dimensional electrophoresis (2-DE), according to their characteristic isoelectric point (pI) and molecular weight (Mr). The use of 2-DE has the advantages of directly presenting the expressed protein profiles and easily detecting proteins with posttranslational modifications (PTMs), such as glycosylation, phosphorylation, and ubiquitination. Although gel-based proteomics easily facilitates start-up of a proteomics

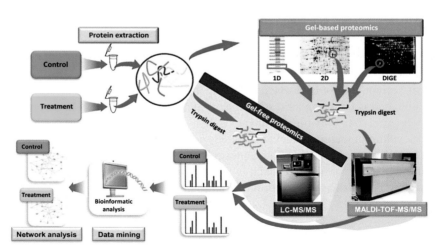

FIGURE 19.3 Two proteomic strategies for buckwheat. In gel-based proteomics, proteins are first separated from the protein mixture by 2-DE gel and then subjected to a second digestion of individual proteins to peptides. Gel-free shotgun proteomics comprises a first digestion of the protein mixture to peptide mixture followed by a separation of peptides by LC. Commonly, the two approaches are followed by the identification of proteins by MS analysis and characterization, that is, the protein network comparison.

laboratory with low cost and minimal labor skills, a large amount of protein (at least 100 µg) is required for loading the 2-DE gel and there are a limited number of proteins that can be identified by mass spectrometry (up to around 300). In addition, harsh proteins, with high Mr and/or extreme pI, and water-insoluble hydrophobic membrane proteins are difficult to identify using gel-based proteomic methods.

In contrast to gel-based proteomics, in the shotgun proteomic methods, the first digestion of proteins is performed by trypsin, producing a mix of peptides that are separated and identified by liquid chromatography tandem mass spectrometry (LC-MS/MS). The multidimensional protein identification technology (MudPIT) was developed to introduce gel-free shotgun proteomics (Washburn et al., 2001). MudPIT is a "first-generation" technology that performs the automation and high throughput of proteome profiling. Second-generation proteomic techniques correspond to quantitative labeling methods, including isotope-coded affinity tags, isobaric tags for relative and absolute quantitation, and targeted mass tags, and were successively developed to address crop adaptation mechanisms to abiotic and biotic stresses (Hu et al., 2015). An innovative label-free quantitative proteomics technique was developed that contributes to understanding a larger set of plant networks and proteome dynamics involved in developmental change and/or environmental stress responses (Matros et al., 2011). Shotgun proteomics combined with the complete genomes obtained using next-generation sequencing (NGS) technologies (Varshney et al., 2009) have provided deeper protein profiling and PTMs mapping, from cellular to subcellular proteome, in major crop species (Thelen and Peck, 2007). The accumulation of cereal and noncereal crop plant species genomes obtained using NGS propels the use of shotgun-based proteomic strategies, allowing new insights into crop science applications and its societal benefits (Agrawal et al., 2012).

PROTEOMIC STUDIES OF MAJOR CEREAL CROPS

To improve crop productivity, it is of great importance to understand a crop's adaptation to climate change and its resistance to abiotic/biotic stress. Based on the available genome sequences of the model plant *Arabidopsis* (Arabidopsis Genome Initiative, 2000) and nonmodel cereal crops, responses to environmental cues in signal transduction pathways and biochemical routes can be scrutinized in the proteomic data of crop plants (Vanderschuren et al., 2013). During the last few decades, genome-wide transcriptomes of major crops were investigated and the gene products produced in response to various external conditions were characterized. Basic information on crop traits and biological perturbations have been gathered and further applied in novel crop breeding. Advanced cultivar breeding programs by interspecies crossing will be achieved through optimization of crop productivity in better fields, to which in-depth proteome quantification and PTM characterization of regulatory proteins will significantly contribute (Agrawal et al., 2013). External factors responsible for crop plant growth and productivity include both abiotic and biotic stresses, which have a marked influence at the stages of seed germination, development, and maturation. The stresses affecting crop plants can be classified into two types: (1) abiotic stress, such as extreme temperatures (freeze and heat), light conditions (light intensity and duration, ultraviolet radiation), water supply (drought and flooding), and nutrients, including toxic heavy metals and pesticides; and (2) biotic stress, including biological pathogens, such as viruses, bacteria, fungi, insects, and other plants and animals (Sunkar, 2010). For instance, oxidative stress initiates an overall proteome modulation, activating proteins that are stress sensors or coordinators for adaptation or resistance to stress. Extensive proteomic analyses of plant responses to abiotic stresses have been conducted on whole extracts of the model plant *Arabidopsis* and on rice, a representative crop plant, under diverse environmental conditions such as increased salinity, cold, flooding, drought, and herbicide use (Wienkoop et al., 2010). When the nuclear subcellular proteome of *Arabidopsis* was subjected to cold stress, different regulatory transcription factors were identified in addition to generic proteins such as heat-shock proteins and calmodulin (Bae et al., 2003). The large-scale plant proteome and genome analysis of *Arabidopsis thaliana* has resulted in the identification of over 13,000 unique proteins in different organs, covering near 50% of the predicted genes (Baerenfaller et al., 2008). Shotgun proteomics combined with prefractionation and MudPIT of rice shoots identified over 3000 nonredundant rice proteins (Lee et al., 2011). Advanced quantitative proteomics have recently powered up investigations of the subcellular proteome in model plants and cereal crop plants for a deeper understanding of biological processes.

Rice (*Oryza sativa* L.) is one of the staple crops feeding half of the world's population. Rice productivity must be increased by at least 25% to keep pace with the expected increase in the world population by 2030. Thus technical advances to obtain better rice grains have led to the development of rice tolerant to several stresses, in particular those imposed by climate change. For the systemic and functional characterization of the rice genome, the International Rice Genome Sequencing Project (2005) was established. Map-based draft genome sequences (Sasaki et al., 2002) and rice proteome studies were subsequently initiated based on this genomic database and, in 2005, a complete and high-quality map-based rice genome sequence was published by IRGSP. The Rice Annotation Project was initiated upon the publication of the complete genome sequence of *O. sativa* ssp. *japonica* cv. Nipponbare. IRGSP accurately and timely annotated the complete rice genome: 389-megabase genome size, 12 chromosomes, and 37,544 protein-coding genes, excluding transposable elements (International Rice Genome Sequencing Project, 2005), were reported. IRGSP provided an updated genome assembly construct named Os-Nipponbare-Reference-IRGSP-1.0 after sequences were reread using NGS techniques and ~44X genome coverage. This high-quality reference rice genome is useful for rice diversity studies and major cereal crop improvement experiments (Kim et al., 2014).

Wheat (*Triticum aestivum*) is a dominant source of global nutrition. Wheat grains are a prevalent source of proteins in our daily diet, such as gluten and storage proteins (Gill et al., 2004). Wheat grains feed about 30% of the worldwide human population. However, climate change decreased wheat production rate by 5.5% between 2000 and 2008 (Lobell et al., 2011). Specialists recommend that wheat production should increase by around 70% to satisfy the demands by 2050 (Foley et al., 2011). The wheat genome is much larger than the rice genome (17 gigabases vs. 389 megabases). This large size is caused by numerous polyploidy events occurring during the ~10,000 years of selective breeding. In fact, the current wheat genome derives from three different genomes entangled to endow human inbred wheat with the capability of adapting to harsh environments. The International Wheat Genome Sequencing Consortium (2014) released a chromosome-based draft sequence of the hexaploid wheat genome. This draft sequence provides new insights into the structural and functional organization and evolution of the complex genome of the world's longest cultivated crop. Cultivated wheat genome displays genomic mosaics reflecting the multiple ancestral hybridizations (Marcussen et al., 2014) and consists of three subgenomes (A, B, and D), which are closely related. A total of 124,201 gene loci are distributed across the chromosomes and subgenomes. Recent wheat proteomics has focused on the study of wheat acclimation and adaptation to various abiotic stresses. Several techniques, such as subcellular proteome extraction, wheat chloroplast proteome profiling (Kamal et al., 2013a), and enhanced sample preparation have been used to better explain the mechanisms of stress tolerance in wheat (Komatsu et al., 2014). At the initial stage of wheat proteomics, particular attention was devoted to water stress response by examining the wheat chloroplast proteome isolated from wheat roots (Kamal et al., 2013b).

Maize (*Zea mays*), also known as corn, is the most widely cultivated crop plant since its domestication by ancient Mexicans about 10,000 years ago. In 2013, According to the Food and Agricultural Organization of the United Nations (2013) total maize production reached 1016 million tons, followed by rice (745 million tons), and wheat (713 million tons). Maize is the most important source of food for both domesticated animals and human populations. In addition, maize has been extensively used for textile adhesives and biofuel production. Maize has also been the model monocotyledon plant for molecular studies, resulting in its establishment as a systematic genetic model (Doebley, 2004). In contrast to rice and wheat, maize is a C4 plant that can fix carbon dioxide efficiently in mesophyll cells through the use of the elaborate enzyme ribulose-1,5-bisphophate carboxylase/oxygenase; the fixed carbon is then sent to bundle-sheath cells via malate or aspartate (Majeran and van Wijk, 2009). The genome of the B73 maize line is 2.3 gigabases. Maize has 10 chromosomes, which are structurally diverse and distributed in chromatin by dynamic changes (Schnable et al., 2009). Genome annotation of maize predicted over 32,000 genes, with 85% of the genome comprising hundreds of transposable element families, nonuniformly dispersed across the maize chromosomes. Currently, two maize proteome databases—Plant GDB (http://www.plantgdb.org/ZmGDB) and Maize Assembled Genomic Island (http://magi.plantgenomics.iastate.edu)—are well established, enabling proteomic studies concerning maize biological phenomena (Pechanova et al., 2013). Maize proteomics includes research on developmental maturation, responses to abiotic/biotic stresses, maize (host)–*Fusarium* (pathogen) interaction, and evaluation on the safety of genetically modified maize.

CURRENT PROTEOMICS OF COMMON AND TARTARY BUCKWHEAT

Unconstrained genome-wide proteomic study of buckwheat is limited by the lack of complete genomic information. Instead, a global transcriptome analysis of aluminum-induced gene expression in common buckwheat was performed using RNA sequencing technology (Yokosho et al., 2014). Complete chloroplast genomic sequences from the buckwheat species *Fagopyrum cymosum, F. esculentum*, and *F. tataricum* were recently released, enabling species distinction based on genetic informative biomarkers and a study of their evolutionary relationships (Cho et al., 2015; Yang et al., 2015b). In 2014 a Russian scientific group introduced a draft genome sequence of Tartary buckwheat in the XXII Plant and Animal Genome: 372 megabases were assembled in Tartary buckwheat corresponding to 70% of the genome. The number of predicted buckwheat genes is 30,000, according to cDNA-based, homology-based, and *ab initio* approaches. Obtaining a complete genome for buckwheat in the near future will allow: systemic -omic studies concerning crop improvement; molecular breeding by identifying novel self-pollinating factors; and metabolic engineering to reinforce physiologically active compounds like rutin. Nonetheless, incomplete buckwheat proteomics is currently possible to some extent, by using the several buckwheat protein sequences deposited in the National Center for Biotechnical Information (NCBI) (134 genes, 270 ESTs, 808 proteins) and establishing homology comparisons to similar sequences of the model plant *A. thaliana*.

We attempted to identify two buckwheat species (*F. esculentum* and *F. tataricum*) proteomes from leaves and stems cultured under light or dark conditions, using 2-DE/matrix-assisted laser desorption ionization-time of flight mass spectrometry (Shin et al., 2010). Samples were obtained from 7-day-old etiolated sprouts, in which the sprouting leaves of *F. esculentum* were larger than those of *F. tataricum* (Fig. 19.4). In addition, the darker red color of the hypocotyledons and cotyledons of *F. tataricum* compared to those of *F. esculentum*, suggested that Tartary buckwheat contains a higher amount of soluble anthocyanin. Light-grown Tartary buckwheat is longer than common buckwheat grown in dark conditions. Likewise, sprouting leaves of dark-grown Tartary buckwheat were larger than those of dark-grown common buckwheat. Gel-based proteomic analysis was performed and compared to plant genome databases in NCBI. A total of 166 unique proteins were identified—79 belonged to common buckwheat and 81 to Tartary buckwheat. These were mainly cytosolic proteins involved in cell metabolism. In common buckwheat, light-dependent protochlorophyllide reductase was identified in light-inhibited leaves as a key regulator of chlorophyll biosynthesis (Blomqvist et al., 2008). Light-inhibited storage protein 13S globulin-3, which induces allergies, was commonly expressed in dark-grown common and Tartary buckwheats (Zhang et al., 2008). One unknown

Common buckwheat   Tartary buckwheat

Light
grown

Dark
grown

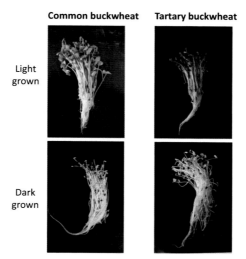

FIGURE 19.4 Morphological differences between common and Tartary buckwheat grown under light or dark conditions for 7 days after seed germination. Under light conditions, the sprouting length of common buckwheat is larger than that of Tartary buckwheat. The dark color of Tartary buckwheat stem under light conditions is caused by the high content of anthocyanin, a major pigment. Under dark conditions, Tartary buckwheat plants are longer than common buckwheat plants.

pentatricopeptide repeat-containing protein was identified in both buckwheat species, suggesting that it acts as a sterility and fertility regulator. Buckwheat proteomics is expected to provide a better understanding of the developmental physiology of buckwheat sprouts.

Using a modified shotgun proteomics method the embryo and the endosperm proteomes from mature common buckwheat seeds were analyzed (Kamal et al., 2011). Protein samples were separated on sodium dodecylsulfate polyacrylamide gel electrophoresis, and the silver-stained bands were excised and subjected to in-gel trypsin digestion. Isolated peptides were identified by ion trap mass spectrometry against a plant genome database. A total of 465 proteins were identified: 270 were embryo proteins, 163 were endosperm proteins, and 33 were common proteins. Sixty-seven of these proteins had database hits with more than two peptides, mainly corresponding to soluble proteins and are involved in cell metabolism. In particular, buckwheat storage proteins such as globulin-11S1, 13S1, 13S2, and 13S3 were identified and all these proteins induce allergenic responses (Nair and Adachi, 2002). Among these, globulin-13S3 was identified in the buckwheat leaf, embryo, and endosperm. It is noteworthy that granule-bound starch synthase 1 was identified in the endosperm, since this enzyme plays a role in the biosynthesis of starch and amylopectin (Hanashiro et al., 2008). The complete genome sequences of buckwheat species will accelerate further understanding of buckwheat seed physiology as well as improvement of buckwheat protein quality.

CHALLENGES AND PERSPECTIVES OF BUCKWHEAT PROTEOMICS

To perform a proteomic study, a buckwheat protein database must be built. So far only a few buckwheat proteins have been deposited in the public databases Uniprot-KB (466 proteins) and SwissProt (45 proteins). Recently, the complete chloroplast genome sequence of Tartary buckwheat was released (Cho et al., 2015). A 159 kilobase sequence of the chloroplast genome was obtained by NGS technology in which 114 genes were annotated to 81 protein-coding genes, 29 transfer RNA genes, and 4 ribosomal RNA genes.

The promising news is that buckwheat draft genome sequencing is in progress (Logacheva et al., 2014). Approximately 70% of the genome sequencing has been done, and a total sequence length of ~530 megabases and ~30 thousands genes are expected when sequencing is complete. Buckwheat genome sequencing and full gene annotation of protein function can be used as a database for genome-wide proteome analysis by large-scale shotgun proteomics and for transcriptome analysis by RNA sequencing.

Based on the genomic database, gel-based and gel-free quantitative proteomic analysis, coupled with high-resolution mass spectrometry, should be carried out to discover valuable biomolecules against environmental stresses (Fig. 19.5). These genetic resources can be further used to develop buckwheat varieties to cope with environmental stresses. In addition, they will also contribute to explore functional biomaterials present in buckwheat such as rutin and D-chiro-inositol via a functional proteomic study.

Proteome studies suggest quantitative proteomics should be preceded by the functional analysis of rutin, amylose, and allergens in buckwheat. Concerning rutin, functional analysis is fundamental for: (1) improving the use of buckwheat

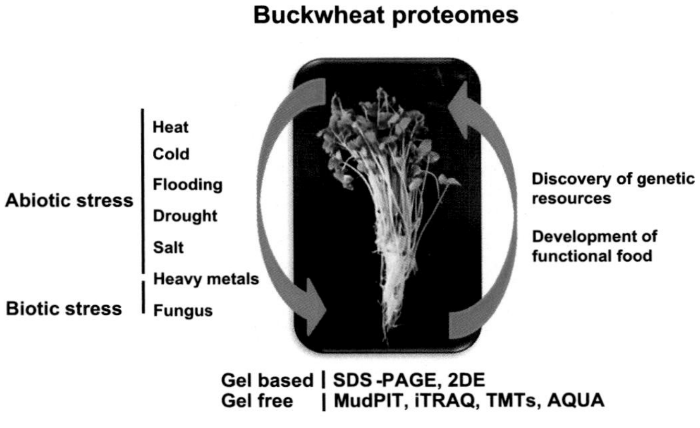

Buckwheat proteomes

Abiotic stress
- Heat
- Cold
- Flooding
- Drought
- Salt
- Heavy metals

Biotic stress
- Fungus

Discovery of genetic resources

Development of functional food

Gel based | SDS -PAGE, 2DE
Gel free | MudPIT, iTRAQ, TMTs, AQUA

FIGURE 19.5 Potential proteomic application for buckwheat. A gel-based and/or gel-free quantitative proteomic analysis conducted on plants under environmental stresses will contribute to the discovery of genetic resources to improve stress tolerance in buckwheat cultivars or explore its functional components as healthy food supplements. *iTRAQ*, isobaric tags for relative and absolute quantitation; *TMTs*, targeted mass tags.

cultivation by-products and developing new functional foods; (2) understanding the major routes of rutin biosynthesis, since related genes have not been cloned and some potential key transcription factors have not yet been identified and characterized; (3) investigating the action of buckwheat-derived pharmaceuticals; and (4) breeding new buckwheat cultivars with high contents of functional metabolites, in particular rutin, based on the information acquired by plant proteomics, metabolomics, and applying genetic engineering technologies.

Starch is a major component of the harvested parts of plants and is used in food and nonfood industries. Demands for natural starch sources that possess novel physical and chemical properties are increasing, because of the desire to develop new products and to reduce the need for postharvest chemical modification processes. A number of naturally occurring mutants with altered proportions of amylose and amylopectin have been studied. Of these, waxy wheat is widely used in food industries. We discovered a waxy mutant whose endosperms are stained brown by iodine. Starch of waxy buckwheat with low amylose may have several applications, including reduction of staling in flour products, such as noodles, cakes, and bread. We expect this unique starch accumulated in the mutant lines to be useful for industrial applications in the near future.

Buckwheat is known to cause allergen reactions in South Korea and Japan and these are also widespread in Asia. Allergy cases have also been reported in Europe (Heffler et al., 2014). However, buckwheat has emerged as a healthy food, since it is a gluten-free crop and gluten is one of the major protein-causing allergies. Likewise, the value of buckwheat as a functional food is increasing. Still, to definitely classify it as a dietary supplement, absolute quantitative proteomics is required for quantifying rutin and allergen proteins among the buckwheat varieties. We recommend the absolute quantitation (AQUA) method of buckwheat allergens as an example, using MS/MS (Fig. 19.6).

First of all the optimization of gel-free quantitative proteomic analysis of buckwheat allergenic proteins must be carried out as follows: (1) allergen protein extraction in a buffer (phenol, urea, urea/thiourea, trichloroacetic acid/acetone, etc.); (2) in-solution digestion; and (3) analysis in an adequate mass spectrometry system (mobile phase gradient condition, MS/MS scan event, etc.).

Absolute quantitative proteome analysis of buckwheat allergen proteins will be employed to synthesize stable isotope-labeled peptides in relation to AQUA peptides (Gerber et al., 2003). The known amount of AQUA peptide will be used prior to the sample as an internal standard, and subjected to multiple reaction monitoring (MRM) by triple-quadrupole LC-MS/MS (Houston et al., 2011). Internal standard candidate peptides are identified from the resultant tryptic peptides, using gel-free proteomic analysis. AQUA peptides should be synthesized for quantification according to the following rules: (1) amino acid length, 8 to 20; (2) amino acid composition, exclusion of cysteine and methionine internal cleavage sites; and (3) redundancy, removal of redundant peptide sequences against buckwheat database. Unique peptides with satisfactory criteria in both label-free mass spectrometry and in silico analyses will finally be selected and

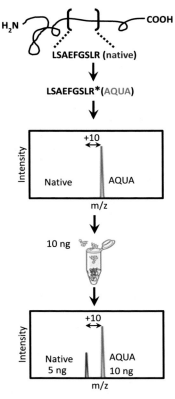

FIGURE 19.6 An overview of AQUA peptide identification. Optimal tryptic peptides for absolute quantitative analysis are chosen from proteins of interest. Isotope-labeled peptide is synthesized and an MRM analysis in LC-MS/MS is optimized. The known amounts of AQUA peptide are analyzed prior to the sample and the native and AQUA peptides are compared to determine native protein amount.

synthesized for AQUA peptides. Stable isotope-labeled arginine or lysine residues will be incorporated into each AQUA peptide at the C-terminal. AQUA peptides can then be compared to their corresponding native allergen peptides (unlabeled), with the labeled peptides possessing a +8 or +10 mass shift in their MS spectrum. For MRM analysis by LC-MS/MS, the following conditions need to be optimized: (1) ion fragmentation (collision energy, ideal transition ions, etc.) and (2) linear quantitation range of native and AQUA peptide concentrations. Finally, the integral peak area ratio of native and AQUA peptides will be compared to calculate the absolute quantitation of allergen protein in the sample using optimal LC-MS/MS followed by MRM. Absolute quantitative proteomics for rutin or allergen contents among the varieties will be useful in screening superior varieties, enabling us to develop a molecular breeding strategy to enhance environmental stress-resistant buckwheat.

ACKNOWLEDGMENTS

This work was financially supported by a grant from the Korea Basic Science Institute (D35403) attributed to J.S. Choi.

ABBREVIATIONS

2-DE	Two-dimensional electrophoresis
LC-MS/MS	Liquid chromatography tandem mass spectrometry
MudPIT	Multidimensional protein identification technology
PTMs	Posttranslational modifications
NGS	Next generation sequencing
MALDI-TOF-MS	Matrix-assisted laser desorption ionization-time of flight mass spectrometry
IRGSP	International Rice Genome Sequencing Project
SDS-PAGE	Sodium dodecylsulfate polyacrylamide gel electrophoresis
ROS	Reactive oxygen species
AQUA	Absolute quantitation
MRM	Multiple reaction monitoring

REFERENCES

Agrawal, G.K., Pedreschi, R., Barkla, B.J., Bindschedler, L.V., Cramer, R., Sarkar, A., Renaut, J., Job, D., Rakwal, R., 2012. Translational plant proteomics: a perspective. J. Proteomics 75, 4588–4601.

Agrawal, G.K., Sarkar, A., Righetti, P.G., Pedreschi, R., Carpentier, S., Wang, T., Barkla, B.J., Kohli, A., Ndimba, B.K., Bykova, N.V., Rampitsch, C., Zolla, L., Rafudeen, M.S., Cramer, R., Bindschedler, L.V., Tsakirpaloglou, N., Ndimba, R.J., Farrant, J.M., Renaut, J., Job, D., Kikuchi, S., Rakwal, R., 2013. A decade of plant proteomics and mass spectrometry: translation of technical advancements to food security and safety issues. Mass Spectrom. Rev. 32, 335–365.

Arabidopsis Genome Initiative, 2000. Analysis of the genome sequence of the flowering plant *Arabidopsis thaliana*. Nature 408, 796–815.

Bae, M.S., Cho, E.J., Choi, E.Y., Park, O.K., 2003. Analysis of *Arabidopsis* nuclear proteome and its response to cold stress. Plant J. 36, 652–663.

Baerenfaller, K., Grossmann, J., Grobei, M.A., Hull, R., Hirsch-Hoffmann, M., Yalovsky, S., Zimmermann, P., Grossniklaus, U., Gruissem, W., Baginsky, S., 2008. Genome-scale proteomics reveals *Arabidopsis thaliana* gene models and proteome dynamics. Science 320, 938–941.

Bai, C.Z., Ji, H.J., Feng, M.L., Hao, X.L., Zhong, Q.M., Cui, X.D., Wang, Z.H., 2015a. Stimulation of dendritic cell maturation and induction of apoptosis in lymphoma cells by a stable lectin from buckwheat seeds. Genet. Mol. Res. 14 (1), 2162–2175.

Bai, C.Z., Feng, M.L., Hao, X.L., Zhao, Z.J., Li, Y.Y., Wang, Z.H., 2015b. Anti-tumoral effects of a trypsin inhibitor derived from buckwheat in vitro and in vivo. Mol. Med. Rep. 12, 1777–1782.

Balzarini, J., Neyts, J., Schols, D., Hosoya, M., Van Danne, E., Peumans, W., De Clercq, E., 1992. The mannose-specific plant lectins from *Cymbidium* hybrid and *Epipactis helleborine* and the (N-acetylglucosamine) n-specific plant lectin from *Urtica dioica* are potent and selective inhibitors of human immunodeficiency virus and cytomegalovirus replication in vitro. Antiviral Res. 18, 191–207.

Blomqvist, L.A., Ryberg, M., Sundqvist, C., 2008. Proteomic analysis of highly purified prolamellar bodies reveals their significance in chloroplast development. Photosyn. Res. 96, 37–50.

Cepkova, P., Dvoracek, V., 2006. Seed protein polymorphism of four common buckwheat varieties registered in the Czech Republic. Fagopyrum 23, 17–22.

Cho, K.S., Yun, B.K., Yoon, Y.H., Hong, S.Y., Mekapogu, M., Kim, K.H., Yang, T.J., 2015. Complete chloroplast genome sequence of Tartary buckwheat (*Fagopyrum tataricum*) and comparative analysis with common buckwheat (*F. esculentum*). PLoS ONE 10, e0125332.

Choi, J.S., Chung, K.Y., Woo, S.H., 2006. Quantitative proteomics towards understanding life and environment. Kor. J. Environ. Agric. 25, 371–381.

Choi, J.S., Cho, S.W., Kim, T.S., Cho, K., Han, S.S., Kim, H.K., Woo, S.H., Chung, K.Y., 2008. Proteome analysis of greenhouse-cultured lettuce with the natural soil mineral conditioner illite. Soil Biol. Biochem. 40, 1370–1378.

Chu, J.X., Li, G.M., Gao, X.J., Wang, J.X., Han, S.Y., 2014. Buckwheat rutin inhibits AngII-induced cardiomyocyte hypertrophy via blockade of CaN-dependent signal pathway. Iran J. Pharm. Res. 13, 1347–1355.

Dhuna, V., Bains, J.S., Kamboj, S.S., Singh, J., Kamboj, S., Saxena, A.K., 2005. Purification and characterization of a lectin from *Arisaema tortuosum* Schott having in-vitro anticancer activity against human cancer cell lines. J. Biochem. Mol. Biol. 38, 526–532.

Doebley, J., 2004. The genetics of maize evolution. Annu. Rev. Genet. 38, 37–59.

Foley, J.A., Ramankutty, N., Brauman, K.A., Cassidy, E.S., Gerber, J.S., Johnston, M., Mueller, N.D., O'Connell, C., Ray, D.K., West, P.C., Balzer, C., Bennett, E.M., Carpenter, S.R., Hill, J., Monfreda, C., Polasky, S., Rockström, J., Sheehan, J., Siebert, S., Tilman, D., Zaks, D.P., 2011. Solutions for a cultivated planet. Nature 478, 337–342.

Food and Agricultural Organization of the United Nations, 2013. Crop prospects and food situation. FAO 2013, no. 2.

Gerber, S.A., Rush, J., Stemman, O., Kirschner, M.W., Gygi, S.P., 2003. Absolute quantification of proteins and phosphoproteins from cell lysates by tandem MS. Proc. Natl. Acad. Sci. USA 100, 6940–6945.

Gill, B.S., Appels, R., Botha-Oberholster, A.M., Buell, C.R., Bennetzen, J.L., Chalhoub, B., Chumley, F., Dvorák, J., Iwanaga, M., Keller, B., Li, W., McCombie, W.R., Ogihara, Y., Quetier, F., Sasaki, T., 2004. A workshop report on wheat genome sequencing: international genome research on wheat consortium. Genetics 168, 1087–1096.

Guozhu, L., Huifan, S., Linhua, F., Honggang, C., 2007. Heredity analysis of flavone and agaromic characters among radiated Tartary buckwheat mutants. Proceedings of the 10th International Symposium on Buckwheat, Yangling, China, pp. 178–185.

Hanashiro, I., Itoh, K., Kuratomi, Y., Yamazaki, M., Igarashi, T., Matsugasako, J., Takeda, Y., 2008. Granule-bound starch synthase I is responsible for biosynthesis of extra-long unit chains of amylopectin in rice. Plant Cell Physiol. 49, 925–933.

Heffler, E., Pizzimenti, S., Badiu, I., Guida, G., Rolla, G., 2014. Buckwheat allergy: an emerging clinical problem in Europe. Allergy & Therapy 5, 1000168.

Herre, J., Willment, J.A., Gordon, S., Brown, G.D., 2004. The role of Dectin-1 in antifungal immunity. Crit. Rev. Immunol. 24, 193–203.

Hilder, V.A., Powell, K.S., Gatehouse, A.M.R., Gatehouse, J.A., Gatehouse, L.N., Shi, W., Hamilton, D.W.O., Merryweather, A., Newell, C.A., Timans, J.C., Peumans, W.J., van Damme, E., Boulter, D., 1995. Expression of snowdrop lectin in transgenic tobacco plants results in added protection against aphids. Transgenic Res. 4, 18–25.

Horbowicz, M., Brenac, P., Obendorf, R.L., 1998. Fagopyritol B1, O-alpha-D-galactopyranosyl-(1→2)-D-chiro-inositol, a galactosyl cyclitol in maturing buckwheat seeds associated with desiccation tolerance. Planta 205, 1–11.

Houston, N.H., Lee, D.G., Stevenson, S.E., Ladics, G.S., Bannon, G.A., McClain, S., Privalle, L., Stagg, N., Herouet-Guicheney, C., MacIntosh, S.C., Thelen, J.J., 2011. Quantitation of soybean allergens using tandem mass spectrometry. J. Proteome Res. 10, 763–773.

Hu, J., Rampitsch, C., Bykova, N.V., 2015. Advances in plant proteomics toward improvement of crop productivity and stress resistance. Front. Plant Sci. 6, 209.

International Rice Genome Sequencing Project, 2005. The map-based sequence of the rice genome. Nature 436, 793–800.

International Wheat Genome Sequencing Consortium, 2014. A chromosome-based draft sequence of the hexaploid bread wheat (*Triticum aestivum*) genome. Science 345, 1251788.

Ishii, S., Katsumura, T., Shiozuka, C., Ooyauchi, K., Kawasaki, K., Takigawa, S., Fukushima, T., Tokuji, Y., Kinoshita, M., Ohnishi, M., Kawahara, M., Ohba, K., 2008. Antiinflammatory effect of buckwheat sprouts in lipopolysaccharide-activated human colon cancer cells and mice. Biosci. Biotechnol. Biochem. 72, 3148–3157.

Kamal, A.H.M., Jang, I.D., Kim, D.E., Suzuki, T., Chung, K.Y., Choi, J.S., Lee, M.S., Park, C.H., Park, S.U., Lee, S.H., Jeong, H.S., Woo S.H., 2011. Proteomics analysis of embryo and endosperm from mature common buckwheat seeds. J Plant Biol. 54, 81–91.

Kamal, A.H.M., Cho, K., Choi, J.S., Bae, K.H., Komatsu, S., Uozumi, N., Woo, S.H., 2013a. The wheat chloroplastic proteome. J. Proteomics 93, 326–342.

Kamal, A.H.M., Cho, K., Choi, J.S., Jin, Y., Park, C.S., Lee, J.S., Woo, S.H., 2013b. Patterns of protein expression in water-stressed wheat chloroplasts. Biol. Plantarum 57, 305–312.

Kayashita, J., Shimaoka, I., Nakajoh, M., Yamazaki, M., Kato, N., 1997. Consumption of buckwheat protein lowers plasma cholesterol and raises fecal neutral sterols in cholesterol-fed rats because of its low digestibility. J. Nutr. 127, 1395–1400.

Kim, S.J., Maeda, T., Sarker, M.Z., Takigawa, S., Matsuura-Endo, C., Yamauchi, H., Mukasa, Y., Saito, K., Hashimoto, N., Noda, T., Saito, T., Suzuki, T., 2007. Identification of anthocyanins in the sprouts of buckwheat. J. Agric. Food Chem. 55, 6314–6318.

Kim, S.J., Zaidul, I.S.M., Suzuki, T., Mukasa, Y., Hashimoto, N., Takigawa, S., Noda, T., Matsuura-Endo, C., Yamauchi, H., 2008. Comparison of phenolic compositions between common and Tartary buckwheat (*Fagopyrum*) sprouts. Food Chem. 110, 814–820.

Kim, S.T., Kim, S.G., Agrawal, G.K., Kikuchi, S., Rakwal R., 2014. Rice proteomics: a model system for crop improvement and food security. Proteomics 14, 593–610.

Komatsu, S., Kamal, A.H.M., Hossain, Z., 2014. Wheat proteomics: proteome modulation and abiotic stress acclimation. Front. Plant Sci. 5, 684.

Kreft, S., Janes, D., Kreft, I., 2013. The content of fagopyrin and polyphenols in common and Tartary buckwheat sprouts. Acta Pharm. 63, 553–560.

Lee, J., Jiang, W., Qiao, Y., Cho, Y.I., Woo, M.O., Chin, J.H., Kwon, S.W., Hong, S.S., Choi, I.Y., Koh, H.J., 2011. Shotgun proteomic analysis for detecting differentially expressed proteins in the reduced culm number rice. Proteomics 11, 455–468.

Li, S.Q., Zhang, Q.H., 2001. Advances in the development of functional foods from buckwheat. Crit. Rev. Food Sci. Nutr. 41, 451–464.

Li, Y., Duan, S., Jia, H., Bai, C., Zhang, L., Wang, Z., 2014. Flavonoids from Tartary buckwheat induce G2/M cell cycle arrest and apoptosis in human hepatoma HepG2 cells. Acta Biochim. Biophys. Sin. 46, 460–470.

Liu, C.L., Chen, Y.S., Yang, J.H., Chiang, B.H., 2008. Antioxidant activity of Tartary [*Fagopyrum tatricum* (L.)] Gaertn.) and common (*Fagopyrum esculentum* Moench) buckwheat sprouts. J. Agric. Food Chem. 56, 173–178.

Lobell, D.B., Schlenker, W., Costa-Roberts, J., 2011. Climate trends and global crop production since 1980. Science 333, 616–620.

Logacheva, M.D., Penin, A.A., Sutormin, R.A., Demidenko, N.V., Naumenko, S.A., Vinogradov, D.V., Mazin, P.V., Kurmangaliev, Y.Z., Gelfand, M.S., Kondrashov, A.S., 2014. A draft genome sequence of Tatary buckwheat, *Fagopyrum tataricum*. Plant & Animal Genome XXII, Jan. 11–15, San Diego, CA, USA.

Ma, M.S., Bae, I.Y., Lee, H.G., Yang, C.B., 2006. Purification and identification of angiotensin I-converting enzyme inhibitory peptide from buckwheat (*Fagopyrum esculentum* Moench). Food Chem. 96, 36–42.

Majeran, W., van Wijk, K.C., 2009. Cell-type specific differentiation of chloroplasts in C4 plants. Trends Plant Sci. 14, 100–109.

Marcussen, T., Sandve, S.R., Heier, L., Spannagl, M., Pfeifer, M., International Wheat Genome Sequencing Consortium, Jakobsen, K.S., Wulff, B.B., Steuernagel, B., Mayer, K.F., Olsen, O.A., 2014. Ancient hybridizations among the ancestral genomes of bread wheat. Science 345, 1250092.

Matros, A., Kaspar, S., Witzel, K., Mock, H.P., 2011. Recent progress in liquid chromatography-based separation and label-free quantitative plant proteomics. Phytochemistry 72, 963–974.

Nair, A., Adachi, T., 2002. Screening and selection of hypoallergenic buckwheat species. Sci. World J. 2, 818–826.

Nile, S.H., Park, S.W., 2014. HPTLC analysis, antioxidant, anti-inflammatory and antiproliferative activities of *Arisaema tortuosum* tuber extract. Pharm. Biol. 52, 221–227.

Pechanova, O., Takac, T., Samaj, J., Pechan, T., 2013. Maize proteomics: an insight into the biology of an important cereal crop. Proteomics 13, 637–662.

Peng, W., Changling, H., Shu, Z., Han, T., Qin, L., Zheng, C., 2015. Antitumor activity of tatariside F isolated from roots of *Fagopyrum tataricum* (L.) Gaertn against H22 hepatocellular carcinoma via up-regulation of p53. Phytomedicine 22, 730–736.

Rubinstein, N., Ilarregui, J.M., Toscano, M.A., Rabinovich, G.A., 2004. The role of galectins in the initiation, amplification and resolution of the inflammatory response. Tissue Antigens 64, 1–12.

Sasaki, T., Matsumoto, T., Yamamoto, K., et al.,2002. The genome sequence and structure of rice chromosome 1. Nature 420, 312–316.

Schnable, P.S., Ware, D., Fulton, R.S., et al.,2009. The B73 maize genome: complexity, diversity, and dynamics. Science 326, 1112–1115.

Shin, D.H., Kamal, A.H.M., Suzuki, T., Yun, Y.H., Lee, M.S., Chung, K.Y., Jeong, H.S., Park, C.H., Choi, J.S., Woo, S.H., 2010. Reference proteome map of buckwheat (*Fagopyrum esculentum* and *Fagopyrun tataricum*) leaf and stem cultured under light or dark. Aus. J. Crop Sci. 4, 633–641.

Steadman, K.J., Burgoon, M.S., Schuster, R.L., Lewis, B.A., Edwardson, S.E., Obendorf, R.L., 2000. Fagopyritols, D-chiro-inositol, and other soluble carbohydrates in buckwheat seed milling fractions. J. Agric. Food Chem. 48, 2843–2847.

Sunkar, R., 2010. Plant stress tolerance: methods and protocols. Meth. Mol. Biol. 639, 365.

Suzuki, T., Honda, Y., Mukasa, Y, 2005. Effects of UV-B radiation, cold and desiccation stress on rutin concentration and rutin glucosidase activity in Tartary buckwheat (*Fagopyrum tataricum*) leaves. Plant Sci. 168, 1303–1307.

Suzuki, T., Kim, S.J., Mohamed, Z.I.S., Mukasa, Y., Takigawa, S., Matsuura-Endo, C., Yamauchi, H., Hashimoto, N., Noda, T., Saito, T., 2007. Structural identification of anthocyanins and analysis of concentrations during growth and flowering in buckwheat (*Fagopyrum esculentum* Moench) petals. J. Agric. Food Chem. 55, 9571–9575.

Suzuki, T., Morishita, T., Noda, T., Ishiguro, K., 2015. Acute and subacute toxicity studies on rutin-rich Tartary buckwheat dough in experimental animals. J. Nurt. Sci. Vitaminol. 61, 175–181.

Thelen, J.J., Peck, S.C., 2007. Quantitative proteomics in plants: choices in abundance. Plant Cell 19, 3339–3346.

Tomotake, H., Yamamoto, N., Yanaka, N., Ohinata, H., Yamazaki, R., Kayashita, J., Kato, N., 2006. High protein buckwheat flour suppresses hypercholesterolemia in rats and gallstone formation in mice by hypercholesterolemic diet and body fat in rats because of its low protein digestibility. Nutrition 22, 166–173.

Vanderschuren, H., Lentz, E., Zainuddin, I., Gruissem, W., 2013. Proteomics of model and crop plant science: status, current limitations and strategic advances for crop improvement. J. Proteomics 93, 5–19.

Varshney, R.K., Nayak, S.N., May, G.D., Jackson, S.A., 2009. Next-generation sequencing technologies and their implications for crop genetics and breeding. Trends Biotechnol. 27, 522–530.

Wang, Z.H., Gao, L., Li, Y.Y., Zhang, Z., Yuan, J.M., Wang, H.W., Zhang, L., Zhu, L., 2007. Induction of apoptosis by buckwheat trypsin inhibitor in chronic myeloid leukemia K562 cells. Biol. Pharm. Bull. 30, 783–786.

Wang, Z., Li, S., Ren, R., Li, J., Cui, X., 2015. Recombinant buckwheat trypsin inhibitor induces mitophagy by directly targeting mitochondria and causes mitochondrial dysfunction in Hep G2 cells. J. Agric. Food Chem. 63, 7795–7804.

Washburn, M.P., Wolters, D., Yates, III, J.R., 2001. Large-scale analysis of the proteome by multidimensional protein identification technology. Nat. Biotechnol. 19, 242–247.

Wienkoop, S., Baginsky, S., Weckwerth, W., 2010. *Arabidopsis thaliana* as a model organism for plant proteome research. J. Proteomics 73, 2239–2248.

Wilkins, M.R., Sanchez, J.C., Gooley, A.A., Appel, R.D., Humphery-Smith, I., Hochstrasser, D.F., Williams, K.L., 1996. Progress with proteome projects: why all proteins expressed by a genome should be identified and how to do it. Biotechnol. Genet. Eng. Rev. 13, 19–50.

Woo, S.H., Kamal, A.H.M., Tatsuro, S., Campbell, C.G., Adachi, T., Yun, S.H., Chung, K.Y., Choi, J.S., 2010. Buckwheat (*Fagopyrum esculentum* Moench.): concepts, prospects and potential. Eur. J. Plant Sci. Biotech. 4 (Special Issue 1), 1–16.

Woo, S.H., Kamal, A.H.M., Park, S.M., Kwon, S.O., Park, S.U., Kumar Roy, S., Lee, J.Y., Choi, J.S., 2013. Relative distribution of free amino acids in buckwheat. Food Sci. Biotechnol. 22, 665–669.

Yang, J., Lu, C., Shen, Q., Yan, Y., Xu, C., Song, C., 2015b. The complete chloroplast genome sequence of *Fagopyrum cymosum*. Mitochon. DNA 29, 1–2.

Yao, Y., Shan, F., Bian, J., Chen, F., Wang, M., Ren, G., 2008. D-Chiro-inositol-enriched Tartary buckwheat bran extract lowers the blood glucose level in KK-Ay mice. J. Agric. Food Chem. 56, 10027–10031.

Yokosho, K., Yamaji, N., Na, J.F., 2014. Global transcriptome analysis of Al-induced genes in an Al-accumulating species, common buckwheat (*Fagopyrum esculentum* Moench). Plant Cell Physiol. 55, 2077–2091.

Zhang, Z., Li, Y., Li, C., Yuan, J., Wang, Z., 2007. Expression of a buckwheat trypsin inhibitor gene in *Escherichia coli* and its effect on multiple myeloma IM-9 cell proliferation. Acta Biochim. Biophys. Sin. (Shanghai) 39, 701–707.

Zhang, X., Yuan, J.M., Cui, X.D., Wang, Z.H., 2008. Molecular cloning, recombinant expression, and immunological characterization of a novel allergen from Tartary buckwheat. J. Agric. Food Chem. 56, 10947–10953.

Zhang, G., Xu, Z., Gao, Y., Huang, X., Zou, Y., Yang, T., 2015. Effects of germination on the nutritional properites, phenolic profiles, and antioxidant activities of buckwheat. J. Food Sci. 80, H1111–H1119.

Chapter twenty

Mineral and Trace Element Composition and Importance for Nutritional Value of Buckwheat Grain, Groats, and Sprouts

P. Pongrac*, K. Vogel-Mikuš*,**, M. Potisek*, E. Kovačec*, B. Budič†, P. Kump**, M. Regvar*, I. Kreft‡

*Department of Biology, University of Ljubljana, Biotechnical Faculty, Ljubljana, Slovenia; **Department of Low and Medium Energy Physics, Jožef Stefan Institute, Jamova, Ljubljana, Slovenia; †Laboratory for Analitical Chemistry, National Institute of Chemistry, Hajdrihova, Ljubljana, Slovenia; ‡Department of Forest Physiology and Genetics, Slovenian Forestry Institute, Ljubljana, Slovenia

INTRODUCTION

Grains are a main component in the human diet and a major source of essential mineral elements. The majority of mineral elements, such as magnesium (Mg), manganese (Mn), iron (Fe), copper (Cu), and zinc (Zn) in grains are tightly bound in phytate (*myo*-inositol hexakisphosphate) salts and are not readily bioavailable. Bioavailability of mineral elements is defined as the fraction of the ingested nutrient that is absorbed and subsequently utilized for normal physiological functions (Fairweather-Tait and Hurrell, 1996). Low concentration and/ or low bioavailability of mineral elements in the diet can result in mineral element malnutrition in humans and associated negative impacts on individual well-being, social welfare, and economic productivity (Stein, 2010). Increase in the concentration and/ or bioavailability of mineral elements in grains can be achieved by agronomic

Molecular Breeding and Nutritional Aspects of Buckwheat. http://dx.doi.org/10.1016/B978-0-12-803692-1.00020-1

approaches (using inorganic fertilizers) or by identifying and/or developing crops with improved mineral element use efficiency (White and Broadley, 2009). In addition, food processing, such as thermal and mechanical processing, soaking, fermentation, and germination/malting of seeds and grain can be applied to improve mineral element concentration and bioavailability (Holtz and Gibson, 2007). These processes are believed to enable or assist in the release of tightly bound mineral elements, especially those bound in phytate. For example, soaking of grains and seeds and germination leads to the remobilization and redistribution of mineral elements, which become available for active processes of the growing embryo until the first photosynthetically active leaves develop. However, few studies address the effect of processing on the concentration, distribution, and bioavailability of mineral elements in grains (Holtz and Gibson, 2007; Nelson et al., 2013). This is the case also for buckwheat (*Fagopyrum* sp.) grains, especially for European cultivars. Such studies are essential to complement the information we have on mineral element composition of buckwheat grains of European cultivars and their milling fractions (Bonafaccia et al., 2003a) and to the spatial distribution of mineral elements in grains of common buckwheat (*Fagoyprum esculentum* Moench; Vogel-Mikuš et al., 2009a; Pongrac et al., 2011) and grains of Tartary buckwheat (*Fagopyrum tataricum* Gaertn.; Pongrac et al., 2013a). The estimation of bioavailability of mineral elements is a complex process. In vitro techniques are not as reliable as in vivo techniques (Fairweather-Tait et al., 2005), but the latter are expensive and time-consuming. In addition, for some mineral elements, techniques to determine their bioavailability are still unavailable. Thus routine and reliable large-scale determination of bioavailability of mineral elements in different foods is still lacking.

The aim of the study was to determine *total* mineral element concentrations in grains and their processed counterparts, namely, groats (hydrothermally processed grains that have the inedible husk removed) and sprouts (7-day-old seedlings) of two buckwheat species from Slovenia. The Tartary buckwheat cultivar used was "Wëllkar," which originates from Luxembourg (Bonafaccia et al., 2003a,b) and was reintroduced to Slovenia in 2010 (Mlin Rangus, Dolenje Vrhpolje at Šentjernej, Slovenia). The common buckwheat cultivar used was "Trdinova," which is cultivated and processed by the same producer. The spatial distribution of mineral elements in groats and sprouts was determined in Tartary buckwheat using synchrotron-based micro-X-ray fluorescence spectroscopy (μ-XRF; at beamline ID21 at European Synchrotron Radiation Facility, Grenoble, France) to complement information on Tartary buckwheat grains (Pongrac et al., 2013a) and to determine the mineral element distribution as affected by hydrothermal processing and germination in this species.

MATERIALS AND METHODS

Grains and groats of the two buckwheat species were obtained from Mlin Rangus (Dolenje Vrhpolje at Šentjernej, Slovenia). Sprouts were cultivated in an automatic sprouter (EasyGreen® MicroFarm System, EasyGreen Factory Inc.,

Nevada, USA) where they were watered by misting every 3 h during the day (five times), and twice during the night (with a 4-h and 5-h gap) with tap water. Three independent experiments were established at room temperature (21°C) and away from direct sunlight.

Seven-day-old sprouts (whole seedlings) were rinsed with tap water, blotted, frozen in liquid nitrogen, and freeze-dried for 5 days at 0.240 mbar and −30°C (Alpha Christ 2-4, Martin Christ Gefriertrocknungsanlagen GmbH, Germany). Grain, groats, and freeze-dried sprouts were homogenized in liquid nitrogen using a pestle and mortar. Powdered material was wet digested in 3 mL HNO_3 and 0.5 mL H_2O_2 and analyzed for concentrations of Mg, phosphorus (P), sulfur (S), potassium (K), calcium (Ca), Mn, Fe, Cu, Zn, and molybdenum (Mo) using inductively coupled plasma-mass spectrometry and inductively coupled plasma-optical emission spectroscopy, as previously described (Pongrac et al., 2013b).

Groats and sprouts of Tartary buckwheat were analyzed for mineral element distributions using μ-XRF. Groats (soaked at 4°C in water for 1 h) and unfolded cotyledons of 7-day-old sprouts were frozen in liquid propane and sectioned at −25°C in cryotome to 25 μm thick sections that were freeze dried for 3 days at 0.240 mbar and −30°C (Vogel-Mikuš et al., 2009b). Freeze-dried sections were analyzed for the distributions of Mg, P, S, K, Ca, Mn, and Fe using 7200 eV at the ID21 beamline (ESRF, Grenoble, France; Koren et al., 2013). Qualitative maps of mineral element spatial distribution were subjected to quantitative analysis (Koren et al., 2013) and quantitative distribution maps were generated with PyMCA software (Solé et al., 2007).

Pairwise comparisons of mineral element concentrations in grains with mineral element concentrations in the two differently processed grains, groats, and sprouts were performed for each buckwheat species separately, while pairwise comparisons of mineral element concentrations in grains, groats, and sprouts were performed between the two buckwheat species. For these purposes, Student's t-test at $p < 0.01$ was used in GenStat (64-bit Release 17.1; VSN International Ltd, Oxford, UK).

RESULTS AND DISCUSSION
Mineral Element Composition of Grains, Groats, and Sprouts

Tartary buckwheat groats had higher concentrations of Mg and lower concentrations of Ca and Cu than Tartary buckwheat grains, while concentrations of P, S, K, Mn, Fe, Zn, and Mo were not affected by hydrothermal processing (Table 20.1). Similarly, groats of common buckwheat had lower concentrations of Ca and Mo than common buckwheat grains, while no difference was seen for the rest of the analyzed mineral elements (Table 20.1). For groat production, grains were hydrothermally processed (soaked in tap water at 95°C for 20 min, and dried to 20% moisture content), which enabled easier removal of firm, inedible husks. Thus grains and groats differ in two characteristics: first, groats are rehydrated in hot water and redried, and, second, groats do not have the husk.

TABLE 20.1 Mineral Element Concentrations (mg/kg dry weight) in Grains, Groats, and 7-Day-Old Sprouts of Tartary Buckwheat and Common Buckwheat

	Tartary Buckwheat (*F. tataricum* cv. Wëllkar)			Common Buckwheat (*F. esculentum* cv. Trdinova)		
	Grains	Groats	Sprouts	Grains	Groats	Sprouts
Mg	2130	2490[a]	5790[a]	2800	2220	5470[a]
	(49)	(136)	(261)	(564)	(755)	(240)
P	3910	3930	7300[a]	3860	3850	7930[a]
	(109)	(189)	(206)	(70)	(179)	(544)
S	1240	1280	2490[a]	1490[b]	1650[b]	3260[a,b]
	(43)	(50)	(85)	(43.9)	(15.4)	(101)
K	5070	5640	7780[a]	6280	5470	7290
	(127)	(444)	(276)	(1230)	(1650)	(436)
Ca	562	194[a]	5830[a]	791[b]	178[a]	8410[a]
	(43)	(5)	(758)	(37.6)	(11.0)	(1090)
Mn	11	9.4	14.5[a]	25.7	18.5	21.1
	(0.45)	(0.57)	(1.23)	(5.21)	(6.43)	(2.86)
Fe	60.6	55.7	100	114	35.3	70.7
	(25.1)	(15.4)	(20.0)	(23.4)	(15.1)	(7.98)
Cu	4.49	3.88[a]	10.1	6.41[b]	5.51[b]	10.8[a]
	(0.09)	(0.10)	(1.83)	(0.19)	(0.30)	(0.14)
Zn	22.6	24.9	220[a]	23.0	26.0	130
	(0.98)	(1.01)	(18.3)	(0.72)	(1.95)	(29.4)
Mo	0.84	0.98	1.32[a]	1.36[b]	0.50[a,b]	1.89[a,b]
	(0.14)	(0.03)	(0.14)	(0.03)	(0.07)	(0.10)

Shown are means from $n = 3$ and standard deviations in parentheses.
[a]Statistically significant differences for pairwise comparisons grains versus groats and grain versus sprouts for each buckwheat species separately at $p < 0.01$.
[b]Statistically significant differences for pairwise comparisons between buckwheat species at $p < 0.01$.

The husk has been shown to contain 71% of total Ca and 29% of total Cu in the Tartary buckwheat grain (Pongrac et al., 2013a) and 85% of total Ca and 17% of total Cu in common buckwheat grain (Vogel-Mikuš et al., 2009). Thus the observed decreases in Ca and Cu concentration in groats can be explained by the removal of the husk. Milling of whole common buckwheat grain (thus including this Ca-rich layer) led to higher Ca concentrations of buckwheat flour than milling from the husk-free groats (Steadman et al., 2001a). Molybdenum concentration in fancy flour (a light-colored flour, mostly central endosperm; Steadman et al., 2001b) milled from common buckwheat groats was reported to be lower than in the fancy flour milled from whole grains (Steadman et al., 2001a), which supports our results in common buckwheat. Although in groats most of the Mo was found in the bran fraction, the highest concentrations were found in the no-hull bran from whole grain millings (Steadman et al., 2001a). This indicates that the largest proportion of Mo is located in the embryo of the grain, but additional detailed studies will be needed to resolve the location of Mo in the grains. The only observed statistically significant increase in groats was seen for Mg concentrations in Tartary buckwheat. Taken together these results suggest that hydrothermal processing in general does not have positive effects on mineral element concentrations in the two buckwheat species. However, there might be changes to the bioavailability of mineral elements because of the processing, but this has not been studied in detail so far. Our results on Fe bioavailability using human intestinal Caco-2 cells in Tartary buckwheat grains and groats do not support the assumption of groats having higher bioavailability of Fe than grains (Pongrac et al., 2016).

Sprouts are vigorously transformed grains as they undergo imbibition, germination, and several steps of seedling development (in buckwheat this comprises hypocotyl elongation, husk shedding, and cotyledon unfolding). During seedling development the loss of dry matter (mostly of nonfibrous carbohydrates) caused by respiration takes place and water content increases. In addition, newly developed roots take up mineral elements, which are available in the solution. Collectively, these events contribute to the concentration of the mineral elements in sprouts. Therefore this was consistent with higher concentrations of mineral elements in both Tartary buckwheat and common buckwheat sprouts compared to grains (Table 20.1). When compared to published reports, higher concentrations of Mg, P, K, and Fe, but lower concentrations of Ca and Zn, were found in Tartary buckwheat sprouts (line KW45 introduced from Japan; Lee et al., 2006) than in our study. In common buckwheat sprouts, higher concentrations of Zn and Mo and comparable concentrations of Mn, Fe, and Cu were found in our study compared to another report on common buckwheat sprouts cultivated in tap water (Lintschinger et al., 1997). Sprouts of Tartary buckwheat did not have higher bioavailability of Fe than grains (Pongrac et al., 2016). Although common buckwheat and Tartary buckwheat sprouts and their antioxidative potential have been used in products that were tested for having positive health effects in rats (eg, Kuwabara et al., 2007; Merendino et al., 2014) and hamsters (Lin et al., 2008), neither their mineral element composition nor mineral element bioavailability has been studied in detail. It would be beneficial to evaluate the contribution of mineral elements, both total and bioavailable portions, to the observed health effects. Likewise, mineral elements play a crucial role in several agronomic traits (eg, yield, appearance, taste, and longevity) as well as in the nutritional value of products. Further studies are required to determine the levels of variation in mineral element composition and their bioavailability in sprouts of different buckwheat species and cultivars.

Comparison of Tartary Buckwheat and Common Buckwheat

Compared to Tartary buckwheat, common buckwheat grains contain larger concentrations of S, Ca, Cu, and Mo, groats contain larger concentrations of S, Cu, and Mo, and sprouts contain larger concentrations of S and Mo (Table 20.1). Therefore the concentrations of S and Mo are constitutively higher in the studied cultivar of common buckwheat than in the studied cultivar of Tartary buckwheat. Analyses of S and Mo concentrations in buckwheat are scarce, thus any strong conclusions are hard to draw. Common buckwheat grains were reported to contain 0.78 mg/kg of Mo (Lintschinger et al., 1997), while groats contained 0.9 mg/kg (Steadman et al., 2001a). In higher plants, Mo is a cofactor for only a few enzymes, such as nitrogenase, nitrate reductase, xanthine dehydrogenase, aldehyde oxidase, and sulfite oxidase (Broadley et al., 2012). The latter is especially important in plants for the detoxification of excess of sulfur dioxide, therefore correlation between S and Mo concentrations can be expected. In addition, molybdate anion $MoSO_4^{2-}$ has similar properties to SO_4^{2-}, which has important implications for Mo availability in soils and uptake by plants (Broadley et al., 2012). Since a large proportion of S can be found in proteins, the measured difference in S concentration between common buckwheat and Tartary buckwheat could be reflected in higher concentrations of proteins in common buckwheat grain. However, this conclusion is not supported based on previous reports on protein concentration in buckwheat (Bonafaccia et al., 2003b). Tartary buckwheat (cultivar "Wëllkar" from Luxembourg) grain fractions contained higher concentrations of proteins and of S-containing amino acid methionine, while cysteine concentrations were higher only in the bran of Tartary buckwheat grain than in corresponding grain parts of common buckwheat (Bonafaccia et al., 2003b). In addition, higher concentrations of Se, Zn, Fe, Co, and Ni were found in grains and in the bran of Tartary buckwheat of the same cultivars (Bonafaccia et al., 2003a). These contrasting results can be explained by (1) the different grain weights between buckwheat species leading to dilution effects in heavier grains and smaller mineral element concentrations (in our study common buckwheat grains were on average 20% heavier than Tartary buckwheat grains), (2) differences in grain weight of the buckwheat cultivars analyzed, and (3) because of the environmental conditions and soil characteristics during plant cultivation. Indeed, several cultivars from Shanxi, China, collected from various locations showed large variability with the following ranges of concentrations found (minimum–maximum; in mg/kg) for common buckwheat ($n = 46$; P 1450–5560; Ca 184–1210; Mn 7–32; Fe 45–737; Cu 6–18; Zn 13–47; Se 0.01–0.56) and for Tartary buckwheat ($n = 36$; P 2270–5150; Ca 175–412; Mn 6–12; Fe 51–195; Cu 4–9; Zn 19–83; Se 0.10–0.41) (Prof. Tao Yunping, personal communications; Geo-database of buckwheat in Shanxi, China: http://zrzy.sxinfo.net/Home_pic.aspx). Based on these data, higher average concentrations of Cu, Fe, Mn, and Ca and lower average concentrations of Se were found in common buckwheat than in Tartary buckwheat, while concentrations of Zn and P were comparable (Prof. Tao Yunping, personal communication; Geodatabase of buckwheat in Shanxi, China: http://zrzy.sxinfo.net/Home_pic.aspx). These results are, however, not a reliable measure of the ranges of mineral element concentrations across the two buckwheat species, as the plants were not cultivated side by side. In contrast, relevant studies on the extent of variability in mineral element concentrations in cereal grains in plants grown in the same environment have already been performed. For example, the level of variability in several mineral element concentrations was evaluated for wheat (eg, Zhao et al., 2009, $n = 175$), rice (eg, Pinson et al., 2015, $n = 1763$), barley (Mamo et al., 2014, $n = 298$), and millet (Bashir et al., 2014, $n = 225$) grain. These studies firmly demonstrated the effect of genetic variability in combination with the environmental conditions (particularly the mineral element composition of the soil and the length of the growing cycle) on the mineral element concentration of the grains. Studies of species-wide variation in mineral element composition of buckwheat species are still lacking. Such studies would contribute considerably to the development of the field, especially as the variability in bioactive compounds (eg, Izydorczyk et al., 2013; Kiprovski et al., 2015) is being reported. Based on thorough dissection of variability of all nutritionally important traits in a particular crop, recommendations on best-fitting cultivars for agronomic or dietary-specific needs could be made.

Mineral Element Distribution in Groats and Sprouts of Tartary Buckwheat

Heterogeneous distribution of mineral elements in cereal (eg, Lombi et al., 2011; Singh et al., 2013), pseudocereal grains (Vogel-Mikuš et al., 2009a; Pongrac et al., 2013a), and seeds (eg, Cvitanich et al., 2010) reflects stringent control over movement of mineral elements within plants. It is also a reason for the observed variability in mineral element concentrations in milling fractions of these grains (eg, Steadman et al., 2001a; Bonafaccia et al., 2003a). To complement knowledge on mineral element distribution in mature Tartary buckwheat grain (Pongrac et al., 2013a) we studied the distribution of mineral elements in processed grains. In cotyledons of groats, densely packed cells with high concentrations of Mg, P, K, Ca, Mn, and Fe were distinguished, while S was more predominant in the vascular bundles (Fig. 20.1). Within these cells, sphere-shaped distribution of P could be assigned to phytate globoids and the corresponding distributions of Mg, K, Ca, Mn, and Fe presumably represent phytate-bound

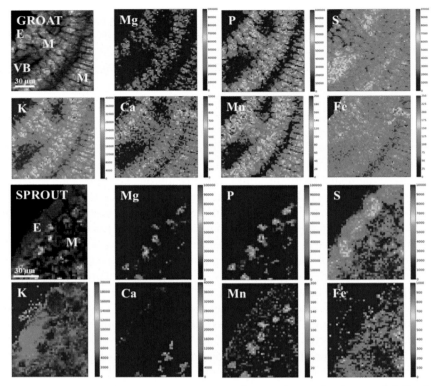

FIGURE 20.1 Morphology of the cross-sections and quantitative spatial distribution maps of Mg, P, S, K, Ca, Mn, and Fe (legends are in mg/kg dry weight) of representative cotyledons of Tartary buckwheat groat and sprout. *E*, Epidermis; *M*, mesophyll; *VB*, vascular bundle.

fractions of these mineral elements. Likewise in wheat bran, phytate was shown to bind to mineral elements in concentration order K > Mg > Ca > Fe (Bohn et al., 2007). The distribution of Mg, P, Mn, and Fe in Tartary buckwheat groat greatly resembles that in the Tartary buckwheat grain (Pongrac et al., 2013a), indicating that hydrothermal processing did not significantly affect the distribution of mineral elements. By contrast, distribution of mineral elements in cotyledons of sprouts was considerably different, especially for S, K, Ca, and Fe (Fig. 20.1). The observed differences in mineral element distribution between the mesophyll of cotyledon in groats and the mesophyll of sprouts can be assigned to morphological and physiological transformations of storage vacuoles in grains and groats into the lytic vacuoles, previously described for root tips of germinating *Nicotiana tabacum* L. (Zheng and Staehelin, 2011), after which some of the mineral elements can accumulate in the lytic compartments. Magnesium, P, and Mn were colocalized as in grains (Pongrac et al., 2013a) and groats, indicating that Mg and Mn are still bound to phytate or other P-containing compounds, but are confined to loosely dispersed hot spots in sprouts. Sulfur was predominantly found in the epidermis where it could be conjugated into flavonoid sulfates such as quercetin, vitexin, isovitexin, orientin, and isoorientin (Harborne, 1975), all of which have been reported in both buckwheat species (Kim et al., 2008). Several of these flavonoids have been allocated to the upper epidermis of cotyledons of buckwheat seedlings (Margna et al., 1990), where they may act as protection against UV-B radiation.

Calcium was present in cell walls, particularly in the surface cell walls of the epidermis. In the inner mesophyll Ca was seen in globular hot spots, which could be designated as Ca oxalate crystals. Oxalic acid concentrations were reported to be very high in buckwheat cotyledons (10,000 mg/kg; Shen et al., 2004) compared to concentrations in the grain (1500 mg/kg; Siener et al., 2006) and in cotyledons it might precipitate in oxalate crystals. Iron was evenly distributed in the mesophyll with hot spots that could only partially be correlated to S distribution in mesophyll. This Fe and S-rich area can be attributed to the formation of non-heme Fe–S proteins, where Fe is coordinated to the thiol group of cysteine or to inorganic S as clusters, or to both (Broadley et al., 2012). The most well-known Fe–S protein is ferredoxin, which acts as an electron transmitter in a number of metabolic processes and in green leaves as the reductant for sulfate (Broadley et al., 2012). The lack of strong colocalization between P and Fe suggested that Fe is released from phytate or other P-containing compounds in sprouts. With the technique used, we were unable to study the distribution of Zn. It is noteworthy to say that around 70% of total Zn in the common buckwheat flour was shown to be available (Ikeda et al., 1990). Since the binding strength of mineral elements plays a role in their bioavailability we anticipate future studies to focus on the chemical complexation of mineral elements as well as direct estimations of bioavailability from different buckwheat products.

ACKNOWLEDGMENTS

This study was financed by the Slovenian Research Agency project (Z4-4113) and programs (P1-0112 and P3-0395), and the IAEA framework of coordinated research projects RC 16796 "Applications of synchrotron radiation for environmental sciences and materials research for development of environmentally friendly resources" (CSI, Katarina Vogel-Mikuš). The European Synchrotron Radiation Facility, Grenoble, France, is acknowledged for provision of synchrotron radiation facilities at beamline ID21 for µ-XRF analyses (project LS-2225). We thank Dr Hiram Castillo-Michel from ID21 for his help with the beamline operation, Prof. Tao Yunping for providing mineral element concentrations in common and Tartary buckwheat from Shanxi, China, Mlin Rangus for providing grains and groats of Tartary buckwheat, and Dr Courtney Giles for reading the original manuscript.

REFERENCES

Bashir, E.M.A., Ali, A.M., Ali, A.M., Ismail, M.I., Parzies, H.K., Haussmann, B.I.G., 2014. Patterns of pearl millet genotype-by-environment interactions for yield performance and grain iron (Fe) and zinc (Zn) concentration in Sudan. Field Crops Res. 166, 82–91.

Bohn, L., Josefsen, L., Meyer, A.S., Rasmussen, S.K., 2007. Quantitative analysis of phytate globoids isolated from wheat bran and characterisation of their sequential dephosphorylation by wheat phytase. J. Agric. Food Chem. 55, 7547–7552.

Bonafaccia, G., Gambelli, L., Fabjan, N., Kreft, I., 2003a. Trace elements in flour and bran from common and Tartary buckwheat. Food Chem. 83, 1–5.

Bonafaccia, G., Marocchini, M., Kreft, I., 2003b. Composition and technological properties of the flour and bran from common and Tartary buckwheat. Food Chem. 80, 9–15.

Broadley, M., Brown, P., Cakmak, I., Rengel, Z., Zhao, F., 2012. Function of nutrients: micronutrients. In: Marschner, P. (Ed.), Marschner's Mineral Nutrition of Higher Plants. third ed. Academic Press, Amsterdam, Netherlands, pp. 191–248.

Cvitanich, C., Przybyłowicz, W.J., Urbanski, D.F., Jurkiewicz, A.M., Mesjasz-Przybyłowicz, J., Blair, M.W., Astudillo, C., Jensen, E.O., Stougaard, J., 2010. Iron and ferritin accumulate in separate cellular locations in *Phaseolus* seeds. BMC Plant Biol. 10, 26.

Fairweather-Tait, S., Hurrell, R.F., 1996. Bioavailability of mineral and trace elements. Nutr. Res. Rev. 9, 295–324.

Fairweather-Tait, S., Lynch, S., Hotz, C., Hurrell, R., Abrahamse, L., Beebe, S., Bering, S., Bukhave, K., Glahn, R., Hambidge, M., Hunt, J., Lonnerdal, B., Miller, D., Mohktar, N., Nestel, P., Reddy, M., Sandber, A.S., Sharp, P., Teucher, B., Rinidad, T.P., 2005. The usefulness of in vitro models to predict the bioavailability of iron and zinc: a consensus statement from the HarvestPlus Consultation. Int. J. Vitam. Nutr. Res. 75 (6), 371–374.

Harborne, J.B., 1975. Flavonoid sulphates: new class of sulphur compounds in higher plants. Phytochemistry 14, 1147–1155.

Holtz, C., Gibson, R.S., 2007. Traditional food-processing and preparation practices to enhance the bioavailability of micronutrients in plant-based diets. J. Nutr. 137, 1097–1100.

Ikeda, S., Edotani, M., Naito, S., 1990. Zinc in buckwheat. Fagopyrum 10, 51–55.

Izydorczyk, M.S., McMillan, T., Bazin, S., Kletke, J., Dushnicky, L., Dexter, J., 2013. Canadian buckwheat: a unique, useful and under-utilized crop. Can. J. Plant Sci. 93, 1–16.

Kim, S.-J., Zaidul, I.S.M., Suzuki, T., Mukasa, Y., Hashimoto, N., Takigawa, S., Noda, T., Matsuura-Endo, C., Yamauchi, H., 2008. Comparison of phenolic compositions between common and tartary buckwheat (*Fagopyrum*) sprouts. Food Chem. 110, 814–820.

Kiprovski, B., Mikulič-Petkovšek, M., Slatnar, A., Veberič, R., Štampar, F., Malenčić, D., Latković, D., 2015. Comparison of phenolic profiles and antioxidant properties of European *Fagopyrum esculentum* cultivars. Food Chem. 185, 41–47.

Koren, Š., Arčon, I., Kump, P., Nečemer, M., Vogel-Mikuš, K., 2013. Influence of CdCl$_2$ and CdSO$_4$ supplementation on Cd distribution and ligand environment in leaves of the Cd hyperaccumulator *Noccaea* (*Thlaspi*) *praecox*. Plant Soil 370, 125–148.

Kuwabara, T., Han, K.-H., Hashimoto, N., Yamauchi, H., Shimada, K.-I., Sekikawa, M., Fukushima, M., 2007. Tartary buckwheat sprout powder lowers plasma cholesterol level in rats. J. Nutr. Sci. Vitaminol. 53, 501–507.

Lee, S.-S., Park, C.-H., Park, B.-J., Kwon, S.-M., Chang, K.-J., Kim, S.-L., 2006. Rutin, catechin derivatives, and chemical components of Tartary buckwheat (*Fagopyrum tataricum* Gaertn.) sprouts. Korean J. Crop Sci. 51(S), 277–282.

Lin, L.-Y., Peng, C.-C., Yang, Y.-L., Peng, R.Y., 2008. Optimization of bioactive compounds in buckwheat sprouts and their effect on blood cholesterol in hamsters. J. Agric. Food Chem. 58, 1216–1223.

Lintschinger, J., Fuchs, N., Moser, H., Jäger, R., Hlebeina, T., Markolin, G., Gössler, W., 1997. Uptake of various trace elements during germination of wheat, buckwheat and quinoa. Plant Foods Hum. Nutr. 50, 223–237.

Lombi, E., Smith, E., Hansen, T.H., Paterson, D., de Jonge, M.D., Howard, D.L., Persson, D.P., Husted, S., Ryan, C., Schjoerring, J.K., 2011. Megapixel imaging of (micro)nutrients in mature barley grains. J. Exp. Bot. 62, 273–282.

Mamo, B.E., Barber, B.L., Steffenson, B.J., 2014. Genome-wide association mapping of zinc and iron concentration in barley landraces from Ethiopia and Eritrea. J. Cereal Sci. 60, 497–506.

Margna, U., Margna, E., Paluteder, A., 1990. Localization and distribution of flavonoids in buckwheat seedling cotyledons. J. Plant Physiol. 136, 166–171.

Merendino, N., Molinari, R., Costantini, L., Mazzacuto, A., Pucci, A., Bonafaccia, F., Esti, M., Ceccantoni, B., Papeschi, C., Bonafaccia, G., 2014. A new "functional" pasta containing Tartary buckwheat sprouts as an ingredient improves the oxidative status and normalizes some blood pressure parameters in spontaneously hypertensive rats. Food Funct. 5, 1017–1026.

Nelson, K., Stojanovska, L., Vasiljevic, T., Mathai, M., 2013. Germinated grains: a superior whole grain functional food? Can. J. Physiol. Pharmacol. 91, 429–441.

Pinson, S.R.M., Tarpley, L., Yan, W., Yeater, K., Lahner, B., Yakubova, E., Huang, X.-Y., Zhang, M., Guerinot, M.L., Salt, D.E., 2015. Worldwide genetic diversity for mineral element concentrations in rice grain. Crop Sci. 55, 294–311.

Pongrac, P., Vogel-Mikuš, K., Regvar, M., Vavpetič, P., Pelicon, P., Kreft, I., 2011. Improved lateral discrimination in screening the elemental composition of buckwheat grain by micro-PIXE. J. Agric. Food Chem. 59, 1275–2128.

Pongrac, P., Vogel-Mikuš, K., Jeromel, L., Vavpetič, P., Pelicon, P., Kaulich, B., Gianoncelli, A., Regvar, M., Eichert, D., Kreft, I., 2013a. Spatial distribution of mineral elements in Tartary buckwheat (*Fagopyrum tataricum*) grain as revealed by micro-imaging techniques. Food Res. Int. 54, 125–131.

Pongrac, P., Kreft, I., Vogel-Mikuš, K., Regvar, M., Germ, M., Grlj, N., Jeromel, L., Eichert, D., Budič, B., Pelicon, P., 2013b. Relevance for food sciences of quantitative spatially resolved element profile investigation in wheat (*Triticum aestivum*) grain. J. Roy. Soc. Interf. 10, 20130296.

Pongrac, P., Scheers, N., Sandberg, A.-S., Potisek, M., Arčon, I., Kreft, I., Kump, P., Vogel-Mikuš, K., 2016. The effects of hydrothermal processing and germination on Fe speciation and Fe bioaccessibility to human intestinal Caco-2 cells in Tartary buckwheat. Food Chem. 199, 782–790.

Shen, R., Iwashita, T., Ma, J.F., 2004. Form of Al changes with Al concentration in leaves of buckwheat. J. Exp. Bot. 55 (394), 131–136.

Siener, S., Hönow, R., Seidler, A., Voss, S., Hesse, A., 2006. Oxalate contents of species of the Polygonaceae, Amaranthaceae and Chenopodiaceae families. Food Chem. 98, 220–224.

Singh, S.P., Vogel-Mikuš, K., Arčon, I., Vavpetič, P., Jeromel, L., Pelicon, P., Kumar, R., Tuli, R., 2013. Pattern of iron distribution in maternal and filial tissues in wheat grains with contrasting levels of iron. J. Exp. Bot. 64 (11), 3249–3260.

Solé, V.A., Papillon, E., Cotte, M., Walter, P., Susini, J., 2007. A multiplatform code for the analysis of energy-dispersive X-ray fluorescence spectra. Spectrochim. Acta Part B 62, 63–68.

Steadman, K.J., Burgoon, M.S., Lewis, B.A., Edwardson, S.E., Obendorf, R.L., 2001a. Minerals, phytic acid, tannin and rutin in buckwheat seed milling fractions. J. Sci. Food Agric. 81, 1094–1100.

Steadman, K.J., Burgoon, M.S., Lewis, B.A., Edwardson, S.E., Obendorf, R.L., 2001b. Buckwheat seed milling fractions: description, macronutrient composition and dietary fibre. J. Cereal Sci. 33, 271–278.

Stein, A.J., 2010. Global impacts of human mineral malnutrition. Plant Soil 335, 133–154.

Vogel-Mikuš, K., Kump, P., Nečemer, M., Pelicon, P., Arčon, I., Pongrac, P., Povh, B., Bothe, H., Regvar, M., 2009b. Micro-PIXE analysis for localisation and quantification of elements in roots of mycorrhizal metal-tolerant plants. Varma, A., Kharkwal, A.C. (Eds.), Symbiotic Fungi: Principles and Practice, Soil Biology, vol. 18, Springer-Verlag, Berlin, Heidelberg, pp. 227–242.

Vogel-Mikuš, K., Pelicon, P., Vavpetič, P., Kreft, I., Regvar, M., 2009a. Elemental analysis of edible grains by micro-PIXE: common buckwheat case study. Nucl. Instrum. Meth. B 267, 2884–2889.

White, P.J., Broadley, M.R., 2009. Biofortification of crops with seven mineral elements often lacking in human diets—iron, zinc, copper, calcium, magnesium, selenium and iodine. New Phytol. 182, 49–84.

Zhao, F.-J., Su, Y.H., Dunham, S.J., Rakszegi, M., Bedo, Z., McGrath, S.P., Shewry, P.R., 2009. Variation in mineral micronutrient concentrations in grain of wheat lines of diverse origin. J. Cereal Sci. 49, 290–295.

Zheng, H., Staehelin, L.A., 2011. Protein storage vacuoles are transformed into lytic vacuoles in root meristematic cells of germinating seedlings by multiple, cell type-specific mechanisms. Plant Physiol. 155, 2023–2035.

Chapter twenty one

The Effect of Environmental Factors on Buckwheat

M. Germ, A. Gaberščik

Department of Biology, Biotechnical Faculty, University of Ljubljana, Jamnikarjeva, Ljubljana, Slovenia

INTRODUCTION

We are living in an era of global changes of environmental conditions. The planet is facing changes of temperatures caused by global warming (Fischlin et al., 2007) and an increased amount of solar UV-B radiation caused by the diminishing ozone layer (Chipperfield et al., 2015). According to computer modeling, climate change will not reduce the average level of precipitation, but is likely to increase the frequency of extreme events such as heavy rainfall and droughts (Albritton and Meira Filho, 2001). These changes will affect food production, since, as pointed out by McKersie and Leshem (1994), crops are more vulnerable to different types of stress because they are being bred outside their ecological boundaries for greater yield and cultivation. Consequently, crops that are more resistant to stress conditions will become very important. Such crops can include different species and cultivars of buckwheat originating in relatively harsh environments (Ohnishi, 1998).

In addition to global changes, crop production is also affected by shortage of nutrients and soil degradation, compaction, and pollution. Today, soil formation is typically 13–40 times slower than soil degradation and loss because of erosion (Duran and Pleguezuelo, 2008). Pimentel (2006) reports that the annual erosion rate from terrestrial ecosystems is about 75 billion tons and this also includes organic matter as well as the nitrogen and phosphorus (P) content of soil. This results in poor nutrient composition of crops, which in turn negatively affects food quality for consumers of plants, including humans. In nutrient-poor areas, fertilizers are used not only to increase crop production and quality but also to increase concentrations of essential mineral elements in the edible portions of plants (White and Brown, 2010). Buckwheat has a wide ecological range and consequently it can grow well in almost all kinds of disadvantageous living environments, including nutrient-poor soil (Kreft, 2007). In spite of this,

273

buckwheat is rich in nutrients and limiting amino acids, as well as flavones, fagopyrin, buckwheat sterols, and thiamine-binding proteins, which have medical effects and also the potential to be used as active components in functional food production (Li and Zhang, 2001).

Because of its high plasticity, buckwheat can be used for food fortification—a process of increasing the bioavailable concentrations of essential elements in edible portions of crop plants through agronomic intervention or genetic selection. A great deal of research has been dedicated to selenium (Se), an essential nutrient for microorganisms, animals, and humans. As a result of low Se concentrations and availability in soils, and the consequently low Se concentrations in crop plants, Se malnutrition in humans is common (White and Broadley, 2009). Cultivation of plants enriched with Se is an effective way to reduce dietary deficiencies and increase health benefits (Pyrzynska, 2009).

ORIGIN OF BUCKWHEAT

Ohnishi systematically and extensively searched for the ancestors of common and Tartary buckwheat and discovered that based on the distribution of their wild ancestors, common buckwheat originated in the northwestern corner of China's Yunnan Province. Tartary buckwheat, on the other hand, originated in the northwest part of Sichuan Province, as determined by the variability of allozymes in the buckwheat (Ohnishi, 1998). Both wild and cultivated types of Tartary buckwheat exist in the same regions of northeastern Pakistan, Tibet, Yunnan, and Sichuan (Tsuji and Ohnishi, 2001). In China and Nepal, Tartary buckwheat grows at higher elevations than common buckwheat. Environmental conditions, including temperatures, the length of the vegetative period, and UV radiation differ significantly along the elevation gradient.

Response of Buckwheat to Radiation Conditions

Buckwheat thrives in open place habitats that experience high levels of solar radiation. In full sunlight *Fagopyrum esculentum*, variety Darja, achieves a photosynthetic level of up to 16.5 $\mu mol/m^2 s$ (Gaberščik et al., 2002), which is in a range of the activity of mesophytes growing in sunny habitats (Larcher, 2003).

An important part of the solar radiation that significantly affects growth and development of plants is UV-B radiation, which depends on the altitude of the location. The increase along the altitudinal gradient ranges from 6–8% (Caldwell et al., 1980) to 20% (Blumthaler et al., 1993) for each 1000 m of elevation. Consequently, buckwheat, a plant originating in high-altitude habitats with elevated UV levels, may be a suitable crop for culturing under enhanced UV-B radiation. Further enhancement of UV-B radiation is expected in the future. For example, Chipperfield et al. (2015) report that models project that recovery of stratospheric ozone and the reduction of UV-B levels to normal levels will take place no earlier than 2050.

Short wave UV-B radiation may exert an adverse effect on plants. UV-B radiation induces synthesis of phenolic substances that provide a variety of benefits for plants and their consumers (Schreiner et al., 2014). Phenolic substances also absorb radiation in the UV range and prevent damage from shortwave radiation (Gaberščik et al., 2002; Kataria et al., 2014). On the other hand, UV-B radiation may damage important biomolecules such as DNA, proteins, and membranes, and affect a variety of processes in plants. Tartary buckwheat has very high potential for production of UV-absorbing compounds. It has been shown that it contains about 100-fold more rutin in comparison to common buckwheat (Fabjan et al., 2003), although this is not necessarily the case for other UV-absorbing compounds. High amounts of rutin along with branched habitus results in less pronounced damaging effects of UV-B radiation (Germ et al., 2013). A study of UV-B radiation effects on common buckwheat revealed that chlorophyll a and b and carotenoid contents were slightly affected by UV-B. Net photosynthesis was disturbed in an early phase of development, but later in the growth season, photosynthetic activity was less affected, because of the effective protection offered by UV-absorbing compounds. Changes in photosynthetic rates as well as transpiration rates led to a decrease in water use efficiency, possibly caused by disturbances in the functioning of stomata (Gaberščik et al., 2002). Disorder in water economy and water deficiency both disturb physiological processes but, above all, water use efficiency and the costly production of UV-B-absorbing compounds in treated common buckwheat plants result in lower biomass production. The number and weight of seeds were also reduced in plants grown under elevated UV-B radiation (Gaberščik et al., 2002).

RESPONSE TO TEMPERATURE CONDITIONS

Buckwheat is very sensitive to low temperatures, but it is relatively tolerant to high temperatures. For two varieties of common buckwheat, cv. Bednja (tetraploid) and cv. Siva (diploid), it was shown that freezing of the leaf tissue occurs between -3 and $-4°C$, while the net photosynthesis reached the lower temperature compensation point at $-1°C$, with no differences among varieties (Kajfež-Bogataj and Gaberščik, 1986; Gaberščik et al., 1986). These values serve to classify buckwheat into the group of freezing-sensitive plants (Larcher, 2003) even though the differences regarding the age of leaf, growing conditions, and genotypes exist. The upper temperature compensation point of photosynthesis for cv. Bednja and cv. Siva was determined to be $43°C$, while the range was very wide with a peak at $15°C$. However, the photosynthetic activity remained relatively high up to temperatures of approximately $40°C$ (Kajfež-Bogataj and Gaberščik, 1986). This relatively wide temperature range for photosynthesis presents an advantage for buckwheat whereby it can benefit from different weather conditions during the vegetation period and is somewhat resistant to global warming. According to the fourth assessment report by the Intergovernmental Panel on Climate Change, global surface average temperature will experience a 1.1–$6.4°C$ range increase by the end of the 21st century (Fischlin et al., 2007). Warming above $3°C$ will significantly lower fixed carbon uptake of global terrestrial vegetation (Xu et al., 2010).

WATER AVAILABILITY

Of the various environmental parameters, drought is the most important abiotic factor limiting crop productivity (Bettaieb et al., 2009). On a worldwide scale, the decrease in average yield because of water deficit is predicted to be more than 50% (Wang et al., 2003).

Plants respond to water limitation in different ways: (1) by avoiding the period of water deficiency, (2) by maintaining high water potential by different protection mechanisms, and (3) by sustaining metabolic activity during periods of low water availability (Larcher, 2003). Common and Tartary buckwheat were exposed to moderate water shortage that had a slight negative effect on the yield of plants (Germ et al., 2013). Water deficit significantly reduces the transpiration rate caused by stomata closing, and this contributes to a reduction in water loss and maintenance of a favorable water potential for the plant (Larcher, 2003). In a majority of higher plants, including agricultural crops, closing of the stomata is the first reaction to drought (Kawakami et al., 2006). Both common and Tartary buckwheat produce lower total biomass in conditions of water limitation, but the inhibitory effect was more evident in common buckwheat (Germ et al., 2013). Lower biomass production is a consequence of the disturbance in the process of gas exchange (Fernandez et al., 2002). Xiang et al. (2013) showed that in Tartary buckwheat, water limitation leads to lower total chlorophyll content and rate of photosynthesis.

NUTRIENT AVAILABILITY

Compared to cereals, buckwheat is more productive in soils with low amounts of nutrients (Kreft, 2007; Clark, 2007). Buckwheat can use phosphorus (P) from a stabile inorganic pool, mainly calcium-bound P, which is utilized by other crops (Clark, 2007; Teboh and Franzen, 2011). A comparison of buckwheat grown in plots with low amounts of plant-available P with plants grown in P-fertilized plots showed that soil-P availability was not affected by buckwheat, but that there was an enhancement of tartrate concentration in the rhizosphere. These changes were significantly higher in low available P in comparison to P-fertilized plots. This suggests exudation of tartrate via the plants' roots into a rhizosphere, affecting the availability of P (Possinger et al., 2013). Teboh and Franzen (2011) also showed that the growing of buckwheat mobilized P from stabile inorganic pools to the P pool available for plant uptake.

BIOFORTIFICATION WITH SELENIUM

Se is an essential nutrient for microorganisms, animals, and humans, but its essential role in plants is still unclear. At low doses, it has positive effects on plants but it is toxic at higher doses, and there is a fine boundary between these concentrations (Kaur et al., 2014). As a result of the chemical similarities between Se and S, uptake, transport, and assimilation of selenate in plants follow the sulfate pathway (Brown and Shrift, 1981; Sors et al., 2005). Foliar spraying of buckwheat plants with sodium selenate solution increases the Se content in leaves (Smrkolj et al., 2006; Vogrinčič et al., 2009; Golob et al., 2015) and foliar spraying with Se [10 mg Se(VI) L^{-1}] of Tartary buckwheat plants showed that Se was effectively assimilated and taken into the seeds, where its concentration was as high as twice as much as in untreated plants (Kreft et al., 2013). Seeds were collected at the end of the season to examine the effect of Se on the subsequent generation of the Se-treated plants. In adult plants the dry mass of the leaves was significantly greater in the Se-treated progeny plants than in the controls. Se promotes the scavenging of H_2O_2 caused by increased GSH-Px activity (Hartikainen et al., 2000) and the growth-stimulating effect of Se is therefore a consequence of Se antioxidative function (Xue et al., 2001). Golob et al. (2015) studied the feasibility of increasing the concentration of Se in grain and in green parts of Tartary buckwheat. Their results are in agreement with those of Li et al. (2008) who reported that, in contrast to selenite, selenate is not readily assimilated in organic forms in roots, but is highly mobile in xylem transportation. Regarding recommended Se concentrations, edible parts of Tartary buckwheat were adjudged safe for human consumption (Golob et al., 2015).

MULTIPLE EFFECTS OF ENVIRONMENTAL FACTORS ON BUCKWHEAT
Drought and UV Radiation

The response of common buckwheat (*Fagopyrum esculentum*) and Tartary buckwheat (*F. tataricum*) to enhanced UV-B radiation and water limitation revealed that both these factors led to lower amounts of chlorophylls *a* and *b* in common buckwheat and a positive interaction was observed between enhanced UV-B radiation and water limitation (Germ et al., 2013). Water limitation induces the production of UV-absorbing compounds and lowered chlorophyll content and effective photochemical efficiency of photosystem II (PSII) in Tartary buckwheat to a degree that is less pronounced than the effects of elevated UV-B radiation. Similarly in the study of Alexieva et al. (2001), UV-B application caused greater membrane damage than drought stress in pea and wheat. The reduction of biomass production in Tartary buckwheat was only moderate, while in common buckwheat exposed to both stresses the decrease was more pronounced. This was probably a consequence of the tradeoff between UV-absorbing compounds and biomass production. It has been reported that the production of UV-absorbing compounds in primary producers demands an additional supply of energy, which results in a lower growth rate (Germ et al., 2006; Tsormpatsidis et al., 2010). The negative effect of elevated UV-B radiation on growth parameters in common buckwheat, which was highly significant in watered plants, was less pronounced in plants exposed to a limited supply of water. In Tartary buckwheat, UV-B radiation mitigated the negative effects of water limitation to such an extent that the biomass production actually increased.

Drought and Se

The exposure of two cultivars (cvs.) of common buckwheat (*F. esculentum*), Pyra and Siva, to water deficit and Se showed that in plants exposed to water deficit, the stomatal conductance was significantly lower, while Se exerted a significantly higher stomatal conductance on Siva plants deficient in water (Tadina et al., 2007). The addition of Se mitigated the negative effect of water limitation in most cases but the effect was significant only in cv. Siva. Similar findings were reported by Kuznetsov et al. (2003) who concluded that Se regulates the water status of wheat plants. A significantly higher effective photochemical efficiency of PSII was obtained in Siva water-deficient plants and in Pyra plants exposed to Se and water deficit. This was possibly because of improvement of plant water management during treatment. Water deficit, Se spraying, and a combination of the two resulted in shorter plants with a reduced number of nodes in both cvs. Foliar application of Se was beneficial to watered buckwheat, as revealed from the photochemical efficiency of PSII and biomass production in both cvs. Biomass production failed to increase when Se treatment was applied under water deficit conditions (Tadina et al., 2007). Yao et al. (2012) proved that Se enhanced the recovery of wheat seedlings after rehydration following drought stress. The same pattern was seen in the results of a study of Nawaz et al. (2014) who demonstrated that Se application mitigated the negative effects of drought stress in wheat seedlings.

Se and UV Radiation

The combined effects of UV-B irradiation and foliar treatment with Se on common and Tartary buckwheat grown outdoors showed that neither enhanced UV-B radiation nor Se treatment affected the potential quantum yield of PSII in either species (Breznik et al., 2005). A similar effect on potential quantum yield was found in a study of barley and strawberry plants (Valkama et al., 2003). The effective quantum yield of PSII was lowered because of UV-B radiation in both buckwheat species and was mitigated by the addition of Se. Se treatment also mitigated the stunting effect of UV-B radiation and the lowering of biomass in common buckwheat plants (Breznik et al., 2005). Yao et al. (2013) reported that Se addition increased wheat yield and protein concentration, and increased most microelement concentration in wheat grains, which improved quality of wheat exposed to stress caused by UV-B.

CONCLUSIONS

This review reveals buckwheat to be a plant with a wide ecological range. Its high tolerance to different environmental factors is also caused by its high potential for production of phenolic substances that have an important role in mitigating different stress conditions. The higher sensitivity of common buckwheat in comparison to Tartary buckwheat reveals that breeding has reduced its tolerance to environmental factors.

ACKNOWLEDGMENTS

This research was financed by the Ministry of Education, Science and Sport, Republic of Slovenia, through the programs "Biology of plants" (P1-0212), and projects J4-4224 and J4-5524. The authors are grateful to Dragan Abram for technical assistance and Bill Milne for revising the English text.

REFERENCES

Albritton, D.L., Meira Filho, L.G., 2001. Technical summary of the working group I report. In: Houghton, J.T., Ding, Y., Griggs, D.J., Noguer, M., van der Linden, P.J., Dai, X., Maskell, K., Johnson, C.A. (Eds.), Climate Change 2001: the Scientific Basis. Cambridge University Press, Cambridge, pp. 71–73.

Alexieva, V., Sergiev, I., Mapelli, S., Karanov, E., 2001. The effect of drought and ultraviolet radiation on growth and stress markers in pea and wheat. Plant Cell Environ. 24, 1337–1344.

Bettaieb, I., Zakhama, N., Aidi Wannes, W., Kchouk, M.E., Marzouk, B., 2009. Water deficit effects on *Salvia officinalis* fatty acids and essential oils composition. Sci. Hort. 120, 271–275.

Blumthaler, M., Ambach, W., Huber, M., 1993. Altitude effect of solar UV radiation dependent on albedo, turbidity and solar elevation. Meteorol. Z. 2, 116–120.

Breznik, B., Germ, M., Gaberščik, A., Kreft, I., 2005. Combined effects of elevated UV-B radiation and the addition of selenium on common (*Fagopyrum esculentum* Moench) and tartary [*Fagopyrum tataricum* (L) Gaertn.] buckwheat. Photosynthetica 43, 583–589.

Brown, T.A., Shrift, A., 1981. Exclusion of selenium from proteins of selenium-tolerant *Astragalus* species. Plant Physiol. 67, 1051–1053.

Caldwell, M.M., Robberecht, R., Billings, W.D., 1980. A steep latitudinal gradient of solar ultraviolet-B radiation in the artic-alpine life zone. Ecology 61 (3), 600–611.

Chipperfield, M.P., Dhomse, S.S., Feng, W., McKenzie, R.L., Velders, G.J.M., Pyle, J.A., 2015. Quantifying the ozone and ultraviolet benefits already achieved by the Montreal Protocol. Nat. Commun. 6, 7233.

Clark, A., 2007. Managing cover crops profitably, third ed. Sustainable Agriculture NetworkBeltsville, MD.

Duran, Z.H.V., Pleguezuelo, C.R.R., 2008. Soil-erosion and runoff prevention by plant covers. A review. Agron. Sustain. Dev. 28, 65–86.

Fabjan, N., Rode, J., Košir, I.J., Wang, Z., Zhang, Z., Kreft, I., 2003. Tartary buckwheat (*Fagopyrum tataricum* Gaertn.) as a source of dietary rutin and quercitrin. J. Agric. Food Chem. 51, 6452–6455.

Fernandez, J.R., Wang, M., Reynolds, F.J., 2002. Do morphological changes mediate plant responses to water stress? A steady-state experiment with two C_4 grasses. New Phytol. 155, 79–88.

Fischlin, A., Midgley, G.F., Price, J.T., Leemans, R., Gopal, B., Turley, C., Rounsevell, M.D.A., Dube, O.P., Tarazona, J., Velichko, A.A., 2007. Ecosystems, their properties, goods, and services. In: Parry, M.L., Canziani, O.F., Palutikof, J.P., van der Linden, P.J., Hanson, C.E. (Eds.), Climate Change 2007 Impacts, Adaptation and Vulnerability. Contribution of Working Group II to the Fourth Assessment Report of the Intergovernmental Panel on Climate Change, Cambridge University Press, Cambridge, pp. 211–272.

Gaberščik, A., Martinčič, A., Kajfež-Bogataj, L., Kreft, I., 1986. Possibility of laboratory determination of resistance of buckwheat plants to freezing. Fagopyrum 6, 10–11.

Gaberščik, A., Vončina, M., Trošt Sedej, T., Germ, M., Björn, L.O., 2002. Growth and production of buckwheat (*Fagopyrum esculentum*) treated with reduced, ambient, and enhanced UV-B radiation. J. Photochem. Photobiol., B Biology 66 (1), 30–36.

Germ, M., Mazej, Z., Gaberščik, A., Trošt Sedej, T., 2006. The response of *Ceratophyllum demersum* L. and *Myriophyllum spicatum* L. to reduced, ambient, and enhanced ultraviolet-B radiation. Hydrobiologia (Den Haag) 570, 47–51.

Germ, M., Breznik, B., Dolinar, N., Kreft, I., Gaberščik, A., 2013. The combined effect of water limitation and UV-B radiation on common and Tartary buckwheat. Cereal Res. Commun. 41 (1), 97–105.

Golob, A., Stibilj, V., Kreft, I., Germ, M., 2015. The feasibility of using Tartary buckwheat as a Se-containing food material. J. Chem. 2015, 1–5.

Hartikainen, H., Xue, T., Piironen, V., 2000. Selenium as an anti-oxidant and pro-oxidant in ryegrass. Plant Soil 225, 193–200.

Kajfež-Bogataj, L., Gaberščik, A., 1986. Analysis of net photosynthesis response curves for buckwheat. Fagopyrum 6, 6–8.

Kataria, S., Baroniya, S.S., Baghel, L., Kanungo, M., 2014. Effect of exclusion of solar UV radiation on plants. Plant Sci. Today 1 (4), 224–232.

Kaur, N., Sharma, S., Kaur, S., Nayyar, H., 2014. Selenium in agriculture: a nutrient or contaminant for crops? Arch. Agron. Soil Sci. 60 (12), 1593–1624.

Kawakami, J., Iwama, K., Jitsuyama, Y., 2006. Soil water stress and the growth and yield of potato plants grown from microtubers and conventional seed tubers. Field Crops Res. 162, 903–911.

Kreft, I., 2007. Buchweizen Slowenien. In: Kreft, I., Ries, C., Zewen, C. (Eds.), Das Buchweizen Buch: mit Rezepten aus aller Welt. 2. überarbeitete und erweiterte Aufl. Islek ohne Grenzen EWIV, Arzfeld, pp. 71–79.

Kreft, I., Mechora, Š., Germ, M., Stibilj, V., 2013. Impact of selenium on mitochondrial activity in young Tartary buckwheat plants. Plant Physiol. Biochem. 63, 196–199.

Kuznetsov, Vas.V., Kholodova, V.P., Kuznetsov, Vl.V., Yagodin, B.A., 2003. Selenium regulates the water status of plants exposed to drought. Doklady Biol. Sci. 390, 266–268.

Larcher, W., 2003. Physiological Plant Ecology, 4th ed. Springer-Verlag, Berlin.

Li, S., Zhang, H., 2001. Advances in the development of functional foods from buckwheat. Crit. Rev. Food Sci. Nutr. 41 (6), 451–464.

Li, H.-F., McGrath, S.P., Zhao, F.-J., 2008. Selenium uptake, translocation and speciation in wheat supplied with selenate or selenite. New Phytol. 178 (1), 92–102.

McKersie, B.D., Leshem, Y.Y., 1994. Stress and Stress Coping in Cultivated Plants. Kluwer Academic Publishers, Dordrecht.

Nawaz, F., Ashraf, Y.M., Ahmad, R., Waraich, E.A., Nauman Shabbir, R., 2014. Selenium (Se) regulates seedling growth in wheat under drought stress. Adv. Chem. 2014, 1–7.

Ohnishi, O., 1998. Search for the wild ancestor of buckwheat. III. The wild ancestor or cultivated common buckwheat, and of Tartary buckwheat. Econ. Bot. 52 (2), 123–133.

Pimentel, D., 2006. Soil erosion: a food and environmental threat. Environ. Dev. Sustain. 8, 119–137.

Possinger, A.R., Byrne, L.B., Breen, N.E., 2013. Effect of buckwheat (*Fagopyrum esculentum*) on soil-phosphorus availability and organic acids. J. Plant Nutr. Soil Sci. 176 (1), 16–18.

Pyrzynska, K., 2009. Selenium speciation in enriched vegetables. Food Chem. 114, 1183–1191.

Schreiner, M., Martínez-Abaigar, J., Glaab, J., Jansen, M., 2014. UV-B induced secondary plant metabolites. Optik Photonik 9 (2), 34–37.

Smrkolj, P., Germ, M., Kreft, I., Stibilj, V., 2006. Respiratory potential and Se compounds in pea (*Pisum sativum* L.) plants grown from Se-enriched seeds. J. Exp. Bot. 57 (14), 3595–3600.

Sors, T.G., Ellis, D.R., Salt, D.E., 2005. Selenium uptake, translocation, assimilation and metabolic fate in plants. Photosynth. Res. 86 (3), 373–389.

Tadina, N., Germ, M., Kreft, I., Breznik, B., Gaberščik, A., 2007. Effect of water deficit and selenium on common buckwheat (*Fagopyrum esculentum* Moench) plants. Photosynthetica 45 (3), 472–476.

Teboh, J.M., Franzen, D.W., 2011. Buckwheat (*Fagopyrum esculentum* Moench) potential to contribute solubilized soil phosphorus to subsequent crops. Commun. Soil Sci. Plant Anal. 42 (13), 1544–1550.

Tsormpatsidis, E., Henbest, R.G.C., Battey, N.H., Hadley, P., 2010. The influence of ultraviolet radiation on growth, photosynthesis and phenolic levels of green and red lettuce: potential for exploiting effects of ultraviolet radiation in a production system. Ann. Appl. Biol. 156, 357–366.

Tsuji, K., Ohnishi, O., 2001. Phylogenetic relationships among wild and cultivated Tartary buckwheat (*Fagopyrum tataricum* Gaertn.) populations revealed by AFLP analyses. Genes Genet. Syst. 76 (1), 47–52.

Valkama, E., Kivimäenpää, M., Hartikainen, H., Wulff, A., 2003. The combined effects of enhanced UV-B radiation and selenium on growth, chlorophyll fluorescence and ultrastructure in strawberry (*Fragaria × ananassa*) and barley (*Hordeum vulgare*) treated in the field. Agric. Forest Meteorol. 120, 267–278.

Vogrinčič, M., Cuderman, P., Kreft, I., Stibilj, V., 2009. Selenium and its species distribution in above-ground plant parts of selenium enriched buckwheat (*Fagopyrum esculentum* Moench). Anal. Sci. 25, 1357–1363.

Wang, W., Vinocur, B., Altman, A., 2003. Plant responses to drought, salinity and extreme temperatures: towards genetic engineering for stress tolerance. Planta 218, 1–14.

White, P.J., Broadley, M.R., 2009. Biofortification of crops with seven mineral elements often lacking in human diets—iron, zinc, copper, calcium, magnesium, selenium and iodine. New Phytol. 182, 49–84.

White, P.J., Brown, P.H., 2010. Plant nutrition for sustainable development and global health. Ann. Bot. 105, 1073–1080.

Xiang, D.-B., Peng, L.-X., Zhao, J.-L., Zou, L., Zhao, G., Song, C., 2013. Effect of drought stress on yield, chlorophyll contents and photosynthesis in Tartary buckwheat (*Fagopyrum tataricum*). J. Food Agric. Environ. 2 (3,4), 1358–1363.

Xu, Z., Zhou, G., Shimizu, H., 2010. Plant responses to drought and rewatering. Plant Signal. Behav. 5 (6), 649–654.

Xue, T., Hartikainen, H., Piironen, V., 2001. Antioxidative and growth-promoting effect of selenium in senescing lettuce. Plant Soil 237, 55–61.

Yao, X., Chu, J., Liang, L., Geng, W., Li, J., Hou, G., 2012. Selenium improves recovery of wheat seedlings at rewatering after drought stress. Russ. J. Plant Physiol. 59 (6), 701–707.

Yao, X., Jianzhou, C., Xueli, H., Binbin, L., Jingmin, L., Zhaowei, Y., 2013. Effects of selenium on agronomical characters of winter wheat exposed to enhanced ultraviolet-B. Ecotoxicol. Environ. Safety 92, 320–326.

Chapter twenty two

The Effect of Habitat Conditions and Agrotechnical Factors on the Nutritional Value of Buckwheat

G. Podolska

Institute of Soil Science and Plant Cultivation—State Research Institute, Puławy, Puławy, Poland

INTRODUCTION

Buckwheat is a crop variety widely utilized in medicine, industry, and food production. Such wide applications are mainly because of its unique chemical composition. The protein content of buckwheat grains ranges from 8.5% to 19%. Water-soluble albumins and globulins constitute the main fraction and represent 50% of the proteins present in the grain. The most remarkable feature of buckwheat grains is a small content of prolamins and a lack of α-gliadin, which makes it suitable as a food source in the diets of celiac sufferers. In terms of nutritional value, the amino acid composition of buckwheat is the most favorable of all cereals. Buckwheat proteins are rich in lysine—the amino acid that is lacking in other cereals, thus limiting the biological value of other cereal proteins (Bonafaccia et al., 2003; Dietrych-Szóstak and Suchecki, 2003; Dziedzic et al., 2010; Krkošková and Mrázová, 2005; Li and Zhang, 2001; Stempińska and Soral-Śmietana, 2006; Steadman et al., 2001; Wei et al., 2003).

Furthermore, starch is the major component of buckwheat grains. It is accumulated in the endosperm. In the whole grain of buckwheat, the starch content varies from 59% to 70% of the dry mass. The amylose content of buckwheat starch granules fluctuates between 15% and 52% and its degree of polymerization varies from 12 to 45 glucose units. Moreover, buckwheat grains contain from 5% to 11% of fiber. The soluble fraction ranges from 3% to 7%, while the

283

Molecular Breeding and Nutritional Aspects of Buckwheat. http://dx.doi.org/10.1016/B978-0-12-803692-1.00022-5

insoluble fraction ranges from 2% to 4% (Qian and Kuhn, 1999; Krkošková and Mrázová, 2005).

Buckwheat grains contain from 1.5% to 4% of total lipids, with the content of raw fat in buckwheat flour exceeding 3%. Free lipids isolated from buckwheat grains constitute 2.5% of dry matter (d.m.), whereas bound lipids constitute about 1.3% of d.m. Buckwheat germ contains the highest amount of lipid (10–22%), while the hull contains the smallest amount (0.8%). Furthermore, triacylglycerides are the main components of the neutral fraction of lipids containing fatty acids from C12 to C22, with a predominating contribution of: oleic (42%), linoleic (32%), and palmitic acids (16%) (Soral-Śmietana et al., 1984; Steadman et al., 2001).

The content of minerals in buckwheat grains and their morphological fractions (dry base) reaches 2–2.5%. Buckwheat is rich in potassium (K), magnesium (Mg), calcium (Ca), and sodium (Na) and is considered an important nutritional source of such microelements as iron (Fe), manganese (Mn), and zinc (Zn). Furthermore, buckwheat seeds contain valuable vitamins, such as thiamin, riboflavin, niacin, pantothenic acid, and pyridoxine. In addition, buckwheat contains vitamins of antioxidative nature, such as vitamin E and trace quantities of β-carotene (Bonafaccia et al., 2003; Li and Zhang, 2001; Wei et al., 1995).

Buckwheat grains and hulls consist of a number of components with healing properties and biological activity, that is, flavonoids and flavons, phenolic acids, condensed tannins, phytosterols, and fagopyrins. Flavonoids consist of six compounds: rutin, quercetin, orientin, vitexin, isovitexin, and isoorientin. Rutin is the main flavonoid occurring in buckwheat. Its content ranges from 4% to 6%. Fruit-seed hulls of buckwheat contain rutin, orientin, isoorientin, vitexin, and quercetin, in quantities ranging from 60 to 74 mg/100 g per hull.

The chemical content of buckwheat nuts is mostly influenced by weather conditions during the growing period, as well as by habitat conditions, the genetic factor—a cultivars, and cultivation technology. It is also determined by factors that affect the growth and development of plants, such as sowing term, sowing density, and nutrients; primarily the amount of nitrogen. The most important factors influencing plant growth and development include: the amount of light, temperature, and the availability of water and/or nutrients. These factors affect the process of photosynthesis, metabolism of plants, and the reactions that occur in the cells, consequently affecting the yield and chemical composition.

Buckwheat is a thermophilous plant, which in the initial growth period, that is, from the four blades phase to early blooming, requires air temperature in the range of 17–19°C. During flowering, the temperature should not exceed 19°C, and in the seed-maturing phase, it should range from 17 to 19°C. Buckwheat is not affected by the length of the day; however, long day conditions promote the development of flowers, fruits, and side shoots. The water requirements of buckwheat are rather high. The processing of 2 tons of grain and 5 tons of straw requires using an average of 3500 tons of water. During flowering and grain formation, buckwheat consumes from 15 to 20 times more water than in the early stages of its development. During flowering, buckwheat is very sensitive to changes in air humidity. If the air humidity is lower than 30–40%, plants, flowers, and buds wither (Ruszkowscy and Ruszkowscy, 1967).

Buckwheat requires a high amount of nutrients. For the production of 2.5 t/ha of yield and 6 t/ha of straw, buckwheat requires about 90 kg N/ha, 60 kg P_2O_5, and 150 kg K_2O. The highest rate of nitrogen uptake occurs in the stages of early plant development to the full flowering phase. At those phases, buckwheat assimilates approximately 60% of the total amount of the required nitrogen and potassium, and only 40% of phosphorus. At the stage of full flowering, the demand for phosphorus increases. The availability of nutrients depends mainly on the soil abundance and weather conditions, while their effectiveness depends on the delivery of this component to buckwheat at the proper time. The impact of nitrogen amount on the yield and chemical composition of buckwheat depends on the habitat conditions and variety (Murakami et al., 2002; Ruszkowscy and Ruszkowscy, 1967; Noworolnik, 1999; Mazurek, 1999).

Sowing date, as a fundamental element for production, is of considerable importance because of its significant effects on how plants exploit its maximum production potential. One of the important factors that may influence the selection of sowing date is the temperature; in other words appropriate planting time of cultivar or a group of similar cultivars depends on the existing set of environmental factors and should be optimized to be suitable for plant germination, establishment, and survival. Various sowing dates in various climatic conditions influence the growth period, composition of the plant effective material, and yield. The term of sowing has an impact on the content of chemical compounds in buckwheat seeds. It is primarily associated with the fact that in different sowing terms, different conditions for the growth of plants, and maturing of seeds occur. The literature frequently reports interactions between the sowing terms and years in shaping the nutrient composition in buckwheat (Hore et al., 2002; Omidbaigi and De Mastro, 2004; Kwiatkowski, 2010; Dietrych-Szostak and Podolska, 2008).

The density of sowing impacts on the amount of light per leaf area unit, hence the efficiency of photosynthesis. Changes of sowing patterns with constant density can alter the conditions for light absorption concerning plant biomass. As a result, higher material production influences the qualitative and quantitative yields.

These factors, by affecting the processes of growth and development, have an influence not only on the level of yields, but also on the chemical composition of buckwheat nuts (Chai et al., 1998; Chang et al., 2003; Feng et al., 2003; Hagels et al., 1995; Noworolnik, 1995; Omidbaigi and De Mastro, 2004; Sobhani et al., 2014; Subedi et al., 2007; Tahir et al., 2006; Zhang et al., 1998, 2001).

This chapter discusses available information showing how habitat conditions, especially weather and soil conditions, affect the nutritional value of buckwheat nuts. In addition, the effect of abiotic stress (water limitation) on nutritional value and the effect of agrotechnical factors (nitrogen fertilization, sowing term, sowing density) on fat, fiber, protein content, and the flavonoids content are also reviewed.

Environmental Conditions

Environmental conditions have significant impacts on the grain quality of buckwheat. Dry and sunny weather promotes the accumulation of proteins. Apart from having an impact on the total protein content, weather conditions significantly affect the amount of amino acids. Under lower rainfall conditions combined with high air temperatures occurring during flowering and ripening, buckwheat grains produce a greater amount of amino acids (Barta et al., 2004; Inoue et al., 2005; Kwiatkowski, 2010; Michalova et al., 1998).

Additionally, meteorological conditions alter the properties of proteins in buckwheat grains. Under drought stress, plants synthesize a set of new proteins or change their fractional and amino acid composition (Henckel, 1975). Podolska et al. (2007) have shown that the amount of albumins and globulins is not significantly affected by soil moisture; however, soil moisture had a significant effect on prolamin content in buckwheat nuts. The drought stress from flowering phase to harvest time resulted in a decrease of prolamin content, mainly of gliadin content (Table 22.1).

Weather conditions occurring during the growing period additionally affect the fat content. For example, Kwiatkowski (2010) claim that grains accumulate higher amounts of fat in dry years compared to the wet ones, which is not consistent with the studies of Kulka and Górecki (1995), who have reported that the fat content in buckwheat grains increases when they mature in high air humidity and low temperature.

The content of antioxidant compounds, including flavonoids, is strongly determined by weather conditions during the growing period. The rutin content, which is the main buckwheat flavonoids, increases in long-day conditions, probably because of higher UV-B radiation (Gabersik et al., 2002; Germ, 2004; Kreft et al., 2002; Ohsawa and Tsutsumi, 1995).

Rutin accumulation in buckwheat is also associated with the response to drought and cold stress. A higher rutin concentration in buckwheat grains is associated with the optimal soil moisture during the growing period and the occurrence of water deficits at the level of 30% of ppw from the period of flowering to grain setting (Kwiatkowski, 2010).

TABLE 22.1 Content of the Protein Fractions in the Buckwheat (The area under the pick stated as mAU*s) (Podolska et al., 2007)

Soil Water Capacity	Albumins + Globulins	Gliadins	Glutenins
60% swc	31,365	3,211	4,362
30% swc—whole vegetation period	31,770	2,915	2,370
60% swc—until full blooming and 30% swc—from full blooming to harvest term	32,472	2,776	3,288

swc, Soil water capacity.

Cultivar

The chemical composition of buckwheat grains largely depends on genetic factor, ie, a cultivar; nonetheless, comparing cultivars in terms of their chemical composition is quite difficult, and conclusions should only be drawn for the cultivation of cultivars under the same habitat conditions and under the same production technology. Comparison of the chemical composition of the cultivars from different habitats is subject to a large error, thus no definite conclusions can be drawn as the researcher cannot be sure if results are affected by growing conditions or a type of cultivar. The content of proteins in Polish buckwheat cultivars amounts to, on average, 12.2% of d.m. and oscillates in a very narrow range, from 11.97% in Kora cultivar to 12.65% in Panda cultivar (Stempińska and Soral-Śmietana, 2006) (Table 22.2). Dietrych-Szostak and Suchecki (2003) showed that the difference in protein content among Polish cultivars amounted to 1.8%. The lowest protein content was recorded in Luba cultivar (12.8%), while the highest was in Hruszowska (14%). Wei et al. (2003) reported that the protein content among the Japanese cultivars of common buckwheat varied only slightly and ranged from 15.36% to 15.55%. Prakash et al. (1987) performed studies that have demonstrated high differences in the protein content among different buckwheat cultivars. Comparing the protein content in Polish, Russian, and Indian buckwheat cultivars growing in the conditions of northern India, he found differences in seed protein content ranging from 9.2% to 13.8%.

Starch is the major storage component of buckwheat grains. The results of starch analysis in buckwheat grains of three Polish cultivars have shown that the starch content lies in a narrow range, from 63% to 66% of d.m. The analysis of the tested grains showed the differences of resistant starch content among different cultivars. The highest content was recorded for the seeds of Kora cultivar (63.64% of d.m.), then for Luba (66.00% of d.m.), and the lowest for Panda (64.23% of d.m.) (Stempińska and Soral-Śmietana, 2006) (Table 22.2).

Stempińska and Soral-Śmietana (2006) reported that total lipid content in three Polish cultivars was from 2.4 to 2.7% of d.m. (Table 22.2). The differences in the contents of fats and fatty acids among the cultivars are not significant, but

TABLE 22.2 Chemical Compositions of Buckwheat Grains (Stempińska and Soral-Śmietana, 2006)

Cultivar	Ash Content (% d.m.)	Protein Content (% d.m.)	Starch Content (% d.m.)	Lipids Content (% d.m.)
Kora	2.32 ± 0.02	11.91 ± 0.19	63.64 ± 1.76	2.71 ± 0.10
Luba	2.21 ± 0.01	11.98 ± 0.05	66.00 ± 0.28	2.68 ± 0.03
Panda	3.42 ± 0.04	12.65 ± 0.05	64.23 ± 0.86	2.46 ± 0.04

% d.m., percentage of dry matter.

Kora cultivar dominates over Luba and Panda cultivars. Triacylglycerides are the main component of the neutral lipid fraction; linoleic (18:2), oleic (18:1), and palmitic (16:0) acids account for 88% of total fatty acids (Soral-Śmietana et al., 1984; Horbowicz and Obendorf, 2005).

Fiber is of special interest among the components performing physiological functions. The analysis of fiber showed that the insoluble fraction dominates over the soluble fraction in the proportion of 3:1 in the case of Kora and Panda cultivars, and 4:1 in the case of Luba (Stempińska and Soral-Śmietana, 2006).

Generally, there are no differences between concentrations of magnesium in Polish cultivars (Dietrych-Szostak and Suchecki, 2003). Magnesium content amounts from 0.14% of d.m. (Emka) to 0.16% of d.m. (Hruszowska, Luba, and Panda). The varieties differ, however, in terms of zinc content. Its highest concentrations were recorded for Kora, Emka, and Luba, while the lowest were recorded for Hruszowska and Panda. Zinc content differed and depended on a cultivar origin. In the Japanese cultivars, zinc content ranged from 0.00129% to 0.00205% of d.m., while in Slovenian ones ranged from from 0.00152% to 0.00273% of d.m. (Ikeda and Yamaguchi, 1993; Ikeda and Yamashita, 1994; Ikeda et al., 2000). Polish cultivars contained from 0.0027% to 0.003% of zinc in their d.m. (Dietrych-Szostak and Suchecki, 2003).

Antioxidant compounds contents and compositions differ depending on the buckwheat cultivars. The flavonoids content of *Fagopyrum esculentum* is on average 10 mg/g. About six flavonoids have been isolated from buckwheat grains. The dehulled nuts of Polish cultivars contain rutin and isovitexin, while the hulls contain as many as six flavonoids such as: rutin, quercetin, orientin, vitexin, isovitexin, and isoorientin. Rutin dominates in the total pool of buckwheat grain flavonoids. In the dehulled nuts of the Polish cultivars, the rutin content oscillated between 9.0 and 19.5 mg/100 g. It was the highest with Emka and Panda cultivars, amounting to 19.5 mg/100 g, lower with Hruszowska and Luba (12.5 mg/100 g), and the lowest with Kora (9.0 mg/100 g). The hulls of these varieties contained a twofold higher amount of rutin than the seeds. The cultivars that had the lowest amounts of rutin in the hulls had the highest amounts of rutin in the seeds, in contrast with the defatted nuts (Dietrych-Szostak and Oleszek, 1999, 2001; Dietrych-Szostak and Suchecki, 2003, 2006).

However, Brunori et al. (2007, 2010) obtained different results, proving significant diversity in the rutin content among the cultivars. The difference in rutin content among the cultivars amounted to 62 mg/100 g of d.m. (Brunori et al., 2007) (Table 22.3) and 53 mg/100 g of d.m. (Brunori et al., 2010). Brunori et al. (2010) examined the rutin content in 35 buckwheat cultivars from different breeding centers from around the world, grown in different habitats and agritechnological conditions. The rutin content in the cultivars grown in the same location ranged from 17 mg/100 g of dry weight (Arakawa Village cultivar) to 47 mg/100 g (Karmen cultivar). The results from both studies have shown that the cultivars react differently to habitat conditions in terms of their rutin content. The most valuable cultivars are the universal ones, which means that habitat

TABLE 22.3 Rutin Content of the Grain *F. esculentum* Cultivars Grown in the Summer 2006 at Two Locations (Brunori et al., 2007)

Cultivar	Matrice Average	Matrice Standard deviation	Camigliatello Silano Average	Camigliatello Silano Standard deviation
Aleksandrina	15	1.59	40	0.37
Anita Belorusskaya	16	1.17	20	0.15
Iliya	77	0.04	47	1.70
Karmen	36	1.38	25	2.44
Lena	23	0.01	58	5.07
Vlada	19	0.72	21	1.75
Zhnayarka	21	0.83	24	0.67

conditions have no major impact on their rutin content. These include Vlada, Zhnayarska, and Anita Belorusskaya (Table 22.3).

Sowing Term

The sowing term affects the content of chemicals in buckwheat seeds. This is primarily because, depending on the sowing term, there are different conditions for plant growth and different conditions for seed maturation. The literature shows the interaction of sowing term and year in shaping the content of nutrients in buckwheat seeds. Sowing term impacts the root development and efficiency of nitrogen absorption. A delay in the sowing term increases the heat stress effects during the grain-filling period, and the changes in the optimal conditions of plant growth may influence the nutritional value of seeds (protein, fat, carbohydrate, and rutin content) (Sobhani et al., 2012, 2014). Additionally, sowing term influences the percentage of protein. Sobhani et al. (2014) recorded the highest protein content, amounting to 14.80%, for the third sowing term, but the lowest for the first (13.68%) Table 22.4. The differences equaled 1.18%. These data are consistent with Dietrych-Szostak and Podolska (2008), who found that the sowing of buckwheat at the beginning of Jun. and Jul. resulted in a higher protein content in the seeds compared with the sowing in the first decade of May. The differences between the highest and the lowest protein content equaled 2.5% in the seeds and 1.06% in the hulls. Interestingly, Liszewski (1998) obtained different results, showing that a 2-week Table 22.5 delay in sowing did not significantly change the total protein content of buckwheat nuts.

TABLE 22.4 Mean Comparison of Starch, Protein, and Rutin Content of Buckwheat (Sobhani et al., 2014)

Factors	Starch (%)	Protein (%)	Stem Rutin (%)	Leaf Rutin (%)	Flower Rutin (%)
P1	47.05a	14.10a	0.10a	0.47a	0.85a
P2	46.99a	14.00a	0.07a	0.47a	0.61b
D					
D1	46.20d	13.68d	0.05b	0.48b	0.47d
D2	46.41c	13.79c	0.04b	0.32c	0.99a
D3	46.86b	14.80a	0.06b	0.32c	0.77b
D4	48.62a	13.95b	0.17a	0.65a	0.68c
N					
N1	47.17b	13.87c	0.065c	0.42c	0.75ab
N2	47.03c	14.08ba	0.088b	0.45bc	0.70cb
N3	47.70a	14.07b	0.10a	0.54a	0.79a
N4	46.21d	14.19a	0.088b	0.48b	0.68c

Means followed by the same letter in each column are not significantly different ($p = 0.05$). P1 and P2 were planting treatment mounds with the width of 50 cm associated with two planting rows regarding the distance intervals of 20 cm and those with the width of 60 cm along with three planting rows which are of the distance intervals of 15 cm. D1, D2, D3, and D4 were sowing date Jun. 20, Jul. 5, Jul. 20, and Aug. 5. N1, N2, N3, and N4 were nitrogen treatments 0, 50, 100, and 150 kg ha^{-1}.

TABLE 22.5 Protein Content in Buckwheat Depending on Sowing Term (% d.m.) (Dietrych-Szostak and Podolska, 2008)

	Protein Content (%)			
Year	2004		2005	
Sowing term	Dehulled seeds	Hulls	Dehulled seeds	Hulls
May 6	14.3	3.12	15.6	2.85
Jun. 30	14.1	2.88	16.9	3.31
Jul. 1	16.6	3.94	16.6	3.40
LSD ($\alpha = 0.05$)	0.24	0.124	0.406	0.105

% d.m., percentage of dry matter; %, percent.

Sobhani et al. (2014) reported that sowing term in Aug. increases the storage of carbohydrates, which equals 48.62% compared with the first sowing term (Jun. 20), equaling 46.20%. Similar results indicated that the highest content of carbohydrates was obtained in the late sowing term, as was confirmed in Sobhani et al. (2012).

Morishita et al. (1995) reported that temperature is the most important environmental factor, which affects the accumulation of flavonoids and rutin in buckwheat. It has been reported that rutin content (%) when sowing terms occur after summer is half of that obtained for sowing terms during summer (Li et al., 2008; Mao et al., 2003). According to the experiments performed in Berlin, Germany, the highest rutin content (%) of buckwheat was achieved 45 days after cultivation. Furthermore, Baumgertel et al. (2010) reported the highest rutin content 58 days after cultivation. Investigating buckwheat characteristics in seven different sowing terms in Iran indicated the best sowing term as Jul. 5, with the highest rutin content (0.60 g/plot in d.m.) and the lowest content (0.31 g/plot in d.m.) from Sep. 5 and Jun. 5 (Omidbaigi and De Mastro, 2004).

Moreover, Sobhani et al. (2014) Table 22.4 investigated the impact of the sowing term on the rutin content in the leaves, stems, and flowers of buckwheat. The highest leaf and stem rutin content was recorded at the sowing term on Aug. 5. The results indicate that a suitable temperature and day length increase the length of growth period and leaf area index. In the first and second sowing terms, high temperature causes the decrease of leaf rutin content (%) as compared to the vegetative stages. Thus leaf rutin content is the highest during the previously mentioned period in comparison with the others. The highest flower rutin content was obtained in the third sowing term (Jul. 20). Similarly, Klykov and Moiseenko (2010) conducted a study that took into account three sowing terms: May 30, Jun. 15 and 30, Jul. 15 and 30. The authors have established that rutin content increases with the delay of the sowing term.

Additionally, Dietrych-Szostak and Podolska (2008) reported that in Polish conditions the sowing term has a significant influence on rutin content in buckwheat nuts and hulls as well. A several-fold higher concentration was found in the hulls from the last sowing term (the beginning of Jul.) Table 22.6. They suggested that such a high concentration was associated with weather conditions during nut formation and ripening, as can also be concluded from the literature. According to Kreft et al. (2002), the synthesis of rutin and other polyphenolic compounds depends on the violet radiation. A small amount of UV-B radiation stimulates rutin synthesis. Oomah and Mazza (1996) believe that the place of cultivation, hence the climate or its selected elements, affects the content of rutin and other flavonoids in buckwheat. Schneider et al. (1996) state that a larger amount of rutin is synthesized at a high temperature of 24.5°C during the day and at 18°C at night compared to 18°C during the day and 12°C during the night. In summary, the review of the literature indicates that the concentration of protein and antioxidant compounds in buckwheat nuts is significantly affected by the sowing term. A 2-month delay in the sowing term changes the conditions of nut maturing drastically, which causes quantitative changes of chemical compounds (Tables 22.4–22.6).

TABLE 22.6 Response of Sowing Term to Flavonoid Content in Buckwheat (% d.m.) (Dietrych-Szostak and Podolska, 2008)

	Year							
	2004				2005			
	Sowing term							
	Start of May	Start of Jun.	Start of Jul.	NIR-LSD	Start of May	Start of Jun.	Start of Jul.	NIR-LSD
Flavonoids	Dehulled Seeds							
Rutin	0.016	0.015	0.014	0.001	0.016	0.016	0.015	n.o
	Hulls							
Rutin	0.061	0.055	0.230	0.017	0.100	0.108	0.235	0.061
Orientin	0.033	0.037	0.076	0.001	0.041	0.046	0.084	0.025
Isoorientin	0.025	0.028	0.133	0.005	0.036	0.047	0.129	0.049
Vitexin	0.014	0.017	0.023	0.001	0.012	0.013	0.015	n.o
Isovitexin	0.024	0.027	0.081	0.003	0.039	0.036	0.099	0.030

n.o, No significant difference; % d.m., percentage of dry matter.

Nitrogen Fertilization

Nitrogen is one of plant's basic elements. Buckwheat's (*F. esculentum*) response to nitrogen depends on the initial nitrogen content of the soil, climatic conditions, and nitrogen additives, as well as the term of their application. Several studies have shown that nitrogen fertilizers promote protein increase. Grain protein content is considerably correlated with the available nitrogen in the soil and the quantity of nitrogen applied to buckwheat (Chai et al., 1998; Zhang et al., 1998, 2001; Feng et al., 2003; Sobhani et al., 2012, 2014). The results obtained during the experiment performed in Poland indicate that nitrogen at the dose 40 kg/ha, compared with 0 kg N/ha, resulted in an increase in the content of protein and amino acids in nuts (Kwiatkowski, 2010). The studies of Dietrych-Szostak and Podolska (2008) Table 22.7 showed that the protein content in buckwheat seeds increased at the nitrogen dose of 60 kg N/ha, but the protein increase was also dependent on weather conditions during the growing season, and ranged from 4% to 17%, depending on the year. Dietrych-Szostak and Suchecki (2006), when determining the protein content in the defatted nuts and hulls of buckwheat cultivars, found by several-fold more protein in the defatted nuts compared to the hulls. The seeds contained from 12.75% to 15.44%, while the hulls contained from 3.06% to 3.88% of protein. The results are consistent with those obtained by the authors, as they prove the existence of differences in the protein content in the seeds (defatted nuts) and hulls depending on N content. Baburkov et al. (1999) claimed that grain protein was significantly affected by nitrogen and phosphorous fertilizers. Sobhani et al. (2014)

TABLE 22.7 Response of Nitrogen Fertilization Doses to Content of Protein and Flavonoids (% d.m.) in Buckwheat (2004) (Dietrych-Szostak and Podolska, 2008)

Traits	Nitrogen Fertilization Doses (kg N/ha)			
	0	30	60	LSD
Dehulled Seeds				
Protein	12.7	13.7	14.7	1.141
Rutin	0.0141	0.0138	0.0109	0.0006
Hulls				
Protein	2.87	3.20	3.46	0.256
Rutin	0.0505	0.0490	0.0441	0.0062
Orientin	0.0326	0.0318	0.0328	n.o
Isoorientin	0.0273	0.0253	0.0228	0.0038
Vitexin	0.0126	0.0125	0.0119	n.o
Isovitexin	0.0311	0.0308	0.0306	n.o

n.o, no significant difference; % d.m., percentage of dry matter; kg N/ha, kilograms of nitrogen per hectare.

recorded an increase in the protein content of buckwheat seeds together with increasing nitrogen doses. The lowest quantity of nitrogen was found in the nonfertilized treatments (13.87%), while the highest was found in the treatments fertilized with 150 kg N/ha (14.19%).

In contrast to the positive effect of nitrogen on the protein content, large doses of nitrogen adversely affect starch content. Sobhani et al. (2014) proved that using 50 kg nitrogen can provide the highest starch content, but nitrogen at the dose of 150 kg N/ha significantly reduces starch content in buckwheat seeds. A negative correlation between protein and starch contents has been discussed also in Kalinova et al. (2005) and Ikeda et al. (1995). However, Kalinova et al. (2006) observed a positive correlation between protein and starch content.

The research has shown that the high levels of nitrogen fertilizer led to the decrease in rutin content in seeds (Hagels et al., 1995; Dietrych-Szostak et al., 2008; Kwiatkowski, 2010), but nitrogen influences rutin content in leaves, branches, and flowers (Klykov and Moiseenko, 2010; Sobhani et al., 2014). In contrast to seeds, it had a positive effect on rutin content in leaves. The results of Sobhani et al. (2014) indicated that nitrogen fertilizer levels had a significant influence on leaf rutin content. It was found that 150 kg/ha of nitrogen can increase the leaf rutin content (%) as compared to doses of 0, 50, and 100 kg N/ha. Kalinova et al. (2005, 2006) obtained the highest leaf rutin content using 50 kg N/ha.

Furthermore, Kreft et al. (1999) reported the effects of nutritional ingredients on the rutin contents of grain, leaf, flower, and stem in buckwheat as the mean rutin contents of leaf, stem, and flower were 300–46,000 ppm. The effects of nutrition on buckwheat rutin content have been investigated and measured in various cultivars (Kim et al., 2004; Kitabayashi et al., 1995; Michalova et al., 1998).

Sowing Density

Sowing density determines the amount of light available for plants, affects the efficiency of photosynthesis, and has an impact on the plant nutrition area. There have been very few studies on the influence of sowing density on the chemical composition of buckwheat, and only several of them have shown interactions with other agritechnological factors. Zhang et al. (1997) reported that sowing density can influence the grain protein content. Additionally, Kwiatkowski (2010) found an impact of sowing density on the protein content, and interactions between density and years. In 2006, the highest crude protein content was recorded for nuts harvested from crops sown at a density of 100 items/m², compared to sowing at 300 items/m². In the following years, sowing density had no impact on the amount of protein in the nuts. Sobhani et al. (2012) reported that the pattern of planting influenced the fat content on the 5% level. Buckwheat grown in a smaller row spacing (20 cm) had a fat content of 1.63%, and in a bigger row spacing (60 cm) the fat content was 1.83%.

Concerning dual mutual effects, the pattern of planting and nitrogen influenced the percentage of protein and starch. Also the pattern of planting and the sowing term influenced the protein content percentage wise.

Sobhani et al. (2014) reported that planting density had no impact on protein content, starch content, and stem and leaf rutin content, but found interaction between agrotechnical factor and sowing density in the protein content and rutin content in flowers. During a 2-year experiment the changes of sowing term and planting pattern, as well as those of sowing term and nitrogen fertilizer, showed that the interactions of year with a planting pattern and of a year with sowing term are meaningful. Therefore the highest amount of rutin was obtained for the planting pattern with 20-cm row spacing while sowing the seeds in the third term and applying 100 kg of nitrogen.

Hagels et al. (1995) and Klykov and Moiseenko (2010) showed that seed sowing rate is one of the most effective factors influencing rutin content. Furthermore, Klykov and Moiseenko (2010) reported that the highest rutin content in seeds was observed in the sowing of 1 million seeds per hectare. Increasing the seed-sowing rate by up to 2 million seeds per hectare decreases the rutin content by 29%.

REFERENCES

Baburkov, A.M., Alinov, A., Moudry, J., 1999. Influence of nitrogen fertilizer application on yield and chemical composition of buckwheat seeds. Series Crop Sci. 16, 35–40.

Barta, J., Kalinova, J., Moudry, J., 2004. Effects of environmental factors on protein content and composition in buckwheat flour. Cereal Res. Comm. 32 (4), 541–548.

Baumgertel, A., Loebers, A., Kreis, W., 2010. Buckwheat as a source for the herbal drug Fagopyriherba: rutin content and activity of flavonoid-degrading enzymes during plant development. Eur. J. Plant Sci. Biotechnol. 4 (1), 82–86.

Bonafaccia, G., Marocchini, M., Kreft, I., 2003. Composition and technological properties of the flour and bran in common and Tartary buckwheat. Food Chem. 80, 9–15.

Brunori, A., Brunon, G., Baviello, E., Marconi, M., Colonna, B., Ricci, M., 2005. The yield of five buckwheat (*Fagopyrum esculentum* Moench) varieties grown in central and southern Italy. Fagopyrum 22, 98–102.

Brunori, A.,Vegvari, G., Kadyrov, R., 2007. Rutin content of the grain of seven buckwheat (*Fagopyrum esculentum* Moench) varieties from Belarus grown in central and southern Italy. Advances in Buckwheat Research Proceedings of the 10th International Symposium on Buckwheat. China, 414–416.

Brunori, A., Baviello, G., Colonna, M., Ricci, M., Izzi, G., Toth, M., Vegvari, G., 2010. Recent insights on the prospect of cultivation in central and southern Italy. Advances in Buckwheat Research Proceedings of the 11th International Symposium on Buckwheat. Russia, 589–600.

Chai, Y., Zhang, X., Feng, S.H., Wang, B., Jiang, J.Y., 1998., Study of characters of grain protein in buckwheat. II. Changes of protein content and composition in seed formation period 1, 20–22.

Chang, Q.T., Wang, S.Q., Wang, J.R., 2003. Introduce and extend a new common buckwheat variety. Ping Qia. Fagopyrum 2, 10–11.

Dietrych- Szóstak, D., Suchecki, S.z., 2003. Wybrane cechy jakościowe nasion polskich odmian gryki. Pam. Puł. 133, 35–42.

Dietrych-Szostak, D., Oleszek, W., 1999. Effect of processing on the flavonoid content in buckwheat (*Fagopyrum esculentum* Moench) grain. J. Agric. Food Chem. 47, 4383–4387.

Dietrych-Szostak, D., Oleszek, W., 2001. Obróbka technologiczna a zawartość antyoksydantów w przetworach gryczanych. Przemysł Spożywczy 1, 42–43.

Dietrych-Szostak, D., Podolska, G., 2008. Wpływ terminu siewu na plon i wybrane cechy jakościowe nasion gryki. Fragm. Agronom. 1 (97), 92–100.

Dietrych-Szostak, D., Podolska, G., Maj, L., 2008. Wpływ nawożenia azotem na plon oraz zawartość białka i flawonoidów w orzeszkach gryki. Fragm. Agronom. 1 (97), 101–109.

Dietrych-Szostak, D., Suchecki, S.z., 2006. Zawartość flawonoidów i niektórych składników mineralnych w nasionach polskich odmian gryki. Fragm. Agrom. 1 (97), 101–109.

Dziedzic, K., Gorecka, D., Kobus-Cisowska, J., Jeszka, M., 2010. Możliwości wykorzystania gryki w produkcji żywności funkcjonalnej. Nauka Przyroda Technologia 4 (2), 1–7.

Feng, B.L., Zhang, B., Zhou, J.M., Gao, X.L., 2003. Progress in fertilization on the performance of the pseudo cereals common and flavonoid compounds as possible regulators of reproductive processes in buckwheat. Bio. Zhurnal. 23, 154–159.

Gabersik, A., Voncina, M., Tros, T., Germ, M., Bjorn, L.O., 2002. Growth and production of buckwheat (*Fagopyrum esculentum*) treated with reduced, ambient and enhanced UV-B radiation. J. Photochem. Photobiol. 66, 30–36.

Germ, M., 2004. Environmental factors stimulate synthesis of protective substances in buckwheat. Advances in Buckwheat Research. Proceedings of the 9th International Symposium on Buckwheat, Prague, 55–60.

Hagels, H., Wagenbrcth, D., Schilcher. H., 1995. Phenol compounds of buckwheat herb and influence of plant and agricultural factors (*Fagopyrum esculentum* Moench and *Fagopyrum tataricum* Gartner). The 6th International Symposium on Buckwheat. Current Advances in Buckwheat Research, 801–809.

Henckel, P.A., 1975. Physiological ways of plant adaptation against drought. Agrochimica 19, 5–7.

Hore, D., Rathi, R.S., Collection, M., 2002. Cultivation and characterization of buckwheat northeastern region of India national bureau of plant genetic resources. Regional Station, Brainpan. 793–798.

Horbowicz, M., Obendorf, R.L., 2005. Fagopyritol accumulation and germination of buckwheat seeds matured at 15, 22, 30°C. Crop Sci. 45, 1–11.

Ikeda, S., Yamashita, Y., Murakami, T., 1995. Minerals in buckwheat. Curr. Adv. Buckwheat Res., 789–792.

Ikeda, S., Yamashita, Y., Kreft, I., 2000. Essential mineral composition of buckwheat flour fractions. Fagopyrum 17, 199–212.

Ikeda, S., Yamaguchi, Y., 1993. Zinc contents in various samples and products of buckwheat. Fagopyrum 13, 11–14.

Ikeda, S., Yamashita, Y., 1994. Buckwheat as a dietary source of zinc, copper and manganese. Fagopyrum 14, 29–34.

Inoue, N., Uehara, S.-H., Fujita, K., Yang, Z., Kato, M., Ujihara, A., 2005. Effects of environmental factors on the chemical characteristics related to flour texture in common buckwheat. Fagopyrum 22, 39–44.

Kalinova, J., Moudry, J., Curn, V., 2005. Yield formation in common buckwheat (*Fagopyrum esculentum* Moench). Acta Agron. Hung. 53 (3), 283–291.

Kalinova, J., Triska, J., Vrchotova, N., 2006. Distribution of vitamin E, squalene. epicatechin, and rutin in common buckwheat plants (*Fagopyrum esculentum* Moench). J Agric. Food Chem. 54, 5330–5335.

Kim, S.U., Kim, S.K., Park, C.H., 2004. Introduction and nutritional evaluation of buckwheat sprouts as a new vegetable. Food Res. Int. 37, 319–322.

Kitabayashi, H., Ujhara, A., Hirose, T., Minami, M., 1995. Varietal differences and heritability for rutin content in common buckwheat *Fagopyrum esculentum* Moench. Breed. Sci. 45, 75–79.

Klykov, A.G., Moiseenko, L.M., 2010. Influence of buckwheat cultivations upon rutin content. Proceedings of the 11th International Symposium on Buckwheat, Orel. 475–483.

Kreft, S., Knapp, M., Kreft, I., 1999. Extraction of rutin from buckwheat (*Fagopyrum esculentum* Moench) seeds and determination by capillary electrophoresis. J. Agric. Food Chem. 47, 4649–4652.

Kreft, S., Strukelj, B., Gaberscik, A., Kreft, I., 2002. Rutin in buckwheat herbs grown at different UV-B radiation levels: comparison of two UV spectrophotometric and an HPLC method. J. Exp. Bot. 53, 1801–1804.

Krkoškova, B., Mrazova, Z., 2005. Prophylactic components of buckwheat. Food Res. Int. 38, 561–568.

Kwiatkowski, J., 2010. Agrotechniczne uwarunkowania produkcji gryki (*Fagopyrum esculentum* Moench) o wysokiej wartości technologicznej, odżywczej i reprodukcyjnej orzeszków. Dissertations and Monographs UWM Olsztyn 153, 111.

Li, H.C., Bozhang, L., Zi, X., 2008. Inter-varietal variations of rutin content in common buckwheat flour (*Fagopyrum esculentum* Moench). Transgenic Res. 17 (1), 121–132.

Li, S., Zhang, G.H., 2001. Advances in the development of functional foods from buckwheat. Crit. Rev. Food Sci. Nutr. 41, 451–464.

Liszewski, M., 1998. Wpływ terminu siewu i gęstości siewu na plon I niektóre cechy biologiczno użytkowe gryki tetraploidalnej uprawianej na glebie kompleksu żytniego dobrego. Fragm. Agron. 2 (58), 13–23.

Mao, C., Cheng, G.X., Chen, L.Z., 2003. Breeding a new Tartary variety, Wei HeiQiao No. 1, with high yield and quality. Fagopyrum 1, 12–14.

Mazurek, J., 1999. Biologia kwitnienia i owocowania gryki w zależności od nawożenia jej azotem. Biuletyn Naukowy, Olsztyn 4, 19–26.

Michalova, A., Dotlaci, l.L., Cejka, L., 1998. Evaluation of common buckwheat cultivars. Roslinna Vyroba 44 (8), 361–368.

Morishita, T., Hajika, M., Sakai, S., Tetsuka, T., 1995. Development of simple spectrophotometer assay for the rutin-degrading enzyme in buckwheat. The 6th International Symposium on Buckwheat. Curr. Adv. Buckwheat Res. 18, 833–837.

Murakami, T., Murayama, S., Uchitsu, M., Yoshida, S., 2002. Root length and distribution on field-grown buckwheat (*Fagopyrum esculentum* Moench). Soil Sci. Plant. Nutr. 48 (4), 609–613.

Noworolnik, K., 1995. Nitrogen fertilization efficiency of buckwheat grown at various soil conditions. Current Advances in Buckwheat Research. Proceedings of the 6th International Symposium on Buckwheat, 601–604.

Noworolnik, K., 1999. Interactions between several agricultural factors in relation to grain yield of buckwheat. Biuletyn Naukowy, Olsztyn, 4, 65–70.

Ohsawa, R., Tsutsumi, T., 1995. Inter-varietal variations of rutin content in common buckwheat flour (*Fagopyrum esculentum* Moench). Euphytica 86, 183–189.

Omidbaigi, R., De Mastro, G., 2004. Influence of sowing time on the biological behavior, biomass production, and rutin content on buckwheat. Ital. J. Agron. 8 (1), 47–50.

Oomah, D.B., Mazza, G., 1996. Flavonoids and antioxidative activities in buckwheat. J. Agric. Food Chem. 44, 1746–1750.

Podolska, G., Konopka, I., Dziuba, J., 2007. Response of grain yield, yield components and allergic protein content of buckwheat to drought stress. Advances in Buckwheat Research. Proceedings of the 10th International Symposium on Buckwheat. China, 323–328.

Prakash, D., Prakash, N., Misra, P.S., 1987. Protein and amino-acid composition of *Fagopyrum* (buckwheat). Plant Foods Hum. Nutr. 36, 341–344.

Qian, J., Kuhn, M., 1999. Physical properties of buckwheat starches from various origins. Starch/Starke 5, 81–85.

Ruszkowscy, B., Ruszkowscy, B., 1967. Gryka. PWRiL, Warszawa, Poland.

Schneider, M., Kuhlmann, H., Marquardt, R., 1996. Investigation on rutin content in Fagopyrum esculentum under specific climatic condition in the phytotrone. Proceedings of International Symposium on Breeding Research on Medicinal and Agromatic Plants, Quedlinburg, pp. 351–354.

Sobhani, M.R., Rahmikhdoev, G., Mazaheri, D., Majidian, M., 2012. The effect of sowing date, pattern of planting and nitrogen on quantitative and qualitative yield in summer sowing buckwheat (*Fagopyrum esculentum* Moench). Adv. Environ. Biol. 6 (1), 440–446.

Sobhani, M.R., Rahmikhdoev, G., Mazaheri, D., Majidian, M., 2014. Influence of different sowing date and planting pattern and N rate on buckwheat yield and its quality. Aust. J. Crop Sci. 8 (10), 1402–1414.

Soral-Śmietana, M., Fornal, Ł., Fornal, J., 1984. Characteristics of lipids in buckwheat grain and isolated starch and their changes after hydrothermal processing. Nahrung 28, 483–492.

Steadman, K.J., Burgoon, M.S., Lewis, B.A., Edwardson, S.E., Obendorf, R.L., 2001. Buckwheat seed milling fraction: description, macronutrient composition and dietary fiber. J. Cereal Sci. 33, 271–278.

Stempińska, K., Soral-Śmietana, M., 2006. Składniki chemiczne i ocena fizykochemiczna ziarniaków gryki—porównanie trzech polskich odmian. Żywność Nauka Technologia Jakość 13 (2/47 Supplement), 348–357.

Subedi, K.D., Ma, B.L., Xue, A.G., 2007. Planting date and nitrogen effects on grain yield and protein content of spring wheat. Crop Sci. 47, 36–44.

Tahir, I.S., Nakata, N., Ali, A.M., Mustafa, H.M., Saad, A.S.I., Takata, K., Ishikawa, N., Abdalla, O.S., 2006. Genotypic and temperature effects on wheat grain yield and quality in a hot irrigated environment. Plant Breed. 125 (4), 323–330.

Wei, Y., Hu, X., Zhang, G., Ouyang, S., 2003. Studies on the amino acid and mineral content of buckwheat protein fractions. Nahrung/Food 47, 114–116.

Wei, Y., Zhang, G.Q., Li, Z.X., 1995. Study on nutritive and physico-chemical properties of buckwheat flour. Nahrung 39, 48–54.

Zhang, X., Chai, Y., Shang, A.J., 2001. Effects of seeding date on grain protein content and composition of buckwheat. Fagopyrum 1, 11–13.

Zhang, X., Chai, Y., Wang, B., Ma, Y.A., Feng, B.L., Liang, L.Y., 1998. Study of grain protein content in buckwheat grain protein content and composition. Fagopyrum 1, 15–19.

Zhang, X., Feng, S.H., Lui, J., Bai, Y.B., Jiang, J.Y., 1997. Effects of plant density on buckwheat grain protein and its compositions. Fagopyrum 2, 14–16.

Chapter | twenty three

Cultivation, Agronomic Practices, and Growth Performance of Buckwheat

S. Farooq*, R. Ul Rehman**, T.B. Pirzadah**, B. Malik**, F. Ahmad Dar**, I. Tahir**

*Department of Botany, University of Kashmir, Srinagar, Jammu and Kashmir, India; **Department of Bioresources, University of Kashmir, Srinagar, Jammu and Kashmir, India

INTRODUCTION

Globalization of agriculture and consequently its industrialization seems unrelenting, with adverse side effects felt throughout the globe, thus diminishing the genetic diversity in the agricultural sector. Besides, adoption of monoculturing practices and biased technological development and its usage is practiced for a few high-energy-demanding plant species only. Consequently, world food security and economic growth are dependent on a handful of crops, which has placed future food supply and rural income at risk. Even though mankind over time has been using more than 10,000 edible species, today about 150 species only are being commercialized significantly at the global level, out of which just 12 species meet the major nutritional (80% dietary energy) requirements, and the 60% protein and calorie requirements are met by just four species: rice, wheat, maize, and potato (FAO, 2005). The limited use of a few crops for maximum economic benefits has profound environmental consequences. The loss of many crop varieties has provoked organizations and scientists worldwide to retrieve, research, and disseminate knowledge regarding production and utilization of many neglected, underexploited, and new alternative crop species. These alternative crops were traditionally used for their food, fiber, fodder, oil, or medicinal properties. These crops possess immense potential in the food, health, and energy sectors besides providing opportunities for employment generation. However, with the modernization of agricultural practices many have become

299

Molecular Breeding and Nutritional Aspects of Buckwheat. http://dx.doi.org/10.1016/B978-0-12-803692-1.00023-7

neglected because they are held in low esteem, and some have been neglected to the extent that erosion of their gene pool has become severe and thus they are often regarded as lost crops.

Among the various underutilized crops, buckwheat is one of the ancient domesticated crops of Asia, Central, and Eastern Europe that has been mainly used as a staple food especially in arid regions of the world. Buckwheat is believed to have originated in central Asia; however, the origin of its domestication dates back about 4000–5000 years in South China (Gondola and Papp, 2010). Therefore China is considered as the original center of buckwheat and is extremely rich in buckwheat genetic resources. Buckwheat is chiefly cultivated in India, Nepal, Bhutan, China, Canada, Mongolia, North Korea, far eastern Russia, and Japan. However, buckwheat production declined during the first half of the 20th century, especially in Russia and France. Cultivation was also falling in Canada until buckwheat production began to increase in the 1960s as a result of export demands. Exports from Canada to Japan have steadily grown, where the flour is utilized for the production of noodles (Campbell, 1997). In India, buckwheat is cultivated in Jammu and Kashmir, Himachal Pradesh, Uttrarakhand, West Bengal, Sikkim, Meghalaya, Arunachal Pradesh, and Manipur (Tahir and Farooq, 1988) (Fig. 23.1). The Scottish coined the term "buckwheat" from two Anglo-Saxon terms, boc (beech) and whoet (wheat), because it is similar to beech nut (Edwardsen, 1995; Berglund and Duane, 2003). However, buckwheat is neither a nut nor a cereal but is included in a separate group called "pseudo-cereals" as it shows both similarities and differences with cereals. Buckwheat is not considered a true cereal, but the seeds contain a cereal-like starchy endosperm. Buckwheat belongs to the family Polygonaceae and genus *Fagopyrum*. It is a broad leaved, annual, and dicotyledonous crop that attains a maximum height of about 2–5 ft. (Skrabanja et al., 2004). The genus *Fagopyrum* includes about 19 species besides four new species that have recently been included,

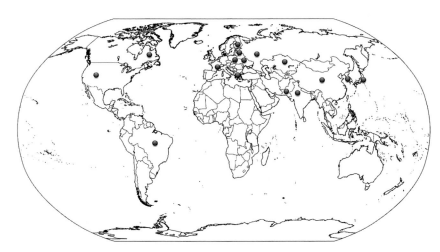

FIGURE 23.1 World buckwheat production sites.

namely, *Fagopyrum crispatifolium* (Liu et al., 2008); *Fagopyrum pugense* (Tang et al., 2010); *Fagopyrum wenchuanense* (Shao et al., 2011); and *Fagopyrum qiangcai* (Shao et al., 2011), and their taxonomic position as well as phylogenetic relationships have been clarified (Zhou et al., 2012; Shao et al., 2011). In Jammu and Kashmir, the genus *Fagopyrum* is represented by four different cultivated species. These include: common buckwheat *Fagopyrum esculentum* Moench, Tartary buckwheat *Fagopyrum tataricum*, coarse buckwheat *Fagopyrum sagittatum*, and Kashmir buckwheat *Fagopyrum kashmirianum* Munshi (Munshi, 1982; Tahir and Farooq, 1988). The perennial species *Fagopyrum cymosum* Meissn grows wild and can be propagated through rhizomes. *F. cymosum* differs from Tartary and common buckwheat in its shoots, branching, and racemes (Tahir and Farooq, 1989a). In addition, all buckwheat species found to date possess eight chromosomes, $2n = 16$, as their base complement except *F. cymosum* and *Fagopyrum gracilipes*, which are tetraploids with $4n = 32$. Buckwheat is characterized by having a single, erect, and hollow stem that shows variation in color from green to red and becomes brown toward maturity. Leaves are triangular or heart shaped, 2–3 in. in length, and are arranged on the stalk in an alternate manner. Inflorescence is formed by seven to nine blossoms. Flowers are showy, attractive, and are generally clustered in racemes at the ends of the branches or on short pedicles arising from the axils of the leaves. The flowers are generally bisexual and consist of distinct variable parts, which include sepals (3–5) and petals (called tepals), stamens (6–9), and a single pistil. The superior ovary of fused carpels has one chamber (locule) with one ovule, which develops into an achene. The color of flowers varies from white or light green to pink or red (Tahir and Farooq, 1989b; Janovska et al., 2009; Cawoy et al., 2009). Buckwheat flowers exhibit sexual dimorphism, ie, they possess two types of flowers. These include pin flowers having long pistils and short stamens while the other type is thrum flowers possessing short pistils and long stamens. Nectar-secreting glands are present at the base of the ovary. Seeds are triangular in shape but sometimes prominently winged and the color of the seeds may be glossy brown, gray brown, silvery gray, or black (Farooq and Tahir, 1987; Krkoskova and Mrazova, 2005). Depending on the variety, seed shape and size also vary. Seed is composed of a thick outer hull and an inner groat that resembles the cereal kernel in its gross chemical composition and structure. Buckwheat hull possesses lesser density than water and thus allows easy removal of the hull from the kernel. However, the hardness of the hull depends on the species of buckwheat, for example, *F. esculentum* possesses a softer hull than its relative species *F. tataricum* (Li and Zhang, 2001). Moreover, buckwheat possesses a dense, fibrous root system with a deep taproot that constitutes about 3–4% of the total weight of the plant. Most of its roots are concentrated in the top 10 in. of the soil. Although buckwheat possesses a shallow root system, its physiological activity is significantly good. In addition, the fibrous root system of buckwheat plays an essential role by promoting earthworm activity, which results in aeration of the soil.

Buckwheat has gained worldwide importance because of the presence of some important bioactive constituents such as rutin, orientin, vitexin, quercetin, isovitexin, and isoorientin, besides other essential components like fago-pyritols find vast potential for glycemic control in type II diabetics, positive cardiovascular effects, treatment of celiac disease, prevention of gall stones, and several hormone-dependent tumors (Tahir and Farooq, 1987; Verardo et al., 2011). Currently, various products of buckwheat like tea, roasted groats, flour, cookies, noodles, pancakes, cakes, and wine are available in the market. Buckwheat has attracted worldwide attention, especially from food scientists, for its healing effects over chronic diseases. The tea prepared from buckwheat groats is considered to be one of the famous health products, especially in various Asian and European countries (Qin et al., 2011). Also, from an agronomical point of view, it is considered as an eco-friendly crop with the need for minimal tilling and chemical fertilizers as well as pesticides. Despite its domestic decline as a staple food and feed source, current research is focused on the neutraceutical aspects of buckwheat, which provides a new perspective for future novel buckwheat products. Currently, Russia is the largest producer of buckwheat. China is the second largest producer of buckwheat with 10.2 million acres under cultivation and contributing about 0.6–0.95 million tons (Li and Zhang, 2001) (Table 23.1). However, over the last two to three decades,

TABLE 23.1 The Current Leading Producers of Buckwheat, Together with Yield and Area (Year 2011; FAOSTAT, 2013)

Countries	Production (tons)	Yield (tons/ha)	Area (ha)
Russian Federation	80,0380	0.949	843,200
China	720,000[a]	0.962	748,000[a]
Poland	92,985	1.227	75,768
France	91,000	2.935	31,000
United States	79,554[b]	1.029	77,244[b]
Brazil	57,000[a]	1.239	46,000[a]
Belarus	44,456	1.091	40,734
Kazakhstan	37,400	1.274	66,780
Japan	32,000	0.567	56,400
Lithuania	26,000	0.955	27,200
World total	2,294,178[c]	0.885	2,327,409[c]

[a]FAO estimates.
[b]Data based on imputation methodology.
[c]Aggregate data (official, semiofficial, and estimated).

the cultivation of buckwheat has declined worldwide because of its low and erratic yield. Several physiological and ecological traits are responsible for its low yield, such as sensitivity to frost, lodging, self-incompatibility, distyly, pollen transfer limitations, indeterminate growth habit, and female sterility. It is in this perspective that buckwheat needs worldwide attention especially from molecular breeders and biotechnologists to improve its yield and revive its cultivation on a large scale.

AGROTECHNIQUES
Site Preparation and Seed Sowing

Buckwheat is a short duration crop, which usually prefers a moist and cool climate and is grown mostly on well-drained sandy soil; however, it also thrives well in acidic soils (pH < 5) (Olson, 2001; Hore and Rathic, 2002). Buckwheat is a cover crop and thus does not require extensive site preparation; it has the ability to grow well even in very poorly tilled soil (Loch and Lazanyi, 2010; Popović et al., 2013a,b; Ikanović et al., 2013). Its cultivation requires the land to be prepared several weeks in advance for weed removal to increase the porosity of the soil. Further, the drainage should be such so that the submerging of seeds is avoided because water lodging will effect germination and thus reduce the crop yield. Sowing date is one of the essential parameters for yield production because of its significant effects on plant genotype. So the appropriate sowing date results in maximum yield. The sowing period of buckwheat depends on various factors such as agroclimatic conditions like temperature (Sobhani et al., 2014). Determination of sowing date aims to find the appropriate planting time of cultivars so that the existing set of environmental factors can be suitable for plant germination and survival (Hore and Rathic, 2002). Lee et al. (2001) reported that the best sowing date for buckwheat is more likely to be late May with maximum production yield of 2059 kg/ha as compared to late Apr., early and late Jun. in China. However, in Western Europe, it is reported that the best sowing date is mid-May to Jul. when there are no risks of late summer frosts (Bernath, 2000; Halbrecq et al., 2005). However, it usually varies from one place to another because of its susceptibility to frost conditions. At lower altitudes, it is usually sown in the months of May to Aug. while at higher altitudes it is generally sown in the months of Apr. to May (Ratan and Kothiyal, 2011). The recommended sowing depth of seeds is about 4–6 cm but during dry climatic conditions seeds are sown deeper for adequate moisture (Björkman, 2012). However, deep sowing delays seeding emergence and decreases crop uniformity. Seeds are usually sown in rows approximately 10 cm apart or scattered randomly in the fields at a rate that ranges from 35 to 40 kg/ha when used as a grain crop but is about 50 kg/ha when it is used as a smoother crop, vegetable crop, or fodder crop and the thinning may be performed after 20 days of sowing (Hore and Rathic, 2002). Seeding rate should be limited as it affects the plant growth resulting in low yield.

Fertilizer Requirements

The efficiency of the crop yield mainly depends on the soil nutrients that are analyzed by soil tests. Buckwheat shows a better response to an equitable fertilizer program, but it is not regarded as an extreme nutrient user. Buckwheat is considered as an efficient green manure crop because it produces high growth rate and biomass, enhances soil nutrient quality especially nitrogen, and has a phosphorus mobilization, high decomposition rate, and high litter quality (Teboh and Franzen, 2011). Thus incorporating buckwheat into the soil improves soil health by enhancing the soil texture of the top soil, improving its tilth and porosity. The recommended fertilizer requirement for buckwheat cultivation includes 47 kg nitrogen, 22 kg phosphorus, and 40 kg potassium to produce a yield of about 1600 kg/ha (Campbell and Gubbels, 1978). Phogat and Sharma (2000) reported that under Indian conditions the higher yield is obtained by applying about 50 kg nitrogen, 20 kg P_2O_5, and 40 kg K_2O or 1500–2000 kg farmyard manure per hectare. In sandy loam soils it is recommended that potash should be added to the soil. However, in fields with high nitrogen content, buckwheat exhibits lodging and results in low yield while a higher dosage of phosphorus alleviates the effect (Inamullah et al., 2012). Edwardsen (1996) reported excessive vegetative development along with increased lodging with higher doses of nitrogen. Saini and Negi (1998) reported that high nitrogen content delays maturity in buckwheat; in addition, the delay or earliness in phenology of buckwheat is also affected by nitrogen and phosphorus (Gardner et al., 1985; Inamullah et al., 2012). Moreover, phosphorus is also reported to strengthen the stem (Gardner et al., 1985; Edwardsen, 1996). In addition to this, phosphorus and nitrogen also influence the grain weight and yield. Omidbaigi et al., (2004) reported a higher number of grains/plant and higher 1000-grain weight with higher doses of nitrogen: 1000-grain weight is also enhanced when the phosphorus level is increased. Moreover, nitrogen and phosphorus used in combination and in high concentration enhance the grain yield drastically (Inamullah et al., 2012). It can be argued that higher doses of nitrogen delay maturity and increase plant height and lodging score while higher doses of phosphorus enhance maturity and decrease lodging score in common buckwheat. Wang and Campbell, 2004 reported that soaking of buckwheat seeds in micronutrients such as manganese and zinc enhances the yield.

Crop Harvesting and Storage

The harvesting period varies according to agroecological conditions. In some high-altitude areas buckwheat is harvested late in the season while at mid- and low-altitudinal areas it is harvested early. Further, because of the indeterminate growth habit of the buckwheat plant, it is difficult to harvest grains at a particular time period. During the harvesting period, seeds of all stages, namely, mature seeds, immature seeds, and a few flowers, are present at the same time period (Farooq and Tahir, 1982). However, this all depends on the grower to determine the particular harvesting period before losses can occur because of the shattering of seeds. However, under normal conditions it can usually be harvested approximately 10 weeks after planting. Harvest should begin when 70–75% of the seeds have reached physiological maturity, and the plants have shed most of their leaves. This stage corresponds with the time the lower seed heads begin to shatter. On a small scale, buckwheat is easy to thresh because most of the seeds fall out if a dry bundle of grain is shaken. The optimum temperature and moisture content for drying is about 45°C and 16%, respectively (Olson, 2001). It is recommended that buckwheat should not be stored for a longer period of time because it is susceptible to rancidity. Generally, in temperate regions, seed maturity reaches an average of 3 months after cultivation. However, the developmental cycle extends to 5 months under unfavorable conditions (Michiyama et al., 1998; Michiyama and Hayashi, 1998). Growing demand for buckwheat seeds and products enhances turnover and storage requirements. Stored grains are exposed to the risk of pest infestation, mainly insects and mites, which cause quantitative as well as qualitative losses (Olejarski and Ignatowicz, 2011). The wheat weevil (*Sitophilus granarius* L.) and the lesser grain borer (*Rhyzopertha dominica* F.) are popular storage pests. However, because of a thick seed coat, high tannin content, and phenolic compounds, buckwheat seed is resistant to storage pests (Zadernowski et al., 1992; Ciepielowska and Fornal, 2004). Common buckwheat with thin silver-gray zebra-lining hulls is more prone to wheat weevil infestations than other species (Kwiatkowski and Nietupski, 2013) (Fig. 23.2).

Crop Diseases and Pests

Buckwheat is relatively free of serious insect and disease infestations, but several pathogenic problems have been reported. The buckwheat shoots are damaged by several scoop caterpillars like *Euxoa segetum*, *Euxoa tritici*, *Phytometra gamma*, *Trachea atriplicis*, and *Barathra brassicae*. Some of these pests are polyphytophages as they also cause damage to underground parts. Several phytophages are responsible for damaging buckwheat stalks and leaves, for example, *Chaetocnema concinna* damages young leaves. Also flowers of buckwheat serve as forage for *Meligethes aeneus* (Naumkin, 2013). Leaf spot is caused by *Ramularia*, root rot by *Rhizoctonia*, downy mildew by *Peronospora ducometi*, and Botrytis rot by *Botrytis cinerea* (Rana et al., 2012). Buckwheat is also attacked by many pests such as aphids, wireworms, birds, and rodents and one nematode (*Ditylenchus dipsaci*) (Tahir et al., 1985; Bhat et al., 1986; Olson, 2001). It has been revealed that buckwheat planted earlier than the recommended dates is at higher risk because aphids may attack and stunt the plants in mid-Jun. Moreover, buckwheat plants also act as host for about 20 viruses including cucumber and tobacco mosaic viruses; in addition, certain bacteria are also associated with it (Jacquemart et al., 2012). Beetles and larvae are considered the most harmful pests. Beetles appear at the end of May or early Jun. and begin eating the shoots of *F. esculentum*. Larvae feed first on the node, thus causing curvature of the stem, lodging, or even breakage, which causes a decline in yield production by damaging the nectar glands (Kuznetsova et al., 2012). It has been reported that

FIGURE 23.2 Two cultivated buckwheat species—common buckwheat (*F. esculentum*) and Tartary buckwheat (*F. tataricum*)—at various developmental stages growing in Gurez Valley. Source: Photographs were taken by the authors during Aug. 2013.

the pest *Rhinoncus sibiricus* causes about 30–50% loss in yield of *F. esculentum* (Kuznetsova and Klykov, 2012; Klykov et al., 2014).

Pollination Requirement

Buckwheat is naturally cross-pollinated to produce seeds, so insect pollinators are necessary for effective fertilization and seed set. The efficiency of pollination mainly depends on insect abundance, flower morphology, nectar production, and the ability of the insects to collect, transport, and deposit pollen on a compatible stigma. Cawoy et al. (2008) reported that nectar production can be influenced by heteromorphy, ploidy level, cultivar type, plant age, inflorescence position, and abiotic factors. Tetraploid cultivars produce more nectar and pollen than diploids (about 30–40% more) and are thus more attractive for insects (Alekseyeva and Bureyko, 2000). Among abiotic factors, light is an essential factor that influences nectar production. It has been reported that nectar production stops when plants or inflorescences are transferred from light to dark conditions (Cawoy et al., 2008). Also nectar volume is light dependent as nectar volume/flower is enhanced by about 41% when the light irradiance doubles (Cawoy et al., 2007a). The available data reveal that nectar production appears to be linked to photosynthesis but when plants undergo defoliation, nectar production persists and the nectar is still highly sugary (defoliation; minimum 30%; control 50%). Photosynthesis may take place in other parts of the plant such as the inflorescence pedicles, the cyme bracts, and the main stem (Cawoy et al., 2008). Bjorkman (1995) reported that in the case of buckwheat approximately 1% pollination is carried out by means of wind. However, depending on the speed of the wind, pollen can reach 600 m if wind speed is lower than 3 m/s and 1000 m when wind speed is higher than 6 m/s. The various pollinators involved in the pollination mechanism of buckwheat include *Hymenoptera*: honey bees (*Apis mellifera* L.), bumble bees, solitary bees, and wasps; *Diptera*; *Syrphidae, Calliphoridae*, and others; and *Lepidoptera, Hemiptera*, and *Nevroptera* (Tahir et al., 1985; Jacquemart et al., 2007). Moreover, the major visitors of buckwheat belong to *Apoidea* (*Hymenoptera, Apis mellifera*, and *Bombus* species) and *Syrphidae* (*Diptera* and *Eristalis* species) (Racys and Montvilienne, 2005). The main pollinators are honey bees as they collect both types of pollen (pin and thrum) on the same trip. Thus buckwheat is considered as an excellent plant for bee pasture and insectary gardens (Mader et al., 2011; Lee-Mader et al., 2014). Besides its foraging and prospecting behavior, collecting nectar and pollen, the honey bee promotes frequent contacts with stigmas (Bjorkman, 1995; Jacquemart et al., 2007). However, the honey bee spends more time on thrum flowers than on pin flowers because there is more nectar production in thrums. Ornelas et al. (2004) observed unusual differences in nectar production between morphs in distylous species. However, the total sugar concentration is similar in both morphs, but sucrose level is significantly higher in thrum flowers (16.8% against 12.9%), so the sucrose/hexose ratio is higher for thrum flowers (Cawoy et al., 2006a). It is recommended that one honey bee colony per acre is sufficient for effective pollination. It has been reported that honey bees are more active during warm and sunny days. In addition, they are also active during the flowering peak and the optimum temperature for effective pollination is 20°C (Sugawara, 1956; Alekseyeva and Bureyko, 2000). Goodman et al. (2001) reported that bumble bees and honey bees visit buckwheat predominantly between 9 and 12 h daily while syrphids are still active in the afternoon. On average a single honey bee visits about 14–20 flowers/min and works on buckwheat for 4–5 h/day (Jacquemart et al., 2007). It has also been observed that under experimental conditions the efficiency of honey bee pollination is good since this insect deposits compatible pollen on most flowers (>90%) without discrimination between floral morphs (Jacquemart et al., 2007). The low temperature influences the rate of pollination

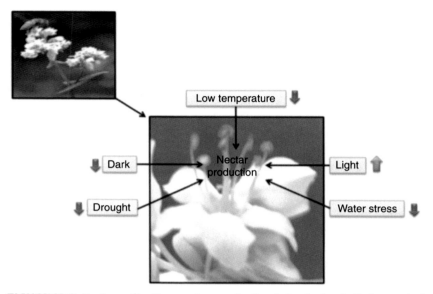

FIGURE 23.3 Factors affecting nectar production in buckwheat. *Red arrows* indicate decline in nectar production while *green arrow* indicates increase in nectar production.

because of the reduction in sucrose production in buckwheat plants, consequently lowering the relative concentration of sugars in the nectar. It has been reported that under cool and wet climatic conditions, sugar production per flower can be about 15 times higher than during dry and drought conditions. Also the optimal soil humidity for nectar production averages 60% but water stress reduces sugar production (Racys and Montvilienne, 2005) (Fig. 23.3).

Buckwheat as Smart Cover Crop in Intercropping System

Buckwheat is considered as a smart crop in intercropping systems and is intercropped with potato, soyabean, sunn hemp, millet, safflower, sunflower, and other warm season annuals (Pirzadah et al., 2013; Pavek, 2014). Many farmers who have used buckwheat as a cover crop refer to its ability to "mellow" the soil, leaving it in a friable and fertile state, ready for planting (Petrich, 2000). Buckwheat acts as a smother crop because of its allelopathic effect to crowd out obnoxious weeds (Tahir and Farooq, 1990; Geneau et al., 2011). It has been reported that rice production was enhanced by about 20% while weed density was reduced to about 80% when 2 tons of buckwheat pellet were applied before rice planting (Jacquemart et al., 2012). It has also been reported that buckwheat contains higher concentrations of calcium, potassium, magnesium, and phosphorus than other plants like sorghum, maize, wheat, and clover (Warman, 1991). Buckwheat used as a cover crop shows significant results in managing weeds by declining weed biomass (75–99%) when compared to bare ground treatments (Iqbal et al., 2003). Incorporation of buckwheat immediately after killing reduces the emergence and biomass of various weeds like shepherd's purse (*Capsella bursa pastoris*), corn chamomile (*Anthemis arvensis*), and redroot pigweed (Kumar et al., 2008). The mechanism behind suppression of weeds is its ability to control soil nitrogen availability that interferes with weed emergence and growth. Creamer and Baldwin (1999) observed that buckwheat possesses a carbon:nitrogen ratio of 34 (C:N = 34), thus immobilizing nitrogen during decomposition. Several weeds exhibit rapid early growth and nitrogen uptake; however, when buckwheat is incorporated into the soil, it reduces the nitrogen level in the soil and thus reduces weed emergence and growth (Kumar et al., 2008). Potassium levels were higher in the soil that was left with buckwheat residues (Kumar et al., 2008). Further, buckwheat has the ability to sequester phosphorus and is considered an efficient crop to uptake phosphorus and use it in various physiological processes. It is found that buckwheat possesses the capacity to take up about 10 times more phosphorus than wheat, thus it is regarded as an efficient phosphorus scavenger (Zhu et al., 2002). The reason behind the high phosphorus uptake is buckwheat's ability to acidify the rhizosphere by oozing out protons (H⁺), which in turn solubilize the calcium-bound phosphorus in an alkaline soil, thus makes phosphorus available to the plant (Zhu et al., 2002) (Fig. 23.4).

FIGURE 23.4 Buckwheat as an efficient phosphorus scavenger. *Red dots* indicate protons (H⁺) oozing out from buckwheat roots and *green dots* indicate forms of phosphorus ($H_2PO_4^-$) available to plants in calcium-bound state, that is, calcium-bound phosphorus (CBP).

LIMITATIONS IN PRODUCTION YIELD
Seed Shattering in the Crop

Although buckwheat finds an important place in the functional food sector it suffers from a problem of seed shattering especially during harvesting time, which is the main cause of its decline (Lee et al., 1996). Some researchers have pointed out that the shattering problem is more predominant in Tartary buckwheat than common buckwheat (Tahir and Farooq, 1988). Because of the grain shattering problem, buckwheat yield is about 40–50% lower by machine than by hand harvesting (Aufhammer et al., 1994; Radics and Mikohazi, 2010). The shattering problem in buckwheat has usually two forms, namely, brittle and weak pedicles. However, brittle pedicles are mainly observed in wild buckwheat, which grows itself and is regulated by two complementary dominant genes, *Sht1* and *Sht2*. The *Sht1* locus is linked to the *S* locus; almost all common buckwheat cultivars possess the allele *Sht1* (Matsui et al., 2004). It has been reported that this grain-shattering problem is mainly observed in wild buckwheat because of the formation of an abscission layer across the pedicle (Oba et al., 1998; Wang et al., 2005). However, cultivated buckwheat also exhibits a shattering problem under unfavorable conditions because of small pedicle diameter. In addition, it has also been reported that tetraploids possess a lower grain-shattering problem than the diploid ones, although the reason for differences of grain shattering are not clear (Hayashi, 1992; Fujimura et al., 2001). Moreover, the seed-shattering and nonshattering forms of buckwheat exhibit distinct geographical distributions. Seed shattering (wild and weedy forms of buckwheat) is distributed from Sichuan Province in China to Tibet to Kashmir and Northern Pakistan (Ohnishi, 1998a,b) while nonshattering Tartary buckwheat types are cultivated in the Himalayan regions and in Southern China at high-altitude areas of 1200–3000 m (Ohnishi, 1992; Wang and Campbell, 2004); also they are distributed as weedy forms in northwest directions to Europe (Kump and Javornik, 2002; Romanova, 2004). Fesenko (2006) carried out genetic analysis between nonshattering (*F. tataricum* Gaertn.) and shattering (*F. tataricum* Potanini) accessions and reported that all the nonshattering lineages carry recessive alleles at two independently segregating loci that are responsible for affecting the development of a functional abscission layer. Recessive homozygotes at any of the loci suppressed the expression of the dominant allele at the other one (recessive epistasis). Moreover, seed shattering exhibits a linked inheritance with the number of vegetative nodes on the main stem. In a wide crossing program it has been revealed that the accessions of *F. esculentum* possess recessive alleles at two or three loci that affect abscission layer development (Wang et al., 2005). Therefore in *F. esculentum*, three or four genes affecting development of the abscission layer have been reported. However, nonshattering populations of common buckwheat carry a single recessive gene providing shattering resistance. In the case of *F. tataricum* at least two such genes have been reported. Mutation of either of these genes leads to a nonshattering phenotype. Currently, several varieties of buckwheat are reported that are resistant to shattering. Suzuki et al. (2012)

reported that the green flower mutant of buckwheat (W/SK86GF) is resistant to grain shattering. Similarly, *Fagopyrum gigantum*, which is the cross-hybridized product of *F. tataricum* and *F. cymosum*, also exhibits resistance to grain shattering. Funatsuki et al. (2000) reported that the Russian variety "Green flower" is considered a useful genetic resource of seed-shattering resistance because of strong pedicles (by increasing number of vessels) than normal buckwheat (Alekseeva et al., 1988). Further, upon genetic analysis of morphological characters it has been observed that homostyly, shattering, and acute achenes are controlled by different dominant single genes. The shattering gene has a linkage to the homostyly locus at a rate of 7.81% and to the allele PS44/PS42.9 at a rate of 22.54% (Li and Chen, 2014).

Flower Abortion

Flower abortion is one of the main limiting factors in *F. esculentum*, which reduces its yield. Even after removing the possible effects of an inadequate pollination by insect visitors through cross-pollination, a proportion of flowers do not set seeds (Cawoy et al., 2007a,b). It has been revealed that there are certain congenital defects such as male or female sterility, embryo abortion, and self-incompatibility that are responsible for failure in seed production (Woo et al., 2004). Further, it has been reported that there are at least three stages at which development to mature seeds may fail in buckwheat: (1) abnormal flower morphogenesis (before anthesis) resulting in undersized and sterile gynoecia, (2) early failure of seed set in pollinated flowers with normal gynoecia (during or shortly after anthesis and pollination), and (3) most importantly, indeterminate growth habit caused by the morphogenetic processes that are involved in the establishment of reproductive structures. However, it has been reported that pollen is not considered as a limiting factor because male sterility seems marginal since pollen viability is high (>90%) at moderate temperatures (Cawoy et al., 2006b). Bjorkman (1995) reported that deficiency in female reproductive structures is considered as a limiting factor in flower abortion. About 35–97% of the flowers do not set seeds even when the possible effect of an inadequate pollination by insects is eliminated by means of hand pollination with exclusively compatible pollen (Cawoy et al., 2007a). In experiments without pollen limitations, Taylor and Obendorf (2001) reported that 76–91% of pollinated flowers did not reveal any sign of fertilization. Two causes can explain the seed production failure: male or female sterility/embryo abortion. The limitations in female reproductive structures depend on their location on the plant as well as age of the plant (Cawoy et al., 2007a). Further, Cawoy et al. (2007a,b) reported that flowers within inflorescences do not possess an equal chance for seed set and also exhibit a declining trend from the proximal to the distal end in the reproductive structures. Moreover, the first flowers in anthesis are also more likely to produce seeds than the later ones (Taylor and Obendorf, 2001). Moreover, it is seen that over time the size of gynoecia declines and there is a presence of abnormal undersized pistils in numerous flowers (Samborborska-Ciana et al., 1989; Nagatomo and Adachi, 1985).

Michiyama et al. (1999) reported that seed sets are enhanced to about 3–10% when all the inflorescence are excised except the terminal one. Furthermore, in plants where half the inflorescences are excised, compensation mechanisms are set up at the level of the meristems with a 25% increment in flower number and about 10–15% enhancement in seed set (Halbrecq et al., 2005).

Abiotic Stress Conditions Responsible for Low Production Yield in Buckwheat

Buckwheat production in the main growing countries has declined during the last few decades. Several reasons for this decline include: (1) failure to benefit from modern agriculture improvements (Adachi, 1990), (2) erratic yield linked to an indeterminate growth habit and the occurrence of defective reproductive organs (Taylor and Obendorf, 2001; Quinet et al., 2004), and (3) sensitivity to abiotic factors such as drought, frost, temperature, water deficit, and photoperiod (Delperee et al., 2003; Jacquemart et al., 2012).

Temperature Stress

Buckwheat cultivation is mostly feasible under temperate climates. The optimal temperature for germination is 10°C (Kalinova and Moudry, 2003). The developmental stage most sensitive to frost is the one to two leaf stage. At this stage, exposure for 4–6 h to temperatures between −1 and −3°C is lethal (Kalinova and Moudry, 2003). Temperature also influences both development and fertility of the reproductive structures especially during the flowering period (Bjorkman, 2000). The optimum temperature range is from 18 to 23°C. Below 15°C, flowering is inhibited with delayed initiation and reduction in flower numbers. Slawinska and Obendorf (2001) reported that a low temperature of 10°C induces early withering of flowers. At the same time high temperatures (30°C) and dry winds have a negative impact causing flower abortion, withering, and embryo sac malformation (Gang and Yu, 1998). Moreover, viability of the pollen grain is also dependent on temperature and relative humidity. Adhikari and Campbell (1998) reported that pollen placed at 23°C under dry air lost viability within an hour. Similarly, it was reported that a 45% reduction in fertilization events occurs with pollen originating from plants grown at 25°C compared to plants grown at 18°C (Slawinska and Obendorf, 2001). Moreover, high temperature (30°C) also enhances lodging as a result of a reduction of stem diameter (Michiyama and Sakaurai, 1999).

Photoperiod

Buckwheat initiates flowering over a wide range of day length (Nagatomo and Adachi, 1985) and even under continuous light (Hao et al., 1995); in addition, it also depends on the particular variety (Romanova and Koshkin, 2010; Hara et al., 2011). It has been reported that the varieties that are photosensitive initiate flowering at lower nodes and produce higher ratios of inflorescence and flowers per inflorescence under short days compared to long days. Also the long days cause a decline in the percentage of fertilized flowers but do not affect the seed weight (Michiyama et al., 2003). This could be because of the alterations in a source–sink relationship causing increasing competition for resources between vegetative and generative organs (Hagiwara et al., 1998).

Water Stress

Buckwheat needs sufficient amounts of water to produce good yields (Gang and Yu, 1998). It has been reported that to produce 1 kg of seeds, buckwheat requires approximately 225–315 kg of water (Gang and Yu, 1998). Because of the shallow root system of buckwheat, it quickly undergoes withering under drought conditions (Campbell, 1997). However, rehydration quickly takes place when normal water conditions are restored and the growth resumes (Delperee et al., 2003). During flowering and seed set, limited water supply is not suitable for endosperm development and may induce embryo abortion (Kalinova et al., 2002). The fertility loss can cause a 50% decline in yield. Sugimoto and Sato (2000) reported that flooding for more than 10 days at the flowering stage or more than 3 days during the maturation stage affects seed weight. Moreover, because of water lodging the main stem remains in contact with the soil, becoming prone to diseases and rotting (Sangma and Chrungoo, 2010). Flooding at the first leaf stage or flowering stage reduces yield by lowering flower production (Jacquemart et al., 2012). Reports revealed that buckwheat undergoes lodging during strong winds and when there is abundant precipitation especially in soils rich in nitrogen content (Schulte auf'mErley et al., 2005).

Conclusion and Future Perspective

Currently, buckwheat is receiving increased attention as a neutraceutical food crop because of certain unique active principal components. However, progress in cultivar breeding and crop management is still needed. Buckwheat has immense potential in crop improvement programs as reported by various research agencies like the International Plant Genetic Resources Institute and the Consultative Group on International Agriculture. Nowadays, research needs to be focused on buckwheat genetic analysis that could help to identify better agronomic traits such as low seed shedding, enhanced nutrient quality, self-compatibility, and homostyly in varieties under development. Moreover, molecular intervention in this crop has a vast scope that will offer vital cues to the factors involved in its low yield. It can also be developed for cultivation on marginal lands because of its tolerance toward abiotic stresses, and this strategy would diminish its competition for land with other high-yielding staple crops.

The genetic improvement of buckwheat is an important area to determine its potential for rapid distribution through seeds, propagules, etc.; it can also contribute toward understanding the diverse aspects of crop biology. These include improved agricultural properties such as resource use efficiency or disease resistance and improvement in harvested products such as nutritional content, storage performance, or processing properties for both food and other uses. In this direction, a nonconventional biotechnological approach has been found quite useful for the improvement of desired traits in a number of crop plants. Similarly, these approaches could be used for addressing the productivity and yield-related problems in buckwheat.

REFERENCES

Adachi, T., 1990. How to combine the reproductive system with biotechnology in order to overcome the breeding barrier in buckwheat. Fagopyrum 10, 7–11.

Adhikari, K.N., Campbell, C.G., 1998. In vitro germination and viability of buckwheat (*Fagopyrum esculentum* Moench) pollen. Euphytica 102, 87–92.

Alekseeva, E.S., Malikov, V.G., Falendysh, L.G., 1988. Green flower form of buckwheat. Fagopyrum 8, 79–82.

Alekseyeva, E.S., Bureyko, A.L., 2000. Bee visitation, nectar productivity and pollen efficiency of common buckwheat. Fagopyrum 17, 77–80.

Aufhammer, W., Fujimoto, F., Yasue, T., 1994. Development and utilization of the seed yield potential of buckwheat (*F. esculentum*). Bodenkultur 45, 37–47.

Berglund, Duane, R., 2003. Buckwheat production. NDSU Extension Service Publication A-687. Retrieved from http://www.ag.ndsu.edu/pubs/plantsci/crops/a687w.htm

Bernath, J., 2000. Medicinal and Aromatic Plants. Mezo Publication, Budapest, p. 667.

Bhat, M.R., Bhali, R.K., Tahir, I., 1986. Predator complex of melon aphid (*Aphis gossypii* Glov.), a serious pest of buckwheat (*Fagopyrum* spp.) in Kashmir (India). Buckwheat Newsletter, Fagopyrum 6, 12.

Bjorkman, T., 1995. The effect of pollen load and pollen grain competition on fertilization success and progeny performance in *Fagopyrum esculentum*. Euphytica 83, 47–52.

Bjorkman, T., 2000. Buckwheat production. Guide to buckwheat production in the northeast. http://www.nysaes.cornell.edu/hort/faculty/Bjorkman/buck/Buck.html

Björkman, T., 2012. Northeast Buckwheat Growers Newsletter, No. 33.

Campbell, C.G., 1997. Buckwheat: *Fagopyrum esculentum* Moench. Promoting the conservation and use of underutilized and neglected crops, 19. Rome: International Plant Genetic Resources Institute.

Campbell, C.G., Gubbels, G.H., 1978. Growing Buckwheat. Agriculture Canada Publication, 1468.

Cawoy, V., Deblauwe, V., Halbrecq, B., Ledent, J.F., Kinet, J.M., Jacquemart, A.L., 2006a. Morph differences and honey bee morph preference in the distylous species *Fagopyrum esculentum* Moench. Int. J. Plant Sci. 167, 853–861.

Cawoy, V., Halbrecq, B., Jacquemart, A.L., Lutts, S., Kinet, J.M., Ledent, J.F., 2007a. Genesis of grain yield in buckwheat (*Fagopyrum esculentum* Moench) with a special attention to the low seed set. In: Yan, C., Zongwen, Z. (Eds.), Current Advances in Buckwheat Research. Proceedings of the 10th International Symposium on Buckwheat. Yangling, China, 14–18, Aug. 2007, pp. 111–119.

Cawoy, V., Kinet, J.M., Jacquemart, A.L., 2008. Morphology of nectaries and biology of nectar production in the distylous species *Fagopyrum esculentum* Moench. Ann. Bot. 102, 675–684.

Cawoy, V., Ledent, J.F., Kinet, J.M., Jacquemart, A.L., 2009. Floral biology of common buckwheat (*Fagopyrum esculentum* Moench). Eur. J. Plant Sci. Biotechnol. 3, 1–9.

Cawoy, V., Lutts, S., Kinet, J.M., 2006b. Osmotic stress at seedling stage impairs reproductive development in buckwheat (*Fagopyrum esculentum* Moench). Physiol. Plant. 128, 689–700.

Cawoy, V., Lutts, S., Ledent, J.F., Kinet, J.M., 2007b. Resource availability regulates reproductive meristem activity, development of reproductive structures and seed set in buckwheat (*Fagopyrum esculentum* Moench). Physiol. Plant. 131, 341–353.

Ciepielowska, D., Fornal, L., 2004. Natural resistance of buckwheat seeds and products to storage pests. In: Faberová, I., Dvořáček, V., Čepková, P., Hon, I., Holubec, V., Stehno, Z. (Eds.), Advances in Buckwheat Research. Proceedings of the 9th International Symposium on Buckwheat. Research Institute of Crop Production, Prague, pp. 639–645.

Creamer, N.G., Baldwin, K.R., 1999. Summer cover crops. Horticulture Information Leaflets (HIL-37). North Carolina Cooperative Extension, N.C. State University, Raleigh, N.C. http://www.ces.ncsu.edu/depts/hort/hil/hil-37.html

Delperee, C., Kinet, J.M., Lutts, S., 2003. Low irradiance modifies the effect of water stress on survival and growth related parameters during the early developmental stages of buckwheat (*Fagopyrum esculentum* Moench). Physiol. Plant. 119, 211–220.

Edwardsen, S.E., 1995. Using growing degree days to estimate optimum windrowing time in buckwheat. In: Matano, T., Ujihara, A. (Eds.), Current Advances in Buckwheat Research: 6th International Symposium on Buckwheat in Shinshu, 24–29. Japan: Shinshu University, Nagano, pp. 509–514.

Edwardsen, S., 1996. Buckwheat: pseudocereal and nutraceutical. In: Janick, J. (Ed.), Progress in New Crops. ASHS Press, Alexandria, VA, USA, pp. 195–207.

FAO, 2005. The state of food insecurity in the world 2004. FAO. Rome.

FAOSTAT, 2013. FAO Statistical Databases. Rome: FAO. Available online at http://faostat. fao.org/

Farooq, S., Tahir, I., 1982. Grain characteristics and composition of some buckwheat (*Fagopyrum* Gaertn.) cultivated in Kashmir. J. Econ. Tax. Bot. 3, 877–881.

Farooq, S., Tahir, I., 1987. Comparative study of some growth attributes in buckwheats (*Fagopyrum* spp.) cultivated in Kashmir. Buckwheat Newsletter, Fagopyrum 7, 9–12.

Fesenko, I.N., 2006. Non-shattering accessions of *Fagopyrum tataricum* Gaertn. carry recessive alleles at two loci affecting development of functional abscission layer. Fagopyrum 23, 7–10.

Fujimura, Y., Oba, S., Horiguchi, T., 2001. Effects of fertilization and poliploidy on grain shedding habit of cultivated buckwheats (*Fagopyrum* spp.). J. Crop Sci. 70, 221–225.

Funatsuki, H., Maruyama-Funatsuki, W., Fujino, K., Agatsuma, M., 2000. Ripening habit of buckwheat. Crop Sci. 40, 1103–1108.

Gang, Z., Yu, T., 1998.A primary study of increasing the production rate of buckwheat. In: Campbell, C., Przybylski, R. (Eds.), Current Advances in Buckwheat Research. Proceedings of the 7th International Symposium on Buckwheat, Winnipeg, Manitoba, Canada, Aug. 12–14, pp. 18–23.

Gardner, F.P., Pearce, R.B., Mitchell, R.L., 1985. Mineral nutrition. Physiology of Crop Plants. Iowa State Univ. Press, Ames, USA.

Geneau, C.E., Wackers, F.L., Luka, H., Daniel, C., Balmer, O., 2011. Selective flowers to enhance biological control of cabbage pests by parasitoids. Basic Appl. Ecol. 13, 85–93.

Gondola, I., Papp, P.P., 2010. Origin, geographical distribution and phylogenic relationships of common buckwheat (*Fagopyrum esculentum* Moench.). In: Dobranszki, J. (Ed.), Buckwheat 2. Eur.J. Plant Sci. Biotechnol. 4 (Special Issue 2), pp. 17–32.

Goodman, R., Hepworth, G., Kaczynski, P., McKee, B., Clarke, S., Bluett, C., 2001. Honeybee pollination of buckwheat (*Fagopyrum esculentum* Moench) cv. Manor. Aust. J. Exp. Agric. 41, 1217–1221.

Hagiwara, M., Inoue, N., Matano, T., 1998. Variability in the length of flower bud differentiation period of common buckwheat. Fagopyrum 15, 55–64.

Halbrecq, B., Romedenne, P., Ledent, J.F., 2005. Evolution of flowering, ripening and seed set in buckwheat (*Fagopyrum esculentum* Moench): quantitative analysis. Eur. J. Agron. 23, 209–224.

Hao, X., Li, G., Yang, W., Zhou, N., Lin, R., Zhou, M., 1995. The difference and classification of light reaction of buckwheat under different treatments of light duration. In: Matano, T., Ujihara, A. (Eds.), Current Advances in Buckwheat Research. Proceedings of the 6th International Symposium on Buckwheat, Aug. 24–29, Shinshu University Press, Shinshu, Japan, pp. 541–549.

Hara, T., Iwata, H., Okuno, K., Matsui, K., Obsawa, R., 2011. QTL analysis of photoperiod sensitivity in common buckwheat by using markers for expressed sequence tags and photoperiod-sensitivity candidate genes. Breed. Sci. 61, 394–404.

Hayashi, H., 1992. Changes of factors involving in decision of harvest of buckwheat during grain maturation. Hokuriku Crop Sci. 28, 81–82.

Hore, D., Rathic, R.S., 2002. Collection, cultivation and characterization of buckwheat in northeastern region of India. Fagopyrum 19, 11–15.

Ikanović, J., Rakić, S., Popović, V., Janković, S., Glamočlija, Đ., Kuzevski, J., 2013. Agroecological conditions and morpho-productive properties of buckwheat. Biotechnol. Anim. Husband. 29 (3), 555–562.

Inamullah, Saqib, G., Ayub, M., Khan, A.A., Anwar, S., Khan, S.A., 2012. Response of common buckwheat to nitrogen and phosphorus fertilization. Sarhad J. Agric. 28 (2), 171–178.

Iqbal, Z., Hiradate, S., Noda, A., Isojima, S., Fujii, Y., 2003. Allelopathic activity of buckwheat: isolation and characterization of phenolics. Weed Sci. 51 (5), 657–662.

Jacquemart, A.L., Gillet, C., Cawoy, V., 2007. Floral visitors and importance of honey bee on buckwheat (*Fagopyrum esculentum* Moench) in central Belgium. J. Hort. Sci. Biotechnol. 82, 104–108.

Jacquemart, A.L., Cawoy, V., Kinet, J.M., Ledent, J.F., Quinet, M., 2012. Is buckwheat (*Fagopyrum esculentum* Moench) still a valuable crop today. Eur. J. Plant Sci. Biotechnol. 6 (Special Issue 2), 1–10.

Janovska, D., Kalinova, J., Michalova, A., 2009. Metodikapěstovánípohanky v ekologickém a konvenčnímzemědělství.Metodika pro praxi.Praha 6—Ruzyně: Výzkumnýústavrostlinnévýroby, v.v.i.

Kalinova, J., Moudry, J., 2003. Evaluation of frost resistance in varieties of common buckwheat (*Fagopyrum esculentum* Moench). Plant Soil Environ. 49, 410–413.

Kalinova, J., Moudry, J., Curn, V., 2002. Technological quality of common buckwheat (*Fagopyrum esculentum* Moench). Rostlinna Vyroba 48, 279–284.

Klykov, A.G., Anisimov, M.M., Moiseenko, L.M., Chaikina, E.L., Parskaya, N.S., 2014. Effect of biologically active substances on morphological characteristics, rutin content and productivity of *Fagopyrum esculentum* Moench. Agric. Sci. Devel. 3 (1), 139–142.

Krkoskova, B., Mrazova, Z., 2005. Prophylactic components of buckwheat. Food Res. Int. 38, 561–568.

Kumar, V., Brainard, D.C., Bellinder, R.R., 2008. Suppression of Powell amaranth (*Amaranthus powellii*), shepherd's-purse (*Capsella bursa-pastoris*), and corn chamomile (*Anthemis arvensis*) by buckwheat residues: role of nitrogen and fungal pathogens. Weed Sci. 56 (2), 271–280.

Kump, B., Javornik, B., 2002. Genetic diversity and relationships among cultivated and wild accession of Tartary buckwheat (*Fagopyrum tataricum* Gaertn.) as revealed by RAPD markers. Genet. Resour. Crop Evol. 49, 565–572.

Kuznetsova, A.V., Klykov, A.G., 2012. Efficiency of chemical and biological preparations in *Rhinoncus sibiricus* Faust control. J. Siberian Mess. Agric. Sci. 3, 25–29.

Kuznetsova, A.V., Klykov, A.G., Timoshinov, R.V., Moiseyenko, L.M., 2012. Harmfulness of *Rhinoncus sibiricus* Faust in Primorskykrai. J. Rep. Russ. Acad. Agric. Sci. 5, 35–38.

Kwiatkowski, J., Nietupski, M., 2013. Susceptibility of various buckwheat genotypes to wheat weevil (*Sitophilus granarius* L.) and lesser grain borer (*Rhizopertha dominica* F.) infestations. Proceedings of the 12th International Symposium on Buckwheat, Laško, Aug. 21–25. Pernica: Fagopyrum, pp. 186–187.

Lee, H.B., Kim, S.L., Park, C.H., 2001. Productivity of whole plant and rutin content under the different quality of light in buckwheat. Proceedings of the 8th Iinternational Symposium on Buckwheat, 84–89.

Lee, J.H., Aufhammer, W., Kubler, E., 1996. Produced, harvested and utilizable grain yield of the pseudocereals buckwheat (*Fagopyrum esculentum*, Moench), quinoa (*Chenopodium quinoa*, Wild) and amaranth (*Amaranthus hypochondriacus*, L × A. hybridus, L.) as affected by production techniques. Bodenkultur 47, 5–14.

Lee-Mader, E., Hopwood, J., Morandin, L., Vaughan, M., Black, S.H., 2014. Farming with Native Beneficial Insects. Storey Publishing, North Adams, MA.

Li, J.H., Chen, Q.F., 2014. Inheritance of seed protein subunits of common buckwheat (*Fagopyrum esculentum* Moench) cultivar Sobano and its homostylous wild type. J. Agric. Sci. 6, 6.

Li, S., Zhang, Q.H., 2001. Advances in the development of functional foods from buckwheat. Crit. Rev. Food Sci. Nutr. 41, 451–464.

Liu, J.L., Tang, Y., Xia, Z.M., Shao, J.R., Cai, G.Z., Luo, Q., 2008. *Fagopyrum crispatifolium* Liu, J L., a new species of Polygonaceae from Sichuan, China. J. Syst. Evol. Res. 46, 929–932.

Loch, J., Lazanyi, J., 2010. Soil nutrient content in buckwheat production. In: Dobranszki, J. (Ed.), Buckwheat 2. Eur. J. Plant Sci. Biotechnol. 4 (Special issue 1), 93–97.

Mader, E., Shepherd, M., Vaughan, M., Black, S.H., LeBuhn, G., 2011. Attracting Native Pollinators. Storey Publishing, North Adams, MA.

Matsui, K., Nishio, T., Tetsuka, T., 2004. Genes outside the S supergene suppress S functions in buckwheat (*Fagopyrum esculentum*). Ann. Bot. 94, 805–809.

Michiyama, H., Arikuni, M., Hirano, T., Hayashi, H., 2003. Influence of day light before and after the start of anthesis on the growth, flowering and seed setting in common buckwheat (*Fagopyrum esculentum* Moench). Plant Prod. Sci. 6, 235–242.

Michiyama, H., Fukui, A., Hayashi, H., 1998. Differences in the progression of successive flowering between summer and autumn ecotype cultivars in common buckwheat (*Fagopyrum esculentum* Moench). Jpn. J. Crop Sci. 67, 498–504.

Michiyama, H., Hayashi, H., 1998. Differences of growth and development between summer and autumn type-cultivars in common buckwheat (*Fagopyrum esculentum* Moench). Jpn. J. Crop Sci. 67, 323–330.

Michiyama, H., Sakaurai, S., 1999. Effect of day and night temperature on the growth and development of common buckwheat (*Fagopyrum esculentum* Moench). Jpn. J. Crop Sci. 68, 401–407.

Michiyama, H., Tachimoto, A., Hayashi, H., 1999. Effect of defloration and restriction of the number of flower clusters on the progression of successive flowering and seed-setting in common buckwheat (*Fagopyrum esculentum* Moench). Jpn. J. Crop Sci. 68, 91–94.

Munshi, A.H., 1982. A new species of *Fagopyrum* from Kashmir Himalaya. J. Econ. Tax. Bot. 3, 627–630.

Nagatomo, T., Adachi, T., 1985. Fagopyrum esculentum. Halevy, A.H. (Ed.), Handbook of Flowering, III, CRC Press, Boca Raton, FL, USA, pp. 1–8.

Naumkin, V., 2013. Complex of insects on buckwheat plantings. Proceedings of the 12th International Symposium on Buckwheat, Laško, Aug. 21–25. Pernica: Fagopyrum, pp. 125–126.

Oba, S., Suzuki, Y., Fujimoto, F., 1998. Breaking strength of pedicels and grain shattering habit in two species of buckwheat (Fagopyrum spp.). Plant Prod. Sci. 1 (1), 62–66.

Ohnishi, O., 1992. Buckwheat in Bhutan. Fagopyrum 12, 5–13.

Ohnishi, O., 1998a. Search for the wild ancestor of common buck wheat. 1. Description of new *Fagopyrum* (Polygonaceae) species and their distribution in China and Himalayan hills. Fagopyrum 15, 18–28.

Ohnishi, O., 1998b. Search for the wild ancestor of common buckwheat. III. The wild ancestor of cultivated common buckwheat, and of Tartary buckwheat. Econ. Bot. 52, 123–133.

Olejarski, P., Ignatowicz, S., 2011. Integrowanametodazwalczaniaszkodnikówmagazyno wychpodstawązapewnieniawysokiejjakościprzechowywanegoziarnazbóż [IPM as a principle to ensure the high quality of stored grain]. Progress in Plant Protection/Postępy w OchronieRoślin 51 (4), 1879–1885.

Olson, M., 2001. Common buckwheat, agri-facts, agriculture, food and rural management. Alberta, Canada. http://www1.agric.gov.ab.ca/$department/deptdocs.nsf/all. agdex103/$file.118_20-2.pdf? Open Element

Omidbaigi, R., Mostra, G.D., Bahrami, K., 2004. Influence of nitrogen and phosphorus fertilization on the grain characteristics of buckwheat (*Fagopyrum esculentum* Moench). Proceedings of the 9th International Symposium on Buckwheat, Prague, pp. 457–460.

Ornelas, J.F., Gonzales, C., Jimenez, L., Lara, C., Martinez, A.J., 2004. Reproductive ecology of distylous *Palicourea padifolia* (Rubiaceae) in a tropical montane cloud forest. II. Attracting and rewarding mutualistic and antagonistic visitors. Amer. J. Bot. 91, 1061–1069.

Pavek, P.L.S., 2014. Evaluation of cover crops and plantings dates for dryland Eastern Washington rotations. Plant Materials Technical Note No. 25. United States Department of Agriculture—Natural Resources Conservation Service. Spokane, WA.

Petrich, C., 2000. Phosphorus mobilization and weed suppression by buckwheat. Cropping systems and soil fertility. Minnesota Department of Agriculture. http://.www.mda.state.mn.us

Phogat, B.S., Sharma, G.D., 2000. Under-utilized food crops: their uses, adaptation and production technology. Technical Bulletin, pp. 10–15, Jun. NBPGR (ICAR), New Delhi.

Pirzadah, T.B., Malik, B., Tahir, I., Rehman, R.U., 2013. Buckwheat: an introspective and future perspective in Kashmir Himalayas. Proceedings of the 12th International Symposium on Buckwheat, Laško, Aug. 21–25. Pernica: Fagopyrum, pp. 212–215.

Popović, V., Sikora, V., Berenji, J., Glamočlija, Đ., Marić, V., 2013a. Effect of agroecological factors on buckwheat yield in convential and organic cropping systems, Institute of PKB Agroeconomik, Belgrade, 19(1–2), pp. 155–165.

Popović, V., Sikora, V., Ikanovic, J., Rajičič, V., Maksimović, L., Katanski, S., 2013b. Production, productivity and quality of buckwheat in organic growing systems in course environmental protection. XVII Eco-Conference, Novi Sad, Sep. 25–28, pp. 395–404.

Qin, P., Tingjun, M., Li, W., Fang, S., Guixing, R., 2011. Identification of Tartary buckwheat tea aroma compounds with gas chromatography-mass spectrometry. J. Food Sci. 76, S401–S407.

Quinet, M., Cawoy, V., Lefevre, I., Van Miegroet, F., Jacquemart, A.L., Kinet, J.M., 2004. Inflorescence structure and control of flowering time and duration by light in buckwheat (*Fagopyrum esculentum* Moench). J. Exp. Bot. 55, 1509–1517.

Racys, J., Montvilienne, R., 2005. Effect of bee-pollinators in buckwheat (*Fagopyrum esculentum* Moench) crops. J. Apicult. Sci. 49, 47–51.

Radics, L., Mikohazi, D., 2010. Principles of common buckwheat production. In: Dobranszki, J. (Ed.), Buckwheat 2. Eur. J. Plant Sci. Biotechnol. 4 (Special issue 1), 57–63.

Rana, J.C., Chauhan, R.C., Sharma, T.R., Gupta, N., 2012. Analyzing problems and prospects of buckwheat cultivation in India. Eur. J. Plant Sci. Biotechnol. 6 (2), 50–56.

Ratan, P., Kothiyal, P., 2011. *Fagopyrum esculentum* Moench (common buckwheat) edible plant of Himalayas: a review. Asian J. Pharm. Life Sci. 1 (4), 426–442.

Romanova, O., 2004. Northern populations of Tartary buckwheat with respect to day length. Proceedings of the 9th International Symposium on Buckwheat at Prague: 173–178.

Romanova, O., Koshkin, V., 2010. Photoperiod response of landraces and improved varieties of buckwheat from Russia and from the main buckwheat cultivating countries. In: Dobranszki, J. (Ed.), Buckwheat 2. Eur. J. Plant Sci. Biotechnol. 4 (Special Issue 1), 123–127.

Saini, J.P., Negi, S.C., 1998. Effect of spacing and nitrogen on Indian buckwheat (*Fagopyrum esculentum* Moench.) under dry temperate conditions. Ind. J. Agron. 43, 351–354.

Samborborska-Ciana, A., Januszewicz, E., Ojczyk, T., 1989. The morphology of buckwheat flowers depending on the course of plant flowering. Fagopyrum 9, 23–26.

Sangma, S.C., Chrungoo, N.K., 2010. Buckwheat gene pool: potentialities and drawbacks for use in crop improvement programmes. In: Dobranszki, J. (Ed.), Buckwheat 2. Eur. J. Plant Sci. Biotechnol. 4 (Special Issue 1), 45–50.

Schulte auf'mErley, G., Kaul, H.P., Kruse, M., Aufhammer, W., 2005. Yield and nitrogen utilization efficiency of the pseudocereals amaranth, quinoa and buckwheat under differing nitrogen fertilization. Eur. J. Agron. 22, 95–100.

Shao, J.R., Zhou, M.L., Zhu, X.M., Wang, D.Z., Bai, D.Q., 2011. *Fagopyrum wenchuanense* and *Fagopyrum qiangcai*, two new species of Polygonaceae from Sichuan, China. Novon 21, 256–261.

Skrabanja, V., Kreft, I., Golob, T., Modic, M., Ikeda, S., Ikeda, K., Kreft, S., Bonafaccia, G., Knapp, M., Kosmel, K.J., 2004. Nutrient content in buckwheat milling fractions. Cereal Chem. 81, 172–176.

Slawinska, J., Obendorf, R.L., 2001. Buckwheat seed set in planta and during in vitro inflorescence culture: evaluation of temperature and water deficit stress. Seed Sci. Res. 11, 223–233.

Sobhani, M.R., Rahmikhdoev, G., Mazaheri, D., Majidian, M., 2014. Influence of different sowing date and planting pattern and N rate on buckwheat yield and its quality. Aust. J. Crop Sci. 8 (10), 1402–1414.

Sugawara, K., 1956. On buckwheat pollen. III. The relation between pollen germination and temperature. Proc. Crop Sci. Soc. Japan 24, 264–265.

Sugimoto, H., Sato, T., 2000. Effects of excessive soil moisture at different growth stages on seed yield of summer buckwheat. Jpn. J. Crop Sci. 69, 189–193.

Suzuki, T., Mukasa, Y., Morishita, T., Takigawa, S., Noda, T., 2012. Traits of shattering resistant buckwheat "W/SK86GF". Breed. Sci. 62, 360–364.

Tahir, I., Farooq, S., 1987. Occurrence of phenolic compounds in four cultivated buckwheats with particular reference to grain quality. Buckwheat Newsletter, Fagopyrum 7, 7–8.

Tahir, I., Farooq, S., 1988. Review article on buckwheat. Buckwheat Newsletter, Fagopyrum 8, 33–53.

Tahir, I., Farooq, S., 1989a. Grain and leaf characteristics of perennial buckwheat (*F. cymosum* Meissn). Buckwheat Newsletter, Fagopyrum 9, 41–43.

Tahir, I., Farooq, S., 1989b. Some morpho-physiological characteristics in four buckwheats (*Fagopyrum* spp.) grown in Kashmir. J. Econ. Tax. Bot. 13 (2), 433–436.

Tahir, I., Farooq, S., 1990. Growth analysis and yield in buckwheats (*Fagopyrum* spp.) grown in Kashmir. Acta Physiol. Plant. 12 (4), 311–324.

Tahir, I., Farooq, S., Bhat, M.R., 1985. Insect pollinators and pests associated with cultivated buckwheat in Kashmir (India). Fagopyrum 5, 3–5.

Tang, Y., Zhou, M.L., Bai, D.Q., Shao, J.R., Zhu, X.M., Wang, D.Z., et al., 2010. *Fagopyrum pugense* (Polygonaceae), a new species from Sichuan, China. Novon 20, 239–242.

Taylor, D.P., Obendorf, R.L., 2001. Quantitative assessment of some factors limiting seed set in buckwheat. Crop Sci. 41, 1792–1799.

Teboh, J.M., Franzen, D.W., 2011. Buckwheat (*Fagopyrum esculentum* Moench) potential to contribute solubilized soil phosphorus to subsequent crops. Commun. Soil Sci. Plant Anal. 42, 1544–1550.

Verardo, V., Arráez-Román, D., Segura-Carretero, A., Marconi, E., Fernández-Gutiérrez, A., Caboni, M.F., 2011. Identification of buckwheat phenolic compounds by reverse phase high performance liquid chromatography-electrospray ionization-time of flight-mass spectrometry (RP-HPLC-ESI -TOF-MS). J. Cereal Sci. 52, 170–176.

Wang, Y., Campbell, C.G., 2004. Buckwheat production, utilization, and research in China. Fagopyrum 21, 123–133.

Wang, Y., Scarth, R., Campbell, G.C., 2005. Inheritance of seed shattering in interspecific hybrids between *Fagopyrum esculentum* and *F. homotropicum*. Crop Sci. 45, 693–697.

Warman, P.R., 1991. Effect of incorporated green manure crops on subsequent oat production in an acid, infertile silt loam. Plant Soil 134 (1), 115–119.

Woo, S.H., Omoto, T., Kim, H.S., Park, C.R., Campbell, C., Adachi, T., Jong, S.K., 2004. Breeding improvement of processing buckwheat: prospects and problems of interspecific hybridation. Edited by Faberova, I., Dvoracek, V., Cepkova, P., Hon, I., Holubec, V., and Stehno, Z. Advances in Buckwheat Research (I). Proceedings of the 9th International Symposium on Buckwheat, Aug. 18–22, 2004. Prague. Czech Republic. 35Q.354.

Zadernowski, R., Pierzynowska-Korniak, G., Ciepielewska, D., Fornal, L., 1992. Chemical characteristic and biological functions of phenolic acids of buckwheat and lentil seeds. Fagopyrum 12, 27–35.

Zhou, M.L., Bai, D.Q., Tang, Y., Zhu, X.M., Shao, J.R., 2012. Genetic diversity of four new species related to southwestern Sichuan buckwheats as revealed by karyotype, ISSR and allozyme characterization. Plant System. Evol. 298, 751–759.

Zhu, Y.G., He, Y.Q., Smith, S.E., Smith, F.A., 2002. Buckwheat (*Fagopyrum esculentum* Moench) has high capacity to take up phosphorus (P) from calcium (Ca)-bound source. Plant Soil 239 (1), 1–8.

Chapter twenty four

Cultivation of Buckwheat in China

F.-L. Li*, M.-Q. Ding**,†, Y. Tang**,†,‡, Y.-X. Tang**, Y.-M. Wu**, J.-R. Shao†,‡, M.-L. Zhou**

*Xichang Institute of Agricultural Science, Alpine Crop Research Station, Xichang, Sichuan, China; **Biotechnology Research Institute, Chinese Academy of Agricultural Sciences, Beijing, China; †School of Life Sciences, Sichuan Agricultural University, Yaan, Sichuan, China; ‡Department of Food Science, Sichuan Tourism University, Chengdu, Sichuan, China

THE CULTURAL HISTORY OF BUCKWHEAT

Buckwheat is native to China. The first description of buckwheat in Chinese books goes back thousands of years. One of China's famous agriculturalists, Ying Ding (1921 and 1928), wrote in his book: there are two types of Chinese buckwheat, the Common Buckwheat (CB) and Tartary Buckwheat (TB). The CB is native to the north of China, being cultivated before the Northern and Southern Dynasties (420–581 AD). It was popular in the central plains during the Tang and Song Dynasties and spread to the south of China. As for the TB, which is native to the southwest of China, since the Yuan and Ming Dynasties, it was widely cultivated in the southwest of China. Rui-Ting Guan (1987) supposed that the initial period of buckwheat cultivation was no later than the West Han Dynasty (1 BC). After gradually developing through the West Han, Wei, Jing, and the Southern and Northern Dynasties, cultivated buckwheat formed a certain scale in the early Tang Dynasty, and by continuously developing in the Song and Yuan Dynasties, it was widely planted in the south and north of China and became a staple food in certain areas.

The history of Chinese buckwheat cultivation goes back more than 1500 years; the ancient people attached importance to buckwheat for its short growth period, high adaptability, and resistance to barren conditions. Buckwheat can make full use of light, heat, water, and soil resources and can be planted in spring, summer, and fall; these characters increase the multiple cropping index and have great benefit to disaster prevention and reduction.

321

Molecular Breeding and Nutritional Aspects of Buckwheat. http://dx.doi.org/10.1016/B978-0-12-803692-1.00024-9

INTRODUCTION OF CHINESE BUCKWHEAT VARIETIES

Before the 1990s, the breeding of buckwheat did not work well in China. The cultivated species in different districts normally adjusted to their own surroundings and their production was hardly improved. Since the 1990s the work toward breeding improvement and different varieties promoted the production of buckwheat. According to the statistics, during 1998–2015, 45 varieties or species of buckwheat have been examined and approved by national or provincial committees. Thirty-one are national level (26 TBs and 5 CBs), 12 are province level (5 TBs and 7 CBs), and 2 are *Fagopyrum cymosum* Meisn species. Next, we will briefly describe the characters of some new varieties (Table 24.1) (Zhang and Lin, 2007).

F. cymosum belongs to *Fagopyrum* Mill, Polygonaceae family, a perennial herb known as a resource plant full of nutrients and medicinal values. *F. cymosum*'s remarkable effects inhibit the invasion and metastasis of malignant tumors, and have antiseptic and antiinflammatory qualities, making it a major component of cancer prevention and anticancer medicine (compound rhizoma fogopyri cymosi kernel, Jizhi syrup, Weimaining capsule, etc.). The aerial part of *F. cymosum* is rich in amino acids and trace elements such as Fe, Mn, Cu, Zn, etc., so it is a high-quality forage. Recently, scientists in Guizhou and Beijing used wild *F. cymosum* as a material to breed new varieties, which was approved by the provincial committee.

THE BUCKWHEAT CULTIVATED TECHNOLOGY IN CHINA

The Rotation System of Buckwheat

Crop rotation is important in agriculture, which means different crops growing in one piece of land in a certain order over a certain year; this behavior adjusts soil fertility, controls diseases, pests, and weeds, and provides high and stable yields. However, continuous cropping of buckwheat has led to a decrease of yield and quality, and is unfavorable regarding making full use of the soil resource. A farmers' proverb says, " with three year planting, buckwheat lost its edges" (in Dingbian, Jingbian in Shaanxi and Yanchi in Ningxia, etc.) and "continuous cropping make buckwheat become beard" (in Sichuan Liangshan); both sayings indicate that continuous cropping of buckwheat leads to serious consumption of soil nutrients, and fertility will be difficult to recover for several years.

Buckwheat can be planted in different rotation systems, but in the pursuit of high production, it is wiser to choose a more popular crop, such as bean, potato, and sweet potato, or millet, broomcorn, maize, wheat, oat, and tobacco. On the contrary, rape and leaf mustard are not good choices for crop rotation with buckwheat, because these crops use too many nutrients, especially phosphorus, which is necessary to increase phosphate fertilizer in buckwheat planting.

Buckwheat rotation systems, because of the different crop distributions, are different in different parts of China. In the high-altitude localities of the northwest, northeast, and North China, buckwheat normally rotates with naked oat and potato. In North China, buckwheat is usually planted as the later crop of winter wheat and potato, double cropping in a year or triple cropping in 2 years.

TABLE 24.1 The Characters of Some New Varieties

No.	Variety	Habitat	Growth Period (day)	Plant Height (cm)	1000-Grain Weight (g)	Resistance
1.	Jiujiang TB	Jiangxi	80	108.5	20.2	Lodging, drought, and barren
2.	Xiqiao No. 1 TB	Sichuan	75–80	90–105	19–20.5	Lodging, drought, shattering, and barren
3.	Chuanqiao No. 1 TB	Sichuan	80	90	20–21	Drought, cold, and lodging
4.	Fenghuang TB	Hunan	85–90	106	22–24	Drought, lodging, water logging, disease, and insects
5.	Xinong 9920 TB	Shaanxi	88	105.5	17.9	Lodging, drought, and barren
6.	Qianku No. 2 TB	Guizhou	80	90.5	21.8	Lodging, drought, shattering, and barren
7.	Qianku No. 4 TB	Guizhou	83	96.2	20.2	Lodging, drought, shattering, and barren
8.	Liuku No. 2 TB	Guizhou	90–100	100–130	20–22	
9.	Xinong 9909 TB	Shaanxi	85–95	110–120	17–20	Lodging, drought, shattering, and barren
10.	Qianku No. 3 TB	Guizhou	90	107.4	23.4	Lodging, drought, cold, disease, and barren
11.	Zhaoku No. 1 TB	Yunnan	90	108	21.9	Lodging, drought, shattering, and barren
12.	Jinqiaomai (ku) No. 2 TB	Shanxi	93	120.3	18	
13.	Qianku No. 5 TB	Guizhou	92	120.7	16.8	Lodging, drought, cold, disease, and barren
14.	Chuanqiao No. 3 TB	Sichuan	81–86	94.9	20.6	Lodging, drought, shattering, and barren
15.	Yunqiao No. 1 TB	Yunnan	83–88	101.7	17.4–20.5	
16.	Xinong 9940 TB	Shaanxi	92–94	106–112	19.2–21.9	Lodging, drought, shattering, and barren
17.	Diku No. 1 TB	Yunnan	87	98.7	20	

(*Continued*)

TABLE 24.1 The Characters of Some New Varieties (*cont.*)

No.	Variety	Habitat	Growth Period (day)	Plant Height (cm)	1000-Grain Weight (g)	Resistance
18.	Zhaoku No. 2 TB	Yunnan	80–89	80.6–122.3	20.3–21	Lodging and drought
19.	Fengku No. 3 TB	Hunan	86	106.1	18.0	
20.	Yunqiao No. 2 TB	Yunnan	84	118.3	20.7	
21.	Qianku No. 7 TB	Guizhou	84	109.3	18.0	
22.	Xiqiao No. 3 TB	Sichuan	79	107.5	20.1	Lodging, drought, disease, and barren
23.	Fengku No. 2 TB	Hunan	82	105.2	19.3	Lodging, drought, shattering, and barren
24.	Dingtianqiao No. 1 CB	Gansu	80	70–90	28	Lodging, drought, shattering, and barren
25.	Pingqiao No. 7 CB	Gansu	67–81	101–109	24.9–26.7	Drought, cold, and disease
26.	Qinghongqiao No. 1 CB	Gansu	70–80	95–106	26–27	Drought, cold, barren, disease, and shattering
27.	Yantianqiao No. 1 CB	Shaanxi	77	104.5	27.6	Lodging, shattering, and disease
28.	Qianjinqiaomai No. 1 *F. cymosum*	Guizhou	180–200 (per year)	100–150	45–51	Drought and disease
29.	Jinqiao No. 1 *F. cymosum*	Beijing	200–210 (per year)			Drought

In the low-altitude localities of South China, buckwheat is planted after spring crops to realize triple cropping in a year. In the subtropical zone, farmers rush to plant winter buckwheat after late-fall crops.

CONCLUDING REMARKS

The planting areas of buckwheat in China are mostly thinly populated areas with poor soil, cold weather, and locked traffic. Geographically speaking, they are the economically less-developed areas on the borders of China. People live in these places and plant buckwheat in traditional ways unmanaged; obviously, it is hard to increase the yield of buckwheat. Recently, with improvements in living standards, people have come to realize the unique nutritive value of buckwheat (Zhang et al., 2012), making the demand for buckwheat produce increase rapidly. Today, people focus on a healthy lifestyle and protection of the environment, so researchers are continuously improving the yield and quality of buckwheat to meet the growing demand in the market. Good agriculture practice could standardize buckwheat production and form a quality management system, so popularizing this technology is an ideal way to expand the buckwheat industry.

ACKNOWLEDGMENTS

This research was supported by the Key Project of Science and Technology of Sichuan, China (grant no. 04NG001-015, "Protection and exploitation of wild-type buckwheat germplasm resource").

REFERENCES

Ding, Y., 1921. The origin of Chinese crop (中国作物原始). Voice of Agriculture (农声), 82–85.

Ding, Y., 1928. The textual criticism of cereal name (谷类名实考). Voice of Agriculture (荞麦动态), 99–115.

Guan, R.T., 1987. Primary investigation of Chinese cultivated buckwheat origination. Res. Prog. Buckwheat (荞麦动态) 2, 2–4.

Zhang, Z.L., Zhou, M.L., Tang, Y., Li, F.L., Tang, Y.X., Shao, J.R., Xue, W.T., Wu, Y.M., 2012. Bioactive compounds in functional buckwheat food. Food Res. Int. 49, 389–395.

Zhang, Z.W., Lin, R.F., 2007. Regulation and Standard of Buckwheat Germplasm Resources, Description and Database. Chinese Agriculture Science and Technology Press, Beijing.

Chapter twenty five

Characterization of Functional Genes in Buckwheat

M.-L. Zhou*, G. Wieslander, Y. Tang†,
Y.-X. Tang*, J.-R. Shao‡, Y.-M. Wu***

*Biotechnology Research Institute, Chinese Academy of Agricultural Sciences, Beijing, China; **Department of Occupational and Environmental Medicine, Uppsala University, Uppsala, Sweden; †Department of Food Science, Sichuan Tourism University, Chengdu, Sichuan, China; ‡School of Life Sciences, Sichuan Agricultural University, Yaan, Sichuan,China

INTRODUCTION

Traditional plant breeding has produced most of the buckwheat varieties that we use today. The most widely grown buckwheat species include common buckwheat (CB, *Fagopyrum esculentum*) and Tartary buckwheat (TB, *Fagopyrum tataricum*) (Zhou et al., 2012). The lack of selection criteria and variation in responses of plants at different developmental stages and environmental stresses have resulted in only limited success, which is no longer sufficiently powerful to satisfy current and future needs for the production of buckwheat in the world. Contrary to the classical breeding and marker-assisted selection approaches, direct introduction of genes by genetic engineering seems a more attractive and quicker solution for breeding new varieties (Zhou et al., 2011). Therefore deciphering the molecular mechanism of buckwheat growth and development and response to environmental signals is of critical importance for the development of rational breeding. According to recent reports, many buckwheat genes are isolated and characterized by scientific researchers from all over the world. These genes are classified into functional and regulatory proteins according to the functions of their encoding products. In this chapter, recent progress on the genes and their roles in regulating buckwheat growth, development, and metabolism are discussed. This opens an excellent opportunity to develop stress tolerant and nutritionally improved buckwheat in the future.

Molecular Breeding and Nutritional Aspects of Buckwheat. http://dx.doi.org/10.1016/B978-0-12-803692-1.00025-0

ENVIRONMENTAL STRESSES TOLERANCE

Phytotoxicity of heavy metals, such as Al^{3+}, Cu^{2+}, and Cd^{2+}, is a serious problem limiting crop production in soils. Buckwheat is a short season crop that grows well on acidic soils and accumulates high Al in the leaves (Yokosho et al., 2014; Zhu et al., 2015). It has been reported that oxalate secretion of the roots is able to chelate ionic Al, which is one of the mechanisms for the high Al tolerance in buckwheat (Ma and Hiradate, 2000; Shen et al., 2002). Although much progress has been made in understanding the physiological mechanisms for Al tolerance, accumulation, and transportation in buckwheat, the corresponding genes have not been identified. One of the main reasons is the lack of genome sequence information. However, the transcriptomics or RNA sequencing (RNA-Seq) offers an efficient tool to facilitate gene discovery. A genome-wide transcriptome analysis of Al-responsive genes using RNA-Seq technology has been conducted in CB (Yokosho et al., 2014) and TB (Zhu et al., 2015), respectively. RNA-Seq of CB reveals that *FeSTAR1* (Sensitive To proton rhizotoxicity 1/Al Resistance transcription factor 1), *FeALS3* (ALuminum Sensitive 3), *FeALS1* (ALuminum Sensitive 1), *FeMATE1*, *FeMATE2* (Multidrug And Toxic Compound Extrusion 1 and 2), and also a number of transporter genes are highly expressed in the roots and leaves and responded to Al stress, indicating they function in Al tolerance and accumulation (Yokosho et al., 2014). Additionally, genes involved in organic acid metabolism are also upregulated in both CB and TB (Zhu et al., 2015). These data provide a reference platform for further characterizing the functions of genes involved in Al tolerance and accumulation in buckwheat. Buckwheat is also known as a Cu^{2+} and Zn^{2+} accumulator (Tani and Barrington, 2005). Metallothioneins are low molecular mass (4–10 kDa) proteins that bind metal ions, such as Zn^{2+}, Cu^{2+}, or Cd^{2+}, via the thiol groups of their cysteine residues (Cobbett and Goldsbrough, 2002). *FeMT3* (Metallothionein type 3) was the first reported gene related to heavy metal homeostasis (Nikolić et al., 2010). Gene expression analysis showed that *FeMT3* was highly induced by Zn^{2+} and Cu^{2+} treatments, indicating possible involvement of *FeMT3* in Zn^{2+} and Cu^{2+} homeostasis (Nikolić et al., 2010).

In addition to heavy metal stress, drought, oxidative stress, and low temperature also greatly reduce the survival and yield of many crop plants (Zhou et al., 2011). Real-time RT-PCR analysis showed that oxidative treatment (H_2O_2), as well as drought treatment, significantly enhanced the accumulation of *FeMT3* transcript in buckwheat leaves. Additionally, *FeMT3*-overexpressing yeast cells were significantly less sensitive to the oxidant (H_2O_2) than the untransformed control cells (Samardžić et al., 2010). These data revealed that *FeMT3* functions in drought stress defense and reactive oxygen species-related cellular processes. MYB transcription factors (TFs) play important roles in the abiotic stress response in plants (Zhou et al., 2015). Our group has isolated a novel nuclear transcription activator *FtMYB12*, which was greatly induced by low temperature (Zhou et al., 2015). Overexpression of *FtMYB12* in *Arabidopsis* plants resulted in enhanced cold tolerance (Zhou et al., 2015). The results revealed that

FtMYB12 plays an important role in the regulation of cold stress responsive signaling in TB. However, the detailed mechanisms of stress tolerances of buckwheat still need further studies.

SECONDARY METABOLISM

Flavonoids are the largest group of secondary metabolites thought to provide health benefits through cell signaling pathways and antioxidant effects in plants. Buckwheat is rich in flavonoids, such as rutin, orientin, vitexin, quercetin, isovitexin, and isoorientin (Zhang et al., 2012). Rutin was recognized as the most health protective and has also been proven to be antiinflammatory and anticarcinogenic (Zhang et al., 2012). The rutin biosynthetic pathway is a branch of a large phenylpropanoid pathway, and the pathway genes are conserved among different plants (Du et al., 2010). Seeds of TB contain 40–50 times higher rutin compared to CB (Zhang et al., 2012). qRT-PCR analysis demonstrated that the amounts of transcripts of four key enzyme genes *phenylalanine ammonialyase (PAL), chalcone synthase (CHS), chalcone isomerase (CHI)*, and *flavonol synthase (FLS)* are relatively higher in TB compared to CB, indicating that the expression of flavonoid biosynthesis genes is strongly correlated with the rutin content (Gupta et al., 2011). Later, the complete open reading frame of *FtFLS1* was cloned and characterized by Li et al. (2012a). The transcripts of *FtFLS1* showed an organ-specific expression pattern, with similar trends in flavonoid content (Li et al., 2012a). The correlating biosynthetic genes with the biosynthesis and accumulation of flavonoids were also observed in TB sprouts (Li et al., 2012b; Thwe et al., 2014b), white- and red-flowered buckwheat cultivars (Kim et al., 2013a), TB (Li et al., 2013; Kim et al., 2014), and hairy root culture of TB (Thwe et al., 2013, 2014a). Two key enzyme genes, *hydroxycinnamoyl coenzyme A quinate hydroxycinnamoyltransferase (HQT)* and *p-coumarate 3'-hydroxylase (C3H)*, which are involved in chlorogenic acid biosynthesis, have been isolated (Kim et al., 2013b). However, their functions in buckwheat have not been studied in much detail. Glycosylation of flavonoids is crucial for multiple physiological properties and functions in plants, and among them flavonoid-C-glycosides are prevalent compounds that show several biological functions, such as antifungal and antibacterial activities (Jay et al., 2005). Two isozymes of flavonoid C-glucosyltransferases (FeCGTa and FeCGTb) from CB cotyledon were identified by Nagatomo et al. (2014), indicating that C-glucosylflavones play an essential role in the defense mechanism of buckwheat during the early growth stage.

MYB TFs play important roles in regulatory networks controlling the flavonoid biosynthetic pathway in plants. In *Arabidopsis*, TT2 (Transparent testa 2, AtMYB123) encoded an R2R3 MYB TF, which can increase the proanthocyanidin accumulation (Nesi et al., 2001). In our group we also identified a TT2/AtMYB123 homologous protein FtMYB123L through bioinformatics analysis from the latest available floral transcriptome data from *F. tataricum* (Logacheva et al., 2011; Zhou et al., 2013). Two TF genes, *FtMYB1* and *FtMYB2*, were isolated from TB (Bai et al., 2014). Bioinformatic analysis showed that *FtMYB1* and *FtMYB2* have a high degree of similarity with *TaMYB14* and *AtMYB123/ TT2* (Bai et al., 2014). These two MYB TFs localize in the nucleus and activate their target genes such as *dihydroflavonol-4-reductase* (*DFR*), thereby enhancing the production of proanthocyanidins (Bai et al., 2014). These results can be useful in planning for improvement in the yield of flavonoids through genetic engineering.

PROTEASE INHIBITORS

Proteinase inhibitors are one of protective proteins that can degrade the proteolytic enzymes secreted from insect pests and pathogenic microorganisms during their penetration into the host plant tissues (Khadeeva et al., 2009). In CB, two types of aspartic proteinase (AP) genes, *FeAP9* and *FeAPL1* (AP-like), have been isolated from the cDNA library of developing seeds (Milisavljevic et al., 2008). The mRNA expression profiles analysis showed that *FeAPL1* is restricted to the seeds only, whereas *FeAP9* is also present in the other plant tissues including leaves, roots, and flowers (Milisavljevic et al., 2008). Additionally, the *FeAP9* mRNA expression is induced by different abiotic stresses, including dark, drought, and UV-B light, as well as wounding and salicylic acid (Timotijević et al., 2010). Bioinformatic analysis of their sequences showed that *FeAPL1* lacks a unique element of plant-specific insert among APs (Milisavljevic et al., 2008). The differential expression pattern and protein sequence of these two unique APs reveal that they may play different roles in physiological processes. Overexpression of the serine proteinase inhibitor *BWI-1a* (ISP) gene from buckwheat seeds in tobacco and potato plants showed that these transgenic plants provide sufficient protection at least against two bacterial phytopathogens, *Pseudomonas syringae* pv. *tomato* and *Clavibacter michiganensis* ssp. *michiganensis* (Khadeeva et al., 2009). A trypsin inhibitor from TB (FtTI) protein was purified by affinity chromatography and centrifugal ultrafiltration, and the *FtTI* gene also was successfully cloned by Ruan et al. (2015). The biotest demonstrated that *FtTI* showed a higher toxic killing effect on *Mamestra brassicae* larvae (Ruan et al., 2015). Therefore it is possible to engineer insect and pathogen resistance in transgenic buckwheat or crop plants by manipulating the expression of protease inhibitors in the future.

SELFINCOMPATIBILITY

Selfincompatibility (SI) is a genetic mechanism to prevent self-fertilization after pollination. CB is a self-incompatible plant with dimorphic flowers, namely, short-styled (SS) and long-styled (LS) flowers, and is associated with a type of SI called heteromorphic SI. The *S*-locus (dominant allele) supergene is thought to govern self-incompatibility, flower morphology, and pollen size in buckwheat (Matsui et al., 2004). Plants with SS flowers are heterozygous (*S/s*) and plants with LS flowers are homozygous recessive (s/s) at the *S*-locus. Differentially expressed gene analysis from both SS and LS floral morphs found that one gene homology to *EARLY FLOWERING 3* is expressed only in SS plants (Yasui et al., 2012). Further genetic linkage analysis showed that this novel gene was completely linked to the *S*-locus, hence this gene was designated as *S-LOCUS EARLY FLOWERING 3* (*S-ELF3*) (Yasui et al., 2012). The ion beam-induced LS mutant plants showed a deletion in the genomic region spanning *S-ELF3*, whereas *S-ELF3* was present in the genome of SS plants (Yasui et al., 2012). Furthermore, independent disruptions of *S-ELF3* were detected in a self-compatible line of buckwheat, indicating that *S-ELF3* is a suitable candidate gene for the control of the SS and SI phenotype of buckwheat plants (Yasui et al., 2012). In *Arabidopsis*, APETALA3 (AP3) encoding MADS-box transcription factors are required to specify petal and stamen identity. An AP3 homolog, *FaesAP3*, from CB was isolated and identified. Expression analysis showed that *FaesAP3* was restricted to the developing stamens only. Additionally, overexpression of *FaesAP3* in the *Arabidopsis ap3* mutant rescued stamen development, indicating that *FaesAP3* functions in the development of stamens in buckwheat (Fang et al., 2014). These results provide some potential for biotechnical engineering to create sterile lines of buckwheat in the future.

ALLERGENS

Buckwheat seeds contain 15–17% protein and are rich in essential amino acids. However, buckwheat seeds also contain proteins designated as allergenic, that is, an immunoglobulin E (IgE)-mediated hypersensitive response capable of causing serious symptoms, including anaphylactic shock, which limits their use as a general food source and additive (Yoshioka et al., 2004; Zhang et al., 2012). In CB, two allergenic protein genes, *FA02* and *Fag e 1*, were cloned (Fujino et al., 2012; Yoshioka et al., 2004). *FA02* consists of two separate components: a 41.3-kDa α-subunit and a 21-kDa β-subunit. Immunoblotting analysis revealed that the 21-kDa β-subunit displayed the greatest reactivity with the antiserum (Fujino et al., 2012). *Fag e 1* is a 22-kDa protein and consists of eight epitopes, mutation of the central amino acids within each epitope led to significantly decrease or complete loss in IgE binding (Yoshioka et al., 2004). In TB, a 24-kDa allergenic protein gene *TBa* was isolated by Wang et al. (2006). Later, her group cloned another 56-kDa allergenic protein gene *TBt* (Zhang et al., 2008). Analysis of their immunological activity showed that *TBt* had a lower IgE binding ability than the recombinant *TBa* (Zhang et al., 2008). Biochemical analysis of these allergenic proteins could help therapeutic efforts and have the potential of developing hypoallergenic buckwheat.

CONCLUDING REMARKS

This chapter summarized the research progress of genes regulating buckwheat growth, development, and metabolism. However, only a few genes from buckwheat have been identified. Thus discovering new genes is still important to understand the molecular mechanism of buckwheat growth and development, and response to environmental signals. Rapid progress in the acquisition of genomics and transcriptomics information and combinatorial systems omics analysis has proven to be a powerful tool to identify new functional genes from buckwheat.

ACKNOWLEDGMENTS

This research was supported by the Key Project of Science and Technology of Sichuan, China (grant no. 04NG001-015, "Protection and exploitation of wild-type buckwheat germplasm resource") and the National Natural Science Foundation of China (grant no. 31572457).

REFERENCES

Bai, Y.C., Li, C.L., Zhang, J.W., Li, S.J., Luo, X.P., Yao, H.P., Chen, H., Zhao, H.X., Park, S.U., Wu, Q., 2014. Characterization of two Tartary buckwheat R2R3-MYB transcription factors and their regulation of proanthocyanidin biosynthesis. Physiol. Plant. 152, 431–440.

Cobbett, C., Goldsbrough, P., 2002. Phytochelatins and metallothioneins: roles in heavy metal detoxification and homeostasis. Annu. Rev. Plant Physiol. 53, 159–182.

Du, H., Huang, Y., Tang, Y., 2010. Genetic and metabolic engineering of isoflavonoid biosynthesis. Appl. Microbiol. Biotechnol. 86, 1293–1312.

Fang, Z.W., Qi, R., Li, X.F., Liu, Z.X., 2014. Ectopic expression of *FaesAP3*, a *Fagopyrum esculentum* (Polygonaceae) AP3 orthologous gene rescues stamen development in an Arabidopsis *ap3* mutant. Gene 550, 200–206.

Fujino, K., Funatsuki, H., Inada, M., Shimono, Y., Kikuta, Y., 2012. Expression, cloning, and immunological analysis of buckwheat (*Fagopyrum esculentum* Moench) seed storage proteins. J. Agric. Food Chem. 49, 1825–1829.

Gupta, N., Sharma, S.K., Rana, J.C., Chauhan, R.S., 2011. Expression of flavonoid biosynthesis genes vis-à-vis rutin content variation in different growth stages of *Fagopyrum* species. J. Plant Physiol. 168, 2117–2123.

Jay, M., Viricel, M.R., Gonnet, J.F., 2005. C-glycosylflavonoids. Flavonoids: chemistry. In: Andersen, O.M., Markham, K.R. (Eds.), Biochemistry and Applications. CRC Press, Boca Raton, FL, pp. 857–915.

Khadeeva, N.V., Kochieva, E.Z., Tcherednitchenko, M.Y., Yakovleva, E.Y., Sydoruk, K.V., Bogush, V.G., Dunaevsky, Y.E., Belozersky, M.A., 2009. Use of buckwheat seed protease inhibitor gene for improvement of tobacco and potato plant resistance to biotic stress. Biochemistry (Mosc.) 74, 260–267.

Kim, Y.B., Park, S.Y., Thwe, A.A., Seo, J.M., Suzuki, T., Kim, S.J., Kim, J.K., Park, S.U., 2013a. Metabolomic analysis and differential expression of anthocyanin biosynthetic genes in white- and red-flowered buckwheat cultivars (*Fagopyrum esculentum*). J. Agric. Food Chem. 61, 10525–10533.

Kim, Y.B., Thwe, A.A., Kim, Y., Li, X., Cho, J.W., Park, P.B., Valan Arasu, M., Abdullah Al-Dhabi, N., Kim, S.J., Suzuki, T., Hyun Jho, K., Park, S.U., 2014. Transcripts of anthocyanidin reductase and leucoanthocyanidin reductase and measurement of catechin and epicatechin in Tartary buckwheat. Sci. World J., 726567.

Kim, Y.B., Thwe, A.A., Kim, Y.J., Li, X., Kim, H.H., Park, P.B., Suzuki, T., Kim, S.J., Park, S.U., 2013b. Characterization of genes for a putative hydroxycinnamoyl coenzyme A quinate transferase and *p*-coumarate 3′-hydroxylase and chlorogenic acid accumulation in Tartary buckwheat. J. Agric. Food Chem. 61, 4120–4126.

Li, C., Bai, Y., Li, S., Chen, H., Han, X., Zhao, H., Shao, J., Park, S.U., Wu, Q., 2012a. Cloning, characterization, and activity analysis of a flavonol synthase gene FtFLS1 and its association with flavonoid content in Tartary buckwheat. J. Agric. Food Chem. 60, 5161–5168.

Li, X., Kim, Y.B., Kim, Y., Zhao, S., Kim, H.H., Chung, E., Lee, J.H., Park, S.U., 2013. Differential stress-response expression of two flavonol synthase genes and accumulation of flavonols in Tartary buckwheat. J. Plant Physiol. 170, 1630–1636.

Li, X., Thwe, A.A., Park, N.I., Suzuki, T., Kim, S.J., Park, S.U., 2012b. Accumulation of phenylpropanoids and correlated gene expression during the development of Tartary buckwheat sprouts. J. Agric. Food Chem. 60, 5629–5635.

Logacheva, M.D., Kasianov, A.S., Vinogradov, D.V., Samigullin, T.H., Gelfand, M.S., Makeev, V.J., Penin, A.A., 2011. De novo sequencing and characterization of floral transcriptome in two species of buckwheat (*Fagopyrum*). BMC Genom. 12, 30.

Ma, J.F., Hiradate, S., 2000. Form of aluminium for uptake and translocation in buckwheat (*Fagopyrum esculentum* Moench). Planta 211(3), 355-360.

Matsui, K., Nishio, T., Tetsuka, T., 2004. Genes outside the S supergene suppress S functions in buckwheat (*Fagopyrum esculentum*). Ann. Bot. 94, 805–809.

Milisavljevic, M.Dj., Timotijevic, G.S., Radovic, S.R., Konstantinovic, M.M., Maksimovic, V.R., 2008. Two types of aspartic proteinases from buckwheat seed—gene structure and expression analysis. J. Plant Physiol. 165, 983–990.

Nagatomo, Y., Usui, S., Ito, T., Kato, A., Shimosaka, M., Taguchi, G., 2014. Purification, molecular cloning and functional characterization of flavonoid C-glucosyltransferases from *Fagopyrum esculentum* M. (buckwheat) cotyledon. Plant J. 80, 437–448.

Nesi, N., Jond, C., Debeaujon, I., Caboche, M., Lepiniec, L., 2001. The Arabidopsis TT2 gene encodes an R2R3 MYB domain protein that acts as a key determinant for proanthocyanidin accumulation in developing seed. Plant Cell 13, 2099–2114.

Nikolić, D.B., Samardžić, J.T., Bratić, A.M., Radin, I.P., Gavrilović, S.P., Rausch, T., Maksimović, V.R., 2010. Buckwheat (*Fagopyrum esculentum* Moench) FeMT3 gene in heavy metal stress: protective role of the protein and inducibility of the promoter region under Cu(2+) and Cd(2+) treatments. J. Agric. Food Chem. 58, 3488–3494.

Ruan, J., Yan, J., Hou, S., Chen, H., Wu, Q., Han, X., 2015. Expression and purification of the trypsin inhibitor from Tartary buckwheat in *Pichia pastoris* and its novel toxic effect on *Mamestra brassicae* larvae. Mol. Biol. Rep. 42, 209–216.

Samardžić, J.T., Nikolić, D.B., Timotijević, G.S., Jovanović, Z.S., Milisavljević, M.Đ., Maksimović, V.R., 2010. Tissue expression analysis of FeMT3, a drought and oxidative stress related metallothionein gene from buckwheat (*Fagopyrum esculentum*). J. Plant Physiol. 167, 1407–1411.

Shen, R.F., Ma, J.F., Kyo, M., Iwashita, T., 2002. Compartmentation of aluminium in leaves of an Al-accumulator, *Fagopyrum esculentum* Moench. Planta 215 (3), 394–398.

Tani, F.H., Barrington, S., 2005. Zinc and copper uptake by plants under two transpiration rates. Part II. Buckwheat (*Fagopyrum esculentum* L.). Environ. Poll. 138, 548–558.

Thwe, A.A., Kim, J.K., Li, X., Kim, Y.B., Uddin, M.R., Kim, S.J., Suzuki, T., Park, N.I., Park, S.U., 2013. Metabolomic analysis and phenylpropanoid biosynthesis in hairy root culture of Tartary buckwheat cultivars. PLoS One 8, e65349.

Thwe, A.A., Kim, Y., Li, X., Kim, Y.B., Park, N.I., Kim, H.H., Kim, S.J., Park, S.U., 2014a. Accumulation of phenylpropanoids and correlated gene expression in hairy roots of Tartary buckwheat under light and dark conditions. Appl. Biochem. Biotechnol. 174, 2537–2547.

Thwe, A.A., Kim, Y.B., Li, X., Seo, J.M., Kim, S.J., Suzuki, T., Chung, S.O., Park, S.U., 2014b. Effects of light-emitting diodes on expression of phenylpropanoid biosynthetic genes and accumulation of phenylpropanoids in *Fagopyrum tataricum* sprouts. J. Agric. Food Chem. 62, 4839–4845.

Timotijević, G.S., Milisavljević, M.Dj., Radović, S.R., Konstantinović, M.M., Maksimović, V.R., 2010. Ubiquitous aspartic proteinase as an actor in the stress response in buckwheat. J. Plant Physiol. 167, 61–68.

Wang, Z.H., Wang, L., Chang, W.J., Li, Y.Y., Zhang, Z., Wieslander, G., Norbäck, D., 2006. Cloning, expression, and identification of immunological activity of an allergenic protein in Tartary buckwheat. Biosci. Biotechnol. Biochem. 70, 1195–1199.

Yasui, Y., Mori, M., Aii, J., Abe, T., Matsumoto, D., Sato, S., Hayashi, Y., Ohnishi, O., Ota, T., 2012. *S-LOCUS EARLY FLOWERING 3* is exclusively present in the genomes of short-styled buckwheat plants that exhibit heteromorphic self-incompatibility. PLoS One 7, e31264.

Yokosho, K., Yamaji, N., Ma, J.F., 2014. Global transcriptome analysis of Al-induced genes in an Al-accumulating species, common buckwheat (*Fagopyrum esculentum* Moench). Plant Cell Physiol. 55, 2077–2091.

Yoshioka, H., Ohmoto, T., Urisu, A., Mine, Y., Adachi, T., 2004. Expression and epitope analysis of the major allergenic protein Fag e 1 from buckwheat. J. Plant Physiol. 161, 761–767.

Zhang, X., Yuan, J.M., Cui, X.D., Wang, Z.H., 2008. Molecular cloning, recombinant expression, and immunological characterization of a novel allergen from Tartary buckwheat. J. Agric. Food Chem. 56, 10947–10953.

Zhang, Z.L., Zhou, M.L., Tang, Y., Li, F.L., Tang, Y.X., Shao, J.R., Xue, W.T., Wu, Y.M., 2012. Bioactive compounds in functional buckwheat food. Food Res. Int. 49, 389–395.

Zhou, M.L., Zhu, X.M., Shao, J.R., Tang, Y.X., Wu, Y.M., 2011. Production and metabolic engineering of bioactive substances in plant hairy root culture. Appl. Microbiol. Biotechnol. 90, 1229–1239.

Zhou, M.L., Bai, D.Q., Tang, Y., Zhu, X.M., Shao, J.R., 2012. Genetic diversity of four new species related to southwestern Sichuan buckwheats as revealed by karyotype ISSR and allozyme characterization. Plant System. Evol. 298 (4), 751–759.

Zhou, M.L., Tang, Y., Zhang, K.X., Li, F.L., Yang, P.Y., Tang, Y.X., Wu, Y.M., Shao, J.R., 2013. Identification of TT2 gene from floral transcriptome in *Fagopyrum tataricum*. Food Res. Int. 54, 1331–1333.

Zhou, M.L., Wang, C.L., Qi, L.P., Sun, Z.M., Tang, Y., Tang, Y.X., Shao, J.R., Wu, Y.M., 2015. Ectopic expression of *Fagopyrum tataricum* FtMYB12 improves cold tolerance in *Arabidopsis thaliana*. J. Plant Growth Reg. 34, 362–371.

Zhu, H., Wang, H., Zhu, Y., Zou, J., Zhao, F.J., Huang, C.F., 2015. Genome-wide transcriptomic and phylogenetic analyses reveal distinct aluminum-tolerance mechanisms in the aluminum-accumulating species buckwheat (*Fagopyrum tataricum*). BMC Plant Biol. 15, 16.

Chapter twenty six

Flavor and Lipid Deterioration in Buckwheat Flour Related to Lipoxygenase Pathway Enzymes

T. Suzuki

National Agriculture and Food Research Organization (NARO), Kyushu Okinawa Agricultural Research Center, Koshi, Kumamoto, Japan

INTRODUCTION

Buckwheat (*Fagopyrum esculentum* Moench) is recognized as an important crop in Japan, China, Korea, and European countries (Ikeda, 2002; Kreft et al., 2003), and is related to their cultures. In the Japanese food industry, buckwheat flour is mainly used for making noodles. The freshness of buckwheat flour is very important in quality. However, buckwheat flour deteriorates easily (Tohyama et al., 1982; Muramatsu et al., 1986; Suzuki et al., 2005). Several reports have shown that lipid degradation and oxidation in buckwheat flour during storage are some of the main changes in measurable indexes of the deterioration of quality (Tohyama et al., 1982; Muramatsu et al., 1986; Suzuki et al., 2005). Therefore lipid degradation pathways in buckwheat flour are important to study the mechanisms of the deterioration of quality.

On the other hand, buckwheat boiled noodles (soba), important in traditional food items in Japan (Ikeda, 2002), have a unique flavor, which is one of the most important characteristics of quality. Buckwheat flour and dough contain a number of volatile compounds (Aoki et al., 1981; Aoki and Koizumi, 1986;

Molecular Breeding and Nutritional Aspects of Buckwheat. http://dx.doi.org/10.1016/B978-0-12-803692-1.00026-2

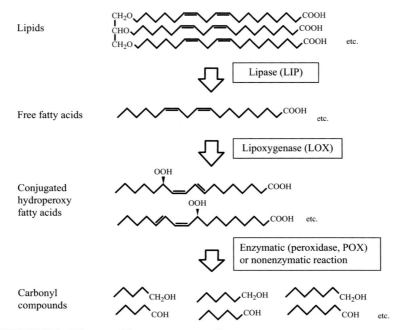

FIGURE 26.1 Scheme of lipoxygenase pathway.

Yajima et al., 1983; Przybylski et al., 1995; Janes et al., 2012). The most important contributors are reported to be carbonyl compounds such as aldehydes and ketones (Aoki and Koizumi, 1986; Ohinata et al., 2001; Kawakami et al., 2008; Janes et al., 2012). Among them, hexanal is known as a cause of the "beany" flavor compound in soybean [*Glycine max* (L.) Merr.] products (Matoba et al., 1975, 1985; Axelrod, 1974; Dahuja and Madaan, 2004). In soybean products, hexanal is generated through the lipoxygenase pathway, which was first proposed in rice bran (Takano, 1993) (Fig. 26.1). Lipase (triacylglycerol lipase EC 3.1.1.3) (LIP) catalyzes the first step of the lipid by hydrolysis. Lipoxygenase (EC 1.13.11.12) (LOX) catalyzes the dioxygenation of polyunsaturated fatty acids in lipids. LOX has a significant effect on flavor generation in some crops including soybean (Fukushima, 1994), rice (*Oryza sativa* L.) (Robinson et al., 1995), and some vegetables (Baardseth and Slinde, 1987). Peroxidase (EC 1.11.1.7) (POX) also has a role in the generation of flavor in soybean (Ashie et al., 1996). From these backgrounds, as a first step to understanding the mechanisms of flavor generation and lipid deterioration, and characterization of these enzymes on buckwheat flour is important. A number of studies have investigated the purification and characterization of LIP (Ohinata et al., 1997; Suzuki et al., 2004), LOX (Suzuki et al., 2007), and POX (Kondo et al., 1982; Suzuki et al., 2006) in buckwheat flour. In this review, we summarize the characteristics of these enzymes and possible mechanisms of flavor generation and lipid deterioration in lipoxygenase pathways in buckwheat.

Characteristics of Lipase, Lipoxygenase, and Peroxidase in Buckwheat Flour

LIP is an important enzyme in the food industry, because lipid hydrolysis sometimes causes deterioration of food quality (Ashie et al., 1996). An increase in free fatty acids indicates deterioration of the quality of buckwheat flour such as increase in water-soluble acids (WSAs). This will be responsible for lipid peroxidation and deterioration of the flavor. From these backgrounds, LIP in buckwheat seed has been purified and characterized (Kondo et al., 1982; Ohinata et al., 1997; Suzuki et al., 2004). LIP consists of at least two isozymes, LIP I and LIP II. The molecular weights of LIP I and II are 150 and 28.4 kDa by SDS-PAGE, and 171 and 26.5 kDa by gel filtration, respectively, indicating that LIP I and II are monomers. Both LIPs are stable below 30°C, and retain activity at 70°C. In addition, both LIP I and II maintain about 50% of their activity at 30°C compared to 10°C. Therefore, LIP would remain active during storage even if stored at 10°C. The optimal pH is determined using triolein as substrate. The optimal pH values are 3.0 (LIP I) and 6.0 (LIP II), respectively. Both LIP I and II show activity between pH 3.0 and pH 7.0, and are inactive below pH 2.0 and above pH 8.0. LIP I reacts in a narrower optimum pH range than LIP II; LIP I activity peaks between pH 3 and pH 4 whereas LIP II peaks between pH3 and pH6. Both isozymes have higher activities in the acidic pH range. Optimal pHs of buckwheat LIP I and LIP II are distinctly different from those of rape (Antonian, 1988), mustard (Antonian, 1988), and rice (Aizono et al., 1976). The optimal pHs of lipase from these plants are between 8 and 9. On the other hand, optimal pHs of buckwheat LIP I and II are very similar to those of castor bean acid lipase (Robert et al., 1962) for which the optimal pH was 4.3. Both LIPs have activity between pH 3.0 and 7.0. The pH of buckwheat flour is around 6.8. Therefore the pH of buckwheat flour is suitable for LIP. It is known that buckwheat flour deteriorates easily, and LIP activity plays an important role in lipid deterioration. To inactivate LIP activity in buckwheat flour, heat treatment would be effective because buckwheat LIP is not stable above 30°C when triolein is used as a substrate. However, heat treatment is costly and would result in deterioration of flavor, color, and a number of physical properties. Therefore it is desirable to breed buckwheat cultivars whose LIP is not reactive in flour or dough. Further, to develop such a cultivar, it is important to clarify which isozyme is more important for the quality of buckwheat flour.

To date, only a few studies have been published on LOX because of the difficulty of measuring LOX activity in buckwheat flour (Axelrod, 1974). From this background, immunoblotting analysis using a LOX-specific antibody raised against soybean LOX is employed to investigate buckwheat LOX (Suzuki et al., 2009). LOX protein content in buckwheat seed is two or four orders of magnitude less than that in other cereals such as rice, wheat, and soybean. Therefore, unlike rice and soybean, where LOX activity plays an important role in lipid degradation and oxidation, lipid degradation and oxidation in buckwheat may not be linked to LOX activity.

POX is also widely distributed in plants (Van and Cairns, 1982; Amako et al., 1994). It plays important roles in food quality, including deterioration of color and flavor (Ashie et al., 1996; Ibaraki et al., 1988; Ibaraki and Hirano, 1989). Carbonyl compounds such as aldehydes and ketones are the major contributors to flavors in soybean (Fukushima, 1994). These compounds are mainly generated by lipid peroxidation by POX (Matoba et al., 1975, 1985; Dahuja and Madaan, 2004). Buckwheat POX in flour has been partially characterized (Kondo et al., 1982; Suzuki et al., 2006, 2012). The POX in common buckwheat consists of at least two isozymes, POX I and POX II. The molecular weights of POX I and II are 46.1 and 58.1 kDa by gel filtration. In buckwheat seed, most of the POX activity was detected in the embryo, similar to LIP, LOX, and oil compounds (Dorrell, 1970). The K_m for tested substrates was different for POX I and II. POX II has a greater affinity than POX I for all substrates tested. In particular, POX I does not catalyze a reaction with ABTS. The K_m of POX II for ascorbic acid, guaiacol, and o-dianisidine are 0.137, 0.038, and 0.202 mM whereas those of POX I are 0.229, 0.043, and 0.288 mM, respectively. In addition, both POX I and II have low K_m for phenolic substrates such as quercetin. Therefore buckwheat POX is also related to the change of color in food products. The lower K_m of POX II and its greater quantity than POX I suggests that POX II is the major POX isozyme in buckwheat seed. The optimal temperature and thermal stabilities of POX I and II are also different. The optimal temperature for POX I is 30°C, whereas POX II is 10°C. In thermal stability, POX I is more stable at higher temperatures than POX II: POX I maintains activity at 0–30°C and only becomes unstable above 40°C, whereas POX II is stable to 20°C and unstable above 30°C. POX I and II are inactivated at 60°C and 50°C, respectively. In POX I, more than 50% of activity is retained in the temperature range 0–50°C. On the other hand, POX II has greatest activity in the lower temperature range 0–20°C and it decreases gradually above 20°C. Therefore POX II rather than POX I would catalyze reactions during storage, even if buckwheat seeds were stored at a low temperature such as 0–10°C.

Effects of Enzymes in Lypoxygenase Pathway on the Deterioration of Quality During Flour Storage

The effects of LIP, LOX, and POX on the deterioration of quality of buckwheat flour have been investigated by storage test of buckwheat flour (Suzuki et al., 2009). A brief experimental procedure is as follows. Prior to the start of the test, they screened 14 of 46 buckwheat cultivars to obtain a wide variation in the enzymes' activity and concentration. Buckwheat flour was placed in polyethylene bags and stored for 0, 4, 10, and 30 days at 5°C or 20°C in a dark room. The enzymes (LIP, POX activity, and LOX protein content) and indexes of flour lipid deterioration (pH, WSA, peroxide value [POV], and carbonyl value [COV]) were then analyzed to assess the correlation between enzymes and indexes of flour lipid deterioration. During storage, the pH decreased at both 5°C and 20°C. The pH decreased more at 20°C than at 5°C. WSA increased at both 5°C and 20°C, and more at the latter temperature. A decrease of pH and increase of WSA generally indicate the accumulation of free fatty acids. Changes in COV, similar to POV, also differed between varieties and storage temperatures. POV is an index of the amount of peroxidative compounds such as conjugated hydroperoxy fatty acids, which will degrade into carbonyl compounds (Takano, 1993). COV indicates the quantity of carbonyl compounds, volatile compounds such as aldehydes or ketones such as hexanal and nonanal (Kumazawa and Oyama, 1965; Takano, 1993). During storage at both 5°C and 20°C, LIP activity showed a significant negative correlation with pH (0, 4, 10, and 30 storage days) and a significant positive correlation with WSA (0, 4, 10, and 30 storage days). The LIP activity also showed a significant correlation with changes in POV and COV. An increase of POV and COV also indicates deterioration of buckwheat flour. Therefore LIP activity plays an important role in the deterioration of quality of buckwheat flour in terms of lipid degradation. The quantity of LOX1 protein concentration was partly correlated negatively to changes in WSA at 5°C and 20°C. On the other hand, for storage at 5°C the concentration of LOX 2 protein did not significantly correlate to any index, whereas for storage at 20°C it partly correlated significantly with changes in pH, POV, and COV. The correlations of POX with pH, WSA, POV, and COV were not observed for storage at 20°C. However, for storage at 5°C, POX showed a significant correlation to changes in pH and POV. These results indicate that mainly LIP but partly POX and LOX may have an influence on flour deterioration during storage. In addition, interestingly, the rutin concentration showed a significant correlation to changes in pH, WSA, and COV. The rutin concentration also showed a negative correlation with WSA at both 5°C and 20°C during the entire storage period. Rutin, a kind of flavonol glycoside, is widely distributed in plants (Sando and Lloyd, 1924; Couch et al., 1946). It has beneficial effects on fragile capillaries (Griffith et al., 1944; Shanno, 1946), and antioxidative (Afanas'ev et al., 1989; Steger-Hartmann et al., 1994), antihypertensive (Matsubara et al., 1985), and antiinflammatory activities (Afanas'ev et al., 2001). Therefore buckwheat has been utilized as a rutin-rich material for food (Kreft et al., 2006). Buckwheat flour contains about 20 mg rutin per 100 g flour. Quantitatively, LIP activity in buckwheat flour can be inhibited by 40% by the presence of rutin (Suzuki et al., 2005). However, the inhibition mechanism of LIP activity by rutin is not clear. Further studies are necessary on the effects of rutin on the deterioration of quality of buckwheat flour together with its effect on enzymes such as LIP.

Effects of Enzymes in Lypoxygenase Pathway on Generation of Noodle Volatile Compounds

Flavor components in buckwheat include many volatile compounds. Carbonyl compounds such as aldehydes and ketones have been focused on important compounds of buckwheat flavor. In soybean products, as described previously, such compounds are generated through the lipoxygenase pathway. In buckwheat, the relationship between enzymes in the lipoxygenase pathway and generation of flavor compounds has been investigated (Suzuki et al., 2010). They quantified LIP, POX activities, and LOX levels in the flour of 12 buckwheat varieties/breeding lines. In addition, using these 12 buckwheat varieties, they also quantified the volatile compounds produced by boiled buckwheat noodles using head space GC/MS, and then investigated correlations of the previously measured parameters. LIP activity in flour showed significant positive correlations with volatile compounds in head space butanal, tentative 3-methyl-butanal, tentative 2-methyl-butanal, and hexanal. POX activity showed a significant positive correlation to 3-methyl-butanal and 2-methyl-butanal. This result indicates that LIP and POX activities are important components in the enzymatic generation of volatile compounds. On the other hand, LOX showed a significant correlation to no volatile compound. In soybean, the enzymic action of LOX is key in generating hexanal, which is the major source of the "beany" flavor (Fukushima, 1994; Robinson et al., 1995), among the enzymes of the lipoxygenase pathway. In addition, LOX is also an important enzyme in the generation of unfavorable volatile compounds during the storage of rice (Suzuki et al., 1999). Therefore in buckwheat the key enzyme that generates volatile compounds such as hexanal may be different from those in soybean and rice. The free fatty acid levels such as C18:1, C18:2, and C18:3 in flour showed significant correlations with the volatile compounds such as pentanal and hexanal. These free fatty acids are the product of LIP activity, and the substrate of POX or other enzymatic and/or nonenzymatic reactions, which result in the generation of volatile compounds (Takano, 1993). From the results mentioned previously it is clear that enzymatic reactions such as LIP and POX are important in generating volatile compounds of boiled buckwheat noodles. Some volatile compounds found in this experiment such as hexanal and certain methyl-butanals are important contributors to the unique flavor of buckwheat. Therefore LIP and POX should also be important factors in generating the organoleptic qualities of boiled buckwheat noodles.

From these results it is clear that the lipid degradation and oxidation mechanism in buckwheat flour is unique compared to rice and soybean because not LOX but LIP is the key enzyme. LIP activity in buckwheat flour apparently plays a role in lipid degradation and the deterioration of quality. On the other hand, rutin tends to prevent flour deterioration (Suzuki et al., 2005). In addition, LIP and POX activity in buckwheat flour is important for flavor generation of boiled buckwheat noodles whereas rutin does not have important roles in this respect. This fact indicates that to develop the variety whose flavor is enhanced but whose flour does not deteriorate easily, increasing LIP activity and rutin concentration would be effective. Studies on the acceptability of flavor in bakery and nonbakery products of buckwheat have been summarized (Antonio et al., 2015). Volatile compounds generated from the lipoxygenase pathway may have a relationship with the flavor of bakery products. Further study is necessary to make this clear.

ACKNOWLEDGMENTS

We thank Dr T. Morishita, Dr Y. Mukasa, Dr T. Noda, Mr S. Takigawa, and Dr H. Yamauch for their useful advice in this study. We also thank Mr N. Murakami, Mr T. Yamada, Mr M. Oizumi, Mr T. Saruwatari, Mr S. Nakamura, Mr K. Suzuki, Mr A. Morizumi, Mr T. Hirao,

T. Takakura, K. Abe, Ms K. Fujii, Ms M. Hayashida, Ms T. Ando, and Mr Y. Honda for their field and laboratory experiments. And we are grateful to Dr A. Horigane and Ms S. Yamada for their useful suggestions.

REFERENCES

Afanas'ev, I.B., Dorozhko, A.I., Brodskii, A.V., Kostyuk, V.A., Potapovitch, A.I., 1989. Chelating and free radical scavenging mechanisms of inhibitory action of rutin and quercetin in lipid peroxidation. Biochem. Pharmacol. 38, 1763–1769.

Afanas'ev, I.B., Ostrakhovitch, E.A., Mikhal'chik, E.V., Ibragimova, G.A., Korkina, G.L., 2001. Enhancement of antioxidant and anti-inflammatory activities of bioflavonoid rutin by complexation with transition metals. Biochem. Pharmacol. 61, 677–684.

Aizono, Y., Funatsu, M., Fuziki, Y., Watanabe, M., 1976. Purification and characterization of rice bran lipase II. Agric. Biol. Chem. 40, 317–324.

Amako, K., Chen, G., Asada, K., 1994. Separate assays specific for ascorbate peroxidase and guaiacol peroxidase and for the chloroplastic and cytosolic isozymes of ascorbate peroxidase in plants. Plant Cell Physiol. 35, 497–504.

Dahuja, A., Madaan, T.R, 2004. Off-flavour development in soybeans: comparative role of some antioxidants and related enzymes. J. Sci. Food Agric. 84, 547–550.

Antonian, E., 1988. Recent advances in the purification, characterization and structure determination of lipase. Lipids 23, 1101–1106.

Antonio, J., Piskuła, M.K., Zieliński, H., 2015. Recent advances in processing and development of buckwheat derived bakery and non-bakery products—a review. Pol. J. Food Nutr. Sci. 65, 9–20.

Aoki, M., Koizumi, N., Ogawa, G., Yoshizaki, T., 1981. Identification of the volatile components of buckwheat flour and their distribution of milling fractions. Nippon Shokuhin Kogyo Gakkaishi 28, 476–481, [in Japanese, English abstract].

Aoki, M., Koizumi, N., 1986. Organoleptic properties of the volatile components of buckwheat flour and their changes during storage after milling. Nippon Shokuhin Kogyo Gakkaishi 33, 769–772, [in Japanese, English abstract].

Ashie, N.A., Simpson, B.K., Smith, J.P., 1996. Mechanisms for controlling enzymatic reactions in foods. Crit. Rev. Food Sci. Nutr. 36, 1–30.

Axelrod, B., 1974. Lipoxygenases. Am. Chem. Soc. Adv. Chem. Series 136, 324–348.

Baardseth, P., Slinde, E., 1987. Enzymes and off-flavors: palmitoyl-CoA hydrolase, lipoxygenase, α-oxidation, peroxidase, catalase activity and ascorbic acid content in different vegetables. Norw. J. Agric. Sci. 1, 111–117.

Couch, J., Naghski, J., Krewson, C., 1946. Buckwheat as a source of rutin. Science 103, 197–198.

Dorrell, O.G., 1970. Fatty acid composition of buckwheat seed. J. Am. Oil Chem. Soc. 48, 693–696.

Fukushima, D., 1994. Recent progress on biotechnology of soybean proteins and soybean protein food products. Food Biotechnol. 8, 83–135.

Griffith, J.Q., Couch, J.F., Lindauer, A., 1944. Effect of rutin on increased capillary fragility in man. Proceedings of the Society for Experimental Biology and Medicine, 55, pp. 228–229.

Ibaraki, T., Hirano, T., Yamashita, N., Baba, N., 1988. Studies on peroxidase activity of butterbur. Bull. Rep. Fukuoka Agric. Res. Center [in Japanese] B-8, 67–72.

Ibaraki, T., Hirano, T., 1989. Control of peroxidase and polyphenoloxidase activities, and the preserve of chlorophylls in processing of waterboiled butterbur. Bull. Rep. Fukuoka Agric. Res. Center [in Japanese] B-9, 85–90.

Ikeda, K., 2002. Buckwheat: composition, chemistry and processing. In: Taylor, S.L. (Ed.), Advances in Food and Nutrition Research. Academic Press, Nebraska, pp. 395–434.

Janes, D., Prosen, H., Kreft, S., 2012. Identification and quantification of aroma compounds of Tartary buckwheat (*Fagopyrum tataricum* Gaertn.) and some of its milling fractions. J. Food Sci. 77, 746–751.

Kawakami, I., Murayama, N., Kawasaki, S., Igasaki, T., Hayashida, Y., 2008. Effects of storage temperature on flavor of stone-milled buckwheat flour. Nippon Shokuhin Kagaku Kogaku Kaishi 55, 559–565.

Kondo, K., Kurogouti, T., Matubashi, T., 1982. Report in research institute of food technology in Nagano prefecture, vol. 10, p. 150. [in Japanese].

Kreft, I., Chang, K.J., Choi, Y.S., Park, C.H., 2003. Ethnobotany of Buckwheat. Jinsol Publishing Co., Seoul, Korea, pp. 91–115.

Kreft, I., Fabjan, N., Yasumoto, K., 2006. Rutin content in buckwheat (*Fagopyrum esculentum* Moench) food materials and products. Food Chem. 98, 508–512.

Kumazawa, H., Oyama, T., 1965. Estimation of total carbonyl content in oil by 2,4-dinitrophenylhydrazine. Yukagaku 14, 167–171, [in Japanese].

Matoba, T., Hidaka, H., Kitamura, K., Kaizuma, N., Kito, M., 1975. Contribution of hydroperoxide lyase activity to *n*-hexanal formation in soybean. J. Agric. Food Chem. 23, 136–141.

Matoba, T., Hidaka, H., Narita, H., Kitamura, K., Kaizuma, N., Kito, M., 1985. Lipoxigenase-2 isozymes are responsible for generation of *n*-hexanal in soybean homogenates. J. Agric. Food Chem. 33, 852–855.

Matsubara, Y., Hiroyasu, K., Yoshitomi, I., Tetsuo, M., Kozo, O., Hideo, M., Katsumi, Y., 1985. Structure and hypotensive effect of flavonoid glycosides in citrus unshiu peelings. Agric. Biol. Chem. 49, 909–914.

Muramatsu, N., Ohinata, H., Obara, T., Matsubashi, T., 1986. Food Packag. 30, 94 [in Japanese].

Ohinata, H., Karasawa, H., Muramatsu, N., Ohike, T., 1997. Properties of buckwheat lipase and depression of free fatty acid accumulation during storage. Nippon Shokuhin Kagaku Kogaku Kaishi 44, 590–593, [in Japanese, English abstract].

Ohinata, H., Karasawa, H., Kurokouchi, H.K., 2001. Influence of milling methods on buckwheat aroma. In: Proceedings of the Eighth International Symposium on Buckwheat pp. 694–697.

Przybylski, R., Woodward, L., Eskin, N., Malcolmson, L.J., Mazza, G., 1995. Effect of buckwheat storage and milling on aroma compounds. Curr. Adv. Buckwheat Res., 783–787.

Robert, L., Allen, J., Aaron, M., 1962. The acid lipase of the castor bean. Properties and substrate specificity. J. Lipid Res. 3, 99–105.

Robinson, D.S., Wu, Z.C., Domoney, C., Casey, R., 1995. Lipoxygenase and the quality of foods. Food Chem. 54, 33–43.

Sando, C., Lloyd, J., 1924. The isolation and identification of rutin from the flowers of elder (*Sambucus canadensis* L.). J. Biol. Chem. 58, 737–745.

Shanno, R.L., 1946. Rutin: a new drug for the treatment of increased capillary fragility. Am. J. Med. Sci. 211, 539–543.

Steger-Hartmann, T., Ulrich, K., Thomas, D., Edgar, W., 1994. Induced accumulation and potential antioxidative function of rutin in two cultivars of *Nicotiana tabacum* L. Z. Naturforsch. 49c, 57–62.

Suzuki, T., Honda, Y., Mukasa, Y., 2004. Purification and characterization of lipase in buckwheat seed. J. Agric. Food Chem. 52, 7407–7411.

Suzuki, T., Honda, Y., Mukasa, Y., Kim, S., 2005. Effects of lipase, lipoxygenase, peroxidase and rutin on quality deteriorations in buckwheat flour. J. Agric. Food Chem. 85, 8400–8405.

Suzuki, T., Honda, Y., Mukasa, Y., Kim, S., 2006. Characterization of peroxidase in buckwheat seed. Phytochemistry 67, 219–224.

Suzuki, T., Honda, Y., Mukasa, Y., 2007. Specificity of anti soybean LOX antibody, and its application to detect LOX-like protein in buckwheat seed. Fagopyrum 24, 15–19.

Suzuki, T., Mukasa, Y., Noda, T., Hashimoto, N., Takigawa, S., Matsuura-Endo, C., Yamauchi, H., 2009. Lipase, peroxidase activity and lipoxygenase-like protein content during ripening in common buckwheat (*Fagopyrum esculentum* Moench cv. Kitawasesoba) and Tartary buckwheat (*F. tataricum* Gaertn. cv. Hokkai T8). Fagopyrum 26, 63–67.

Suzuki, T., Kim, S.J., Mukasa, Y., Morishita, T., Noda, T., Takigawa, S., Hashimoto, N., Yamauchi, H., Matsuura-Endo, C., 2010. Effects of lipase, lipoxygenase, peroxidase and free fatty acids on volatile compound found in boiled buckwheat noodles. J. Sci. Food Agric. 90, 1232–1239.

Suzuki, T., Shin, D.H., Woo, S.H., Mukasa, Y., Morishita, T., Noda, T., Takigawa, S., Hashimoto, N., Yamauchi, H., Matsuura-Endo, C., 2012. Characterization of peroxidase in Tartary buckwheat seed. Food Sci. Technol. Res. 18, 571–575.

Suzuki, Y., Ise, K., Li, C., Honda, I., Iwai, Y., Matsukura, U., 1999. Volatile components in stored rice [*Oryza sativa* (L.)] of varieties with and without lipoxygenase-3 in seeds. J. Agric. Food Chem. 47, 1119–1124.

Takano, K., 1993. Mechanism of lipid hydrolysis in rice bran. Cereal Food World 38, 695–698.

Tohyama, R., Sekizawa, N., Murai, K., Ishiya, T., 1982. Quality change in packaged buckwheat during storage. Nippon Shokuhin Kagaku Kogaku Kaishi 29, 501, [in Japanese, English abstract].

Van, H., Cairns, W., 1982. Progress and prospects in the use of peroxidase to study cell development. Phytochemistry 21, 1843–1847.

Yajima, I., Yanai, T., Nakamura, M., Sakakibara, H., Uchida, H., Hayashi, K., 1983. Volatile flavor compounds of boiled buckwheat flour. Agric. Biol. Chem. 47, 729–738.

Chapter twenty seven

Bitterness Generation, Rutin Hydrolysis, and Development of Trace Rutinosidase Variety in Tartary Buckwheat

T. Suzuki*, T. Morishita**

*National Agriculture and Food Research Organization (NARO), Kyushu Okinawa Agricultural Research Center, Koshi, Kumamoto, Japan; **National Agriculture and Food Research Organization (NARO) Hokkaido Agricultural Research Center, Hokkaido, Japan

INTRODUCTION

Rutin, a kind of flavonol, is widely distributed throughout the plant kingdom (Sando and Lloyd, 1924; Couch et al., 1946; Haley and Bassin, 1951; Fabjan et al., 2003). Rutin has a number of bioeffects, such as strengthening of blood capillaries, and antioxidative, antihypertensive, and alpha-glucosidase inhibitory activities. Buckwheat is the only cereal known to contain rutin in its seeds. Among buckwheat species, Tartary buckwheat (*F. tataricum* Gaertn.) contains an approximately 100-fold greater amount of rutin in its seeds than that of common buckwheat. From these backgrounds, buckwheat has been identified as a rutin-rich material for food products. However, Tartary buckwheat seed also contains extremely high levels of rutinosidase activity (Fig. 27.1). This activity is sufficient to hydrolyze the rutin present in buckwheat flour [approximately 1–2% (w/w)] within a few minutes after the addition of water (Yasuda et al., 1992; Yasuda and Nakagawa, 1994; Suzuki et al., 2002). Tartary buckwheat rutinosidase activity consists of at least two isozymes with very similar characteristics (Yasuda et al., 1992; Yasuda and Nakagawa, 1994; Suzuki et al., 2002, 2004). To date, a number of researchers have investigated rutin hydrolysis in foods such as bread and confectionaries. However, the main part

345

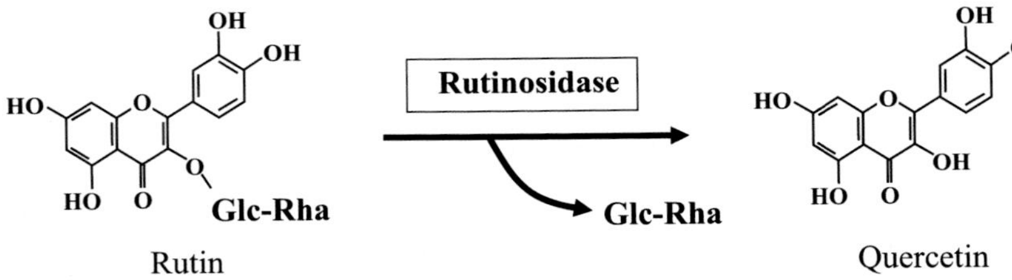

FIGURE 27.1 Catalysis of rutin decomposition in Tartary buckwheat.

of rutin contained in these foods is hydrolyzed as a result of rutinosidase activity (Brunori and Baviello, 2010; Vogrincic et al., 2010; Brunori et al., 2013). In addition, Tartary buckwheat is also known as "bitter buckwheat." Tartary buckwheat flour and products generally have a strong, bitter taste, thereby limiting their use in food products. In Tartary buckwheat dough, at least three bitter compounds have been reported (Kawakami et al., 1995). However, the mechanism responsible for generating the bitter compounds is not clear. In this context, the development of a Tartary buckwheat variety with low rutinosidase activity has been anticipated for the production of rutin-rich and nonbitter foods.

Our research group has identified a trace rutinosidase Tartary buckwheat (Suzuki et al., 2014a) and developed a variety named "Manten-Kirari" (Suzuki et al., 2014b). The rutinosidase activity of the flour is about two or three orders of magnitude less than the common variety such as "Hokkai T8." Moreover, "Manten-Kirari" flour is not bitter. In this review, we summarize the characteristics of rutinosidase, the discovery of trace rutinosidase Tartary buckwheat, and the development and characterization of the trace rutinosidase variety.

CHARACTERIZATION OF RUTINOSIDASE IN TARTARY BUCKWHEAT

To understand rutin decomposition in Tartary buckwheat, it is important to characterize the enzymes that catalyze rutin hydrolysis. Rutinosidase catalyzes the hydrolysis of the 3-glycoside of flavonols such as rutin or isoquercitrin [speculated to be the precursor of rutin (Barber, 1963; Barber and Behrman, 1991; Suzuki et al., 2005)]. Rutinosidase activity has been found in a number of plants (Suzuki, 1962; Yasuda and Nakagawa, 1994; Suzuki et al., 2002; Baumgertel et al., 2003), and microorganisms (Hendson et al., 1992; Narikawa et al., 2000). Although common buckwheat has only a little rutinosidase activity, Tartary buckwheat contains huge amounts of rutinosidase. The first report regarding purification and characterization of rutinosidase in Tartary buckwheat was a rutin-degrading enzyme (RDE) (Yasuda and Nakagawa, 1994). In 2002, as the second report focused on flavonol 3-glycosidase (f3g), also purified and characterized from Tartary buckwheat seeds (Suzuki et al., 2002). Both RDE and f3g have very similar characteristics except for kinetic properties and molecular weight. Both RDE and f3g consist of two major isozymes, which were separated using ion-exchange chromatography. The f3gs were purified to homogeneity from Tartary buckwheat seeds using ammonium sulfate precipitation, ion-exchange chromatography, and gel-filtration chromatography. The molecular weights of each isozyme were 58,200 (f3gI) and 57,400 (f3gII) on sodium dodecyl sulfate polyacrylamide gel electrophoresis (SDS-PAGE) and 89,000 on gel filtration. For both isozymes, rutin 3-glycosidase activity and isoquercitrin 3-glycosidase activity were optimal at pH 5.0 and 40°C. The f3g also had strong activity around the pH of Tartary buckwheat dough (about pH 6.4), therefore f3gs are suitable to hydrolyze rutin in Tartary buckwheat dough. The kinetic constants and V_{max} with rutin and isoquercitrin as substrates were also similar in both isozymes. The optimal pH and optimum temperature were very similar to that of RDE (Yasuda and Nakagawa, 1994). But the molecular weight (RDE: 68,000 on SDS-PAGE and 70,000 by gel filtration) and kinetic constants (RDE: K_m for rutin were 130 mM and 120 mM) were quite different. The amino acid sequences of both f3g isozymes in the amino terminus were identical. These sequences shared identity with other flavonoid glycosidases. The f3g can catalyze rutin (K_m for rutin is about 0.12 mM and V_{max} is about 620 nkat/mg protein) and isoquercitrin 3-glycosidase activity (K_m for isoquercitrin is about 1.1 mM and V_{max} is about 67 nkat/mg protein), though the V_{max} value for isoquercitrin is one-tenth of that for rutin. Isoquercitrin was speculated to be the precursor of rutin (Suzuki, 1962; Barber and Behrman, 1991; Suzuki et al., 2005). From these results, rutinosidase in Tartary buckwheat seed has suitable characteristics to hydrolyze rutin in Tatary buckwheat flour. On the other hand, buckwheat leaves also contain rutinosidase (Baumgertel et al., 2003). The rutinosidase consists of two isozymes: FGHI and FGHII. The molecular weights of rutinosidase are different from that of f3g and RDE; 74.5 kDa for FGHI and 85.3 kDa for FGHII, respectively. The V_{max} value of FGH for rutin was 745 nkat/mg protein, and optimal enzyme activity was seen at pH 4.8; these are almost the same as for f3g. However, K_m for rutin is 0.561 µM, which is quite a lot lower than that of f3g. In Tartary buckwheat leaves and cotyledons, rutinosidase activity, and rutin may have roles for plant defense mechanisms such as UV irradiation and cold stress. It was reported that an antifungal agent, 3,4-dihydroxybenzoic acid, is formed by peroxidase-dependent oxidation of quercetin on browning onion scales (Takahama and Hirota, 2000). Buckwheat also contains peroxidase activity (Kondo et al., 1982; Suzuki et al., 2009). Therefore, in buckwheat leaves, the FGHs might catalyze the first step in the generation of an antifungal agent during germination.

DISCOVERY OF A TRACE RUTINOSIDASE MUTANT OF TARTARY BUCKWHEAT

The trace rutinosidase individuals were first screened from approximately 300 ethyl methanesulfonate mutant lines (about 8400 plants) and 200 genetic resources collected from Nepal, Russia, Europe, and Japan (Suzuki et al., 2014a). As described earlier, rutinosidase in Tartary buckwheat has almost the same characteristics and molecular weight, indicating it is difficult to detect the presence or absence of each isozyme. However, the mobility of each isozyme on native PAGE gel is different. They developed an original staining method to detect each rutinosidase isozyme on gel using a rutin–copper complex (Suzuki et al., 2002). They identified four genetic resources containing seeds, in which no rutinosidase signals were detected using the in-gel detection method. They also investigated in vitro rutinosidase activity in the seeds of individuals within these resources, and divided seeds into two groups: those with high enzyme activity (>400 nkat/g seed) and those will trace enzyme activity (<1.5 nkat/g seed). Among the examined genetic resources, the frequency of individuals with trace rutinosidase characteristics in seeds ranged from 7.1 to 46.4%, whereas the other individuals had normal rutinosidase levels. The geographic location where each genetic resource was collected by Namai and Gotoh (1994) was the foot of the mountain located near Chamaita, south of the Hewa River, in eastern Nepal. The screened individuals were sown, grown for approximately, and individually harvested. After harvesting the seeds, the rutinosidase isozyme composition was again investigated using the in-gel detection method. Individuals that did not have both rutinosidase activities using the in-gel detection method were selected as trace rutinosidase lines, and called "f3g-162," the most promising individual.

They performed an analysis of the progeny from hybridization between "f3g-162" (trace rutinosidase line) and "Hokkai T8" (normal rutinosidase variety). The F_1 seeds of a cross between "f3g-162" (as seed parent) and "Hokkai T8" (as pollen parent) were planted, and F_2 seeds were harvested from each F_1 plant. A total of 157 seeds of the F_2 population were planted, and segregation of rutinosidase activity, isozyme composition of rutinosidase, rutin concentration, and bitterness in F_3 seeds (testa is derived from F_2 progeny) were investigated. The progeny could be clearly divided into two groups: those with activity under 1.5 nkat/g seed (trace rutinosidase individuals) and those with activity over 400 nkat/g seed (normal rutinosidase individuals). A segregation pattern corresponded to a ratio of 1:3 (trace rutinosidase:normal rutinosidase); no individuals with rutinosidase activities between 1.5 and 400 nkat/g seed were found. Among the 157 F_2 progeny, 42 were trace rutinosidase individuals and 115 were normal rutinosidase individuals. This suggested that the trace rutinosidase trait is dominated by a single recessive gene, which was named *rutinosidase-trace A* (*rutA*).

They also evaluated the bitterness of flour prepared from seeds of both trace and normal rutinosidase groups. In trace rutinosidase individuals, none of the panelists detected bitterness in any of the flour samples. On the other hand, all flour samples from normal rutinosidase individuals were uniformly found to have strong bitterness. They also confirmed that rutin was completely hydrolyzed in dough samples from individuals in the normal rutinosidase group, whereas almost no rutin had been hydrolyzed in dough prepared from the flour of trace rutinosidase individuals. The results of progeny analysis indicate that the trace rutinosidase characteristic is conferred by a single recessive gene.

However, Tartary buckwheat seeds contain at least two rutinosidase isozymes (Yasuda and Nakagawa, 1994; Suzuki et al., 2002), and more than two cDNAs corresponding to rutinosidase were cloned and sequenced (Fujino et al., 2012). Their present finding suggests that both isozyme genes are controlled at the transcriptional level by a single regulatory gene and/or are tightly linked. Quercetin, the hydrolysis product of rutin by rutinosidase, has been identified as the principal bittering compound in several fruits (Hladik and Sinmen, 1996; Peterson and Dwyer, 1998). However, Kawakami et al. (1995) reported that Tartary buckwheat dough contains at least three bitter compounds: quercetin and the unidentified compounds "F3" and "F4." Although it remains unclear which of these three compounds is the major cause of bitterness in Tartary buckwheat, the present findings clearly demonstrate that rutin hydrolysis leads to the generation of bitterness in seeds. Based on these results, "f3g-162" is a promising breeding material because of its nonbitter and trace rutinosidase characteristics. However, the agronomical characteristics of "f3g-162," such as maturing time and yield, are not suitable for cultivation in the Hokkaido region, which is the largest production area of Tartary buckwheat in Japan. To develop a nonbitter variety of Tartary buckwheat for practical agronomic cultivation, the cross-breeding of "f3g-162" with a leading variety of buckwheat was required.

BREEDING AND CHARACTERISTICS OF TRACE RUTINOSIDASE VARIETY IN TARTARY BUCKWHEAT

After identification of rutinosidase trace line "f3g-162," breeders tried to improve the agronomic characteristics in terms of maturing time, plant height, and yield. Crosses between the trace rutinosidase line "f3g-162" (as a seed parent) and the normal rutinosidase variety "Hokkai T8" (as a pollen parent) using hot-water emasculation (Mukasa et al., 2009), harvested F_1 seeds were sown, and plants were individually harvested to obtain F_2 seeds, respectively. In the F_2 progeny, plants with trace rutinosidase activity were selected for further propagation. The agronomical characteristics of progeny up to the F_7 generation were evaluated, and the most promising line was named "Mekei T27," and was continually propagated. In 2014, "Manten-Kirari" was officially registered as a variety of Tartary buckwheat with the Japanese Ministry of Agriculture, Forestry, and Fisheries.

Although Tartary buckwheat is recognized as a plant in which it is difficult to employ the cross-breeding method because of its small flower size, "Manten-Kirari" is the first to be bred using this method, which involved hot-water emasculation (Mukasa et al., 2009). This "hot-water emasculation" had a 90% success rate; therefore it is very effective for the crossing of Tartary buckwheat. Tartary buckwheat materials for breeding "semidwarf lines" have been developed by gamma ray-induced mutations (Morishita et al., 2010). This line is promising for improving the agronomic characteristics of "Manten-Kirari" because of strong lodging resistance and low plant height (Morishita et al., 2010; Kasajima et al., 2012, 2013). Further breeding to cross a semidwarf cultivar with "Manten-Kirari" would be effective to develop a new practical cultivar of Tartary buckwheat.

In "Manten-Kirari," many agronomic characteristics such as shortening of the growth period, smaller plant height, and increased grain yields of "f3g-162" (seed parent) were improved. Among these characteristics, they considered that the shortening of the growth period is the most important factor for large-scale cultivation of "Manten-Kirari." In the Hokkaido region, typhoons most frequently occur after Aug., and the nonfrost period extends from late May to late Nov. Therefore, the optimal cultivation period of buckwheat in Hokkaido is limited to between late May and late Aug. "f3g-162" was originally collected in eastern Nepal, where Tartary buckwheat is generally cultivated as an autumn ecotype and is typically sown in late Jun. (Namai and Gotoh, 1994). Therefore, it is difficult for "f3g-162" to reach maturity if cultivated in the Hokkaido region. On the other hand, "Manten-Kirari" reaches maturity in middle Aug., the same as "Hokkai T8." Therefore, "Manten-Kirari" is suitable for cultivation in the Hokkaido region. In addition to the maturation period, the other agronomic characteristics of "Manten-Kirari" were similar to those of "Hokkai-T8." On the other hand, several characteristics, such as t1000-seed weight and 1-L weight of seeds, significantly differed between the two varieties. However, these differences are not expected to limit the practical use of "Manten-Kirari" in agriculture.

"Manten-Kirari" seeds have only a trace rutinosidase activity. Therefore, they hypothesized that foods prepared from "Manten-Kirari" would contain markedly higher levels of rutin with only trace bitterness compared to those of the previous variety. They clarified that only a small amount of rutin in "Manten-Kirari" dough was hydrolyzed to quercetin. Because rutinosidase cannot become active in food under dried conditions or after heat treatment, rutin-rich and nonbitter foods can be made with "Manten-Kirari" dough if it is dried and/or heated within several hours of being prepared. The rutin-hydrolyzed rate is affected by the time course after addition of water to the flour, and by the temperature and blending ratio of Tartary buckwheat flour. The dough rutin content of "Manten-Kirari" was investigated in a time course study of rutin hydrolysis with various water contents and blending ratios of Tartary buckwheat flour (Suzuki et al., 2015). In the normal rutinosidase variety, "Hokkai T8," most of the rutin was hydrolyzed within 30 min after the addition of water, whereas about 90% of rutin remained in "Manten-Kirari." They also investigated the residual rutin ratio of foods such as white bread, butter-enriched roll, pound cake, and galette. In the normal rutinosidase variety, "Hokkai T8," rutin was hydrolyzed completely in all foods tested. On the other hand, 88.5, 49.8, 31.0, and 26.2% of rutin remained in "Manten-Kirari"-containing pound cake, white bread, butter-enriched roll, and galette, respectively. In addition, "Hokkai T8" bread contained strong bitterness, whereas "Manten-Kirari" foods showed minimal bitterness. These results indicate that "Manten-Kirari" is a promising material for the production of rutin-rich and nonbitter food products.

To date, there are no food materials like "Manten-Kirari," which can be used as a staple food with large amounts of flavonoids. Therefore, the toxicity of "Manten-Kirari" should be studied. To investigate the toxicity of rutin-rich dough of "Manten-Kirari," subacute and acute toxicity studies (5,000 and 10,000 mg/kg flour, respectively) were performed using rats (Suzuki et al., 2015). As a control, they employed and compared common buckwheat flour whose rutin concentration is about 1% of that of Tartary buckwheat. In the subacute toxicity study, no toxic symptoms were observed. In addition, no rats died during the test. Food intake and body weight in the "Manten-Kirari"-treated and common buckwheat groups were not significantly different when compared with the control group. On the other hand, some investigated properties, such as urine protein and serum albumin, were significantly different in the "Manten-Kirari" and common buckwheat groups compared with the control group. However, these changes were not caused by toxicity, but by transient changes. On pathologico-anatomic observation, certain abnormalities were observed in some organs of a number of rats. However, the incidental rates in the "Manten-Kirari" and common buckwheat groups did not differ when compared to controls. In addition, no unusual symptoms were observed in the "Manten-Kirari"-treated group when compared with the control group on pathologico-anatomic observation. Therefore, they concluded that these abnormalities may be caused by natural generation. In the acute toxicity study, no rats died and no toxic symptoms were observed during the test. Body weight in the "Manten-Kirari"-treated group was not significantly different when compared with the control group. From these results about examination for safety, they concluded that dough at a dose of 5000 mg flour/kg is a noneffect level.

The commercial cultivation of "Manten-Kirari" was started in 2012, and confectionary products and several noodles containing "Manten-Kirari" flour are commercially available. They confirmed that these products have nonbitterness and most of the rutin has remained without hydrolysis. On the other hand, heat treatment of flour or seeds also inactivates rutinosidase activity in normal rutinosidase varieties. However, the heat inactivation of rutinosidase requires a long treatment duration and high temperature, which deteriorates product quality such as flavor, color, and physical properties as well as increasing production costs. Therefore, the "Manten-Kirari" variety is promising for the production of nonbitter foods compared to varieties with normal rutinosidase activity, because it does not require heat treatment.

In Japan, Tartary buckwheat is recognized as a crop that can be cultivated at the northern limit of the upland farming area. In this upland area, the number of unused and abandoned fields is rapidly increasing because of a decline in field workers; the previous generation of farmers have advanced in age and are unable to work the land. The commonly farmed upland crops such as wheat, potato, sugar beet, and beans cannot grow sufficiently because of the shortened growth season. However, because Tartary buckwheat has a relatively very short growth period of only 80–90 days, it can be cultivated in these farming areas. In addition, Tartary buckwheat can be cultivated with relatively little labor

(Sharama, 2005) and is also suitable for repeated cultivation as it is resistant to replant disease. The yield of common buckwheat is low because the activity of pollen-disseminating insects is markedly reduced by low temperatures often occurring during the flowering period. However, Tartary buckwheat can still be pollinated because of its self-pollinating characteristics even in such low temperature conditions. Because of these advantages, "Manten-Kirari" is regarded as a new crop that can grow in this region in Japan and is also a material for rutin-rich and nonbitter foods.

ACKNOWLEDGMENTS

We thank Dr Y. Mukasa, Dr T. Noda, Dr K. Ishiguro, Mr S. Takigawa, and Dr H. Yamauch for their useful advice in the study. We also thank Mr N. Murakami, Mr T. Yamada, Mr M. Oizumi, Mr T. Saruwatari, Mr S. Nakamura, Mr. K. Suzuki, Mr A. Morizumi, Mr T. Hirao, T. Takakura, K. Abe, Ms K. Fujii, Ms M. Hayashid., and Ms T. Ando for their field and laboratory experiments.

REFERENCES

Barber, G.A., 1963. The formation of uridine diphosphate L-rhamnose by enzymes of the tobacco leaf. Arch. Biochem. Biophys. 103, 276–282.

Barber, G.A., Behrman, E.J., 1991. The synthesis and characterization of uridine 5′-(β-L-rhamnopyranosyl diphosphate) and its role in the enzymatic synthesis of rutin. Arch. Biochem. Biophys. 288, 239–242.

Baumgertel, A., Grimm, R., Eisenbeiss, W., Kreis, W., 2003. Purification and characterization of a flavonol 3-O-β-heterodisaccharidase from the dried herb of *Fagopyrum esculentum* Moench. Phytochemistry 64, 411–418.

Brunori, A., Baviello, G., 2010. The use of Tartary buckwheat whole flour for bakery products: recent experience in Italy. The Annals of the University Dunarea de Jos of Galati, Fascicle VI. Food Technol. 34, 33–38.

Brunori, A., Antonini, A., Baviello, G., Rossi, P., Fiore, A., Nobili, C., 2013. Different treatments applied to Tartary buckwheat whole flour to preserve rutin. The Proceedings of Papers on 12th International Symposium on Buckwheat, ISBN 978-961-93535-1-6.

Couch, J., Naghski, J., Krewson, C., 1946. Buckwheat as a source of rutin. Science 103, 197–198.

Fabjan, N., Rode, J., Kosir, I.J., Wang, Z., Zhang, Z., Kreft, I., 2003. Tartary buckwheat (*Fagopyrum tataricum* Gaertn.) as a source of dietary rutin and quercitrin. J. Agric. Food Chem. 51, 6452–6455.

Fujino, K., Matsui, K., Suzuki, T., Morishita, T., 2012. Analysis of rutinosidase in the new breeding line of *Fagopyrum tartaricum*. Jpn. J. Crop Sci. 81, 170–171.

Haley, T., Bassin, M., 1951. The isolation, purification and derivatives of plant pigments related to rutin. J.Am. Pharm. Assoc. Sci. Ed. 40, 111–112.

Hendson, M., Hildebrand, D.C., Schroth, M.N., 1992. Distribution among pseudomonads of sequences homologous to rutin glycosidase and β-glucosidase genes of *Pseudomonas viridiflava*. Phytopathology 82, 1230–1233.

Hladik, C.M., Sinmen, B., 1996. Taste perception and feeding behavior in nonhuman primates and human populations. Evol. Anthropol. 5, 58–71.

Kasajima, S., Endo, A., Itoh, H., Yoshida, H., Suzuki, T., Mukasa, Y., Morishita, T., Shimizu, A., 2013. Internode elongation patterns in semi dwarf and standard-height genotypes of Tartary buckwheat. Fagopyrum 30, 15–18.

Kasajima, S., Itoh, H., Yoshida, H., Suzuki, T., Mukasa, Y., Morishita, T., Shimizu, A., 2012. Growth, yield, and dry matter production of a gamma ray-induced semi dwarf mutant of Tartary buckwheat. Fagopyrum 29, 7–12.

Kawakami, A., Kayahara, H., Hjihara, A., 1995. Properties and elimination of bitter components derived from Tartary buckwheat (*Fagopyrum tataricum*) flour. Nippon Shokuhin Kagaku Kogaku Kaishi 42, 892–898.

Kondo, K., Kurogouti, T., Matubashi, T., 1982. Report in Research Institute of Food Technology in Nagano Prefecture, vol. 10, p. 150.[in Japanese]

Morishita, T., Mukasa, Y., Suzuki, T., Shimizu, A., Yamaguchi, H., Degi, K., Aii, J., Hase, Y., Shikazono, N., Tanaka, A., Miyazawa, Y., Hayashi, Y., Abe, T., 2010. Characteristics and inheritance of the semi dwarf mutants of Tartary buckwheat (*Fagopyrum tataricum* Gaertn.) induced by gamma ray and ion beam irradiation. Breed. Res. 12, 39–43.

Mukasa, Y., Suzuki, T., Honda, Y., 2009. Suitability of rice-Tartary buckwheat for crossbreeding and for utilization of rutin. JARQ 43, 199–206.

Namai, H., Gotoh, T., 1994. Title in Japanese. Norin Suisan Gijutu Joho Kyokai Norinsuisan Gene Bank no Kisyohseibutsu toh no Idenshigen Chousa Shuushuu Itaku Jigho Seika Hokokusho IV: 1–26.

Narikawa, T., Hirofumi, S., Takaaki, F., 2000. A β-rutinosidase from *Penicillium rugulosum* IFO 7242 that is a peculiar flavonoid glycosidase. Biosci. Biotechnol. Biochem. 64, 1317–1319.

Peterson, J., Dwyer, J., 1998. Flavonoids: dietary occurrence and biochemical activity. Nutr. Res. 18, 1995–2018.

Sando, C., Lloyd, J., 1924. The isolation and identification of rutin from the flowers of elder (*Sambucus canadensis* L.). J. Biol. Chem. 58, 737–745.

Sharama, V.K., 2005. A preliminary study on fertilizer management in buckwheat. Fagopyrum 22, 95–97.

Suzuki, H., 1962. Hydrolysis of flavonoid glycosides by enzymes (rhamnodiastase) from *Rhamnus* and other sources. Arch. Biochem. Biophys. 99, 476–783.

Suzuki, T., Honda, Y., Funatsuki, W., Nakatsuka, K., 2002. Purification and characterization of flavonol 3-glucosidase, and its activity during ripening in Tartary buckwheat seeds. Plant Sci. 163, 417–423.

Suzuki, T., Honda, Y., Mukasa, Y., 2004. Purification and characterization of lipase in buckwheat seed. J. Agric. Food Chem. 52, 7407–7411.

Suzuki, T., Honda, Y., Mukasa, Y., 2005. Effects of UV-B radiation, cold and desiccation stress on rutin concentration and rutin glucosidase activity in Tartary buckwheat (*Fagopyrum tataricum*) leaves. Plant Sci. 168, 1303–1307.

Suzuki, T., Morishita, T., Mukasa, Y., Takigawa, S., Yokota, S., Ishiguro, K., Noda, T., 2014a. Discovery and genetic analysis of non-bitter Tartary buckwheat (*Fagopyrum tataricum* Gaertn.) with trace rutinosidase activity. Breed. Sci. 64, 339–343.

Suzuki, T., Morishita, T., Mukasa, Y., Takigawa, S., Yokota, S., Ishiguro, K., Noda, T., 2014b. Breeding of "Manten-Kirari", a non-bitter and trace-rutinosidase variety of Tartary buckwheat (*Fagopyrum tataricum* Gaertn.). Breed. Sci. 64, 344–350.

Suzuki, T., Morishita, T., Noda, T., Ishiguro, K., 2015. Acute and subacute toxicity studies on the rutin-rich Tartary buckwheat dough in experimental animals. J. Nutr. Sci. Vitaminol. 61, 175–181.

Suzuki, T., Mukasa, Y., Noda, T., Hashimoto, N., Takigawa, S., Matsuura-Endo, C., Yamauchi, H., 2009. Lipase, peroxidase activity and lipoxygenase-like protein content during ripening in common buckwheat (*Fagopyrum esculentum* Moench cv. Kitawasesoba) and Tartary buckwheat (*F. tataricum* Gaertn cv. Hokkai T8). Fagopyrum 26, 63–67.

Takahama, U., Hirota, S., 2000. Deglucosidase of quercetin glucosides to the aglycone and formation of antifungal agents by peroxidase-dependent oxidation of quercetin on browning of onion scales. Plant Cell Physiol. 41, 1021–1029.

Vogrincic, M., Timoracka, M., Melichacova, S., Vollmannova, A., Kreft, I., 2010. Degradation of rutin and polyphenols during the preparation of Tartary buckwheat bread. J. Agric. Food Chem. 58, 4883–4887.

Yasuda, T., Masaki, K., Kashiwagi, T., 1992. An enzyme degrading rutin in Tartary buckwheat seeds. J. Jpn. Soc. Food Sci. Technol. 39, 994–1000.

Yasuda, T., Nakagawa, H., 1994. Purification and characterization of rutin-degrading enzymes in Tartary buckwheat seeds. Phytochemistry 37, 133–136.

Chapter twenty eight

Protease Inhibitors in Buckwheat

H. Chen, J. Ruan

Sichuan Agriculture University, College of Life Sciences, Sichuan, People's Republic of China

INTRODUCTION

Protein inhibitors of proteases are widely distributed in nature, having been found in animals, plants, and microorganisms. The most researched class of inhibitors is the plant protease inhibitors (PIs), which are distributed in many different plant tissues and organs. Generally, plant PIs can be found at high concentrations in storage tissues, such as seeds, where their content can vary from 1% to 15% of total soluble proteins (Wang et al., 2006). The PIs form stable complexes with specific proteases, thereby blocking, changing, or preventing access to the enzyme's active site. Many studies have been published on PI functions in plant defense against bacterial, fungal, and pest attacks (Tsybina et al., 2004). PIs are known to be involved in many biological activities, as anticarcinogenic, antiparasitic, antipest, and antimicroorganism agents, and in blood coagulation and platelet aggregation (Zhou et al., 2015). Research has applied PIs as new medicines in highly active antiretroviral combination therapy to improve the life expectancy of HIV-positive patients (Li et al., 2015). PIs also have the capacity to offset many heritable diseases, such as epilepsy and emphysema. Buckwheat PIs can also disturb the life cycles of many viruses and may help suppress many viral disorders (Bai et al., 2015). The plant serine PIs, especially trypsin inhibitors, have been widely researched, and have been grouped into 18 different families. The best studied of these are the Kunitz, Bowman–Birk, Potato I and II, and Squash families of inhibitors. Kunitz-type trypsin inhibitors are often heterogeneous, 18–24-kDa proteins consisting of many isoforms, with low cysteine content and one or two polypeptide chains; generally, four cysteine residues make up two disulfide linkages with a single trypsin active site, usually an arginine residue situated in one of the protein loops (Oparin et al., 2012). The Bowman–Birk-type PIs are usually 8–10 kDa, with high cysteine content, seven disulfide bridges, and two independent active sites for trypsin and chymotrypsin binding (Yuan et al., 2015).

Molecular Breeding and Nutritional Aspects of Buckwheat. http://dx.doi.org/10.1016/B978-0-12-803692-1.00028-6

Buckwheat is widely distributed at high latitudes and in the colder climatic regions of China, Russia, and some European countries. Common buckwheat (*Fagopyrum esculentum* Moench) and Tartary buckwheat (*Fagopyrum tataricum* L. Gaertn.) are the two main cultivated species that are most commonly grown for food. Common buckwheat is grown mainly in cold plateau and mountainous districts, in Asia, North America, Eastern Europe, and Australia. Tartary buckwheat, an important functional food and medicinal plant product, is planted in mountainous regions and is a common dietary component in East Asian and Central European countries. Many PIs have been isolated from common buckwheat seeds. An 86-amino acid long trypsin inhibitor was isolated and characterized by Ruan et al. (2011) from Tartary buckwheat seeds. Homology analysis showed that this buckwheat trypsin inhibitor is a Potato type I family PI.

CLASSIFICATION AND CHARACTERIZATION OF BUCKWHEAT PIs

According to Kiyohara and Iwasaki (1985), there are two types of buckwheat trypsin inhibitors: temporary and permanent. The temporary trypsin inhibitors have serine as the N-terminal amino acid and are composed of 85–99 amino acid residues. The permanent inhibitors have leucine as the N-terminal amino acid, with a peptide length of 51–67 amino acid residues. The common buckwheat PIs BWI-1 and BTI-1, which both start with leucine, are permanent inhibitors (Belozersky et al., 1995; Pandya et al., 1996). However, the common buckwheat PIs BWI-2a and BWI-2b, which start with serine, are temporary inhibitors (Park et al., 1997). Serine PIs from common buckwheat seeds are separated into two main groups—anionic and cationic, based on their behavior on an ion-exchange column and their isoelectric point (pI). Four cationic (BWI-1c, BWI-2c, BWI-3c, and BWI-4c), and three anionic (BWI-1a, BWI-2a, and BWI-4a) inhibitors have been purified and characterized (Park and Ohba, 2004). The molecular masses of the cationic inhibitors were in the range 6.0–8.5 kDa and of the anionic inhibitors, 7.7–9.2 kDa. The buckwheat trypsin inhibitors showed high pH stability in the pH range of 2–12, and were quite thermostable. On the other hand, germination efficiently degrades buckwheat trypsin inhibitors. Alignment analysis of the amino acid sequences of the examined PIs from common buckwheat seeds suggested that BWI-1c, BWI-3c, and BWI-4c belong to the Potato type I PI family. The common buckwheat trypsin inhibitors showed high sequence similarity (up to 65%) with members of this family. The active site of BWI-1 was suggested to be Arg[45]–Asp[46], based on its sequence homology to inhibitors of this family (Tsybina et al., 2001). The three-dimensional structure of the trypsin inhibitor BWI-2c from common buckwheat seeds was identified by nuclear magnetic resonance spectroscopy, showing two antiparallel α-helices connected by disulfide bonds. BWI-2c constitutes a new family of PIs with a particular α-helical hairpin fold. The linker sequence between the helices is the so-called trypsin inhibitory loop for direct binding to the active site of the enzyme that cleaves BWI-2c at the functionally important residue Arg[19]. The inhibition constant was determined for BWI-2c against trypsin to be 1.7×10^{-10} M, and BWI-2c was investigated with other enzymes, including those from various insect digestive systems: it showed high specificity to trypsin-like proteases (Oparin et al., 2012). Li et al. (2006) cloned and identified a new trypsin inhibitor gene from common buckwheat. Sequence analysis demonstrated that its 392-bp cDNA has an open reading frame of 216 bp, encoding 72 amino acids. The inferred amino acid sequence displayed 96% and 93% homology with BWI-1 and BTI-2, innate trypsin inhibitors from common buckwheat seeds. Ruan et al. (2015) designed degenerate primers for the conserved Tartary buckwheat trypsin inhibitor amino acid sequence, and cloned a novel trypsin inhibitor gene (*FtTI*). The full-length sequence of *FtTI* was obtained by rapid amplification of cDNA ends and genome walking, and contained 644 bases. The *FtTI* gene had an open reading frame of 264 bp, encoding a protein of 87 amino acid residues. FtTI had the typical inhibitory motif of the serine superfamily, namely, the central signature area XENXXV and WPELVG-like motif in the N-terminal region, and a conserved RCDRV motif in the C-terminal region. Dixon plot analysis showed that the Tartary buckwheat trypsin inhibitor (FtTI) is a competitive inhibitor, as reflected by the intersection of the two curves corresponding to the substrates above the x-axis. The dissociation constant (K_i) was 1.6×10^{-9} M, indicating that FtTI is a highly effective inhibitor of bovine trypsin. According to sequence homology analysis with PIs of the Potato type I PI family, the reactive site of FtTI also contains Asp[66]–Arg[67] (Ruan et al., 2011).

INSECT-RESISTANT ACTIVITY OF BUCKWHEAT PIs

PIs are natural resistance-related proteins that are often found in seeds and are induced in some plant tissues by wounding or herbivory. PIs play a vital protective role in plants against insect and pest attack. Great progress has been made in recent years toward the design of extremely effective broad-spectrum inhibitors and, in the field, of PI-expressing transgenic plants resistant to major herbivorous pests. Buckwheat PIs also play an important role in the plant's natural protection against pathogenic fungi, as well as insect pests. Moreover, PIs do not have any proven deleterious or toxic effects on humans or domestic animals. These merits make buckwheat serine PIs the perfect candidate for application in transgenic crops for resistance to insect pests. Wang et al. (2006) confirmed that the trypsin inhibitor TBTI2 from Tartary buckwheat seeds can suppress the growth and development of cotton bollworm (*Helicoverpa armigera*) larvae. To a certain extent this PI might be bioactive and a valuable antiinsect factor. The average daily weight gain of cotton bollworm larvae fed an artificial diet containing different concentrations of TBTI2 was significantly reduced after 12 days relative to controls fed only water. The relative growth rate was suppressed to 50.4% of controls when food contained 180 IU of TBTI2. The development of all tested insects was delayed to some extent. TBTI2 transgenic plants were resistant to the greenhouse; white wing butterfly: a considerable number of eggs were laid on the control plant leaves and full value insects incubated from these eggs. Only single ovipositors of 1–2 eggs were observed on transgenic plants, and no progeny emerged from them (Khadeeva et al., 2009). Significant genotypic variations in defense efficiency were observed between PI members belonging to different genera of the same family (potato and tobacco), as well as among different lines of the same species. Adoption of a single gene of common buckwheat serine PI into plants of heterologous groups has demonstrated a protective effect against insects. Compared to control plants those grown by us were able to synthesize some functional materials that display a protective effect and restrict growth and development of insects (Zhou et al., 2015). The *FtTI* gene encoding the trypsin inhibitor from Tartary buckwheat was successfully cloned, expressed in *Pichia pastoris*, and inspected for regulatory effects on insect development and growth. FtTI had a higher toxic effect on *Mamestra brassicae* larvae than soybean trypsin inhibitor. The median lethal concentration for the larvae was 15 µg/mL (Ruan et al., 2015).

THE ANTIMICROBIAL AND ANTIFUNGAL ACTIVITIES OF BUCKWHEAT PIs

PIs have also been investigated for their important role in the plant's natural defense against fungal infections. Studies of varieties of *Leguminosae*, *Solanaceae*, and *Poaceae* cultures have shown that they have endogenous inhibitors specific to proteases of the phytopathogenic fungi *Botrytis*, *Colletotrichum*, *Fusarium*, and *Helminthosporium*. It is very important that plants contain many inhibitors, which act not only on animal and bacterial proteases, but also on those of filamentous fungi. Spores of various plant-pathogenic fungi may germinate, permeate, and cause degradation of the plant cell wall. The plant responds to these injuries by secreting protective compounds (eg, PIs), thereby restraining and weakening extracellular fungal enzymes. These protective measures prevent plant cell degradation. At least a few PIs present in seeds of buckwheat serve as a component of the buckwheat defense system. Buckwheat PIs have been shown to possess marked potential for protecting plants against a wide range of microbial plant pathogens. The ability of PIs from common buckwheat seeds to suppress enzymes secreted by plant-pathogenic fungi is correlated with suppression of seed germination and mycelial growth. BWI-1c and BWI-2c, which are cationic PIs from buckwheat seeds, showed a broad spectrum of antifungal activity, but had no effect on bacterial enzymes. BWI-3c and BWI-4c inhibitors protect the plants from both fungal and bacterial infection (Tsybina et al., 2004). According to the obtained data it was recommended that cationic inhibitors from buckwheat seeds take part in the defense of plants against fungal and bacterial infection. The ability of the PIs from common buckwheat seeds to repress germination of spores of the filamentous phytopathogenic fungi *Fusarium oxysporum* and *Alternaria alternata* and to suppress the proteases secreted by these fungi is worthy of attention. The anionic trypsin inhibitors from common buckwheat seeds suppress growth of these fungi's mycelia and inhibit proteases of animal and bacterial origin. Through structural similarity analysis (Oparin et al., 2012), common buckwheat trypsin inhibitor BWI-2c was capable of inhibiting infection by a microorganism. Ruan et al. (2011) reported that a 14-kDa Tartary buckwheat trypsin inhibitor (FtTI) fully inhibits the growth of *Mycosphaerella melonis*, *Alternaria cucumerina*, *Alternaria solani*, *Phytophthora capsici*, and *Colletotrichum gloeosporioides*, even at concentrations below 30 µg/mL. PIs from Tartary buckwheat seeds inhibited the proteases of fungal pathogens and suppressed fungal spore germination and mycelial growth. These results imply that protein PIs are involved in the defensive reaction of plants under stress conditions. Another member of the Potato type I PI family, BWI-1, was shown to suppress spore germination and mycelial growth of the phytopathogenic fungi *A. alternata* and *F. oxysporum* (Dunaevsky et al., 1997). PIs from common buckwheat seeds dramatically suppressed the activity of proteases secreted by mycelial fungi, exogenous mammalian proteolytic enzymes, and bacterial proteases (Dunaevskii et al., 2005). When the gene encoding BWI-1 was transformed into tobacco and potato plants, it significantly suppressed at least two bacterial phytopathogens, *Clavibacter michiganensis* ssp. *michiganensis* and *Pseudomonas syringae* pv. *tomato* (Khadeeva et al., 2009). The gene encoding the common buckwheat trypsin inhibitor BWI-1 was expressed in transgenic tobacco plants under the control of the constitutive 35S promoter of cauliflower mosaic virus and displayed numerous alterations in microsporogenesis (Khadeeva and Yakovleva, 2010).

ANTITUMOR AND ANTI-HIV ACTIVITIES OF BUCKWHEAT PIs

PIs from buckwheat seeds can also inhibit HIV-1 reverse transcriptase (Park and Ohba, 2004). Aside from their proteolytic enzyme inhibitory and antimicroorganism activities, PIs from buckwheat seeds can also inhibit HIV-1 reverse transcriptase and the proliferation of various kinds of cancer cells (Park and Ohba, 2004; Leung and Ng, 2007). The inhibitory effects of BWI-1 and BWI-2a on human tumor T-ALL cell lines CCRF-CEM and JURKAT and normal blood lymphocytes were tested by tetrazolium/formazan (MTT) assay. BWI-1 and BWI-2a showed significant repression, decreasing cellular dehydrogenase activity to 30% and 10%, respectively, at concentrations of up to 5 µg/mL. The inhibitory effect of BWI-2a seemed to be a little stronger than that of BWI-1, in contrast to the case at 50 µg/mL. The anticancer activity of recombinant buckwheat trypsin inhibitor (rBTI) could be, in part, because of its induction of tumor cell apoptosis via upregulation of Bak and Bax, and downregulation of Bcl-xl and Bcl-2, causing mitochondrial release of cytochrome C into the cytosol, activation of caspases 3 and 9, and disruption of mitochondrial transmembrane potential. Zhang et al. (2007) demonstrated the biological effect of rBTI on multiple myeloma IM-9 cells; not only did these cells undergo apoptosis, but also IM-9/Bcl-2 cells were markedly restrained in the presence of rBTI. How rBTI restricts the multiplication of IM-9 cells and whether other plant trypsin inhibitors have an inductive effect on apoptosis of tumor cells have yet to be elucidated. Trypsin inhibitors may influence the expression of one or more cancer-promoting factors. Because some swelling tumor cells have a particular receptor for trypsin inhibitors on their surface, when binding occurs, tumor cell incursion and proliferation can be suppressed and the cellular matrix can also be destabilized. Khadeeva et al. (2009) found that rBTI can specifically suppress the growth of K562 cells in a dose-dependent manner, but there were also slight effects on ordinary human peripheral blood mononuclear cells by flow cytometry analysis and MTT assays. The rBTI can induce apoptosis in some types of human solid tumor cells (HeLa, EC9706, and HepG2) through the mitochondrial apoptotic pathway (Li et al., 2009). Buckwheat PIs inhibited proliferation of liver embryonic WRL68 cells, HepG2 (hepatoma) cells, L1210 (leukemia) cells, and breast cancer (MCF-7) cells with an IC_{50} of 37, 33, 25, and 4 mM, respectively. On the other hand, rBTI was unable to induce nitric oxide production by macrophages or promote a mitogenic reaction from splenocytes (Li et al., 2009). Li et al. (2015) demonstrated that rBTI could decrease growth of cell and metastasis, and induce cell apoptosis by preventing the cell cycle, giving rise to loss of mitochondrial transmembrane potential and consequent caspase-dependent apoptosis in breast cancer cells. The rBTI could irreversibly stimulate NFκB in MCF-7 cells by increasing the phosphorylation of IKKα/β and IκBα, actuating p65 nuclear migration and promoting the generation of reactive oxygen species. rBTI could also facilitate nuclear glucocorticoid receptor expression by promoting the NFκB/p65 pathway. This implied that NFκB/p65 activation is involved in the adjustment of rBTI-induced glucocorticoid acceptor expression in MCF-7 cell lines. Bai et al. (2015) found that rBTI especially suppressed growth of the H22 hepatic carcinoma cell line in vitro and in vivo in a time-dependent and concentration-dependent manner, while there were only slight effects on normal liver cell line 7702. Apart from this, rBTI-induced apoptosis in H22 cells was, at least in part, mediated by the mitochondrial pathway via caspase 9 revealed by DNA electrophoresis, morphological observation of the nuclei, flow cytometry, MTT assay, measurement of caspase activation, and assessment of cytochrome C. rBTI-induced apoptosis occurred in association with mitochondrial dysfunction, leading to the release of cytochrome C from the mitochondria to the cytosol, as well as activation of caspases 3, 8, and 9. Further studies demonstrated that buckwheat PIs can also suppress HIV-1 reverse transcriptase with an IC_{50} of 5.5 mM. Trypsin inhibitors from large brown Japanese buckwheat seeds slightly restrained the activity of HIV-1 reverse transcriptase and the proliferation of hepatoma HepG2 cells and breast cancer MCF-7 cells with IC_{50} values of 48, 79.2, and 63.8 μM, respectively (Yuan et al., 2015).

THE ALLERGENIC ACTIVITY OF BUCKWHEAT PIs

Food allergy is a common complaint that often leads to self-imposed food avoidance. Food allergies can be simplistically categorized into three main types: IgE mediated, mixed (IgE /non-IgE mediated), and non-IgE mediated (cellular, delayed type). Trypsin inhibitors are known to be major allergens in buckwheat. Currently, four main allergy proteins have been found in this plant. As already noted, buckwheat trypsin inhibitors BWI-1 and BWI-2b have IgE-binding activity, albeit to a low extent, as determined by radioallergosorbent test, indicating that they might be minor allergenic proteins in buckwheat seeds (Park et al., 1997). The allergens of 9, 16, 19, and 24 kDa are strong candidates as main allergens, and N-terminal sequence analyses demonstrated that the 9, 16, and 19 kDa allergens have some homology with buckwheat PIs. Attenuation of IgE binding to the 9-kDa allergen was analyzed with periodate oxidative treatment. IgE binding to the 9-kDa allergen transferred to a nitrocellulose membrane showed a clear decrease after periodate oxidative treatment, and that to the 16-kDa allergen showed a moderate decrease. N-terminal amino acid sequence analyses of 16-kDa and 19-kDa allergens showed low and medium homology to the amylase/trypsin inhibitor of millet and buckwheat trypsin inhibitor, respectively. N-terminal sequence analysis of the major 9-kDa buckwheat allergen indicated that it is the isoallergenic trypsin inhibitor of buckwheat, BWI-2a/BWI-2b (Park et al., 2000). The 16-kDa antidigestion buckwheat allergenic protein appeared to be responsible for immediate hypersensitivity reactions, including anaphylactic reaction after buckwheat ingestion, while the 24-kDa allergen was sensitive to pepsin digestion and responsible for CAP system fluorescein-enzyme immunoassay but not immediate hypersensitivity reactions (Tanaka et al., 2002). The 24-kDa major allergen protein from Tartary buckwheat seeds was purified and identified by Superdex-75 gel filtration and Resource Q ion-exchange column chromatography. Its pI was estimated at approximately 5.6, with 15% arginine and 30% glutamine. The allergen protein was a trypsin inhibitor as determined by sequence alignment analysis and showed intense IgE-binding activity in an immunoblotting analysis experiment of sera collected from many patients that were allergic to buckwheat (Wang et al., 2004).

CONCLUDING REMARKS AND FUTURE PROSPECTS FOR BUCKWHEAT PIs

PIs have long been recognized as significant participants in the plant's exogenous security. Many transgenic plants overexpressing particular PIs have been produced, with resistance to many different pathogenic microflora. However, this phenomenon has yet to be completely understood, and its ramifications may go far beyond its current uses. Buckwheat PIs are generally believed to prevent tumor development and metastasis. Today, synthetic buckwheat PIs make up part of the combination therapies against AIDS, and have the potential for application as medicines against many other illnesses. Future research to find additional enzyme–inhibitor interactions at the plant–pathogen junction requires cross-disciplinary methods including structural biology, advanced proteomics, and genomics. Genomic methods will contribute to identifying candidate inhibitors and enzymes and define the sites of positive selection. Structural biology will help understand the pattern of repression, the selective pressures at the interactive surface, and the evolutionary source of the inhibitors. To better understand the mechanism governing the interactions between host plant and phytopathogen, it is essential to strengthen the search for targets of inhibitors in pathogenic fungi. We believe that further studies are needed to more precisely characterize the major buckwheat trypsin inhibitor allergens and to determine the presence of cross-reactivity.

REFERENCES

Bai, C.Z., Feng, M.L., Hao, X.L., Zhao, Z.J., Li, Y.Y., Wang, Z.H., 2015. Anti-tumoral effects of a trypsin inhibitor derived from buckwheat in vitro and in vivo. Mol. Med. Rep. 12 (2), 1777–1782.

Belozersky, M.A., Dunaevsky, Y.E., Musolyamov, A.X., Egorov, T.A., 1995. Complete amino acid sequence of the protease inhibitor from buckwheat seeds. FEBS Lett. 371 (3), 264–266.

Dunaevskii, Y.E., Tsybina, T.A., Belyakova, G.A., Domash, V.I., Sharpio, T.P., Zabreiko, S.A., Belozerskii, M.A., 2005. Proteinase inhibitors as antistress proteins in higher plants. Appl. Biochem. Microbiol. 41 (4), 344–348.

Dunaevsky, Y.E., Gladysheva, I.P., Pavlukova, E.B., Beliakova, G.A., Gladyshev, D.P., Papisova, A.I., Larionova, N.I., Belozersky, M.A., 1997. The anionic protease inhibitor BWI-1 from buckwheat seeds. Kinetic properties and possible biological role. Physiol. Plant. 101 (3), 483–488.

Khadeeva, N.V., Kochieva, E.Z., Tcherednitchenko, M.Y., Yakovleva, E.Y., Sydoruk, K.V., Bogush, V.G., Dunaevsky, Y.E., Belozersky, M.A., 2009. Use of buckwheat seed protease inhibitor gene for improvement of tobacco and potato plant resistance to biotic stress. Biochemistry 74 (3), 260–267.

Khadeeva, N.V., Yakovleva, E.Y., 2010. Inheritance of marker and target genes in seed and vegetative progenies of transgenic tobacco plants carrying the buckwheat serine protease inhibitor gene. Russ. J. Genet. 46 (1), 50–56.

Kiyohara, T., Iwasaki, T., 1985. Chemical and physicochemical characterization of the permanent and temporary trypsin inhibitors from buckwheat. Agric. Biol. Chem. 49 (3), 589–594.

Leung, E.H.W., Ng, T.B., 2007. A relatively stable antifungal peptide from buckwheat seeds with antiproliferative activity toward cancer cells. J. Peptide Sci. 13 (11), 762–767.

Li, Y., Wu, Y., Cui, X., Wang, Z., 2015. NFκB/p65 activation is involved in regulation of rBTI-induced glucocorticoid receptor expression in MCF-7 cell lines. J. Funct. Foods 15, 376–388.

Li, Y., Zhang, Z., Liang, A., Wang, Z., 2006. Cloning and characterization of a novel trypsin inhibitor (BTIw1) gene from Fagopyrum esculentum. Mitochondr. DNA 17 (3), 203–207.

Li, Y.Y., Zhang, Z., Wang, Z.H., Zhang, L., Zhu, L., 2009. rBTI induces apoptosis in human solid tumor cell lines by loss in mitochondrial transmembrane potential and caspase activation. Toxicol. Lett. 189 (2), 166–175.

Pandya, M.J., Smith, D.A., Yarwood, A., Gilroy, J., Richardson, M., 1996. Complete amino acid sequences of two trypsin inhibitors from buckwheat seed. Phytochemistry 43 (2), 327–331.

Park, S.S., Abe, K., Kimura, M., Urisu, A., Yamasaki, N., 1997. Primary structure and allergenic activity of trypsin inhibitors from the seeds of buckwheat (Fagopyrum esculentum Moench). FEBS Lett. 400 (1), 103–107.

Park, J.W., Kang, D.B., Kim, C.W., Ko, S.H., Yum, H.Y., Kim, K.E., Hong, C.-S., Lee, K.Y., 2000. Identification and characterization of the major allergens of buckwheat. Allergy 55 (11), 1035–1041.

Park, S.S., Ohba, H., 2004. Suppressive activity of protease inhibitors from buckwheat seeds against human T-acute lymphoblastic leukemia cell lines. Appl. Biochem. Biotechnol. 117 (2), 65–74.

Oparin, P.B., Mineev, K.S., Dunaevsky, Y.E., Arseniev, A.S., Belozersky, M.A., Grishin, E.V., Egorov, T.A., Vassilevski, A.A., 2012. Buckwheat trypsin inhibitor with helical hairpin structure belongs to a new family of plant defence peptides. Biochem. J. 446 (1), 69–77.

Ruan, J.J., Chen, H., Shao, J.R., Wu, Q., Han, X.Y., 2011. An antifungal peptide from *Fagopyrum tataricum* seeds. Peptides 32 (6), 1151–1158.

Ruan, J., Yan, J., Hou, S., Chen, H., Wu, Q., Han, X., 2015. Expression and purification of the trypsin inhibitor from Tartary buckwheat in *Pichia pastoris* and its novel toxic effect on *Mamestra brassicae* larvae. Mol. Biol. Rep. 42, 209–216.

Tanaka, K., Matsumoto, K., Akasawa, A., Nakajima, T., Nagasu, T., Iikura, Y., Saito, H., 2002. Pepsin-resistant 16-kD buckwheat protein is associated with immediate hypersensitivity reaction in patients with buckwheat allergy. Int. Arch. Allerg. Immunol. 129 (1), 49–56.

Tsybina, T.A., Dunaevsky, Y.E., Musolyamov, A.K., Egorovb, T.A., Belozersky, M.A., 2001. Cationic inhibitors of serine proteinases from buckwheat seeds. Biochemistry 66 (9), 941–947.

Tsybina, T.A., Dunaevsky, Y.E., Popykina, N.A., Larionova, N.I., Belozersky, M.A., 2004. Cationic inhibitors of serine proteinases from buckwheat seeds: study of their interaction with exogenous proteinases. Biochemistry 69 (4), 441–444.

Wang, Z., Zhang, Z., Zhao, Z., Wieslander, G., Norbäck, D., Kreft, I., 2004. Purification and characterization of a 24 kDa protein from Tartary buckwheat seeds. Biosci. Biotechnol. Biochem. 68 (7), 1409–1413.

Wang, Z., Zhao, Z., Zhang, Z., Yuan, J., Noback, D., Wieslander, G., 2006. Purification and characterization of a protease inhibitor from *Fagopyrum tataricum* Gaertn. seeds and its effectiveness against insects. Chin. J. Biochem. Mol. Biol. 22 (12), 960–965.

Yuan, S., Yan, J., Ye, X., Wu, Z., Ng, T., 2015. Isolation of a ribonuclease with antiproliferative and HIV-1 reverse transcriptase inhibitory activities from Japanese large brown buckwheat seeds. Appl. Biochem. Biotechnol. 175 (5), 2456–2467.

Zhang, Z., Li, Y., Li, C., Yuan, J., Wang, Z., 2007. Expression of a buckwheat trypsin inhibitor gene in *Escherichia coli* and its effect on multiple myeloma IM-9 cell proliferation. Acta Biochim. Biophys. Sinica 39 (9), 701–707.

Zhou, X., Wen, L., Li, Z., Zhou, Y., Chen, Y., Lu, Y., 2015. Advance on the benefits of bioactive peptides from buckwheat. Phytochem. Rev. 14 (3), 381–388.

Chapter twenty nine

Buckwheat Tissue Cultures and Genetic Transformation

G. Suvorova

All-Russia Research Institute of Legumes and Groat Crops,
Laboratory of Genetics and Biotechnology, Orel, Russia

INTRODUCTION

The growth of isolated organs, cells, and tissues on nutrient media in glass (in vitro) is called tissue culture. Since the middle of the 20th century the tissue culture technique has developed rapidly for many plant species, opening up great opportunities in the study of cell biology as well as for practical use. As for buckwheat, unfortunately the crop has never been a favorite plant for biotechnological experiments. Nevertheless, for almost 40 years tissue culture approaches for buckwheat have been developing. This chapter overviews the history of buckwheat tissue culture in vitro from the first successful protocols of plantlet regeneration up to genetic engineering manipulations.

CALLUS FORMATION AND MORPHOGENESIS

The first differentiation of buckwheat plants from the long-term cultivated callus was induced by Yamane (1974). Callus was obtained from the young seedlings on White's medium supplemented with 2,4-dichlorophenoxiacetic acid (2,4-D). Plant regeneration spontaneously occurred when the calli were transferred to the fresh RM-1964 medium containing coconut milk. The five plants that developed from the callus tissue after 48 months of subculture were diploid and showed normal morphology. The weakness in this work was an undefined composition of the morphogenic medium with respect to growth regulators.

The classic protocol that described all stages of buckwheat explant growth in vitro was developed by Srejovic and Neskovic (1981). To regenerate the whole plantlets from the cotyledon fragments they provided three different media, which were subsequently used on a specific phase of growth. They used a richer

365

mineral base of B5 medium instead of White's and suggested the application of specific compositions and concentrations of growth regulators. On the callus induction phase a medium with high auxin and low cytokinin content was used (5.0 mg/L 2,4-D + 0.1 mg/L kinetin). Shoot regeneration was achieved on a medium with a high cytokinin:auxin ratio [10^{-5} M 6-benzylaminopurine (BAP) + 10^{-6} M indole-3-acetic acid] and the root formation was induced by indole-3-butiric acid at a concentration of 1.0 mg/L.

At the same time attempts were made in plant regeneration from the isolated protoplast of common (Adachi et al., 1989) and Tartary buckwheat (Lachmann and Adachi, 1990) and their fusion (Lachmann et al., 1994). The best results were achieved with common buckwheat. The protoplast culture technique was developed for further somatic hybridization, which later was replaced by embryo rescue and genetic transformation.

Next, the newly synthesized compounds were tested in buckwheat tissue culture. The phenyl urea derivatives (thidiazuron) were found to be active as the growth regulators in most cases as cytokinins (Guo et al., 1992; Berbec and Doroszewska, 1999), but sometimes they possessed both auxin and cytokinin activity (Zhang and Chen, 1992).

Plant regeneration via callus formation and subsequent morphogenesis from the cotyledon or hypocotyl segments were obtained in many experiments with some modifications of culture media (Takahata and Jumonji, 1985; Rajbhandari et al., 1995; Hao et al., 1998; Woo et al., 2000; Jin et al., 2002). The induction of multiple shoot organogenesis from cotyledons of buckwheat seedlings was obtained by bypassing the callus formation stage (Park and Park, 2001). The regeneration capacity of buckwheat explant was influenced by the plant genotype (Luthar and Marchetti, 1994; Berbec and Doroszewska, 1999).

One of the pathways of differentiation in buckwheat explants can be the induction of somatic embryogenesis. A series of experiments on the study of somatic embryogenesis in somatic tissues of buckwheat has been carried out by Gumerova et al. (1998, 2001, 2003). An indirect somatic embryogenesis was induced via the development of proembryogenic cell complexes, which formed in the hypocotyl explants of 4–5-day-old seedlings.

Rumyantseva et al. (1989a, 1992) have established the long-term culture of morphogenic calli of Tartary buckwheat. Kostyukova and Rumyantseva (2010) found that the formation of embryoidogenic callus is connected with certain competent cells in procambial and subepidermal layers in cotyledons of immature embryos of Tartary buckwheat.

Protocols of somatic embryogenesis from hypocotyls of Tartary and common buckwheat inducing callus formation followed by plantlet regeneration with the use of a complex combination of growth regulators have been developed (Han et al., 2011; Kwon et al., 2013).

Thus for several decades, efficient methods of somatic cell dedifferentiation in the primary explants of buckwheat followed by subsequent tissue and organ differentiation have been developed in detail, giving efficient and reproducible protocols, which have been realized with more or less success.

MICROPROPAGATION

The micropropagation technique is considered to be well developed for buckwheat when seeds, seedlings, or their fragments are used as the primary explants. Because micropropagation presumes the obtaining of uniform clones of the same genotype, the preferred way of clonal propagation should be direct shoot regeneration without callus formation, which avoids somaclonal variation.

Neskovic et al. (1990a) demonstrated the efficiency of clonal propagation of buckwheat plants from the apical meristem of aseptic young seedlings using the determinate genotype as a marker of genetic stability. The regenerated plantlets as well as their progenies had a determinate growth habit with no signs of segregation.

Since micropropagation often finishes all in vitro manipulations, a number of experiments were designed to optimize the method. Bohanec (1987) suggested removing radicula from the germinated seeds after a few days to ensure close contact between medium and explant. Ramchatogoen et al. (1998) demonstrated multiple soot proliferation from a single shoot excised from the maternal tissue for the next cycle of multiplication.

Comparing the different approaches to buckwheat micropropagation Klčová and Gubišová (2008) found that direct regeneration in the nodal segment was the most efficient for rapid multiplication. They did not observe any differences in the explant growth between the basal media mineral composition, but found that agar was the best gelling substance for buckwheat compared to Phytagel.

Five types of cytokinins have been studied by Dobránszki (2009) in an attempt to influence the shoot multiplication rate of seedling segments of common buckwheat. The best results were obtained when meta-Topoline (TOP) was used as a cytokinin and shoot tips were used as an initial explant. The applied BAP or TOP mostly stimulated the growth of shoot rosettes while 6-(γ,γ-dimethylallylamino)purine and kinetin induced a long stem growth.

At the same time it was pointed out that for plant-breeding purposes, it would be of much greater value if it were possible to induce micropropagation from parts of plants at the adult stage when the breeding status of individual plants could be estimated (Bohanec, 1987). While attempts to use stem or leaf petiole cuttings failed because of explant contamination, Bohanec (1987) used an inflorescence as an initial organ for successful induction of micropropagation to start from adult plants. Emphasizing that the maintenance of valuable genotypes of common buckwheat was difficult because of heterostylic self-incompatibility, Takahata (1988) came to the same conclusion and established the direct regeneration system from immature inflorescence both for common and perennial (*Fagopyrum cymosum*) buckwheat.

Suvorova et al. (1998) was successful in buckwheat micropropagation via axillary meristem culture developing the double sterilization procedure for primary explants. The axillary buds or segments of secondary branches were taken out of the field-grown mature plants. The buds were cleaned from the surrounding ocrea, washed and sterilized, then partially cleaned from the leaf primordia and sterilized again. After the double sterilization the leaf primordia

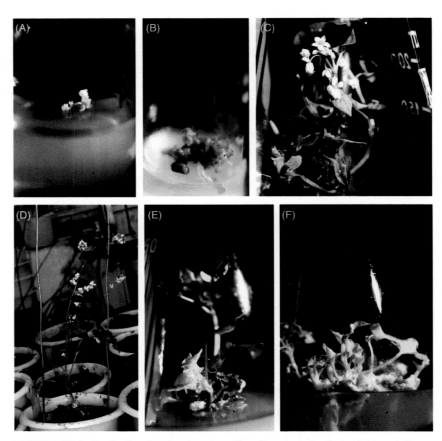

FIGURE 29.1 Buckwheat micropropagation by means of meristem culture.
(A,B) Meristem development in vitro; (C) in vitro flowering; (D) R_0 plant of determinate variety Dicul; (E) mixed morphogenic tissue derived from meristem of the chlorophyll deficiency mutant; (F) chlorophyll deficiency morphogenic tissue.

were removed completely under stereomicroscope and an axillary meristem 0.5–1.0 mm in size was inoculated on the regeneration medium followed by subculture on MS hormone-free medium. The main phases of meristem development are shown in Fig. 29.1(A,B,C).

The method was applied to certain breeding populations of buckwheat (Suvorova et al., 2004). Eight plants out of the 50 large-seeded genotypes selected in the determinate cultivar Dicul were successfully cloned. All the R_0 plants regenerated from the eight tissue clones were characterized by determinate growth (Fig. 29.1D) as well as all plants of the next generations that demonstrated the absence of somaclonal variation. As a result of cloning of a chlorophyll deficiency mutant plant the mixed morphogenic tissue with green, yellow, and variegated sectors was obtained. During subculture the mixed tissue segregated in the green chlorophyll-containing structures and the chlorophyll-deficient tissue (Fig. 29.1E,F) most likely indicated the cytoplasmic nature of mutation. The advantages of the developed micropropagation technique are that the biotechnological approach could be used together with classic buckwheat breeding allowing to reproduce again at the juvenile stage the selected genotype that was considered lost because of the cross-pollination.

The original method of plant regeneration from the young-expanding-leaf-petiole explants taken out from the young plants grown in the greenhouse was developed by Slavinska et al. (2009). In the experiment an effect of triiodobenzoic acid on shoot regeneration was shown. The developed system gave the chance to compare the regenerants and the plants grown from seeds.

The most efficient clonal propagation of the medically valuable mutant plant of perennial buckwheat was performed by Chen et al. (2012). After 2 years of micropropagation through the axillary meristems of adult plants the regenerants were planted into the field and none showed any morphological variants or biochemical and genetic changes. So the unique red-stem mutant of perennial buckwheat with high levels of epicatechin was successfully cloned. Direct plant regeneration using axillary buds was recommended for large-scale production of elite plants.

All developed techniques of buckwheat micropropagation avoided callus induction in the isolated explants and obtained clones genetically identical to the parental plants. But if the regeneration processes are stimulated by any strong factors such as the high concentration of growth regulators causing cell dedifferentiation, the long-term culture, or special selective agents added to the medium, the chance of somaclonal variation emergence is increased. Somaclonal variants of buckwheat were described by Barsukova (2004). The regenerants obtained from the callus tissue of cotyledons, hypocotyls, or immature embryos had characteristics that were different from the initial cultivar Prymorskaya. Variants with winged seeds and black and dark-brown seed color were found in the R_2–R_3 generations. The cell selection for resistance to the copper ions is now being developed (Barsukova, 2013).

But, as for buckwheat, sometimes it is difficult to determine what are the somaclonal variants and what are the recessive homozygotes obtained as a result of the limited cross-pollination in the process of the seed propagation of regenerants.

CULTURE OF IMMATURE EMBRYOS

The immature embryos are very favorable objects for in vitro manipulation because of their high regeneration capacity. The embryos could be used both for callus induction and multiple shoot formation, but the main objective of embryo culture has been an embryo rescue technique because the hybrid ovaries that failed to grow in planta could continue their development by being planted in vitro.

Neskovic et al. (1987) demonstrated the high regeneration ability of immature buckwheat embryos and found three tissue types selected in later subcultures: morphogenic callus tissue, embryogenic tissue, and unorganized nonmorphogenic tissue. Rumyantseva et al. (1989) concluded that in isolated buckwheat embryo culture, all pathways of morphogenesis could be observed: embryogenesis, embryoidogenesis, gemmogenesis, and histogenesis. Suvorova and Kostrubin (1992) found that embryo response depended on the embryo age and recommended a differentiated approach to the different age groups of immature ovules. The young ovules of 4–9 days old were recommended to be grown on a medium containing both auxin and cytokinin. If ovules were isolated later than 9 days after pollination they could be grown on a medium with auxin only. The nearly maturing embryos could germinate on a medium without growth regulators. Wang and Campbell (1999) showed that mature buckwheat embryos could germinate in Petri dishes with ½ MS medium and recommended the method to speed up the breeding cycles.

The first results of the in vitro growing of immature buckwheat embryos were useful for the further development of a method of embryo rescue and its application in the overcoming of postzygotic incompatibility. The interspecific hybrids between the common buckwheat *Fagopyrum esculentum* and its relatives *Fagopyrum cymosum*, *Fagopyrum tataricum*, and *Fagopyrum homotropicum* species have been created by means of ovule or embryo rescue. The results of interspecific hybridization in the genus *Fagopyrum* were presented in the previous chapter.

ANTHER CULTURE

Anther culture means plant regeneration from the haploid microspore cells with the aim of haploid and dihaploid plant production. There are few reports on plant regeneration in the anther culture of buckwheat. Adachi et al. (1988) the first regenerated whole plants in anther culture of the diploid cultivar of common buckwheat Miyazaki-Zarai. The regenerants were all diploid but their origin was unclear. The authors did not exclude the chromosome duplication in plant regenerants. In the experiment of Kong et al. (1992) regenerants were obtained in all six cultivars used. The adopted plants grew well but did not set the seeds. Unfortunately, the number of chromosomes was not counted in this case.

Bohanec et al. (1993) were the first who proved the presence of haploid plants among the regenerants of diploid cv. Daria. They noted the strong tendency toward endoduplication because they detected the presence of diploid, triploid, tetraploid, and aneuploid cells. Berbec and Doroszewska (1998) induced regeneration in anther culture of one tetraploid and two diploid cultivars. The most prevalent class of regenerants was tetraploids. No haploids were found among regenerants derived from the diploid cultivars; likewise no diploid regenerants were obtained from the tetraploid cultivar. The authors assumed that polyploid regenerants must have arisen through multiplication of the initial chromosome number in microspores or somatic tissues. Yui and Yoshida (2001) obtained a number of plantlets regenerated in anther culture of three diploid Japanese cultivars. Most of the plants showed the same amount of DNA as that of the control and were considered to be diploid.

Wang and Campbell (2006) tried to apply the anther culture in the buckwheat breeding process. Among the seven genotypes tested, three were the self-pollinated lines developed from interspecific hybridization between common buckwheat and wild *F. homotropicum*, and two genotypes were F_1 hybrids. Tetraploid and somatic diploid regenerants were predominant and no haploid plants were found. Based on the somatic origin of the regenerants the authors concluded that buckwheat is a difficult species in which to produce haploids or doubled haploids via anther culture.

GENETIC TRANSFORMATION AND BIOACTIVE COMPOUNDS

Since Neskovic et al. (1990b) have demonstrated high susceptibility of common buckwheat to both *Agrobacterium tumefaciens* and *Agrobacterium rhizogenes* the different methods of buckwheat genetic transformation have been developed. The first kanamycin-resistant buckwheat plants were obtained by the same group (Miljus-Djukic et al., 1992) based on *Agrobacterium*-mediated transformation through the cocultivation of buckwheat explants and bacteria on a nutrient medium followed by the selective regeneration of transgenic plants. Kanamycin-resistant and β-glucoronidase-positive transformants were produced later with the *A. tumefaciens* strain LBA 4404 harboring the binary vector pBI121 (Kim et al., 2001). An in planta transformation method based on needle infiltration of *A. tumefaciens* into the apical meristem of buckwheat seedlings was used by Kojima et al. (2000a,b). Bratić et al. (2007) showed that the vacuum infiltration method of in planta transformation was more efficient for buckwheat than infiltration by a syringe.

All of the mentioned studies have used the marker genes of neomycin phosphotransferase II and β-glucuronidase being transferred into the genome of common buckwheat. Only two results are known when buckwheat was transformed with the functional genes. Kojima et al. (2000b) used the pBI121 vector where the β-glucuronidase gene was replaced with cDNA of a rice MADS box gene. The plants transformed with cDNA in a sense orientation were stimulated in branching, while the plants transformed with cDNA in an antisense orientation were inhibited in both branching and growth. The next result was obtained when the *Arabidopsis thaliana* antiporter AtNHX1 gene was transferred into buckwheat by the classic *Agrobacterium*-mediated method (Cheng et al., 2007). Transgenic plants accumulated the higher concentration of Na^+ and proline than the control plants under stress treatment with different concentrations of NaCl.

Buckwheat transformation with *A. tumefaciens* has been developed mostly as a model system, but the transformation with *A. rhizogenes* caused hairy root formation to develop mainly in the applied direction. Hairy root culture of common buckwheat established by infection with *A. rhizogenes* had the highest content of flavonoids (Trotin et al., 1993). The content of rutin in the hairy root clone of common buckwheat was 2.6 times more than that of wild type (Kim et al., 2010). Antioxidant activity in transformed hairy root culture of common buckwheat was increased by three times compared to control (Gabr et al., 2012). The thin hairy root transgenic phenotype of Tartary buckwheat had a higher content of phenolic compounds than the thick phenotype (Park et al., 2011). The red Tartary buckwheat hairy root culture transformed with a wild strain R1000 of *A. rhizogenes* had a several-fold higher concentration of rutin (Thwe et al., 2013). In this way the transgenic hairy root culture of buckwheat may be used as an alternative approach to the production of phenolic compounds.

In recent years, nontransgenic buckwheat tissue culture has been considered as an excellent source of bioactive compounds. The high flavonoid-producing red callus line of *F. cymosum* has been established and used for the molecular cloning of the isoflavone reductase-like gene *FcIRL* (Zhu et al., 2009).

It was found that the production of rutin from the in vitro buckwheat plant culture of both *F. esculentum* and especially *F. tataricum*, as well as the expression level of the rutin biosynthesis-related genes, was increased by exogenous treatment with methyl jasmonate and salicylic acid (Hou et al., 2015).

The suspension and callus cultures of Tartary buckwheat have been widely used as a model object to study the morphogenic processes, intracellular compounds, and tolerance of cultured plant cells to oxidative stress (Gumerova et al., 2008, 2010; Maksyutova et al., 2009; Sibgatullina et al., 2012). Methyl jasmonate treatment activated the accumulation of phenolic compounds in the suspension culture of Tartary buckwheat (Gumerova et al., 2015).

Summarizing this review we may conclude that buckwheat tissue culture evolved in different ways. Plant regenerants were obtained in various explants such as seedling segments, apical and axillary meristems, immature embryos, anthers, and microspores. Micropropagation and embryo rescue techniques may be routinely used in buckwheat breeding. Protocols of plant regeneration are approved in *Agrobacterium*-mediated transformation. Callus and cell cultures are established not only as model objects but as the system for bioactive compound synthesis. Therefore tissue culture approaches are considered to be well developed for buckwheat for use both in practical breeding and basic research.

REFERENCES

Adachi, T., Suputitada, S., Miike, Y., 1988. Plant regeneration from anther culture in common buckwheat (*F. esculentum*). Fagopyrum 8, 5–9.

Adachi, T., Yamaguchi, A., Miike, Y., Hoffmann, F., 1989. Plant regeneration from protoplast of common buckwheat (*F. esculentum*). Plant Cell Rep. 8, 247–250.

Barsukova, E., 2004. Utilization of tissue culture method in buckwheat selection. In: Proceedings of the 9th International Symposium on Buckwheat, Prague, 87–92.

Barsukova, E., 2013. Cellular selection of buckwheat in the conditions of ionic stress. Agrarian Russia 10, 2–4.

Berbec, A., Doroszewska, T., 1998. Callus formation and plant regeneration in anther culture of three buckwheat (*Fagopyrum esculentum* Moench) cultivars. In: Proceedings of the 5th International Symposium on Buckwheat, Winnipeg, pp. V-48–V-52.

Berbec, A., Doroszewska, T., 1999. Regeneration in vitro of three cultivars of buckwheat (*Fagopyrum esculentum* Moench) as affected by medium composition. Fagopyrum 16, 49–52.

Bohanec, B., 1987. Improvement in buckwheat micropropagation procedures. Fagopyrum 7, 13–15.

Bohanec, B., Neskovic, M., Vujicic, R., 1993. Anther culture and androgenetic plant regeneration in buckwheat (*Fagopyrum esculentum* Moench). Plant Cell Tiss. Org. Cult. 35, 259–266.

Bratić, A., Majić, D., Miljuš-Djukić, J., Jovanović, Ž., Maksimović, V., 2007. In planta transformation of buckwheat (*Fagopyrum esculentum* Moench). Arch. Biol. Sci. 59 (2), 135–138.

Chen, C., Lan, J., Xie, S., Cui, S., Li, A., 2012. In vitro propagation and quality evaluation of long-term micro-propagated and conventionally grown *Fagopyrum dibotrys* Hara mutant, an important medicinal plant. J. Med. Plants Res. 6 (15), 3003–3012.

Cheng, L., Zhang, B., Zq, X., 2007. Genetic transformation of buckwheat (*Fagopyrum esculentum* Moench) with AtNHX1 gene and regeneration of salt tolerant transgenic plants. Chin.J. Biotechnol. 23 (1), 51–60.

Dobránszki, J., 2009. Role of cytokinins and explant type in shoot multiplication of buckwheat (*Fagopyrum esculentum* Moench) in vitro. Eur. J. Plant Sci. Biotechnol. 3 (1), 66–70.

Gabr, A., Sytar, O., Ahmed, A., Smetanska, I., 2012. Production of phenolic acid and antioxidant activity in transformed hairy root cultures of common buckwheat (*Fagopyrum esculentum* M.). Aust. J. Basic Appl. Sci. 6 (7), 577–586.

Guo, F., Zhou, J., Luo, X., Ma, H., 1992. Plant regeneration of tetraploid plants of *Fagopyrum esculentum* Moench in tissue culture. In: Proceedings of the 5th International Symposium on Buckwheat, China, pp. 291–300.

Gumerova, E., Akulov, A., Rumyantseva, N., 2015. Effect of methyl jasmonate on growth characteristics and accumulation of phenolic compounds in suspension culture of Tartary buckwheat. Russ. J. Plant Physiol. 62 (2), 195–203.

Gumerova, E., Galeeva, E., Chuenkova, S., Rumyantseva, N., 2003. Somatic embryogenesis and bud formation on cultured *Fagopyrum esculentum* hypocotyls. Russ. J. Plant Physiol. 50 (5), 640–645.

Gumerova E., Gatina, E., Chuenkova, S., Rumyantseva, N., 2001. Somatic embryogenesis in common buckwheat *Fagopyrum esculentum* Moench. In: Proceedings of the 8th International Symposium on Buckwheat, Korea, pp. 377–381.

Gumerova, E., Rumyantseva, N., Gatina, E., 1998. PECC formation in hypocotyl tissues of common buckwheat (*Fagopyrum esculentum* Moench). In: Proceedings of the 7th International Symposium on Buckwheat, Winnipeg, pp. V-53–V-57.

Gumerova, E., Utina, D., Rumyantseva, N., 2008. Establishment and characterization of morphogenic suspension culture of Tatary buckwheat *Fagopyrum tataricum* (L.) Gaertn. Uchenye Zapiski Kazanskogo Gosudarstvennogo Universiteta 150 (2), 126–135.

Gumerova, E., Utina, D., Rumyantseva, N., 2010. The study of morphogenic response in suspension culture of Tartary buckwheat *Fagopyrum tataricum* (L.) Gaertn. In: Proceedings of the 10th International Symposium on Buckwheat, Orel, pp. 160–166.

Han, M., Kamal, A., Huh, Y., Jeon, A., Bae, J., Chung, K., Lee, M., Park, S., Jeong, H., Woo, H., 2011. Regeneration of plantlet via somatic embryogenesis from hypocotyls of Tartary buckwheat (*Fagopyrum tataricum*). Aust. J. Crop Sci. 5 (7), 865–869.

Hao, J., Pel, Y., Qu, Y., Zheng, C., 1998. Study on callus differentiation conditions of common buckwheat. In: Proceedings of the 5th International Symposium on Buckwheat, Winnipeg, pp. V-33–V-37.

Hou, S., Sun, Z., Linghu, B., Wang, Y., Huang, K., Xu, D., Han, Y., 2015. Regeneration of buckwheat plantlets from hypocotyl and the influence of exogenous hormones on rutin content and rutin biosynthetic gene expression in vitro. Plant Cell Tiss. Org. Cult. 120 (3), 1159–1167.

Jin, H., Jia, J.F., Hao, J.G., 2002. Efficient plant regeneration in vitro in buckwheat. Plant Cell Tiss. Org. Cult. 69 (3), 293–295.

Kim, H., Kang, H., Lee, Y., Lee, S., Ko, J., Rha, E., 2001. Direct regeneration of transgenic buckwheat from hypocotyl segment by agrobacterium-mediated transformation. Kor. J. Crop Sci. 46 (5), 375–379.

Kim, Y., Woo, H., Park, T., Park, N., Lee, S., Park, S., 2010. Genetic transformation of buckwheat (*Fagopyrum esculentum* M.) with *Agrobacterium rhizogenes* and production of rutin in transformed root cultures. Aust. J. Crop Sci. 4 (7), 485–490.

Klčová, L., Gubišová, M., 2008. Evaluation of different approaches to buckwheat (*Fagopyrum esculentum* Moench) micropropagation. Czech J. Genet. Plant Breed. 44 (2), 66–72.

Kong, F., Shong, Y., Wang, Z., Yang, L., 1992. Study on plant regeneration from anther culture in common buckwheat (*Fagopyrum esculentum*). In: Proceedings of the 5th International Symposium on Buckwheat, China, pp. 309–314.

Kojima, M., Arai, Y., Iwase, N., Shirotori, K., Shiori, H., Nozue, M., 2000a. Development of a simple and efficient method for transformation of buckwheat plants (*Fagopyrum esculentum*) using *Agrobacterium tumefaciens*. Biosci. Biotechnol. Biochem. 64 (4), 845–847.

Kojima, M., Hihahara, M., Shiori, H., Nozue, M., Yamomoto, K., Sasaki, T., 2000b. Buckwheat transformed with cDNA of a rice MADS box gene is stimulated in branching. Plant Biotechnol. 17 (1), 35–42.

Kostyukova, Y., Rumyantseva, N., 2010. Histological study of embryoidogenic callus induction in immature embryos of Tartary buckwheat *Fagopyrum tataricum* (L.) Gaertn. In: Proceedings of the 10th International Symposium on Buckwheat, Orel, pp. 167–174.

Kwon, S., Han, M., Huh, Y., Roy, S., Lee, C., Woo, S., 2013. Plantlet regeneration via somatic embryogenesis from hypocotyls of common buckwheat (*Fagopyrum esculentum* Moench). Kor. J. Crop Sci. 58 (4), 331–335.

Lachmann, S., Adachi, T., 1990. Callus regeneration from hypocotyl protoplasts of Tartary buckwheat (*Fagopyrum tataricum* Gaertn.). Fagopyrum 10, 62–64.

Lachmann, S., Kishima, Y., Adachi, T., 1994. Protoplast fusion on buckwheat: preliminary results of somatic hybridization. Fagopyrum 14, 7–12.

Luthar, Z., Marchetti, S., 1994. Plant regeneration from mature cotyledons in a buckwheat (*Fagopyrum esculentum* Moench) germplasm collection. Fagopyrum 14, 65–69.

Maksyutova, N., Galeevam, E., Mukhitov, A., Rumyantseva, N., Viktorova, L., 2009. Changes in the growth of Tartary buckwheat (*Fagopyrum tataricum* (L.) Gaertn.) calli with different ability for morphogenesis induced by salicylic acid. Eur. J. Plant Sci. Biotechnol. 3 (1), 71–74.

Miljus-Djukic, J., Neskovic, M., Ninkovic, S., Crkvenjakov, R., 1992. *Agrobacterium*-mediated transformation and plant regeneration of buckwheat (*Fagopyrum esculentum* Moench). Plant Cell Tiss. Org. Cult. 29, 101–108.

Neskovic, M., Vinterhalter, B., Miljus-Djukic, J., Ghalawenji, N., 1990a. Micropropagation of recessive determinate genotypes of buckwheat (*Fagopyrum esculentum* Moench) as an alternate approach to uniform seed production. Plant Breed. 105, 337–340.

Neskovic, M., Vinterhalter, B., Miljus-Djukic, J., Ninkovich, S., Vinterhalter, D., Jovanovic, V., Knezevic, J., 1990b. Susceptability of buckwheat (*Fagopyrum esculentum* Moench.) to *Agrobacterium tumefaciens* and *A. rhizogenes*. Fagopyrum 10, 57–61.

Neskovic, M., Vujicic, R., Budimir, S., 1987. Somatic embryogenesis and bud formation from immature embryos of buckwheat (*Fagopyrum esculentum* Moench). Plant Cell Rep. 6, 423–426.

Park, N., Li, O., Uddin, R., Park, S., 2011. Phenolic compound production by different morphological phenotypes in hairy root cultures of *Fagopyrum tataricum* Gaertn. Arch. Biol. Sci. 63 (1), 193–198.

Park, S., Park, C., 2001. Multiple shoot organogenesis and plant regeneration from cotyledons of buckwheat (*Fagopyrum esculentum* Moench). In: Proceedings of the 8th International Symposium on Buckwheat, Korea, pp. 427–430.

Rajbhandari, B.P., Dhanbhadel, S., Gantam, D.M., Gantam, B.R., 1995. Plant regeneration via calli of leaf and stem explants in common buckwheat ecotypes. In: Proceedings of the 6th International Symposium on Buckwheat, Japan, 1, pp. 191–196.

Ramchatgoen, S., Kachonpadungkitti, Y., Hisajima, S., 1998. Micropropagation of buckwheat (*Fagopyrum esculentum* Moench) plant in vitro. J. Soc. High Technol. Agric. 10 (4), 231–236.

Rumyantseva, N., Salnikov, V., Fedoseeva, N., Lozovaya, V., 1992. Peculiarities of morphogenesis in long-term cultivated buckwheat calluses. Russ. J. Plant Physiol. 39, 143–151.

Rumyantseva, N., Sergeeva, N., Hakimova, L., Gumerova, E., Lozovaya, V., 1989a. Morphogenesis and plant regeneration in buckwheat tissue culture. In: Proceedingsof the 7th International Symposium on Buckwheat, USSR, Orel, 322–328.

Rumyantseva, N., Sergeeva, N., Khakimova, L., Salnikov, V., Gumerova, E., Lozovaya, V., 1989b. Organogenesis and somatic embryogenesis in the culture of two buckwheat species. Russ. J. Plant Physiol. 36, 187–194.

Sibgatullina, G., Rumyantseva, N., Khaertdinova, L., Akulov, A., Tarasova, N., Gumerova, E., 2012. Establishment and characterization of the line of *Fagopyrum tataricum* morphogenic callus tolerant to aminotriazole. Russ. J. Plant Physiol. 59 (5), 662–669.

Slavinska, J., Kantartsi, K., Obendorf, R., 2009. In vitro organogenesis of *Fagopyrum esculentum* Moench (Polygonaceae) as a method to study seed set in buckwheat. Eur. J. Plant Sci. Biotechnol. 3, 75–78.

Srejovic, V., Neskovic, M., 1981. Regeneration of plants from cotyledon fragments of buckwheat (*F. esculentum* Moench). Z. Pflanzenphysiol. 104 (1), 37–42.

Suvorova G., Fesenko, N., Fesenko, M., 1998. Buckwheat micropropagation by means of meristem culture. In: Proceedings of the 7th International Symposium on Buckwheat, Winnipeg, pp. V-25–V-31.

Suvorova G., Fesenko, N., Fesenko, M., 2004. Possibilities of micropropagation technique application in buckwheat breeding. In: Proceedings of the 9th International Symposium on Buckwheat, Prague, pp. 375–378.

Suvorova, G., Kostrubin, M., 1992. Culture of immature buckwheat embryos (*Fagopyrum esculentum* Moench). In: Proceedings of the 5th International Symposium on Buckwheat, China, pp. 140–148.

Takahata, Y., 1988. Plant regeneration from cultured immature inflorescence of common buckwheat (*Fagopyrum esculentum* Moench) and perennial buckwheat (*F. cymosum* Meisn.). Jap. J. Breed. 38, 409–413.

Takahata, Y., Jumonji, E., 1985. Plant regeneration from hypocotyl section and callus in buckwheat (*Fagopyrum esculentum* Moench). Ann. Rep. Fac. Educ. Iwate Univ. 45, 137–142.

Thwe, A., Kim, J., Li, X., Kim, Y., Uddin, M., Kim, S., Suzuki, T., Park, N., Park, S., 2013. Metabolomic analysis and phenylpropanoid biosynthesis in hairy root culture of Tartary buckwheat cultivars. PLoS One 8 (6), e65349.

Trotin, F., Moumou, Y., Vasseur, J., 1993. Flavanol production by *Fagopyrum esculentum* hairy and normal root cultures. Phytochemistry 32, 929–931.

Wang, Y., Campbell, C., 1999. Culture of buckwheat embryos in Petri dishes to speed up the breeding cycle. Fagopyrum 16, 37–41.

Wang, Y., Campbell, C., 2006. Effect of genotypes, pretreatments and media in anther culture of common (*Fagopyrum esculentum*) and self-pollinated buckwheat. Fagopyrum 23, 29–35.

Woo, S., Nair, A., Adachi, T., Campbell, C., 2000. Plant regeneration from cotyledon tissues of common buckwheat (*Fagopyrum esculentum* Moench). In Vitro Cell. Dev. Biol.—Plant 36 (5), 358–361.

Yamane, Y., 1974. Induced differentiation of buckwheat plants from subcultured calluses in vitro. Jap. J. Genet. 49 (3), 139–146.

Yui, M., Yoshida, T., 2001. Callus induction and plant regeneration in anther culture of Japanese buckwheat cultivars (*Fagopyrum esculentum* Moench). Fagopyrum 18, 27–35.

Zhang, J.T., Chen R. 1992. Effect of phenyl urea derivatives on organ formation and plantlet regeneration of hypocotyls of Tartary buckwheat. In: Proceedings of the 5th International Symposium on Buckwheat, China, pp. 301–305.

Zhu, Q., Guo, T., Sui, S., Liu, G., Lei, X., Luo, L., Li, M., 2009. Molecular cloning and characterization of a novel isoflavone reductase-like gene (FcIRL) from high flavonoids-producing callus of *Fagopyrum cymosum*. Acta Pharm. Sinica 44 (7), 809–819.

Chapter thirty

Flavonoid Biosynthesis in Buckwheat

G. Taguchi

Department of Applied Biology, Faculty of Textile Science and Technology, Shinshu University, Ueda, Nagano, Japan

INTRODUCTION

Buckwheat (*Fagopyrum* sp.) is cultivated worldwide for its seeds (achenes), which are consumed and often referred to as pseudocereal. Buckwheat is known to accumulate large amounts of rutin (quercetin 3-O-rutinoside) and other flavonoids (Fig. 30.1), thus it is considered a good dietary source of flavonoids (Couch et al., 1946; Fabjan et al., 2003).

There are two commonly cultivated species of buckwheat, that is, common buckwheat (*Fagopyrum esculentum* Moench) and Tartary buckwheat (*Fagopyrum tataricum* Gaertn.). These two species differ in their flavonoid composition; common buckwheat accumulates mainly rutin and other glycosides, and rarely aglycons such as quercetin in the flowers and seeds, while Tartary buckwheat accumulates quercetin in addition to rutin (Dadáková and Kalinová, 2010). The rutin content in seeds of Tartary buckwheat is much higher than that of common buckwheat (Dadáková and Kalinová, 2010; Kim et al., 2008). Buckwheat also accumulates C-glycosylflavones in the cotyledons of common buckwheat and in trace quantities in those of Tartary buckwheat (Kim et al., 2008).

Flavonoids are reported to have several biological activities, such as antioxidant, hypotensive, and antiinflammatory activity (Afanas'ev et al., 1989; Harborne and Williams, 2000). Thus characterization of flavonoid biosynthesis is important for the utilization of buckwheat as a health-promoting food. Biosynthesis genes of rutin and other flavonols in buckwheat have been isolated, except the genes coding for enzymes involved in the final glycosylation steps (Li et al., 2010). In contrast, those of C-glycosylflavones are still being elucidated; C-glucosyltransferase genes from common buckwheat were identified by Nagatomo et al. (2014). In this chapter, we summarize the information about flavonol O-glycosides and C-glycosylflavone biosynthesis in buckwheat.

377

Molecular Breeding and Nutritional Aspects of Buckwheat. http://dx.doi.org/10.1016/B978-0-12-803692-1.00030-4

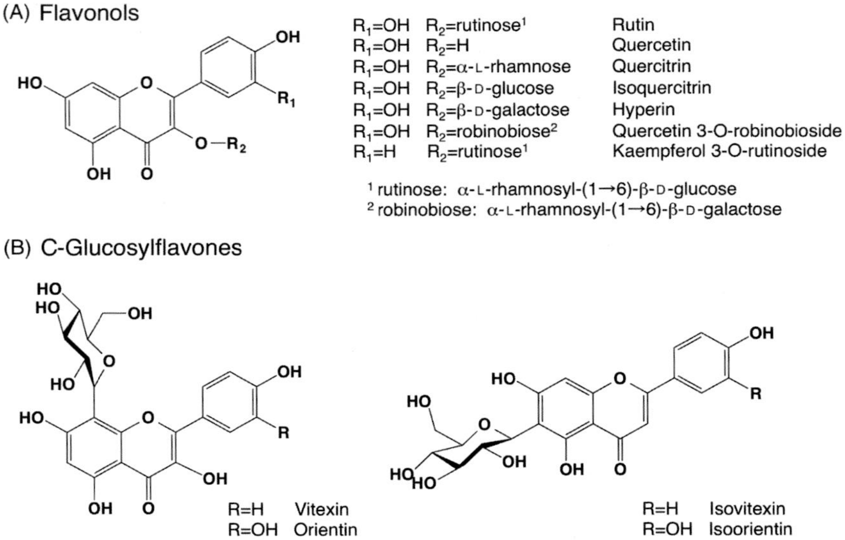

(A) Flavonols

R_1=OH	R_2=rutinose[1]	Rutin
R_1=OH	R_2=H	Quercetin
R_1=OH	R_2=α-L-rhamnose	Quercitrin
R_1=OH	R_2=β-D-glucose	Isoquercitrin
R_1=OH	R_2=β-D-galactose	Hyperin
R_1=OH	R_2=robinobiose[2]	Quercetin 3-O-robinobioside
R_1=H	R_2=rutinose[1]	Kaempferol 3-O-rutinoside

[1] rutinose: α-L-rhamnosyl-(1→6)-β-D-glucose
[2] robinobiose: α-L-rhamnosyl-(1→6)-β-D-galactose

(B) C-Glucosylflavones

R=H	Vitexin
R=OH	Orientin

R=H	Isovitexin
R=OH	Isoorientin

FIGURE 30.1 Chemical structure of major flavonoids found in buckwheat.

FLAVONOLS AND THEIR O-GLYCOSIDES

The most abundant flavonoid compound in buckwheat is rutin. As an example, high-performance liquid chromatography (HPLC) profiles of methanol extracts of common buckwheat (diploid cultivar Shinano No. 1) are shown in Fig. 30.2. Rutin (peak 6) was detected in the methanol fraction mainly in flowers and leaves [3–10% of dry weight (DW)] (Dadáková and Kalinová, 2010; Li et al., 2010; Zielińska et al., 2012). Immature seeds also contain rutin (Fig. 30.2C), but its content decreases during seed ripening (Fig. 30.2D). Generally, the amount of rutin in seeds of common buckwheat (0.01–0.02% of DW) is lower than that in other organs, and the amount increases up to 0.6% of DW in the sprouts (cotyledons) during germination (Fig. 30.2E; Kim et al., 2008). Seeds of Tartary buckwheat accumulate much higher amounts of rutin and quercetin (0.8–1.8% and 0.05–0.09% of DW, respectively) than common buckwheat (Fabjan et al., 2003; Dadáková and Kalinová, 2010). Besides rutin, buckwheat accumulates several flavonol O-glycosides, such as quercitrin, isoquercitrin, hyperoside, quercetin 3-O-robinobioside, and kaempferol 3-O-rutinoside (Fig. 30.1; Dadáková and Kalinová, 2010; Kalinova and Vrchotova, 2009; Kiprovski et al., 2015; Nam et al., 2015). Among them, quercitrin (Fig. 30.2, peak 7) is one of the major flavonoids in the flowers of common buckwheat, whereas quercetin 3-O-robinobioside (Fig. 30.2, peak 5) is found in the cotyledon and immature seeds.

The biosynthetic pathway of these flavonols has been well established in higher plants (Fig. 30.3; Davies and Schwinn, 2004). The biosynthesis of quercetin starts with deamination of phenylalanine by phenylalanine ammonia-lyase (PAL), followed by reactions catalyzed by cinnamate 4-hydroxylase (C4H), 4-coumarate coenzyme A ligase (4CL), chalcone synthase (CHS), chalcone

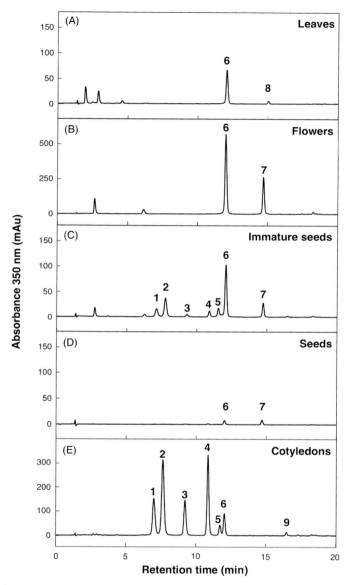

FIGURE 30.2 Phenolic composition of common buckwheat. Phenolic fraction of leaves (A), flowers (B), immature seeds (C), seeds (D), and etiolated cotyledons at 3–4 days after sowing (E) of common buckwheat were extracted with methanol and analyzed by HPLC (Shimadzu LC10Avp System, Kyoto, Japan) using an ODS column (Kinetex EVO C18 5 μm, 150 × 4.6 mm, Phenomenex, Torrance, CA, USA). The column was eluted with 25% methanol containing 0.1% formic acid for 3 min, followed by a linear gradient of 25–55% methanol containing 0.1% formic acid for 20 min at a flow rate of 1.0 mL min⁻¹, 40°C. The eluate was monitored using a diode array detector at 350 nm. Peak identifications: *1*, orientin; *2*, isoorientin; *3*, vitexin; *4*, isovitexin; *5*, quercetin 3-O-robinobioside; *6*, rutin; *7*, quercitrin; *8*, kaempferol 3-O-rutinoside; *9*, unidentified flavonoid compounds.

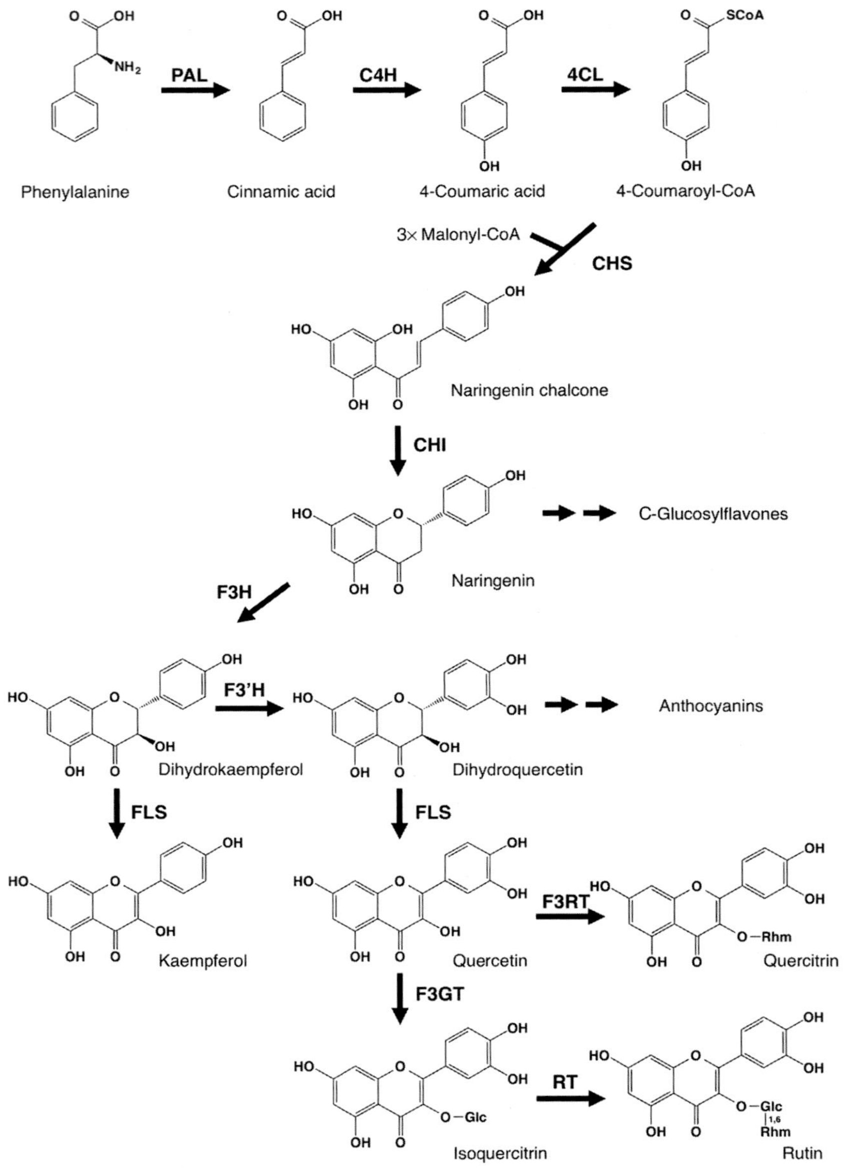

FIGURE 30.3 Biosynthetic pathways of rutin and flavonoids in buckwheat. *PAL,* Phenylalanine ammonia-lyase; *C4H*, cinnamate 4-hydroxylase; *4CL*, 4-coumarate CoA ligase; *CHS*, chalcone synthase; *CHI*, chalcone isomerase; *F3H*, flavanone 3β-hydroxylase; *F3'H*, flavonoid 3'-hydroxylase; *FLS*, flavonol synthase; *F3GT*, flavonol 3-O-glucosyltransferase; *F3RT*, flavonol 3-O-rhamnosyltransferase; *RT*, flavonol 3-O-glucoside 6-O-rhamnosyltransferase.

isomerase (CHI), flavone 3-hydroxylase (F3H), flavonoid 3-hydroxylase (F3′H), and flavonol synthase (FLS). Because buckwheat accumulates large amounts of rutin, this plant was used as a good model organism for the studies of flavonoid biosynthesis especially in the early days. Watkin et al. (1957), in their experiments with Tartary buckwheat, using several radiolabeled compounds, showed that acetic acid and phenylalanine are incorporated into the A ring and B ring of quercetin, respectively. Amrhein and Zenk (1971) reported that PAL activity and the following de novo synthesis of rutin are induced by light illumination of the hypocotyl of common buckwheat. Saito (1974) showed that acetyl coenzyme A was incorporated into the A ring of quercetin in the reaction using a cell-free extract of buckwheat sprouts. Hrazdina et al. (1986) purified and characterized CHS from the buckwheat hypocotyl and proposed that flavonoid biosynthetic enzymes, including soluble enzymes such as PAL, CHS, and glycosyltransferase, form enzyme complexes on the cytoplasmic face of the endoplasmic reticulum membranes that interact with the membrane-embedded enzymes such as C4H and F3′H, channeling the substrates (Hrazdina and Wagner, 1985; Hrazdina et al., 1987). This, now commonly accepted, metabolic channeling concept in the flavonoid biosynthetic pathway has been supported by molecular analyses such as immnoprecipitation, yeast two-hybrid assays, and by using transgenic *Arabidopsis* (Burbulis and Winkel-Shirley, 1999). Despite these extensive studies on flavonoids in buckwheat, isolation of the genes responsible for the biosynthesis of quercetin followed years later after they were isolated in other plants; they were isolated from common buckwheat in this decade (Li et al., 2010).

The final step of rutin biosynthesis is glycosylation of quercetin involving quercetin 3-O-glucosyltransferase (F3GT) and isoquercitrin rhamnosyltransferase. F3GT, whose activity was associated with the accumulation of rutin, was purified partially from the cotyledons of common buckwheat (Suzuki et al., 2005). However, the coding gene (*F3GT*) is yet to be identified because, similar to *Arabidopsis* (Lim et al., 2004), it is expected that many candidate glycosyltransferase (UGT) genes will show F3GT activity in buckwheat and they will be difficult to identify only from in vitro activities of the recombinant enzymes. The rhamnosyltransferase is a kind of UGT that transfers rhamnose moiety of UDP-rhamnose to isoquercitrin, but it has not been reported from buckwheat at present.

The degradation of rutin in buckwheat is catalyzed by rutin-degrading enzymes, which hydrolyze glycosidic bonds of rutin located in the seeds and dried vegetative materials (Baumgertel et al., 2003; Suzuki et al., 2002). In the seeds of Tartary buckwheat these enzymes are accumulated in the testa and separated from their substrate, rutin, which accumulates mainly in the embryo; the enzymes are activated by the addition of water (Suzuki et al., 2002). Regulation of the enzyme activities would be important to keep the quality of flavonoids in buckwheat; however, the physiological roles of these rutin-degrading enzymes in planta have not been clarified in detail.

The accumulation of rutin and the expression patterns of the biosynthetic enzymes in buckwheat have been studied in several developmental stages.

The content of rutin and the key biosynthetic genes of flavonoids—PAL, CHS, CHI, and FLS—was the highest in the flowering stage of both common buckwheat and Tartary buckwheat (Gupta et al., 2011). The content of rutin also increased in the sprouts during the germination process (Kim et al., 2008; Zhang et al., 2015), and the expression of biosynthesis genes such as PAL, 4CL, and F3H were also induced (Li et al., 2012). Interestingly, the expressions of these genes in common buckwheat was the highest in the stem and roots, which differed from the accumulation of flavonoids (primarily in flowers and leaves), suggesting that flavonols might be transported from the stem and roots into flowers and leaves (Li et al., 2010). Light irradiation is another factor affecting rutin and anthocyanin biosynthesis in buckwheat (Amrhein and Zenk, 1971). Tsurunaga et al. (2013) reported that the amount of rutin in the sprouts of common buckwheat was increased by light irradiation, especially by the UV-B light at 300–320 nm, which amount was 1.6 times higher than that under dark conditions. This induction was also observed in the accumulation of anthocyanin. However, the content of rutin in dark-grown sprouts was high (0.6% of DW), while anthocyanin was not detected. Other studies also reported that rutin content and the expression of the biosynthesis enzymes were high in dark-grown sprouts (Li et al., 2010; Suzuki et al., 2005), indicating that light irradiation may not be essential for rutin biosynthesis in contrast to anthocyanin biosynthesis.

Molecular breeding of buckwheat related to flavonoid biosynthesis was reported only in a few studies to date. An overexpression of a transcriptional factor AtMYB12 that upregulates flavonoid biosynthesis in *Arabidopsis* resulted in increased expression of flavonoid biosynthesis genes and rutin content in the hairy root culture of common buckwheat (Park et al., 2012). Gene expression of flavonoid biosynthesis is regulated by these transcription factors (Davies and Schwinn, 2004; Mehrtens et al., 2005). Therefore studies targeting those genes, such as the analysis of the transcription factor of flavonoids in buckwheat, would be required.

C-GLUCOSYLFLAVONES

C-Glucosylflavones are one of the major flavonoids in buckwheat (Dietrych-Szostak and Oleszek, 1999). The glucose moiety of C-glucosides directly binds to the aromatic carbon of aglycon at the anomeric carbon by C—C bond; thus this glucosidic bond is stable for enzyme or acid hydrolysis and shows distinct metabolic and pharmaceutical properties (Talhi and Silva, 2012; Xiao et al., 2014). C-glucosylflavones found in buckwheat are orientin (luteolin 8-C-glucoside), isoorientin (luteolin 6-C-glucoside), vitexin (apigenin 8-C-glucoside), and isovitexin (apigenin 6-C-glucoside) (Figs. 30.1 and 30.2, peaks 1–4).

Most of C-glucosylflavones are accumulated in the cotyledons of common buckwheat (Kerscher and Franz, 1987). The amount of C-glucosylflavones in cotyledons increases during the germination process from trace level at zero days to 2.6–3.2% of DW at 9 days after sawing (Kim et al., 2008). By contrast, cotyledons of Tartary buckwheat accumulate only about 0.1–0.2% of DW (Kim et al., 2008; Li et al., 2012). C-Glucosylflavones are also accumulated in the immature seeds of common buckwheat (about 1% of DW; Zielińska et al., 2012), but this amount decreases during ripening to trace levels in mature seeds (about 0–0.07% of DW; Dietrych-Szostak and Oleszek, 1999; Kim et al., 2008). C-Glucosylflavones are not detected in leaves and flowers (Fig. 30.2). In addition, O-glucosides of flavones have not been reported from buckwheat.

The biosynthetic pathway of C-glucosylflavones in buckwheat was proposed in the 1980s (Kerscher and Franz, 1987), and a partial purification of flavonoid C-glucosyltransferase (CGT) was performed (Kerscher and Franz, 1988). However, the genes responsible for CGTs and details of the enzymatic reactions catalyzed by these enzymes had not been clarified for a long time. CGTs have been purified close to homogeneity from sprouts of common buckwheat and their corresponding genes (*FeCGTs*) were isolated (Nagatomo et al., 2014), which was the first identification of *CGT* genes from a dicot plant. FeCGT enzymes (UGT708C1 and UGT708C2) showed C-glucosylation activity toward 2-hydroxyflavanones and related compounds having a 2′,4′,6′-trihydroxyacetophenone-like structure but not toward flavones or flavonols, indicating that the open-circular form of 2-hydroxyflavanones is their substrate. The enzyme activities of FeCGTs were similar to those of CGT reported from rice (OsCGT; Brazier-Hicks et al., 2009), although the identities of amino acid sequences between FeCGT and OsCGT were not high (41%).

The proposed biosynthetic pathway of C-glucosylflavones in buckwheat is shown in Fig. 30.4 (Kerscher and Franz, 1987; Nagatomo et al., 2014). It starts with the hydroxylation of the 2-position of flavanone, which is the intermediate product of flavonol biosynthesis. In our preliminary results, 2-hydroxylation activity against naringenin was detected in the cell-free extract of the common buckwheat cotyledon (unpublished), suggesting that this 2-hydroxylation would be catalyzed by flavone synthase II, a P450 enzyme, which shows the flavone 2-hydroxylase activity similar to the enzyme reported in rice (Du et al., 2010). The produced 2-hydroxyflavanone is in equilibrium with one of its open-circular forms (2′,4′,6′-trihydroxyacetophenone-like structure, dibenzoylmethane form), which is C-glucosylated by CGT (Nagatomo et al., 2014). The formed C-glucoside is also in equilibrium with its two closed-circular forms, that is, 2-hydroxyflavanone 6-C-glucoside and 8-C-glucoside, because the sugar moiety of the open-circular form of C-glucoside can be rotated. They are then dehydrated to produce two C-glucosylflavones. This dehydration process is considered to be proceeded enzymatically, although the enzyme has not been identified (Nagatomo et al., 2014).

FeCGT genes are expressed in immature cotyledons (1–5 days after sawing) but not in developed cotyledons of common buckwheat, suggesting that C-glucosylflavones are synthesized specifically in the cotyledons during germination (Nagatomo et al., 2014). Interestingly, different from the rutin biosynthesis, light irradiation did not affect the accumulation of C-glucosylflavones (Watanabe and Ito, 2003). Further studies will be required to address why these C-glucosylflavones are accumulated specifically in the cotyledons and immature seeds in buckwheat.

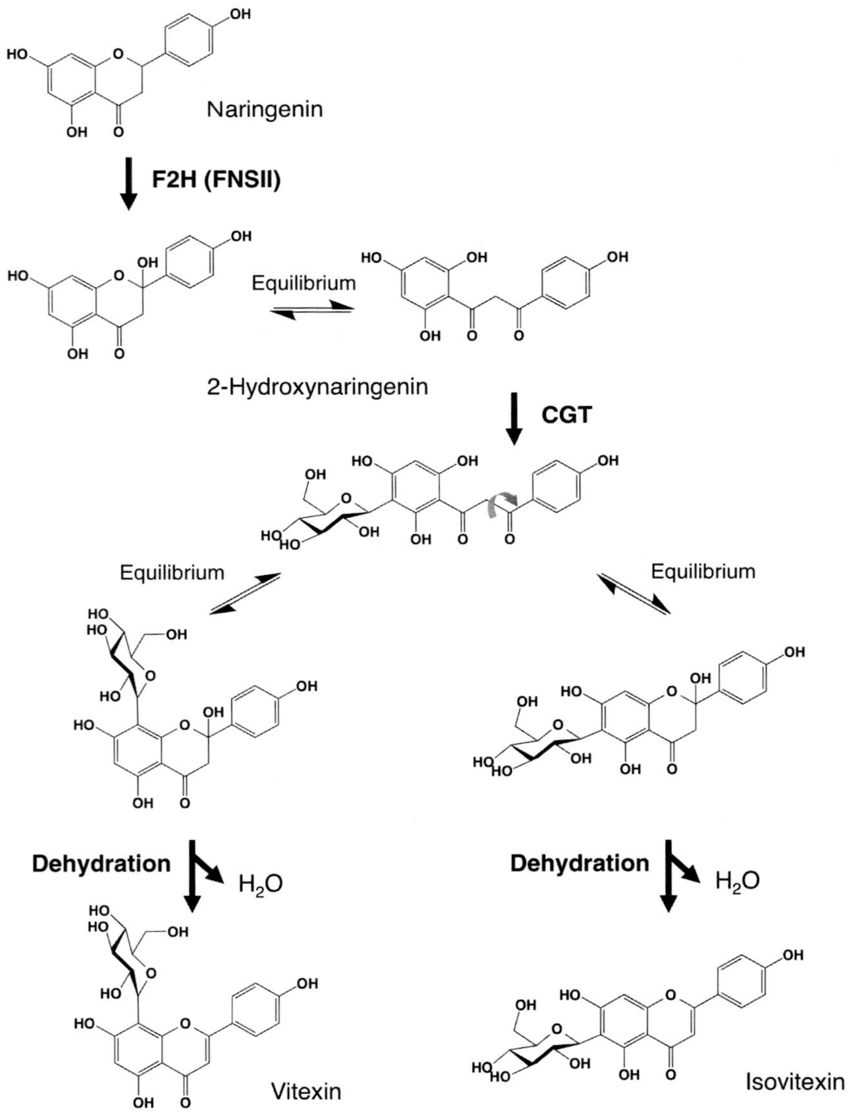

FIGURE 30.4 Biosynthetic pathways of vitexin and isovitexin in buckwheat. *F2H (FNSII)*, Flavanone 2-hydroxylase (flavone synthase II); *CGT*, 2-hydroxyflavanone C-glucosyltransferase.

Flavonoids are important components of buckwheat that show several biological activities bringing health-promoting effects. Although the accumulation of these flavonoids in buckwheat has been studied extensively, several biosynthesis genes responsible for these flavonoids have yet to be identified. Further studies should include isolation of the unidentified biosynthetic genes and elucidation of the regulation systems of flavonoid biosynthesis in buckwheat.

ACKNOWLEDGMENT

This work is supported in part by JSPS KAKENHI grant number 26450120 to GT.

REFERENCES

Afanas'ev, I.B., Dorozhko, A.I., Brodskii, A.V., Koostyuk, V.A., Potapovitch, A.I., 1989. Chelating and free radical scavenging mechanisms of inhibitory action of rutin and quercetin in lipid peroxidation. Biochem. Pharmacol. 38, 1763–1769.

Amrhein, N., Zenk, M.H., 1971. Untersuchungen zur Rolle der Phenylalanin-Ammonium-Lyase (PAL) bei der Regulation der Flavonoidsynthese in Buchweizen (*Fagopyrum esculentum* Moench). Z. Pflanzenphysiol. 64, 145–168.

Baumgertel, A., Grimm, R., Eisenbeiss, W., Kreis, W., 2003. Purification and characterization of a flavonol 3-*O*-beta-heterodisaccharidase from the dried herb of *Fagopyrum esculentum* Moench. Phytochemistry 64, 411–418.

Brazier-Hicks, M., Evans, K.M., Gershater, M.C., Puschmann, H., Steel, P.G., Edwards, R., 2009. The *C*-glycosylation of flavonoids in cereals. J. Biol. Chem. 284, 17926–17934.

Burbulis, I.E., Winkel-Shirley, B., 1999. Interactions among enzymes of the *Arabidopsis* flavonoid biosynthetic pathway. Proc. Natl. Acad. Sci. USA 96, 12929–12934.

Couch, J.F., Naghski, J., Krewson, C.F., 1946. Buckwheat as a source of rutin. Science 103, 197–198.

Dadáková, E., Kalinová, J., 2010. Determination of quercetin glycosides and free quercetin in buckwheat by capillary micellar electrokinetic chromatography. J. Sep. Sci. 33, 1633–1638.

Davies, K.M., Schwinn, K.E., 2004. Molecular biology and biotechnology of flavonoid biosynthesis. In: Andersen, Øyvind M., Markham, Kenneth R. (Eds.), Flavonoids: Chemistry, Biochemistry and Applications. CRC Press, Boca Raton, FL, pp. 143–218.

Dietrych-Szostak, D., Oleszek, W., 1999. Effect of processing on the flavonoid content in buckwheat (*Fagopyrum esculentum* Möench.) grain. J. Agric. Food Chem. 47, 4384–4387.

Du, Y., Chu, H., Chu, I.K., Lo, C., 2010. CYP93G2 is a flavanone 2-hydroxylase required for C-glycosylflavone biosynthesis in rice. Plant Physiol. 154, 324–333.

Fabjan, N., Rode, J., Kosir, I.J., Wang, Z., Zhang, Z., Kreft, I., 2003. Tartary buckwheat (*Fagopyrum tataricum* Gaertn.) as a source of dietary rutin and quercitrin. J. Agric. Food Chem. 51, 6452–6455.

Gupta, N., Sharma, S.K., Rana, J.C., Chauhan, R.S., 2011. Expression of flavonoid biosynthesis genes vis-à-vis rutin content variation in different growth stages of *Fagopyrum* species. J. Plant Physiol. 168, 2117–2123.

Harborne, J.B., Williams, C.A., 2000. Advances in flavonoid research since 1992. Phytochemistry 55, 481–504.

Hrazdina, G., Lifson, E., Weeden, N.F., 1986. Isolation and characterization of buckwheat (*Fagopyrum esculentum* M.) chalcone synthase and its polyclonal antibodies. Arch. Biochem. Biophys. 247, 414–419.

Hrazdina, G., Wagner, G.J., 1985. Metabolic pathways as enzyme complexes: evidence for the synthesis of phenylpropanoids and flavonoids on membrane associated enzyme complexes. Arch. Biochem. Biophys. 237, 88–100.

Hrazdina, G., Zobel, A.M., Hoch, H.C., 1987. Biochemical, immunological, and immunocytochemical evidence for the association of chalcone synthase with endoplasmic reticulum membranes. Proc. Natl. Acad. Sci. USA 84, 8966–8970.

Kalinova, J., Vrchotova, N., 2009. Level of catechin, myricetin, quercetin and isoquercitrin in buckwheat (*Fagopyrum esculentum* Moench), changes of their levels during vegetation and their effect on the growth of selected weeds. J. Agric. Food Chem. 57, 2719–2725.

Kerscher, F., Franz, G., 1987. Biosynthesis of vitexin and isovitexin: enzymic synthesis of the *C*-glucosylflavones vitexin and isovitexin with an enzyme preparation from *Fagopyrum esculentum* M. seedlings. Z. Naturforsch. C. J. Biosci. 42, 519–524.

Kerscher, F., Franz, G., 1988. Isolation and some properties of an UDP-glucose: 2-hydroxy-flavanone-6(or 8)-C-glucosyltransferase from *Fagopyrum esculentum* M. cotyledon. J. Plant Physiolol. 132, 110–115.

Kim, S.J., Zaidul, I.S., Suzuki, T., Mukasa, Y., Hashimoto, N., Takigawa, S., Noda, T., Matsuura-Endo, C., Yamauchi, H., 2008. Comparison of phenolic compositions between common and Tartary buckwheat (*Fagopyrum*) sprouts. Food Chem. 110, 814–820.

Kiprovski, B., Mikulic-Petkovsek, M., Slatnar, A., Veberic, R., Stampar, F., Malencic, D., Latkovic, D., 2015. Comparison of phenolic profiles and antioxidant properties of European *Fagopyrum esculentum* cultivars. Food Chem. 185, 41–47.

Li, X., Park, N.I., Xu, H., Woo, S.H., Park, C.H., Park, S.U., 2010. Differential expression of flavonoid biosynthesis genes and accumulation of phenolic compounds in common buckwheat (*Fagopyrum esculentum*). J. Agric. Food Chem. 58, 12176–12181.

Li, X., Thwe, A.A., Park, N.I., Suzuki, T., Kim, S.J., Park, S.U., 2012. Accumulation of phenylpropanoids and correlated gene expression during the development of Tartary buckwheat sprouts. J. Agric. Food Chem. 60, 5629–5635.

Lim, E.K., Ashford, D.A., Hou, B., Jackson, R.G., Bowles, D.J., 2004. *Arabidopsis* glycosyltransferases as biocatalysts in fermentation for regioselective synthesis of diverse quercetin glucosides. Biotechnol. Bioeng. 87, 623–631.

Mehrtens, F., Kranz, H., Bednarek, P., Weisshaar, B., 2005. The *Arabidopsis* transcription factor MYB12 is a flavonol-specific regulator of phenylpropanoid biosynthesis. Plant Physiol. 138, 1083–1096.

Nagatomo, Y., Usui, S., Ito, T., Kato, A., Shimosaka, M., Taguchi, G., 2014. Purification, molecular cloning and functional characterization of flavonoid C-glucosyltransferases from *Fagopyrum esculentum* M. (buckwheat) cotyledon. Plant J. 80, 437–448.

Nam, T.G., Lee, S.M., Park, J.H., Kim, D.O., Baek, N.I., Eom, S.H., 2015. Flavonoid analysis of buckwheat sprouts. Food Chem. 170, 97–101.

Park, N.I., Li, X., Thwe, A.A., Lee, S.Y., Kim, S.G., Wu, Q., Park, S.U., 2012. Enhancement of rutin in *Fagopyrum esculentum* hairy root cultures by the *Arabidopsis* transcription factor AtMYB12. Biotechnol. Lett. 34, 577–583.

Saito, K., 1974. The incorporation of the carboxyl carbon from acetyl coenzyme A into ring A of quercetin by cell-free extracts prepared from buckwheat seedlings. Biochim. Biophys. Acta 343, 392–398.

Suzuki, T., Honda, Y., Funatsuki, W., Nakatsuka, K., 2002. Purification and characterization of flavonol 3-glucosidase, and its activity during ripening in Tartary buckwheat seeds. Plant Sci. 163, 417–423.

Suzuki, T., Kim, S.J., Yamauchi, H., Takigawa, S., Honda, Y., Mukasa, Y., 2005. Characterization of a flavonoid 3-*O*-glucosyltransferase and its activity during cotyledon growth in buckwheat (*Fagopyrum esculentum*). Plant Sci. 169, 943–948.

Talhi, O., Silva, A., 2012. Advances in C-glycosylflavonoid research. Curr. Org. Chem. 16, 859–896.

Tsurunaga, Y., Takahashi, T., Katsube, T., Kudo, A., Kuramitsu, O., Ishiwata, M., Matsumoto, S., 2013. Effects of UV-B irradiation on the levels of anthocyanin, rutin and radical scavenging activity of buckwheat sprouts. Food Chem. 141, 552–556.

Watanabe, M., Ito, M., 2003. Effects of light on the content of phenolic compounds in buckwheat seedlings (in Japanese). Nippon Shokuhin Kagaku Kogaku Kaishi, 50, 32–34.

Watkin, J.E., Uuderhill, E.W., Neish, A.C., 1957. Biosynthesis of quercetin in buckwheat. II. Can. J. Biochem. Physiol. 35, 229–237.

Xiao, J., Muzashvili, T.S., Georgiev, M.I., 2014. Advances in the biotechnological glycosylation of valuable flavonoids. Biotechnol. Adv. 32, 1145–1156.

Zielińska, D., Turemko, M., Kwiatkowski, J., Zieliński, H., 2012. Evaluation of flavonoid contents and antioxidant capacity of the aerial parts of common and Tartary buckwheat plants. Molecules 17, 9668–9682.

Zhang, G., Xu, Z., Gao, Y., Huang, X., Zou, Y., Yang, T., 2015. Effect of germination on the nutritional properties, phenolic profiles, and antioxidant activities of buckwheat. J. Food Sci. 80, H1111–H1119.

Chapter thirty one

Diversity in Seed Storage Proteins and Their Genes in Buckwheat

N.K. Chrungoo*, L. Dohtdong**, U. Chettry**

*Department of Botany, North Eastern Hill University, Shillong, India;
**Plant Molecular Biology Laboratory, UGC-Centre for Advanced Studies in Botany,
North-Eastern Hill University, Shillong, India

INTRODUCTION

Seed storage proteins, intended as a source of nitrogen during the initial stages of germination and seedling growth, constitute an important source of dietary proteins for human consumption. Cereals and pulses have been contributing to the nutritional requirements of man and constitute an important source of proteins in the human diet. Although cereal grains and legume seeds are the major sources of vegetarian dietary proteins for human consumption, the nutritional quality of the proteins in both does not match the WHO standards. While the major amino acid deficiency in legume seed proteins is their low content of sulfur-containing amino acids cysteine and methionine, cereal proteins have low levels of lysine (Boulter, 1981; Shotwell and Larkins, 1989). Legumins, or 11–13S globulins, are a major class of seed storage proteins that were first characterized in Fabaceae and are widespread in both angiosperms as well as gymnosperms. The proteins are generally classified according to their solubility as albumins, globulins, prolamins, and glutelins, which are, respectively, soluble in water, saline solutions, alcohol, and alkaline solutions (Osborne, 1924). However, other classifications such as site of deposition, metabolic properties (Hills, 2004), and to a lesser extent allergen properties (Breiteneder et al., 2005), are also used to separate and classify seed storage proteins. While the albumins have sedimentation coefficients of 1.7–2S, globulins have sedimentation coefficients ranging from 7S to 8S for vicilins and 11 to 13S for legumins. Seed storage proteins homologous to legumins are, however, widely distributed in both

Molecular Breeding and Nutritional Aspects of Buckwheat. http://dx.doi.org/10.1016/B978-0-12-803692-1.00031-6

monocots and dicots and are selectively known as 11S globulins or legumin-like proteins (Margoliash et al., 1970). Legumin generally exists as hexamers, the monomers of which are composed of a larger, more acidic α-polypeptide and a smaller, more basic β-polypeptide. The two polypeptides are derived from a single preproprotein that is co- and posttranslationally cleaved to yield a characteristic, disulfide-linked α/β-structure (Derbyshire et al., 1976).

Seed storage proteins are stored in storage vacuoles or specialized aggregates called protein bodies, which are assembled within the endoplasmic reticulum (Herman and Larkins, 1991; Casey, 1999). They accumulate exclusively throughout seed development and form the majority of total seed proteins at the end of seed maturation. Thus they have very complex temporal and spatial regulations of expression, which makes them good models for fundamental research on how gene expression is regulated during specific stages of plant development. However, considering the ever-increasing demand for food materials, it is not only necessary to target the seed storage proteins as tools for fundamental research but also to use them for improvement of nutritional quality of conventional crops, wherever applicable. In particular the rich diversity of underexploited crops with a high nutritional value of their seed proteins can prove to be a boon for improvement of nutritional quality of the conventional crops currently under domesticated agriculture. These crops could also constitute an important genetic base to look for suitable heterologous proteins and their genes, which could be used as tools in crop improvement programs. Among the existing known plant resources, the International Plant Genetic Resources Institute and the Consultative Group on International Agriculture have identified common buckwheat, grain amaranth, and *Chenopodium* as important but underutilized nutraceutical crops, which could be used as a genetic base for the identification and isolation of suitable heterologous genes coding for biomolecules of potential economic importance. Because of the high protein content of their grains and the nutritionally balanced amino acid composition of their grain proteins, such minor crops are considered to have great potential in contributing toward crop improvement programs.

Common buckwheat (*Fagopyrum esculentum* Moench) belongs to the family Polygonaceae. Because of its short growth span, the capability to grow at high altitudes, and the high-quality protein content of its grains it is an important crop in mountainous regions of India, China, Russia, Ukraine, Kazakhstan, parts of Eastern Europe, Canada, Japan, Korea, and Nepal. The protein content is higher than that reported for any other cereals and the amino acid composition matches the WHO recommended values for a nutritionally rich protein with a balanced amino acid composition (Rout and Chrungoo, 1996). The nutritionally rich component of protein is a 26-kDa basic subunit, which has more than 6% lysine and nearly 2% methionine (Rout and Chrungoo, 1996; Rout et al., 1997). Because of the balanced amino acid composition, high nutrient value, and homology with seed storage proteins of a leguminous group of plants, this protein could be an important candidate for compensating plants deficient in amino acids.

SEED STORAGE PROTEIN CLASSIFICATION

Seed proteins can be broadly grouped into (1) housekeeping and (2) storage proteins. While the housekeeping proteins are responsible for maintaining normal cell metabolism, the storage proteins not only act as a source of nitrogen during early seedling growth but also have a role in dehydration tolerance during embryo maturation. These proteins are nonenzymatic and have, on the basis of their solubility properties, been empirically classified by Osborne (1924) into water-soluble albumins (1.6S–2S), salt-soluble globulins (7S–13S), aqueous alcohol-soluble prolamins, and weakly acid or alkali-soluble glutelins. While albumins and globulins are the major storage proteins in dicots, prolamins, and glutelins are the major storage proteins in grains of monocots.

The main components of buckwheat seed proteins are the salt-soluble globulins, which constitute 43% of the total seed protein component of the grains. The major seed storage protein of buckwheat is a 13S globulin having a molecular mass of 280 kD (Belozersky, 1975). On the basis of its sedimentation constant, subunit structure, and sequence homologies the protein has been classified as a legumin-like storage protein (Derbyshire et al., 1976; Belozersky and Dunaevsky, 1983; Rout and Chrungoo, 1996). The protein is a hexamer of nonidentical subunits, each consisting of one larger acidic and one smaller basic polypeptide, linked by disulfide bonds. The nutritionally rich component of the protein is a 26-kDa basic subunit, which has more than 6% lysine and nearly 2% methionine (Rout and Chrungoo, 1996; Rout et al., 1997). The protein has 22% α-helix, 36% β-sheet, 12% β-turn, and 30% random coil secondary structure. In comparison with the basic subunits of other legumin-type proteins, the buckwheat legumin subunit has a high content of lysine and methionine (Rout and Chrungoo, 1999). Sequence analysis of the 26-kDa basic subunit revealed 93 and 75% sequence homology with 11S globulin of *Coffea arabica* and β-subunit of 11S globulin of *Cucurbita pepo*, respectively. The subunit has the "globally conserved" N-terminal sequence consisting of "Gly-Ile-Asp-Glu" and the conserved 7 residue domain of "Pro-His-Trp-Asn-Ile-Asn-Ala," characteristic of basic subunits of legumins from nonleguminous angiosperms. Radovic et al. (1996, 1999) have also reported the presence of two more major seed storage proteins in common buckwheat. While one of the proteins was identified as an 8S globulin, the other was identified as a 2S albumin.

The globulin gene sequence encodes two cupin domains, and the gene product forms a radially symmetric homodimer. Each of these homodimers combines with two others to form a radially symmetric trimer (Hirano et al., 1985). This is accomplished via noncovalent bonding between hydrophobic regions in a pocket formed by a helical region at either end of the bicupin structure. While the radially symmetric trimer is the final quaternary product in the case of the 7S globulin family, the trimers stack in pairs to form hexamers in the 11S legumins. The cupin domains found in 7S and 11S globulins share remarkable structural similarities at the tertiary level. Though the primary sequence structure seems to deviate significantly in several areas, the general cupin motif can be found in the peptide sequence of both domains. Each domain has a cupin motif followed by a helix and turn. The cupin motif is composed of two β-strands separated by an internal motif spacer; the three elements together forming the main cupin β-barrel with the two β-strands forming an antiparallel jelly roll, with the inter motif spacer forming a hairpin turn at its center point. The internal motif spacer shows considerable variations in its length between the 11S and 7S globulins, as well as across orthologous and paralogous copies of the globulins. The differences in the bonding pattern between cupin domains in the 11S and 7S globulin has been ascribed to the variations between the multimeric structure of these two classes of proteins and the need of cysteine disulfide bridges between separate cupin domains to stabilize the overall structure (Rodin and Rask, 1990; Adachi et al., 2003). The high level of similarity in the tertiary structure of between-cupin domains of various proteins has led to the postulation that all cupin domain-containing genes share a point of common origin (Dunwell et al., 2000). The proposed models of domain duplication near the origin of the plant storage globulins also hint at a connection with some more ancient relationships with single cupin proteins (Dunwell et al., 2001). The cupin fold, which describes the beta coil domain of the globulins, as well as the single cupin germin-like proteins, is found in a wide array of proteins ranging from bacterial oxalate decarboxylases, to fungal phosphatase isomerases. While some proteins with cupin domains have been reported to be associated with DNA binding and retrotransposition genes, functions of many others have not been elucidated as yet (Khuri et al., 2001).

Even though seed storage protein polymorphism has been revealed as useful genetic markers for cultivar identification, crop origin, and evolutionary studies (Rogl and Javornik, 1996), there is little information available on buckwheat seed proteins compared to that on legumes or cereal storage proteins. Zeller et al. (2004) reported a scheme to identify common buckwheat cultivars and lines grown in different countries on the basis of sodium dodecyl sulfate polyacrylamide gel electrophoresis (SDS-PAGE) of their seed storage proteins. The report inferred a multiallelic nature of the gene locus that coded for the 13S buckwheat globulin in common buckwheat. However, on the basis of studies carried out to investigate the inheritance pattern of the globulins in common buckwheat, Luthar et al. (2007) have suggested that genes controlling globulin subunits are tightly linked and are inherited as a single locus. While we observed a significantly high level of inter- as well as intraaccession variation in the SDS-PAGE profiles of grain proteins in different accessions of *F. esculentum*, we could not detect significant variations in the SDS-PAGE profiles of grain proteins isolated from different accessions of *Fagopyrum tataricuum* (Rout and Chrungoo, 2007). Cluster analysis of the endosperm protein profiles of the accessions revealed three broad clusters with a moderate level of intraspecific variability within the accessions of *F. esculentum*. The results also indicated that "*Fagopyrum himalianum*" belonged to the *esculentum* group and should not be regarded as a different species. We could also demonstrate that *Fagopyrum cymosum*, which to a greater extent resembles *F. esculentum* morphologically, is closer to *F. tataricum* than *F. esculentum*. These observations are in conformity with earlier findings reported on species relationships in *Fagopyrum* using a different marker approach (Kishima et al., 1995; Sharma and Jana, 2002).

BUCKWHEAT LEGUMIN GENE STRUCTURE

The two major reasons for research focus on seed storage proteins have been their application in studies on developmental biology of plants and their importance in human nutrition. On the basis of a reconstructed sequence of two overlapping clones isolated from a cDNA library prepared from grains at mid-maturation stage of development, Samardzic et al. (2004) identified a legumin gene that coded for a methionine-rich subfamily of proteins. The fact that a methionine-rich legumin coexists together with methionine-poor legumins in buckwheat should be an important element regarding the evolutionary position of buckwheat. This may also be supporting evidence that the methionine-rich genes were not lost in evolution but were protected under pressure of an increased need for sulfur. We have cloned a gene from buckwheat that codes for a protein having 6% lysine and 2% methionine (Rout et al., 1997). The firstly identified two-intron structure of buckwheat legumin gene represented an important advancement in knowledge of methionine-rich legumins in lower angiosperms. The two-intron model of legumin genes, first characterized for *Vicia faba LeB4* (Bäumlein et al., 1986) is an exception to the rule. Most angiosperm legumin genes possess three introns with strictly conserved positions (Maraccini et al., 1999). The two-intron structure of the buckwheat legumin gene apparently does not conform to the typical three-intron organization of legumin genes as typified by the *Pisum sativum*-type legumin structure. Shutov and Bäumlein (1999) speculated that the two-intron structure is an ancient feature of the storage globulin single domain ancestor, which predates its duplication.

The synthesis of seed storage proteins is primarily controlled at the transcriptional level, and the seed-specific expression has been shown to be conferred upon the promoter regions of many storage protein genes (Devic et al., 1996; Moreno-Risueno et al., 2008; Lee et al., 2007). Unraveling the molecular basis of seed-specific gene expression has been mainly focused on the identification of specific *cis*-elements or elements within the control regions and the identification of transcription factors associated with them. On the basis of a comparison of the nucleotide sequences of 11S genes of broad bean, pea, and soybean a region in the 5′ flanking segments of these genes has been found to be more conserved than most of the coding region with a 28 bp sequence, designated as "legumin box," located within 250 bp upstream of the transcription start site (Bäumlein et al., 1986). The core motif of the legumin box represents an alternating succession of purine and pyrimidine nucleotides known as "RY" motif, which occurs in multiple copies in the promoter region of SSP genes of most legumes (Dickinson et al., 1988) and cereals (Forde et al., 1985). The motif has been the target of investigations for *cis* analysis of the seed storage protein gene promoters and many workers have identified RY repeat motif as a key *cis*-acting element for seed-specific gene expression (Baumlein et al., 1992; Chamberland et al., 1992; Fujiwara and Beachy 1994; Sakata et al., 1997; Bobb et al., 1997). While the 5'UTR of genes coding for 11S–13S legumin-type proteins are known to have a legumin box, that of 7S genes coding for vicilin-type proteins have the "vicilin box" located within 150 bp of the TSS (Gatehouse et al., 1986). The core sequence (GCCACCTCAT) of the vicilin box has been shown to be necessary for embryo-specific expression of the gene (Gatehouse et al., 1986). Chandrasekharan et al. (2003) have shown that of the four RY elements found in *phas* promoter, mutation in the three distal RY elements led to the expression of the reporter gene in the entire embryo including the radical. This could indicate that the three distal RY elements had a role in restricting the expression of the gene in the radical tissues of the seeds. While mutation in the proximal RY motif led to reduced expression in embryos it abolished the expression of the gene in the radical. These results indicated the involvement of the three distal RY elements as a negative controller of legumin gene expression in the radical. Similarly, deletion of the RY motif in the legumin and napin gene promoter abolished most of the seed-specific promoter activity while its deletion in the *V. faba* legumin promoter resulted in low-level expression of the gene in leaves (Baumlein et al., 1992; Stalberg et al., 1993; Ellerstrom et al., 1996; Reidt et al., 2000). Nunberg et al. (1995) have identified at least three distinct sequences in the proximal promoter region (PPR) of *helianthinin* gene "*Ha*G3A," which interact with nuclear proteins thereby regulating the expression of the gene. Of these two motifs AGATGT ("A" motif) and TGATCT ("T" motif) occur twice in the *helianthinin* PPR. The "A" motif occurs at −111 (A_2) and −58 (A_1) and the "T" motif is situated at −83 (T_2) and −41 (T_1). Mutation of the A and T motifs in the PPR of the *helianthinin* promoter resulted in ectopic expression in non-embryonic tissues (Bogue, 1990). These results indicate an important negative regulatory component in the tissue-specific expression of the gene. Presumably the A and T motifs are involved in such a negative regulatory loop. The third DNA-binding motif identified in the *helianthinin* PPR is the CCAAAT (Y box). This box is similar to the C/EBP-binding motif involved in communication between upstream enhancers and basal promoter elements (Rorth and Montell, 1992). These results suggest a bipartite structure for seed protein regulatory ensemble. The proximal promoter elements direct tissue-specific expression while the more distal elements enhance and modulate this basic pattern. Ellerstrom et al. (1996) have identified a region between −309 and −152 of TSS in the napA promoter, which was involved in regulating the quantitative expression of the gene and had similarity to ABRE. On the other hand, deletion of the region containing the $(CA)_n$ element increased promoter activity in both leaves and endosperm and decreased its activity in the embryo indicating that this element is important for conferring seed-specific expression by serving both as an activator and as a repressor element.

Comparisons of the promoter sequences of various cereal prolamin genes have identified a conserved region comprising two conserved motifs, namely, a 7 bp 5'TGTAAAG3' element and a 9bp 5'(G/C)TGA(G/C)TCA(T/C)3' element located 300 bp upstream of the transcriptional start site (Kreis et al., 1985; Forde et al., 1985). While the 5′TGTAAAG3′ element has been designated as the prolamin-box or endosperm, or E motif, the 5′(G/C)TGA(G/C)TCA(T/C)3′ element has been designated as the "GCN4"-like motif (Muller et al., 1995). Deletion and point mutation analysis has revealed that the prolamin-box is important for the regulation of expression of endosperm-specific genes (Mena et al., 1998; Diaz et al., 2005).

Using the gene-walking approach we isolated a 1-kb 5′UTR (acc. no. EU595873) of the buckwheat 13S globulin. NNPP predicted three probable transcription start sites out of which the TSS at P'_{733} was located closest to the predicted ATG start codon (P'_{774}) and followed the YR rule. Other *cis*-elements identified in the sequence included the −30 TATA box, an endosperm-specific motif identified as prolamin-box in cereals "TGTAAAG," three RY elements "CATGCA," core of "AACA" motif for regulating endosperm-specific and quantitative expression, "CAAT" promoter consensus sequence, and "CACA" box (unpublished data of the authors). Fifteen accessions of common buckwheat were used for generating the nucleotide sequences of the same region. Multiple sequence alignment of these sequences showed the highly conserved position of *cis*-elements in all the accessions.

Functional analysis of SSP gene promoters by expressing the complete genes or promoter–reporter gene fusion in transgenic systems (Sengupta-Gopalan et al., 1985; Bäumlein et al., 1987, 1991; Murai et al., 1983; Ellerstrom et al., 1996) has led to the conclusion that the promoters of SSP genes are directly involved in seed-specific expression of the SSP genes. 5′ serial deletion and nested internal deletion analyses of the upstream regions of SSP gene promoters have revealed that approximately one hundred to a few hundred base pairs are sufficient for tissue-specific regulation and temporal control (Chen et al., 1986, 1989; Colot et al., 1987; Robert et al., 1989; Shirsat et al., 1989). Collectively, these studies suggest that the expression of seed storage protein genes depends on a combinatorial array of distinct regulatory modules and a specific complement of *trans*-acting factors. Thus each gene has a unique combination of *cis*-acting DNA sequences that function to direct its expression. Although the exact information to delineate a promoter and its regulatory elements requires experimental approaches like promoter deletions, substitutions, and linker scanning, prior computational analysis of the sequence can serve as a guide to establish a platform for further promoter analysis. As a first step, potentially useful promoters need to be evaluated in view of their developmental stage specificity, seed specificity, and expression levels. The study and increased understanding of gene promoters, their structure, function, and mechanism of gene regulation, will open up the possibility of modulation of gene expression in homologous as well as heterologous systems. The phenomenon of homology-based gene silencing, frequently encountered in many transgenic systems (Jorgensen, 1995; Matzke and Matzke; 1995 and Meyer, 1995), suggests that it is important to have an available selection of promoters, offering a range of alternative expression patterns.

FUTURE PROSPECTS

The discovery of the cupin superfamily epitomizes the value of an integrated approach toward analysis of storage protein structure, function, and diversity. Comparative analysis of primary sequence information with 3D structure data and its functional implications can throw much light on the role of seed storage proteins in the functional biology of a seed. Research on multifunctional cupins has led to the identification of a particularly stable metal-binding β-sheet structure that confers thermostability to the protein. It is possible that such domains may be important for the stability of the proteins in a heterologous/transgenic system or may even be looked upon as an important tool in taxonomic classification. Since many of these proteins have been found to be responsive to various types of abiotic stresses they could be of interest to plant breeders interested in improving the growth of crops under limiting environments. In a more fundamental context, resolution of cupin structures at the atomic level will confirm the exact significance of such domains in conferring integrity to the proteins under extremophilic conditions. Such information will prove of great potential value, for example, in applied agricultural, medical, or environmental projects utilizing oxalate-degrading enzymes and in nutritional projects that aim to improve the quality of seed proteins. As buckwheat legumin type 13S globulin possesses features that are common to monocots as well as dicots, the protein could work as a model system for an integrated approach in understanding functional biology of seed storage proteins (Figs 31.1 and 31.2)

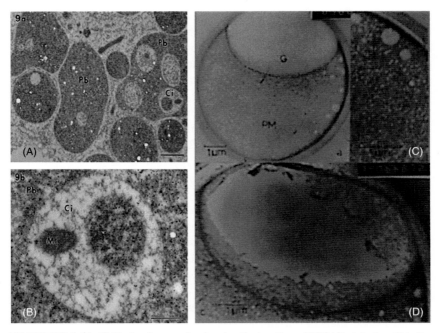

FIGURE 31.1 (A, B) Photomicrographs showing the location of 280 kD globulin within the protein bodies in the endosperm of grains of common buckwheat. The protein bodies are depicted by the mark "*Pb*"; (C, D) Enlarged view of protein bodies showing the presence of amorphous protein within the protein bodies enclosed by a membrane bilayer.

A	B	C	D	E	F	G	H	
1.	Kidney bean	Phaseolin	CTCTTATAATAATA	43	TTCATCA	77	TCTACTCATGATG	Dicots
2.	Pea	legumin(leg A)	TCTCTATAAATTA	33	CGCATCA	33	CTCTTCATGGCT	
3.	Pea	legumin(leg B)	TCTCTATAAATTA	33	CGCATCA	33	CTCTTCATGGCT	
4.	Pea	legumin(leg C)	TCTCTATAAATTA	33	CGCATCA	33	CTCTTCATGGCT	
5.	Vicia faba	legumin(LeB4)	TTCCTATAAATCA	32	TTCACCA	56	GTCACAATGTCC	
6.	Soybean	lectin(Le1)	CTAGTATAAATAG	27	TGCATAC	30	AAAGCAATGGCT	
7.	Kidney bean	lectin(pPVL134)	GTTGTATAAATAG	29	AATGCAT	10	GAATGCATGATC	
8.	Castor bean	lectin(ricin)	TCTGTATTAATTT	48	GACAGCC	34	TCAAGGATGAAA	
9.	Kidney bean	lectin(Lec1)	GTTGTATAAATAG	31	TGCATGA	12	GCATACAATGGCT	
10.	Kidney bean	lectin(Lec2)	TCTCTATAAATAG	33	TGCATGA	12	GCATACAATGGCA	
11.	Buckwheat	legumin-like	TCAGTATAAAATC	30	TTCAATC	41	TCCACCATGTCA	
12.	Wheat	gliadin (pW8233)	CAACTATAAATAG	27	ATGATCC	77	TCCACCATGAAG	Monocots
13.	Wheat	gliadin (~β)	GAGCTATAAAAG	37	CTCACCC	67	TCCACCATGAAG	
14.	Wheat	gliadin (γ)	TAGCTATAAAAAG	29	TCCATAC	76	TCCACCATGAAG	
15.	Wheat	gliadin (yam-2)	CAACTATAATAG	27	ATCATCC	77	TCCACCATGAAG	
16.	Wheat	glutenin (pHSB-26)	TCCCTATAAAGC	30	CTCATCA	61	ATCGAAATGGCT	
17.	Wheat	glutenin (Ac11)	TCCCTATAAAAGC	30	TTCATCA	62	ACCGAGATGGCT	
18.	Maize	glutenin	AAGCTATAATAA	34	TCCATCA	70	GACACCATGAGG	
19.	Barley	hordein (B1)	CTACTATAATAG	27	ATCATCA	52	TCCACCATGAAG	
20.	Maize	Zein (Z4)	TGTGTATAAATAT	29	AATATAT	59	CCAATAATGGCA	
21.	Maize	Zein (19kD)	TGTGTATAAATAT	28	CTAATAT	61	CCAATAATGGCA	
22.	Maize	Zein (ZA1)	CCTATATAAATAG	31	TCCATCA	68	ACAACAATGGCT	
23.	Maize	Zein (pMl1)	AAAATATATATGA	63	ATCACCT	65	ACACAATGGCT	

A. Serial Numbers,
B. Plant
C. Gene
D. Flanking region of TATA sequence (Highlighted), 4 bases on the left and 5 bases on the right.
E. Distance between first 'T' of TATA sequence and nucleotide prior to TSS
F. TSS (highlighted)flank by 3 bases on each side
G. Distance between TSS and nucleotide prior to 'A' of ATG
H. ATG, proposed or observed translation start site (highlighted) flanked by 6 bases on left and 3 bases on the right.

FIGURE 31.2 Alignment of the context sequence around TATA, TSS, and ATG-start codon of buckwheat seed storage protein gene with the corresponding regions of seed storage protein genes from dicot and monocot plants.

REFERENCES

Adachi, M., Kanmori, J., Masuda, T., Yagasaki, K., Kitamura, K., Mikami, B., Utsumi, S., 2003. Crystal structure of soybean 11S globulin: glycinin A3B4 homohexamer. Proc. Natl. Acad. Sci. USA 100, 7395–7400.

Bäumlein, H., Boerjan, W., Nagy, I., Panitz, R., Inzé, D., Wobus, U., 1991. Upstream sequences relating legumin gene expression in heterologous transgenic plants. Mol. Genet. Genom. 225, 121–128.

Bäumlein, H., Müller, A.J., Schiemann, J., Helbing, D., Manteuffel, R., Wobus, U., 1987. A legumin B gene of *Vicia faba* is expressed in developing seeds of transgenic tobacco. Biol. Zentral. 106, 569–575.

Baumlein, H., Nagy, I., Villarroel, R., Inze, D., Wobus, U., 1992. *Cis*-analysis of a seed protein gene promoter: the conservative RY repeat CATGCATG within the legumin box is essential for tissue-specific expression of a legumin gene. Plant J. 2, 233–239.

Bäumlein, H., Wobus, U., Pustell, J., Kalatos, F.C., 1986. The legumin gene family: structure of a B type gene of *Vicia faba* and a possible legumin gene specific regulatory. Elem. Nucl. Acids Res. 14, 2707–2720.

Belozersky, M.A., 1975. Isolation and characterization of buckwheat seed 13S globulin. Biosynthesis of Storage ProteinsNauka, Moscow, pp. 152–156.

Belozersky, M.A., Dunaevsky, Y.E., 1983. Initial changes of the main storage protein in germinating buckwheat seed. Biochimiya 48, 508–511.

Bobb, A.J., Chern, M.S., Bustos, M.M., 1997. Conserved RY-repeats mediate transactivation of seed-specific promoters by the developmental regulator PvALF. Nucl. Acids Res. 251, 641–647.

Bogue, M.A., 1990. Analysis of *cis*-regulatory elements controlling the expression of two sunflower 11s globulin genes. PhD Dissertation. Texas A & M University, USA.

Boulter, D., 1981. Biochemistry of storage protein synthesis and deposition in the developing legume seed. Woolhouse, H.W. (Ed.), Advances in Botanical Research, vol. 9, Academic Press, London.

Breiteneder, H., Shearer, W., Rosenwasser, L., 2005. Molecular properties of food allergens. J. Allerg. Clin. Immunol. 115, 14–23.

Casey, R., 1999. Distribution and some properties of seed globulins. In: Shewry, P.R., Casey, R. (Eds.), Seed Proteins. Kluwer Academic Publishers, Dordrecht, the Netherlands, pp. 159–169.

Chamberland, S., Daigle, N., Bernier, F., 1992. The legumin boxes and the 3′ part of a soybean beta-conglycinin promoter are involved in seed gene expression in transgenic tobacco plants. Plant Mol. Biol. 19, 937–949.

Chandrasekharan, M.B., Bishop, K.J., Hall, T.C., 2003. Module-specific regulation of the beta-phaseolin promoter during embryogenesis. Plant J. 33, 853–866.

Chen, Z.L., Schuler, M.A., Beachy, R.N., 1986. Functional analysis of regulatory elements in a plant embryo-specific gene. Proc. Natl. Acad. Sci. USA 83, 8560–8564.

Chen, Z.G., Stauffacher, C., Li, Y., Schmidt, T., Bomu, W., Kamer, G., Shanks, M., Lomonossoff, G., Johnson, J.E., 1989. Protein–RNA interactions in an icosahedral virus at 3.0. A resolution. Science 245, 154–159.

Colot, V., Robert, L.S., Kavanagh, T.A., Bevan, M.W., Thompson, R.D., 1987. Localization of sequences in wheat endosperm protein genes which confer tissue-specific expression in tobacco. Eur. Mol. Biol. Org. J. 6, 3559–3564.

Derbyshire, E., Wright, D.J., Boulter, D., 1976. Legumin and vicilin storage proteins of legume seeds. Phytochemistry 15, 3–24.

Devic, M., Albert, S., Delseny, M., 1996. Induction and expression of seed specific promoters in *Arabidopsis* embryo-defective mutants. Plant J. 9, 205–215.

Diaz, I., Martinez, M., Isabel-LaMoneda, I., Rubio-Somoza, I., Carbonero, P., 2005. The DOF protein, SAD, interacts with GAMYB in plant nuclei and activates transcription of endosperm-specific genes during barley seed development. Plant J. 42, 652–662.

Dickinson, C.D., Evans, R.P., Nielsen, N.C., 1988. RY repeats are conserved in the 5′-flanking region of legume seed protein genes. Nucl. Acids Res. 16, 371.

Dunwell, J.M., Culham, A., Carter, C.E., Sosa-Aguirre, C.R., Goodenough, P.W., 2001. Evolution of functional diversity in the cupin superfamily. Trends Biochem. Sci. 26, 740–746.

Dunwell, J.M., Khuri, S., Gane, P.J., 2000. Microbial relatives of the seed storage proteins of higher plants: conservation of structure and diversification of function during evolution of the cupin superfamily. Microbiol. Mol. Biol. 64, 153–179.

Ellerstrom, M., Stalberg, K., Ezcurra, I., Rask, L., 1996. Functional dissection of a napin gene promoter: identification of promoter elements required for embryo and endosperm-specific transcription. Plant Mol. Biol. 32, 1019–1027.

Forde, B.G., Heyworth, A., Pywell, J., Kreis, M., 1985. Nucleotide sequence of a Bi Hordein gene and the identification of possible upstream regulatory elements in endosperm storage protein genes from barley, wheat and maize. Nucl. Acids Res. 13, 7327–7337.

Fujiwara, T., Beachy, R.N., 1994. Tissue-specific and temporal regulation of a β-conglycinin gene-roles of the RY repeat and other cis-acting elements. Plant Mol. Biol. 241, 261–272.

Gatehouse, J.A., Evans, I.M., Croy, R., Boutler, D., 1986. Differential expression of genes during legume seed development. Phil. Trans. R. Soc. Lond. B. 314, 367–384.

Herman, R.G., Larkins, B., 1991. Physical mapping of DNA sequences on plant chromosomes by light microscopy and high resolution scanning electron microscopy. Plant Molecular Biology. Plenum Press, New York, pp. 277–284.

Hills, M.J., 2004. Control of storage-product synthesis in seeds. Curr. Opin. Plant Biol. 7, 302–308.

Hirano, H., Fukazawa, C., Harada, L., 1985. The primary structures of the A4 and A5 subunits are highly homologous to that of the A3 subunit is the glycinin seed storage protein of soybean. Fed. Eur. Biochem. Soc. J. 181, 124–128.

Jorgensen, R.A., 1995. Co-suppression, flower color patterns, and metastable gene expression States. Science 268, 686–691.

Kishima, Y., Ogura, K., Mizukami, K., Mikami, T., Adachi, T., 1995. Chloroplast DNA analysis in buckwheat species: phylogenetic relationships, origin of reproductive systems and extended inverted repeats. Plant Sci. 108, 173–179.

Khuri, S., Bakker, F.T., Dunwell, J.M., 2001. Phylogeny, function and evolution of the cupins, a structurally conserved, functionally diverse superfamily of proteins. Mol. Biol. Evol. 18, 593–605.

Kreis, M., Forde, B.G., Rahman, S., Miflin, B.J., Shewry, P.R., 1985. Molecular evolution of the seed storage proteins of barley, rye and wheat. J. Mol. Biol. 183, 499–502.

Lee, S.K., Hwang, S.K., Han, M., Eom, J.S., Kang, H.G., Han, Y., Choi, S.B., Cho, M.H., Bhoo, S.H., Thomas, T.R., Okita, W., Jeon, J.S., 2007. Identification of the ADP-glucose pyrophosphorylase isoforms essential for starch synthesis in the leaf and seed endosperm of rice (Oryza sativa L.). Plant Mol. Biol. 65, 531–546.

Luthar, Z.L., Rogl, S., Kump, B., Avornik, B.J., 2007. 38–48 kDa subunits of buckwheat 13S globulins are controlled by a single locus. Plant Breed. 127, 322–324.

Maraccini, P., Deshayes, A., Petiard, V., Rogers, W.J., 1999. Molecular cloning of the complete 11S seed storage protein gene of Coffea arabica and promoter analysis in transgenic tobacco plants. Plant Physiol. Biochem. 37, 273–282.

Margoliash, E., Nisonoff, A., Reichlin, M., 1970. Immunological activity of cytochrome c. I. Precipitating antibodies to monomeric vertebrate cytochrome c. J. Biol. Chem. 245, 931–939.

Matzke, M.A., Matzke, A.J.M., 1995. How and why do plants inactivate homologous (trans) genes? Plant Physiol. 107, 679–685.

Mena, M., Vicente-Carbajosa, J., Schmidt, R.J., Carbonero, P., 1998. An endosperm-specific DOF protein from barley, highly conserved in wheat, binds to and activates transcription from the prolamin-box of a native β-hordein promoter in barley endosperm. Plant J. 16, 53–62.

Meyer, P., 1995. DNA methylation and transgene silencing in Petunia hybrida. Curr. Top. Microbiol. Immunol. 197, 15–28.

Moreno-Risueno, M.A., González, N., Díaz, I., Parcy, F., Carbonero, P., Vicente-Carbajosa, J., 2008. FUSCA3 from barley unveils a common transcriptional regulation of seed-specific genes between cereals and Arabidopsis. Plant J. 53, 882–894.

Müller, C., Kowenz-Leutz, E., Grieser-Ade, S., Graf, T., Leutz, A., 1995. NF-M (chicken C/ EBPb) induces eosinophilic differentiation and apoptosis in a hematopoietic progenitor cell line. EMBO J. 14, 6127–6135.

Murai, N., Sutton, D.W., Murray, M.G., Slightom, J.L., Merlo, D.J., Reichert, N.A., Sengupta-Gopalan, C., Stock, C.A., Barker, R.F., Kemp, J.D., Hall, T.C., 1983. Phaseolin gene from bean is expressed after transfer to sunflower via tumor-inducing plasmid vectors. Science 222, 476–482.

Nunberg, A.N., Li, Z., Chung, H.J., Reddy, A.S., Thomas, T.L., 1995. Proximal promoter sequences of sunflower helianthinin genes confer regionalized seed-specific expression. J. Plant Physiol. 45, 600–605.

Osborne, T.B., 1924. The Vegetable Proteins, Monographs in Biochemistry. Longmans Green and Co., London.

Radovic, S., Maksimovic, V., Brkljacic, J., Varkonji-Gasic, E., Savic, A., 1999. 2S albumin from buckwheat (*Fagopyrum esculentum* Moench) seeds. J. Agric. Food Chem. 47, 1467–1470.

Radovic, R.S., Makstmoyic, V., Varkonji, E., 1996. Characterisation of buckwheat seed storage proteins. J. Agric. Food Chem. 44, 972–974.

Reidt, W., Wohlfarth, T., Ellerstrom, M., Czihal, A., Tewes, A., Ezcurra, I., Rask, L., Baumlein, H., 2000. Gene regulation during late embryogenesis: the RY motif of maturation-specific gene promoters. Plant J. 21, 401–408.

Robert, L.S., Thompson, R.D., Flavell, R.B., 1989. Tissue-specific expression of a wheat high molecular weight glutenin gene in transgenic tobacco. Plant Cell 1, 569–578.

Rodin, J., Rask, L., 1990. Characterization of a cDNA encoding a *Brassica napus* 12S protein (cruciferin) subunit. J. Biol. Chem. 265, 2720–2723.

Rogl, S., Javornik, B., 1996. Seed protein variation for identification of common buckwheat (*Fagopyrum esculentum* Moench) cultivars. Euphytica 87, 111–117.

Rorth, P., Montell, D.J., 1992. Drosophila C/EBP: a tissue-specific DNA-binding protein required for embryonic development. Genes Dev. 6, 2299–2311.

Rout, M.K., Chrungoo, N.K., 1996. Partial characterization of the lysine rich 280kD globulin from common buckwheat (*Fagopyrum esculentum* Moench): its antigenic homology with seed proteins of some other crops. Biochem. Mol. Biol. Int. 40, 587–595.

Rout, M.K., Chrungoo, N.K., 1999. The lysine and methionine rich basic subunit of buckwheat grain legumin: some results of a structural study. Biochem. Mol. Biol. Int. 47, 921–926.

Rout, A., Chrungoo, N.K., 2007. Genetic variation and species relationships in Himalayan buckwheats as revealed by SDS PAGE by endosperm proteins extracted from single seeds and RAPD based DNA fingerprints. Genet. Resour. Crop Evol. 54, 767–777.

Rout, M.K., Chrungoo, N.K., Rao, K.S., 1997. N-terminal amino acid sequence of the lysine rich subunit of 13S globulin from grains of common buckwheat (*Fagopyrum esculentum* Moench). Phytochemistry 45, 865–867.

Sakata, Y., Chiba, Y., Fukushima, H., Matsubara, N., Habu, Y., Naito, S., Ohno, T., 1997. The RY sequence is necessary but not sufficient for the transcription activation of a winged bean chymotrypsin inhibitor gene in developing seeds. Plant Mol. Biol. 341, 191–197.

Samardzic, J.T., Milisavljevic, M.D., Brkljacic, J.M., Konstantinovic, M.M., Maksimovic, V.R., 2004. Characterization and evolutionary relationship of methionine-rich legumin like protein from buckwheat. Plant Physiol. Biochem. 42, 157–163.

Sengupta-Gopalan, C., Reichert, N.A., Barker, R.F., Hall, T.L., Kemp, J.O., 1985. Developmentally regulated expression of the bean beta-phaseolin gene in tobacco seed. Proc. Natl. Acad. Sci. USA 82, 3320–3324.

Sharma, T.R., Jana, S., 2002. Species relationships in *Fagopyrum* revealed by PCR-based DNA fingerprinting. Theor. Appl. Genet. 105, 306–312.

Shirsat, A., Wilford, N., Croy, R., Boulter, D., 1989. Sequences responsible for the tissue specific promoter activity of a pea legumin gene in tobacco. Mol. Genet. Genom. 215, 326–331.

Shotwell, M., Larkins, B.A., 1989. The biochemistry and molecular biology of seed storage proteins. In: Marcus, E. (Ed.), The Biochemistry of Plants. Academic Press, San Diego, pp. 296–345.

Shutov, A.D., Bäumlein, H., 1999. Origin and evolution of seed storage globulins. In: Casey, R., Shewry, P.R. (Eds.), Seed Proteins. Kluwer Academic Publishers ITP, London, pp. 543–561.

Stalberg, K., Ellerstrom, M., Josefsson, L.G., Rask, L., 1993. Deletion analysis of a 2S seed storage protein promoter of *Brassica napus* in transgenic tobacco. Plant Mol. Biol. 23, 671–683.

Zeller, F.J., Weishaeupl, H., Hsam, S.L.K., 2004. Identification and genetics of buckwheat (*Fagopyrum*) seed storage protein. Proceedings of the 9th International Symposium on Buckwheat. Prague, Czech Republic, pp. 195–201.

Chapter thirty two

Waxy Locus in Buckwheat: Implications for Designer Starches

N.K. Chrungoo*, N. Devadasan*, I. Kreft**

*Department of Botany, North Eastern Hill University, Shillong, Meghalaya, India;
**University of Ljubljana, Biotechnical Faculty, Ljubljana, Slovenia

INTRODUCTION

Starch is the main energy reservoir of higher plants and also a major source of dietary energy for humans and animals. Besides its nutritive value, starch is a very useful raw material with a wide range of applications in both food and nonfood industries. Starch is synthesized in the form of granules with a partially crystalline texture, the morphology, chemical composition, and molecular structure of which are characteristic to each species (Banks and Greenwood, 1975). Starch owes much of its functionality to two major high molecular weight molecules, namely, amylose and amylopectin as well as to the physical organization of these molecules within the granule structure (French, 1984). Even though amylose and amylopectin represent approximately 98–99% of the dry weight of starch granules, the relative proportion of amylose to amylopectin may vary from species to species, within species as well as between tissues from the same species (Jane and Chen, 1992; Shujun et al., 2006; Peroni et al., 2006; Zhu et al., 2008).

The physicochemical properties of starches, namely, gelatinization temperature, retrogradation behavior, swelling power, viscosity, pasting properties, freeze/thaw stability, acid stability, and gel strength, have been shown to have an important bearing on the functional properties of starch (Dedeh and Sackey, 2002; Chen et al., 2003; Amani et al., 2004; Pérez et al., 2005; Peroni et al., 2006; Shujun et al., 2006; Riley et al., 2006). Therefore characterization of starches for their physicochemical, functional, and structural properties is

Molecular Breeding and Nutritional Aspects of Buckwheat. http://dx.doi.org/10.1016/B978-0-12-803692-1.00032-8

essential to unravel their potential for use in industry. Since amylose-free or low amylose starch is considered desirable for certain food and nonfood industries (Morell et al., 1995), most of the approaches aimed at modifying the quality of starch have focused on alteration of the amylose/ amylopectin ratio of the starch granules. Besides other strategies, approaches used to achieve this goal also involve manipulation of the "waxy" locus, which codes for granule-bound starch synthase-I (GBSS-I), the key enzyme involved in amylose biosynthesis in starch (Shure et al., 1983; Klosgen et al., 1986; Hovenkamp-Hermelink et al., 1987; Visser et al., 1989; Van der Leij et al., 1991; Rohde et al., 1988; Wang et al., 1990; Hirano and Sano, 1991; Okagaki, 1992; Dry et al., 1992; Hsieh et al., 1996; Mason-Gamer et al., 1998; Murai et al., 1999). High amylose starches, having 40–80% amylose content, have been reported in maize, barley, rice, and buckwheat (Song and Jane, 2000; Zhu et al., 2008; Christa and Śmietana-Soral, 2008). Similarly, *waxy* (amylose-free) mutants have been found in rice, maize, sorghum, barley, and potato (Denyer et al., 2001; Deschamps et al., 2008). These mutants lack a starch granule-bound protein, known as waxy (Wx) protein, thereby indicating the linkage between amylose biosynthesis and the presence of waxy protein in the starch grains. The protein has been identified as a 59.6-kD protein in pea (Sivak et al., 1993), a 60-kDa protein in rice (Sano, 1984), a 56-, 58-, or 60-kDa protein in maize (Gibbon et al., 2003), and a 58-kDa protein in mungbean (Ko et al., 2009).

While much work has been done on the characterization of starch from cereals and regulation of its biosynthesis, not much information is available about the characteristics of starches in many other potentially important crops including pseudocereals such as buckwheat, which has high potential for expanded utilization as a functional food (Kreft et al., 2002). With up to 52% amylose content, buckwheat starch has 1.6–2.0 times more apparent amylose than corn and wheat (Campbell, 1997; Qian et al., 1998). While the high amylose content gives poor dough-making quality to buckwheat starch, it makes the starch a good raw material for fried foods because of better filming properties.

STARCH

Starch is a naturally occurring polymer of α-D-glucose. It is deposited in the form of granules whose morphology, chemical composition, and supermolecular structure are characteristic of each plant species (Banks and Greenwood, 1975; Whistler and BeMiller, 1997; Tester and Karkalas, 2002). Even though both amylose and amylopectin are polymers of D-glucose, the properties of starch are, to a large extent, a function of structural differences between the two. While amylose comprises largely unbranched α-1,4-linked glucan chains, amylopectin is composed of small size α-1,4-linked chains that are clustered together by α-1,6 linkages between adjoining straight glucan chains (Tetlow et al., 2004). The granules themselves can also vary with respect to shape and structure. While rice has polyhedral starch granules that range in diameter from 3 to 8 μm, potato has ovoid granules that can be up to 100 μm in diameter (Jane et al., 1994). Starch granules from the endosperm of grains of common buckwheat vary in shape from round to polygonal with sizes ranging from 4 to 10 μm. Starch granules of Tartary buckwheat are, however, larger in size and have lower solubility at 70–80°C (Gao et al., 2016). Tartary buckwheat starch granules have been reported to present a typical type "A" X-ray diffraction pattern with crystallinity of 29.9%, RVU gelatinization peak viscosity of 258.3, and final viscosity of 389.6 (Li et al., 1997; Zheng et al., 1998; Qian et al., 1998; Liu et al., 2013). Gao et al. (2014) have shown that compared to corn or wheat starches, buckwheat starches swelled faster, exhibited a greater set back viscosity and formed stiffer and harder gels. These features could indicate a strong potential of Tartary buckwheat starch for producing functional foods with retrograded starch with a low glycemic index. While high amylose starches, having 40–80% amylose, have been reported in maize, barley, rice, and buckwheat (Song and Jane, 2000; Zhu et al., 2008; Christa and Śmietana-Soral, 2008), "*waxy*" (mutants with amylose-free starch) have been found in rice, maize, sorghum, barley, and potato (Denyer et al., 2001; Deschamps et al., 2008). These mutants lack a starch granule-bound protein, known as waxy (Wx) protein, thereby indicating the linkage between amylose biosynthesis and the presence of waxy protein in the starch grains. The protein has been identified as a 59.6 kD protein in pea (Sivak et al., 1993), 60 kDa protein in rice (Sano, 1984), 56, 58, or 60 kDa protein in maize (Gibbon et al., 2003) and 58 kDa protein in mungbean (Ko et al., 2009). Chrungoo et al. (2013) have reported the presence of a 59.7-kDa GBSS-I in the endosperm of grains of common buckwheat and confirmed its localization of GBSS-I within the matrix of the starch grain in the form of discrete internal rings.

Starch biosynthesis involves a coordinated activity of a large number of enzymes having several distinct isoforms (Ball and Morell, 2003; Ral et al., 2004). Although this increases complexity of the process and makes it more difficult to interpret the role of different enzymes, it also increases the range of possibilities for bioengineering. By targeting specific forms of an enzyme it is possible to exercise control over the types of starch molecules that are produced. In addition to its role in amylose biosynthesis, GBSS-I is also responsible for extension of long glucans within the amylopectin fraction (Delrue et al., 1992; Maddelein et al., 1994; Van der Wal et al., 1998). Experiments with potato have shown that, in certain genetic backgrounds, GBSS-I can have a profound influence on granule structure and tuber morphology (Fulton et al., 2002). Besides GBSS-I, GBSS-II, a 77–79-kDa protein, has been shown to be responsible for amylose synthesis in nonstorage tissues (Denyer et al., 1997; Nakamura et al., 1998; Edwards et al., 2002). In contrast to GBSS-I, the role of GBSS-II remains ambiguous. A reduction in its activity in potato tubers did not affect the ratio of amylose to amylopectin (Edwards et al., 1995). Other forms of starch synthases, namely, SSI, II, III, and IV, belonging to a minimum of four sequence families (Li et al., 2003), have been detected in the genomes of various species ranging from the picophytoplanktonic green alga *Ostreococcus tauri* (Ral et al., 2004) to rice (Baba et al., 1993; Tanaka et al., 1995). Due to their conservation throughout the plant kingdom, it is reasonable to assume that they play specific and conserved functions in polysaccharide metabolism.

THE WAXY LOCUS

The Wx gene, also referred to as Waxy locus, codes for a granule bound starch synthase (GBSS-I), which has a molecular mass of 56–60 kDa). Even though amino acid sequence analysis of GBSS-I shows clear diversification into monocotyledonous and dicotyledonous groups, protein from common buckwheat has similarities with dicots as well as monocots (Chrungoo et al., 2013). Wang et al. (2014) have made similar observations for the coding sequence of GBSS-I from *Fagopyrum tataricum*. They have reported a genomic sequence of 3947 bases for the Waxy locus of *F. tataricum* and on the basis of its comparison with the cDNA sequence, suggested it to be comprised of 14 exons and 13 introns. The coding sequence was reported to extend from exon 2–14 with exon 1 and a part of exon 2 forming part of the 5′UTR of the gene. The 14 exon–13 intron architecture of the Waxy locus of *F. tataricum* is in conformity with the architecture of the Waxy locus from other plants (Van der Leij et al., 1991; Okagaki, 1992). Xu et al. (2009) have, however, reported 11 exons and 10 introns for the Waxy locus of *Secale cereale*. Even though several reports have indicated a correlation between CT repeat length with apparent amylose content of starch (Bao et al., 2002, 2006; Jayamani et al., 2007), our investigations on the Waxy locus of common buckwheat did not reveal any correlation between the two. Dobo et al. (2010) have made similar observations about the absence of a strong correlation between CT repeat length with apparent amylose content of starch in rice. They have demonstrated the significance of "G/T" polymorphism in the 5′ leader intron splice site as a marker for differentiating "*waxy*" and "low amylo-" cultivars from the "intermediate and high amylo-" cultivars. They suggested that, regardless of any other base changes, the "T" haplotype effectively distinguished low amylose varieties from the other classes in rice. Biselli et al. (2014) have, however, associated the "T" haplotype even with varieties having up to 22% AAC. Our observations associate the "T" allele with "*waxy*" and "low" amylo- cultivars having starch with <12% AAC. These observations indicate that "G/T" polymorphism in the 5′ leader intron splice site alone may not be the only marker for defining the AAC content of starch in rice. While "G" → "T" mutation in 5′ leader intron splice site is known to reduce the efficiency of splicing of the first intron, leading to reduction in the level of mature GBSS-I mRNA (Cai et al., 1998; Bligh et al., 1998), the process of splicing is sensitive to temperature (Larkin and Park, 1999). While the reasons for low amylose content in cultivars with the sequence AGTTAT in the 5′ leader intron splice site can be ascribed to inefficient splicing of the first intron, the reasons for a wide range of amylose in cultivars with the "G" allele at the 5′ leader intron splice site puts a question mark on the applicability of "G/T" polymorphism in the 5′ leader intron splice site as the sole marker for describing various AAC class starches. Besides the

G/T substitution detected in the 5′ leader intron splice site, other substitutions detected in the Waxy locus include "A/C" polymorphism in exon 6 and "C/T" polymorphism in exon 10. "A" → "C" mutation in exon 6 has been suggested to cause nonconservative change of a tyrosine to serine in the amino acid sequence of GBSS-I (Dobo et al., 2010). Since tyrosine is known to contribute significantly to protein stability through hydrogen bonding (Pace et al., 2001), change of tyrosine to serine in the GBSS-I could lead to destabilization of the protein and consequently reduced catalytic efficiency.

ALTERING STARCH COMPOSITION THROUGH BIOTECHNOLOGICAL APPROACHES

Given the important role starch plays in food and nonfood uses, several efforts are being addressed to the manipulation of its composition through modification of the amylose/amylopectin ratio in different crops in general. Most of the research associated with GBSS expression is concerned with amylose production in crop species and with mutations that generate phenotypes of economical impact. Amylose-free or low amylose starch is considered desirable for certain food and nonfood industries (Morell et al., 1995). Approaches used to achieve this goal are being pursued through the manipulation of the genes involved in the starch biosynthetic pathway using natural or induced mutations and transgenic methods.

While overexpression/decreased expression of some of the genes involved in starch biosynthesis (Itoh et al., 2003; Regina et al., 2006; Sestili et al., 2010) can be one of the means to alter the composition of amylose and amylopectin in starch granules, modification of catalytic activity through enzymes involved in the synthesis of amylose or amylopectin or generation of null mutants through retrotransposon insertion or chemical mutagenesis could be other ways of achieving the same goals. One of the approaches used to alter the amylose/amylopectin ratio of starch was the introduction of Wxᵃ cDNA, which has the sequence AG<u>G</u>TAT, or Wxᵇ cDNA, which has the sequence AG<u>T</u>TAT at the 5′ leader intron splice site, into a null-mutant (*wx*). Such a strategy to obtain high amylose starches was carried out by Itoh et al. (2003) who introduced Wxᵃ cDNA into null-mutant Japonica rice (*wx*). This approach led to 6–11% more amylose content in the transgene than the wild type, which carried the Wxᵃ allele of the Waxy gene. Research aimed at finding different strategies to produce starches with new properties has gained momentum in the past few years. One of these strategies has been the evaluation of a possibility of using microbial starch-binding domains as a universal tool for synthesis of starch with altered properties. One of the functions of SBD is to attach amylolytic enzymes to the insoluble starch granule. Several studies have shown that these enzymes lose most of their catalytic activity toward raw starch granules upon removal of the SBD, while their activity on soluble substrates remains unaltered. Results of the work carried out by Ji et al. (2003), who explored the possibility of engineering artificial granule-bound proteins for incorporation in starch granules during their formation, have revealed localization of SBD domains of granule-bound proteins inside starch granules and not at their surface. Further, amylose-free granules contained 8 times more SBD than the amylose-containing ones. However, no consistent differences in physicochemical properties between transgenic SBD starches and their corresponding controls were found, suggesting that SBD can be used as an anchor for effector proteins without any side effects. On the other hand, evaluation of the possibility of producing amylose-free starch by displacing GBSSI from the starch granule by engineering multiple-repeat CBM20 SBD revealed SBDn expression can affect the physical process underlying granule assembly in both potato genetic backgrounds, without altering the primary structure of the constituent starch polymers and the granule melting temperature. These results have unequivocally established the possibility of manipulation of the starch synthesis pathway through the modification of enzymes belonging to this route, and/or the use of CBM, more specifically SBD, of both microbial and plant origin.

ACKNOWLEDGMENTS

NKC is grateful to the Department of Biotechnology, Government of India, for supporting the work under its Biotech Hub program vide grant no. BT/04/NE/2009.

REFERENCES

Amani, N.G., Buleon, A., Kamenan, A., Colonna, P., 2004. Variability in starch physico-chemical and functional properties of yam (*Dioscorea* sp.) cultivated in Ivory Coast. J. Sci. Food Agric. 84, 2085–2096.

Baba, T., Nishihara, M.K., Kawasaki, T., Shimada, T., Kobayashi, E., Ohnishi, S., Tanaka, K., Arai, Y., 1993. Identification, cDNA cloning, and gene expression of soluble starch synthase in rice (*Oryza sativa* L.) immature seeds. Plant Physiol. 103, 565–573.

Ball, S.G., Morell, M.K., 2003. From bacterial glycogen to starch: understanding the biogenesis of the plant starch granule. Ann. Rev. Plant Biol. 54, 207–233.

Banks, W., Greenwood, C.T., 1975. Fractionation of the starch granule, and the fine structure of its components. In: Banks, W., Greenwood, C.T. (Eds.), Starch and Its Components. Edinburgh University Press, Edinburgh, pp. 5–66.

Bao, J.S., Wu, Y.R., Hu, B., Wu, P., Cui, H.R., Shu, Q.Y., 2002. QTL for rice grain quality based on a DH population derived from parents with similar apparent amylose content. Euphytica 128, 317–324.

Bao, J.S., Corke, H., Sun, M., 2006. Microsatellites, single nucleotide polymorphisms and a sequence tagged site in starch-synthesizing genes in relation to starch physicochemical properties in non waxy rice (*Oryza sativa* L.). Theor. Appl. Genet. 113, 1185–1196.

Biselli, C., Cavalluzzo, D., Perrini, R., Gianinetti, A., Bagnaresi, P., Urso, S., Orasen, G., Desiderio, F., Lupotto, E., Cattivelli, L., Valè, G., 2014. Improvement of Marker-Based Predictability of Apparent Amylose Content in Japonica Rice Through GBSSI Allele Mining. Springer Rice, New York.

Bligh, H.F.J., Larkin, P.D., Roach, P.S., Jonea, P.A., Fu, H., Park, W.D., 1998. Use of alternate splicing sites in granule-bound starch synthase mRNA from low-amylose rice varieties. Plant Mol. Biol. 38, 407–416.

Cai, X.L., Wang, Z.Y., Xing, Y.Y., Zhang, J.L., Hong, M.M., 1998. Aberrant splicing of intron 1 leads to the heterogeneous 5′ UTR and decreased expression of *waxy* gene in rice cultivars of intermediate amylose content. Plant J. 14, 459–465.

Campbell, C.G., 1997. Buckwheat (*Fagopyrum esculentum* Moench). Promoting the conservation and use of underutilized and neglected crops. 19. International Plant Genetic Resources Institute, Rome, Italy.

Chen, Z., Schols, H.A., Voragen, A.G.J., 2003. Physicochemical properties of starches obtained from three different varieties of Chinese sweet potatoes. J. Food Sci. 68, 431–437.

Christa, K., Śmietana-Soral, M., 2008. Buckwheat grains and buckwheat products—nutritional and prophylactic value of their components. Czech J. Food Sci. 26, 153–162.

Chrungoo, N.K., Devadasan, N., Kreft, I., Gregori, M., 2013. Identification and characterization of granule bound starch synthase (GBSS-I) from common buckwheat (*Fagopyrum esculentum* Moench). J. Plant Biochem. Biotechnol. 22, 269–276.

Dedeh, S., Sackey, K.A.E., 2002. Starch structure and some properties of cocoyam (*Xanthosoma sagittifolium* and *Colocasia esculenta*) starch and raphides. Food Chem. 79, 435–444.

Delrue, B., Fontaine, T., Routier, F., Decqm, A., Wieruszeski, J.M., Ball, S., 1992. Waxy *Chlamydomonas reinhardtii*: monocellular algal mutants defective in amylose biosynthesis and granule-bound starch synthase activity accumulate a structurally modified amylopectin. J. Bacteriol. 174, 3612–3620.

Denyer, K., Barber, L.M., Edward, E.A., Smith, A.M., Wang, T.L., 1997. Two isoforms of the GBSS-I class of granule-bound starch synthase are differentially expressed in the pea plant (*Pisum sativum* L.). Plant Cell Environ. 20, 1566–1572.

Denyer, K., Johnson, P., Zeeman, S., Smith, A.M., 2001. The control of amylase synthesis. J. Plant Physiol. 158, 479–487.

Deschamps, P., Moreau, H., Worden, A.Z., Dauvillée, D., Ball, S.G., 2008. Early gene duplication within *Chloroplastida* and its correspondence with relocation of starch metabolism to chloroplasts. Genetics 178, 2373–2387.

Dry, I., Smith, A.M., Edwards, E.A., Bhattacharyya, M., Dunn, P., Martin, C., 1992. Characterization of cDNAs encoding two isoforms of granule bound starch synthase which show differential expression in developing storage organs. Plant J. 2, 193–202.

Dobo, M., Ayres, N., Walker, G., Park, W.D., 2010. Polymorphism in the GBSS gene affects amylose content in US and European rice germplasm. J Cereal Sci. 52, 450–456.

Edwards, A., Marshall, J., Denyer, K., Sidebottom, C., Visser, R.G.F., Martin, C., Smith, A.M., 1995. Biochemical and molecular characterization of a novel starch synthase from potato tubers. Plant J. 8, 283–294.

Edwards, A., Vincken, J.P., Suurs, L.C., Visser, R.G., Zeeman, S., Smith, A., Martin, C., 2002. Discrete forms of amylose are synthesized by isoforms of GBSSI in pea. Plant Cell 14, 1767–1785.

French, D., 1984. Organisation of starch granules. In: Whistler, R.L., Bemiller, J.N., Paschal, E.F. (Eds.), Starch Chemistry and Technology. Academic Press, New York, pp. 182–247.

Fulton, D.C., Edwards, A., Pilling, E., Robinson, H.L., Fahy, B., Seale, R., Donald, A.M., Geigenberger, P., Martin, C., Smith, A.M., 2002. Role of granule bound starch synthase in determination of amylopectin structure and starch granule morphology in potato. J. Mol. Biol. 277, 10834–10841.

Gao, H., Cai, J., Han, W., Huai, H., Chen, Y., Wei, C., 2014. Comparison of starches isolated from three different *Trapa* species. Food Hydrocolloids 37, 174–181.

Gao, J., Kreft, I., Chao, G., Wang, Y., Liu, X., Wang, L., Wang, P., Gao, X., Feng, B., 2016. Tartary buckwheat (*Fagopyrum tataricum* Gaertn.) starch, a side product in functional food production, as a potential source of retrograded starch. Food Chem. 190, 552–558.

Gibbon, B.C., Wang, X., Larkins, B.A., 2003. Altered starch structure is associated with endosperm modification in quality protein maize. Proc. Natl. Acad. Sci. USA 100, 15329–15334.

Hirano, H., Sano, Y., 1991. Molecular characterization of the *waxy* locus of rice (*Oryza sativa*). Plant Cell Physiol. 32, 989–997.

Hovenkamp-Hermelink, J.H.M., Jacobsen, E., Ponstein, A.S., Visser, R.G.F., Vos-Scheperkeuter, G.H., Bijmolt, E.W., de Vries, J.N., Witholt, B., Feenstra, W.J., 1987. Isolation of an amylose-free starch mutant of the potato (*Solanum tuberosum* L.). Theor. Appl. Genet. 75, 217–221.

Hsieh, J., Liu, C., Hsing, Y.C., 1996. Molecular cloning of a sorghum cDNA (acc. no. U23945) encoding the seed *waxy* protein (PGR96-119). Plant Physiol. 112, 1735.

Itoh, K., Ozaki, H., Okada, K., Hori, H., Takeda, Y., Mitsui, T., 2003. Introduction of Wx transgene into rice *wx* mutants leads to both high- and low-amylose rice. Plant Cell Physiol. 44, 473–480.

Jane, J.L., Chen, F.J., 1992. Effect of amylose molecular size and amylopectin branch chain length on paste properties of starch. Cereal Chem. 69, 60–65.

Jane, J.L., Kasemsuwan, T., Leas, S., Zobel, H., Robyt, J.F., 1994. Anthology of starch granule morphology by scanning electron microscopy. Starch-Strake 46, 121–129.

Jayamani, P., Negrao, S., Brites, C., Oliviera, M.M., 2007. Potential of *Waxy* gene microsatellite and single-nucleotide polymorphisms to develop japonica varieties with desired amylose levels in rice (*Oryza sativa*). J. Cereal Sci. 46, 178–186.

Ji, Q., Vincken, J.P., Suurs, L.C., Visser, R.G., 2003. Microbial starch-binding domains as a tool for targeting proteins to granules during starch biosynthesis. Plant Mol. Biol. 51, 789–801.

Klosgen, R.B., Gierl, A., Schwartz-Sommer, Z., Seadler, H., 1986. Molecular analysis of the *waxy* locus of maize. Mol. Genet. Genom. 203, 237–244.

Ko, Y.T., Dong, Y.L., Hsieh, Y.F., Kuo, J.C., 2009. Morphology, associated protein analysis and identification of 58 kDa starch synthase in mungbean (*Vigna radiate* L. cv. KPS1) starch granule. J. Agric. Food Chem. 57, 4426–4432.

Kreft, S., Strukelj, B., Gaberscik, A., Kreft, I., 2002. Rutin in buckwheat herbs grown at different UV-B radiation levels: comparison of two UV spectrophotometric and HPLC method. J. Exp. Bot. 53, 1801–1804.

Larkin, P.D., Park, W.D., 1999. Transcript accumulation and utilization of alternate and non consensus splice sites in rice granule bound starch synthase is temperature-sensitive and controlled by a single nucleotide polymorphism. Plant Mol. Biol. 40, 719–727.

Li, W., Lin, R., Corke, H., 1997. Physicochemical properties of common and Tartary buckwheat starch. Cereal Chem. 74, 79–82.

Li, Z., Sun, F., Xu, S., Chu, X., Mukai, Y., Yamamoto, M., Ali, S., Rampling, L., Kosar-Hashemi, B., Rahman, S., Morell, M.K., 2003. The structural organization of the gene encoding class II starch synthase of wheat and barley and the evolution of the genes encoding starch synthases in plants. Funct. Integr. Genom. 3, 76–85.

Liu, H., Guo, X., Ma, Y., Wang, M., 2013. Preparation technology and properties study of Tartary buckwheat starch. J. Chin. Inst. Food Sci. Technol. 13, 43–49.

Maddelein, M.L., Libessart, N., Bellanger, F., Delrue, B., D'Hulst, C., Ball, S., 1994. Towards an understanding of the biogenesis of the starch granule: determination of granule-bound and soluble starch synthase functions in amylopectin synthesis. J. Biol. Chem. 269, 25150–25157.

Mason-Gamer, R.J., Weil, C.F., Kellogg, E.A., 1998. Granule bound starch synthase: structure, function and phylogenetic utility. Mol. Biol. Evol. 15, 1658–1673.

Morell, M.K., Rahman, S., Abrahams, S.L., Appels, R., 1995. The biochemistry and molecular biology of starch synthesis in cereals. Aust. J. Plant Physiol. 22, 647–660.

Murai, J., Taira, T., Ohta, D., 1999. Isolation and characterization of the three *Waxy* genes encoding the granule-bound starch synthase in hexaploid wheat. Gene 234, 71–79.

Nakamura, T., Virnten, P., Hayakawa, K., Ikeda, J., 1998. Characterization of a granule bound starch synthase isoform found in the pericarp of wheat. Plant Physiol. 118, 125–132.

Okagaki, R.J., 1992. Nucleotide sequence of a long cDNA from the rice waxy gene. Plant Mol. Biol. 19, 1453–1462.

Pace, J.M., Kuslich, C.D., Willing, M.C., Byers, P.H., 2001. Disruption of one intra-chain disulphide bond in the carboxyl-terminal propeptide of the pro alpha1(I) chain of type I procollagen permits slow assembly and secretion of overmodified, but stable procollagen trimers and results in mild osteogenesis imperfecta. J. Med. Genet. 38, 443–449.

Pérez, E., Schultz, F.S., de Delahaye, E.P., 2005. Characterization of some properties of starches isolated from *Xanthosoma sagittifolium* (tannia) and *Colocasia esculenta* (taro). Carbohyd. Polym. 60, 139–145.

Peroni, F., Rocha, T., Franco, C., 2006. Some structural and physicochemical characteristics of tuber and root starches. Food Sci.Technol. Int. 12, 505–513.

Qian, J., Duarte, P.R., Grant, L., 1998. Partial characterization of buckwheat (*Fagopyrum esculentum*) starch. Cereal Chem. 75, 365–373.

Ral, J.P., Derelle, E., Ferraz, C., Wattebled, F., Farinas, B., Corellou, F., Buléon, A., Slomianny, M.C., Delvalle, D., d'Hulst, C., Rombauts, S., Moreau, H., Ball, S., 2004. Starch division and partitioning. A mechanism for granule propagation and maintenance in the picophytoplanktonic green alga *Ostreococcus tauri*. Plant Physiol. 136, 3333–3340.

Regina, A., Bird, A., Topping, D., Bowden, S., Freeman, J., Barsby, T., Kosar-Hashemi, B., Li, Z., Rahman, S., Morell, M., 2006. High-amylose wheat generated by RNA interference improves indices of large-bowel health in rats. Proc. Natl. Acad. Sci. USA 103, 3546–3551.

Riley, C.K., Wheatley, A.O., Asemota, H.N., 2006. Isolation and characterization of starches from eight *Dioscorea alata* cultivars grown in Jamaica. Afr. J. Biotechnol. 5, 1528–1536.

Rohde, W., Becker, D., Salamini, F., 1988. Structural analysis of the *waxy* locus from *Hordeum vulgare*. Nucleic Acids Res. 16, 7185–7186.

Sano, Y., 1984. Differential regulation of *waxy* gene expression in rice endosperm. Theor. Appl. Genet. 68, 467–473.

Sestili, F., Janni, M., Doherty, A., Botticella, E., D'Ovidio, R., Masci, S., Jones, H.D., Lafiandra, D., 2010. Increasing the amylose content of durum wheat through silencing of the SBEIIa genes. BMC Plant Biol. 10, 144.

Shujun, W., Hongyan, L., Wenyuan, G., Haixia, C., Jiugao, Y., Peigen, X., 2006. Characterization of new starches separated from different Chinese yam (*Dioscorea opposite* Thumb.) cultivars. Food Chem. 99, 30–37.

Shure, M., Wessler, S., Federoff, N., 1983. Molecular identification and isolation of the *waxy* locus in maize. Cell 35, 225–233.

Sivak, M.N., Wagner, M., Preiss, J., 1993. Biochemical evidence for the role of the waxy protein from pea *Pisum sativum* L. as a granule-bound starch synthase. Plant Physiol. 103, 1355–1359.

Song, Y., Jane, J., 2000. Characterization of barley starches of waxy, normal, and high amylose varieties. Carbohyd. Polym. 41, 365–377.

Tanaka, K., Ohnishi, S., Kishimoto, N., Kawasaki, T., Baba, T., 1995. Structure, organization, and chromosomal location of the gene encoding a form of rice soluble starch synthase. Plant Physiol. 108, 677–683.

Tester, R.F., Karkalas, J., 2002. Polysaccharides. II. Polysaccharides from eukaryotes. Vandamme, E.J., De Baets, S., Steinbuchel, A. (Eds.), Starch in Biopolymers, 6, Weinheim, Wiley-VCH, pp. 381–438.

Tetlow, I.J., Morell, M.K., Emes, M.J., 2004. Recent developments in understanding the regulation of starch metabolism in higher plants. J. Exp. Bot. 55, 2131–2145.

Van der Wal, M., D'Hulst, C., Vincken, J.P., Buléon, A., Visser, R., Ball, S., 1998. Amylose is synthesised in vitro by extension of and cleavage from amylopectin. J. Biol. Chem. 273, 22232–22240.

Van der Leij, F.R., Visser, R.G.F., Ponstein, A.S., Jacobsen, E., Feenstra, W.J., 1991. Sequence of the structural gene for granule-bound starch synthase of potato (*Solanum tuberosum* L.) and evidence for a single point deletion in the *amf* allele. Mol. Genet. Genom. 228, 240–248.

Visser, R.G.F., Hergersberg, M., Van Der Leij, F.R., Jacobsen, E., Witholt, B., Feenstra, W.J., 1989. Molecular cloning and partial characterization of the genes for granulebound starch synthase from a wildtype and an amylose-free potato (*Solanum tuberosum* L.). Plant Sci. 64, 185–192.

Wang, Z., Wu, Z., Xing, Y., Zheng, F., Guo, X., Zhang, W., Hong, M., 1990. Nucleotide sequence of rice *waxy* gene. Nuc. Acids Res. 18, 5898.

Wang, X., Feng, B., Xu, Z., Sestili, F., Zhao, G., Xiang, C., Lafiandra, D., Wang, T., 2014. Identification and characterization of granule bound starch synthase I (GBSSI) gene of Tartary buckwheat (*Fagopyrum tataricum* Gaertn.). Gene 534, 229–235.

Whistler, R.L., BeMiller, J.N., 1997. Carbohydrate chemistry for food scientists. St. Paul: Eagan Press. Starch, 117–152.

Xu, J., Frick, M., Laroche, A., Ni, Z.F., Li, B.Y., Lu, Z.-X., 2009. Isolation and characterization of the rye *Waxy* gene. Genome 52, 658–664.

Zheng, G.H., Sosulski, F.W., Tyler, R.T., 1998. Wet-milling, composition and functional properties of starch and protein isolated from buckwheat groats. Food Res. Int. 30, 493–502.

Zhu, T., Jackson, D.S., Wehling, R.L., Geera, B., 2008. Comparison of amylase determination methods and the development of a dual wavelength iodine binding technique. Cereal Chem. 85, 51–58.

Chapter | thirty three

Genetic Analyses of the Heteromorphic Self-Incompatibility (S) Locus in Buckwheat

M. Ueno*, Y. Yasui*, J. Aii**, K. Matsui[†], S. Sato**, T. Ota[‡]

*Kyoto University, Graduate School of Agriculture, Kitashirakawa Oiwake-cho, Sakyou-ku, Kyoto, Japan; **Niigata University of Pharmacy and Applied Life Science, Faculty of Applied Life Science, Akiha-ku, Niigata, Japan; [†]National Agriculture and Food Research Organization, Kyushu Okinawa Agricultural Research Center, Suya, Koshi, Japan; [‡]SOKENDAI (The Graduate University for Advanced Studies), School of Advanced Sciences, Department of Evolutionary Studies of Biosystems, Hayama, Japan

INTRODUCTION

Self-incompatibility (SI), which prevents self-fertilization, is widespread in the plant kingdom and an important feature in nature. It is a way to avoid inbreeding depression, ill and/or detrimental effects often elicited by a homozygous state of recessive alleles at some loci (Charlesworth and Charlesworth, 1987; Charlesworth and Willis, 2009). Based on morphological heterogeneity of flowers, SI can be grouped into homomorphic and heteromorphic SI, where only a single flower morph is present in the former whereas more than two flower morphs, two (distyly) in general and three (tristyly) in a few, are present in the latter. SI response takes place upon interactions between pollens and pistils, and SI is also grouped by the nature of male determinant into gametophytic and sporophytic SI, where the haploid genotype of pollen and the diploid genotype of the father plant are the principal determinants of compatibility, respectively. So far, three homomorphic SI systems have been extensively studied (reviewed in Iwano and Takayama, 2012): the gametophytic SI based on SLFs (SFBs)/S-RNase systems in Solanaceae, Rosaceae, and Plantaginaceae; the gametophytic SI based on PrpS/PrsS systems in Papaveraceae; and the sporophytic SI based on SP11 (SCR)/SRK systems in Brassicaceae. On the other hand, little is disclosed for molecular bases of heteromorphic SI except that at least three types of sporophytic systems exist with stigma, style, or ovule as a site of inhibition

411

Molecular Breeding and Nutritional Aspects of Buckwheat. http://dx.doi.org/10.1016/B978-0-12-803692-1.00033-X

(Dulberger, 1992). Notable progress has been made lately, however, for *Primula* (Li et al., 2011, 2015), *Turnera* (Labonne and Shore, 2011; Chafe et al., 2014), *Linum* (Ushijima et al., 2012, 2015), and buckwheat (Yasui et al., 2012).

Buckwheat, *Fagopyrum esculentum*, exhibits distylous SI and there exist two forms of flower: a short-styled flower with long anther (thrum type) and a long-styled flower with short anther (pin type) (Hilderbrand, 1867; Darwin, 1877). With respect to pollen size, pollen grains of thrum-type plants are larger than those of pin-type plants (Schoch-Bodmer, 1934; Tatebe, 1949). Prezygotic response is critical and it has been shown that inhibition of pollen tube growth in thrum-type plants occurred in the upper part of styles with hypertrophy at the tips, whereas inhibition in pin-type plants occurred in the middle part of styles with no hypertrophy (Hirose et al., 1995). The traits, intramorph SI and flower morphology, are controlled by a single locus, that is, the *S* locus, and thrum-type plants are in the heterozygous state (*Ss*) whereas pin-type plants are in the homozygous state of *s* allele (*ss*) (Garber and Quisenberry, 1927; Lewis and Jones, 1992). It has been postulated but not been proven that the buckwheat heteromorphic SI is controlled by the *S* supergene as proposed for dystylous *Primula* (Dowrick, 1956), which consists of at least five genes of the traits: style length (*G*), styler incompatibility (*IS*), pollen incompatibility (*IP*), pollen size (*P*), and anther height (*A*) (Sharma and Boyes, 1961).

Outcrossing can be advantageous for wild species to increase genetic heterogeneity (Goldberg et al., 2010) but not for crops because it results in genetic instability of agronomic traits. Introduction of self-compatibility (SC) trait into SI plants is one of the important breeding objects, and in buckwheat, efforts have been directed to generate SI plants. For example, Campbell (1995) has developed SC lines of buckwheat, using an interspecific cross between *F. esculentum* and wild SC species *Fagopyrum homotropicum* discovered by Ohnishi (1998).

SELF-COMPATIBILITY AND THE *SH* ALLELE OF *F. HOMOTROPICUM*

The *S*-locus allele of the SC line derived from *F. homotropicum* is called *Sh* and it has been shown that the *Sh* allele is dominant over the *s* allele but recessive to the *S* allele (Woo et al., 1999). To obtain further insights, Matsui et al. (2003) have done the crossing between *F. esculentum* cv. "Botansoba" (BTN) and *F. homotropicum* line C9255, and analyzed morphological traits as well as SI of their progenies. A line with long homostyle flowers (LH), which was derived from a cross between a pin-type plant of *F. esculentum* and a plant of *F. homotropicum* line C9255, had been successively self-pollinated and named as Kyukei SC2 (KSC2) at F8 generation (Fig. 33.1). KSC2 is self-compatible and can set seeds, even if flowers are kept in a closed bag, suggesting the SC feature of *F. homotropicum* has been successfully introduced into the *F. esculentum* background. As expected from the dominance of the *Sh* allele over the *s* allele, the pollen diameter of F1 plants derived from the cross between a pin-type plant of BTN and KSC2 was not significantly different from that of KSC2. Furthermore, F2

FIGURE 33.1 Flower morphology in buckwheat (A) pin type, (B) thrum type, and (C) long homostyle.

population derived from self-fertilization of the F1 plant segregated into 106 LH plants and 34 pin-type plants, where self-fertility and pollen size were segregated in accordance with flower morphology. The observed Mendelian segregation has confirmed that SC of KSC2 is controlled by the single allele *Sh*. In addition, cross-compatibility analyses among LH and pin- or thrum-type plants has shown seemly pollen tube growth to ovule in KSC2 pistils for pollens derived from thrum- but not pin-type plants, even though pollens have successfully germinated on recipient stigma in any combinations. On the other hand, pollen tube growth to ovule in pistils of pin- but not thrum-type plants was observed for pollens derived from KSC2. If the *S* supergene is present as proposed for distylous *Primula*, the results would be indicative of KSC2 possessing thrum-type factors for pollen incompatibility and anther height and pin-type factors for styler incompatibility and style length, and it raised the possibility that the *Sh* allele might be produced by a recombination between *S* and *s* alleles as illustrated in Fig. 33.2. Furthermore, a locus controlling pollen size could be located near or at the loci controlling pollen incompatibility and anther height, since pollen size of KSC2 was as large as that of thrum-type plants. The presence of at least two factors, one controlling styler incompatibility and morph of pistils and the other controlling pollen incompatibility and morph of pollen and stamen, is suspected for the *S* supergene, but the details need to be scrutinized by genomic analyses of the *S*-locus (see next section).

GENES PRESENT IN THE S-LOCUS

The dominance of the *S* allele over the *s* allele raises a possibility that unique genes exist in the genomes of thrum-type plants or at least that specific genes are expressed only in the flowers of thrum-type plants. Thus one promising approach to identify *S*-locus gene(s) is to search those expressed only in flowers of thrum- but not pin-type plants, that is, *S* allele-(or haplotype-) specific or short-style-specific genes (SSGs). With the development and advance of next-generation sequencers, an exhaustive search of expressed genes becomes feasible, and we have effectively utilized it to detect SSGs (see Yasui et al., 2012 for details). Using the progenies of the sib-mating line that was used to construct a bacterial

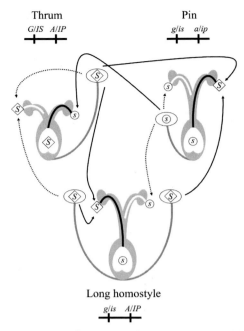

FIGURE 33.2 Compatibility *(solid lines)* and incompatibility *(dotted lines)* reactions of Kyukei SC2 with the *Sh* allele.

artificial chromosome (BAC) library (Yasui et al., 2008), mRNAs were isolated separately from pistils of thrum- and pin-type plants. High-throughput sequencing analyses of expressed genes were conducted using an Illumina Genome Analyzer and short-reads obtained were assembled by the Velvet program (Zerbino and Birney, 2008), yielding 41,599 contigs of various lengths (61–5,334 bps). By excluding contigs with any 32 bp fragments derived from pistils of pin-type plants (*s/s* homozygotes) in silico, 15 contigs were extracted as SSG candidates. Since RNA-seq analyses were conducted only once for pin- and thrum-type plants, statistical significance of the difference in expression was not examined. We then conducted RT-PCR analyses for confirmation, from which four short-style-specific transcripts (*SSG1–SSG4*) were identified.

In major crops or model plants, it is common to generate transgenic plants with overexpressing or knocking-out of the gene of interest and to investigate traits or gene expressions in order to examine the functions of a gene. No reliable methods, however, were established for generating transgenic plants in buckwheat, and we took alternative approaches. One approach was to generate mutants by ion-beam irradiation, and we fortunately obtained a chimeric mutant which possessed a branch setting pin-type flowers on a plant with thrum-type flowers in the background (Fig. 33.3). In this plant, only *SSG1* and *SSG4*, but neither *SSG2* nor *SSG3*, were amplified by PCR when the genomic DNA derived from the section with pin-type flowers was used as a template, whereas all four were amplified when the genomic DNA from the background section with

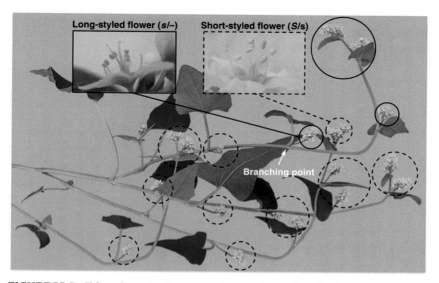

FIGURE 33.3 Chimeric mutant produced by ion-beam irradiation.

thrum-type flowers was used as a template. In addition, complete genetic linkage between the gene(s) controlling the flower morph and *SSG2* or *SSG3* markers was shown in the analyses of 1373 buckwheat plants of the sib-mating line. Furthermore, the association between the *SSG2* and/or *SSG3* marker presence and flower morph was detected for worldwide buckwheat landraces. Southern blot analyses using *SSG3* as a probe also detected hybridization signals only from thrum- but not from pin-type plants. These results collectively indicate that *SSG2* and *SSG3* locate at or very near the *S*-locus and that ion beam-induced deletion around *SSG2* and *SSG3* would have caused a phenotype conversion from *S/s* genotype (generating thrum-type flowers) to *s/–* genotype (generating pin-type flowers) in the chimeric plant. Complete nucleotide sequences of *SSG2* and *SSG3* were obtained with 5′- and 3′-RACE and their homology searches by BLAST programs (Altschul et al., 1990) identified *Arabidopsis EARLY FLOWRING3 (ELF3)* as a homolog of *SSG3* but failed to identify any significant homologs for *SSG2*. Subsequent phylogenetic analyses revealed that *SSG3* was a paralog but not an ortholog of *ELF3*, and *SSG3* was named as *S LOCUS EARLY FLOWRING 3 (S-ELF3)*. In *Arabidopsis thaliana*, ELF3 plays a role in circadian rhythm and interacts with various molecules, such as ELF4 and LUX, to affect the timing of flowering, the regulation of morphological structures, such as hypocotyl length, and the shade avoidance (Hicks et al., 1996, 2001; Coluccio et al., 2011). Involvement of buckwheat *S-ELF3* in circadian rhythm or in morphological development still remains uncertain. RT-PCR analyses revealed that *S-ELF3* was expressed specifically in the pistils and stamens of thrum-type plants, but not in vegetative tissues, such as leaves, roots, and stems, indicating their role in the reproductive system. In contrast, *SSG2* transcript was detected in all the tissues investigated.

Plant SI is generally controlled by multiple factors and their genes locate in a limited genomic region. Therefore the genomic region surrounding *S-ELF3* and *SSG2* may harbor additional genes important for the buckwheat heteromorphic SI. Initial investigation of the genomic region surrounding *S-ELF3* and *SSG2* by chromosome walking of BAC and transformation-competent bacterial artificial chromosome (TAC) genomic libraries has yielded a contig consisting of nine BACs and one TAC, and analyses of the artificial chromosomes using the Roche Genome Sequencer FLX detected a few partial fragments homologous to known genes in 610-kb regions sequenced. No genes of apparent functional importance, however, were identified except *S-ELF3*, *SSG2*, and a homolog of putative uncharacterized protein encoded by the At2g26520 gene in *A. thaliana*. Genes amplifiable by the primers for the homolog of At2g26520 were expressed in both thrum- and pin-type plants and little indication was given for their functional involvement in heteromorphic SI. Lately we have constructed a draft genome of buckwheat and applied a genotyping by sequencing (GBS) approach (Elshire et al., 2011) to identify an *S* haplotype-specific region from the draft genome. GBS reads obtained from 18 thrum-type plants and 18 pin-type plants of worldwide landraces were mapped to the genomic scaffolds and extracted the regions to which any reads from 18 pin-type plants were unmapped. To be conservative, we limited the sites shared by more than 10 thrum-type plants and regarded them as "*S* allele specific." Finally, 88,031 *S* allele-specific sites residing on 332 scaffolds encompassing nucleotide sequences of about 5.4 Mb in total were obtained. Detailed characterization of the regions, including the residing genes other than *S-ELF3*, will be reported in the near future (Yasui et al., 2016).

The functional significance of *S-ELF3* and *SSG2* in heteromorphic SI remains uncertain. Yet the analyses of *Fagopyrum* species may indicate their involvement in heteromorphic SI, as different species of the genus *Fagopyrum* exhibit different SI responses. The genus *Fagopyrum* is known to consist of two distantly related groups, namely, the *cymosum* and *urophyllum* groups (Ohnishi and Matsuoka, 1996), where *F. esculentum* belongs to the *cymosum* group. The representative species of the two groups, that is, *Fagopyrum cymosum* and *Fagopyrum urophyllum*, exhibits heteromorphic SI as in most *Fagopyrum* species, and their analyses revealed that *S-ELF3* orthologs were absent in the pin- but exclusively present in the thrum-type plants. The complete association between flower morph and *S-ELF3* presence in the diverged *Fagopyrum* species favors the significant role of *S-ELF3* in the thrum-type plants of heteromorphic SI species, ever since or prior to the emergence of the genus *Fagopyrum*. In contrast, PCR amplification of *SSG2* was not successful even from thrum-type plants of *F. cymosum*, and Southern blot analysis using buckwheat *SSG2* as a probe failed to detect any hybridization signals for *F. cymosum*, indicating that *SSG2* was dispensable and not significant in the heteromorphic SI of *Fagopyrum* species. Furthermore, the genus *Fagopyrum* is known to contain a few homomorphic SC species including Tartary buckwheat, *Fagopyrum tataricum*. Investigations of *S-ELF3* orthologs in *F. tataricum* in addition to the *Sh* allele derived from *F. homotropicum* are intriguing. PCR amplification of *S-ELF3* was successful from *F. tataricum* and a homomorphic SC buckwheat line (Kyushu PL4) derived from KSC2, and their detailed examinations revealed the following. While the genomic regions of *S-ELF3* in three heteromorphic SI species comprise five exons, including 1983–2010 bp encoding a polypeptide of 661–670 amino acids, the one of Kyushu PL4 contains a single nucleotide deletion resulting in a frameshift in the fifth exon with keeping the basic genomic structure as the others, and the one of *F. tataricum* contains a nonsense mutation in the fifth exon with conspicuous structural alterations, such as an inverted duplication of the 5′-region and an insertion of retrotransposon of 3255 bp in the third intron (Fig. 33.4). In both cases, the protein-coding region was seriously damaged and no functional products could have been generated, because *S-ELF3* was essentially a single-copy gene. The disruptions of *S-ELF3* genes in Kyushu PL4 and *F. tataricum* occurred later and independently in each lineage, as suggested in the evolutionary tree shown in Fig. 33.4. This is in sharp contrast to the long persistence of the *S-ELF3* gene in thrum plants in SI species and supports a role of *S-ELF3* in the heteromorphic SI of *Fagopyrum* species. It should be remembered that this provides an explanation for the molecular mechanism of the dominance of the *S* allele and the diallelism in buckwheat heteromorphic SI; the heteromorphic SI can be attained by the presence and absence of functional genes including *S-ELF3* in the *S* and *s* haplotypes, respectively. Despite much circumstantial evidence, the definitive answer to functional involvement of *S-ELF3* in the heteromorphic SI needs to wait until the introduction of *S-ELF3* in pin-type plants or the generation of a defective mutant of *S-ELF3* in thrum-type plants.

AN INFERENCE OF EVOLUTIONARY HISTORY USING GENETIC DATA OF THE *S*-LOCUS

From the latest studies of the *S*-locus, it becomes apparent that a few genes, including *S-ELF3*, are present solely in the dominant *S* haplotype with complete linkage at the locus. This unique feature provides a tool to study evolutionary relationships of individuals and populations in buckwheat. In plants, nuclear genes generally evolve faster than mitochondrial or chloroplast genes (Wolfe et al., 1987; Drouin et al., 2008) and nuclear genes provide higher resolution in evolutionary analyses. Meanwhile, genetic study using nuclear genes requires caution: as fragments with different genealogy can be brought nearby, analyses need to take the effects of recombination into account. Otherwise, erroneous pictures of evolutionary history would be drawn. From this viewpoint, analyses of extended gene fragments free of genetic recombination have certain advantages, in particular for organisms with limited resources, and can clarify the history of populations and/or individuals at ease, as exemplified in the analyses of Y-chromosomes used to elucidate modern human history (Underhill et al., 2000).

Accordingly, as a first attempt, we have analyzed *S-ELF3* and its closely linked *SSG2* genes to understand the relationships of cultivated buckwheat (Fig. 33.5). The preliminary result illustrates that those derived from Bhutan and

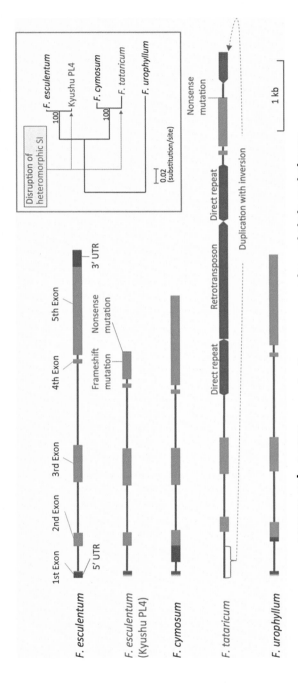

FIGURE 33.4 *S-ELF3* genes in *Fagopyrum* species and their evolutionary relationships.

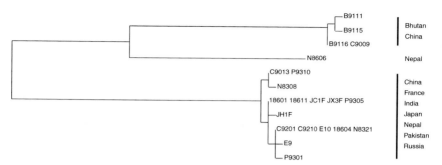

FIGURE 33.5 Relationship of *S* haplotypes inferred from evolutionary analyses of *S-ELF3* and *SSG2* genes.

China as well as one from Nepal are distinct from the others and that most cultivated buckwheat from different parts of the world is close to that from China, though details remain unclear because of the short length of nucleotide sequences analyzed. The current result is consistent with the previous results that have shown that the origin of cultivated buckwheat is located at the area of Yunnan and Sichuan provinces of China (Konishi et al., 2005) and that some buckwheat individuals and/or populations from Bhutan and China are distinct from others (Murai and Ohnishi, 1996).

As mentioned earlier, the *S*-locus can be extended over 5 MB of region. The region is apparently rich in transposable elements and repetitive elements, which are difficult to use for phylogenetic analyses, but still contain much information to answer questions related to the history of buckwheat populations. Archeological remains of buckwheat are so far rare and an early stage of buckwheat cultivation remains unclear despite its interest (Weisskopf and Fuller, 2014). Analyses of the extended region of *S*-locus would provide and shed light on the prehistorical age of buckwheat cultivation and its dispersal. Rapid progress in nucleotide sequencing technology for the past decade makes it possible to analyze long nucleotide sequences of specific genomic regions for a large number of individuals, and even research for an organism of minor crops, such as buckwheat, can take advantage of this. We are close to grasping and understanding the genetic relationship of buckwheat populations and the general picture on the spread of buckwheat cultivation to the world from its origin.

REFERENCES

Altschul, S.F., Gish, W., Miller, W., Myers, E.W., Lipman, D.J., 1990. Basic local alignment search tool. J. Mol. Biol. 215, 403–410.

Campbell, C., 1995. Inter-specific hybridization in the genus *Fagopyrum*. Proceedings of the Sixth International Symposium on Buckwheat, pp. 255–263.

Chafe, P.D.J., Lee, T., Shore, J.S., 2014. Development of a genetic transformation system for distylous *Turnera joelii* (Passifloraceae) and characterization of a self-compatible mutant. Plant Cell Tiss. Org. Cult. 120, 507–517.

Charlesworth, D., Charlesworth, B., 1987. Inbreeding depression and its evolutionary consequence. Annu. Rev. Ecol. System. 18, 237–268.

Charlesworth, D., Willis, J.H., 2009. The genetics of inbreeding depression. Nat. Rev. Genet. 10, 783–796.

Coluccio, M.P., Sanchez, S.E., Kasulin, L., Yanovsky, M.J., Botto, J.F., 2011. Genetic mapping of natural variation in a shade avoidance response: *ELF3* is the candidate gene for a QTL in hypocotyl growth regulation. J. Exp. Bot. 62, 167–176.

Darwin, C., 1877. The Different Forms of Flowers on Plants of the Same Species. John Murray, London.

Dowrick, V.P.J., 1956. Heterostyly and homostyly in *Primula obconica*. Heredity 10, 219–236.

Drouin, G., Daoud, H., Xia, J., 2008. Relative rates of synonymous substitutions in the mitochondrial, chloroplast and nuclear genomes of seed plants. Mol. Phylogenet. Evol. 49, 827–831.

Dulberger, R., 1992. Floral polymorphisms and their functional significance in the heterostylous syndrome. In: Barrett, S.C.H. (Ed.), Evolution and Function of Heterostyly. Springer-Verlag, Berlin, Heidelberg, pp. 41–84.

Elshire, R.J., Glaubitz, J.C., Sun, Q., Poland, J.A., Kawamoto, K., Buckler, E.S., Mitchell, S.E., 2011. A robust, simple genotyping-by-sequencing (GBS) approach for high diversity species. PLoS One 6, e19379.

Garber, R.J., Quisenberry, K.S., 1927. The inheritance of length of style in buckwheat. J. Agric. Res. 34, 181–183.

Goldberg, E.E., Kohn, J.R., Lande, R., Robertson, K.A., Smith, S.A., Igic, B., 2010. Species selection maintains self-incompatibility. Science 330, 493–495.

Hicks, K.A., Alberston, T.M., Wagner, D.R., 2001. *EARLY FLOWERING3* encodes a novel protein that regulates circadian clock function and flowering in *Arabidopsis*. Plant Cell 13, 1281–1292.

Hicks, K.A., Millar, A.J., Carré, I.A., Somers, D.E., Straume, M., Meeks-Wagner, D.R., Kay, S.A., 1996. Conditional circadian dysfunction of the *Arabidopsis early-flowering 3* mutant. Science 274, 790–792.

Hilderbrand, F., 1867. Die Geschlechter-vertheilung bei den pelanzen und das gesetz der vermiedene und unvortheilhaften stetigen selbstbefruchtung. Verlag von Wilhelm Engelmann, Leipzig.

Hirose, T., Ujihara, A., Kitabayashi, H., Minami, M., 1995. Pollen tube behavior related to self-incompatibility in interspecific crosses of *Fagopyrum*. Breed. Sci. 45, 65–70.

Iwano, M., Takayama, S., 2012. Self/non-self discrimination in angiosperm self-incompatibility. Curr. Opin. Plant Biol. 15, 78–83.

Konishi, T., Yasui, Y., Ohnishi, O., 2005. Original birthplace of cultivated common buckwheat inferred from genetic relationships among cultivated populations and natural populations of wild common buckwheat revealed by AFLP analysis. Genes Genet. Syst. 80, 113–119.

Labonne, J.D.J., Shore, J.S., 2011. Positional cloning of the *s* haplotype determining the floral and incompatibility phenotype of the long-styled morph of distylous *Turnera subulata*. Mol. Genet. Genom. 285, 101–111.

Lewis, D., Jones, D.A., 1992. The genetics of heterostyly. In: Barrett, S.C.H. (Ed.), Evolution and Function of Heterostyly. Springer-Verlag, Berlin, Heidelberg, pp. 129–150.

Li, J., Webster, M.A., Smith, M.C., Gilmartin, P.M., 2011. Floral heteromorphy in *Primula vulgaris*: progress towards isolation and characterization of the *S* locus. Ann. Bot. 108, 715–726.

Li, J., Webster, M.A., Wright, J., Cocker, J.M., Smith, M.C., Badakshi, F., Heslop-Harrison, P., Gilmartin, P.M., 2015. Integration of genetic and physical maps of the *Primula vulgaris S* locus and localization by chromosome *in situ* hybridization. New Phytol. 208, 137–148.

Matsui, K., Tetsuka, T., Nishio, T., Hara, T., 2003. Heteromorphic incompatibility retained in self-compatible plants produced by a cross between common and wild buckwheat. New Phytol. 159, 701–708.

Murai, M., Ohnishi, O., 1996. Population genetics of cultivated common buckwheat, *Fagopyrum esculentum* Moench. X. Diffusion routes revealed by RAPD markers. Genes Genet. Syst. 71, 211–218.

Ohnishi, O., 1998. Search for the wild ancestor of buckwheat. I. Description of new *Fagopyrum* (Polygonaceae) species and their distribution in China and Himalayan hills. Fagopyrum 15, 18–28.

Ohnishi, O., Matsuoka, Y., 1996. Search for the wild ancestor of buckwheat. II. Taxonomy of *Fagopyrum* (Polygonaceae) species based on morphology, isozymes and cpDNA variability. Genes Genet. Syst. 71, 383–390.

Schoch-Bodmer, H., 1934. Zum Heterostylieproblem: Griffelbeschaffenheit und Pollenschlauchwachstum bei *Fagopyrum esculentum*. Planta 22, 149–152.

Sharma, K.D., Boyes, J.W., 1961. Modified incompatibility of buckwheat following irradiation. Can. J. Bot. 39, 1241–1246.

Tatebe, T., 1949. Physiological researches on the fertility of the buckwheat. (I) Flower morphology and fertility. Ikushukenkyu 3, 91–95.

Underhill, P.A., Shen, P., Lin, A.A., Jin, L., Passarino, G., Yang, W.H., Kauffman, E., Bonné-Tamir, B., Bertranpetit, J., Francalacci, P., Ibrahim, M., Jenkins, T., Kidd, J.R., Mehdi, S.Q., Seielstad, M.T., Wells, R.S., Piazza, A., Davis, R.W., Feldman, M.W., Cavalli-Sforza, L.L., Oefner, P.J., 2000. Y chromosome sequence variation and the history of human populations. Nat. Genet. 26, 358–361.

Ushijima, K., Ikeda, K., Nakano, R., Matsubara, M., Tsuda, Y., Kubo, Y., 2015. Genetic control of floral morph and petal pigmentation in *Linum grandiflorum* Desf., a heterostylous flax. Horticult. J. 84, 261–268.

Ushijima, K., Nakano, R., Bando, M., Shigezane, Y., Ikeda, K., Namba, Y., Kume, S., Kitabata, T., Mori, H., Kubo, Y., 2012. Isolation of the floral morph-related genes in heterostylous flax (*Linum grandiflorum*): the genetic polymorphism and the transcriptional and post-transcriptional regulations of the *S* locus. Plant J. 69, 317–331.

Weisskopf, A., Fuller, D., 2014. Buckwheat: origins and development. In: Smith, C. (Ed.), Encyclopedia of Global Archaeology. Springer, New York, pp. 1025–1028.

Wolfe, K.H., Li, W.-H., Sharp, P.M., 1987. Rates of nucleotide substitution vary greatly among plant mitochondrial, chloroplast and nuclear DNAs. Proc. Natl. Acad. Sci. USA 84, 9054–9058.

Woo, S.H., Adachi, T., Jong, S.K., Campbell, C.G., 1999. Inheritance of self-compatibility and flower morphology in an inter-specific buckwheat hybrid. Can. J. Plant Sci. 79, 483–490.

Yasui, Y., Hirakawa, H., Ueno, M., Matsui, K., Katsube-Tanaka, T., Yang, S.J., Aii, J., Sato, S., Mori, M., 2016. Assembly of the draft genome of buckwheat and its applications in identifying agronomically useful genes. DNA Research, Article DOI:dsw012.

Yasui, Y., Mori, M., Aii, J., Abe, T., Matsumoto, D., Sato, S., Hayashi, Y., Ohnishi, O., Ota, T., 2012. S-LOCUS EARLY FLOWERING 3 is exclusively present in the genomes of short-styled buckwheat plants that exhibit heteromorphic self-incompatibility. PLoS One 7, e31264.

Yasui, Y., Mori, M., Matsumoto, D., Ohnishi, O., Campbell, C.G., Ota, T., 2008. Construction of a BAC library for buckwheat genome research—an application to positional cloning of agriculturally valuable traits—. Genes Genet. Syst. 83, 393–401.

Zerbino, D.R., Birney, E., 2008. Velvet: algorithms for de novo short read assembly using de Bruijn graphs. Genome Res. 18, 821–829.

Chapter | thirty four

Biochemical and Technological Properties of Buckwheat Grains

S. Bobkov
Laboratory of Plant Physiology and Biochemistry, All-Russia Research Institute of Legume and Groat Crops, Orel, Streletsky, Russia

INTRODUCTION

Common buckwheat (*Fagopyrum esculentum* Moench) is one of the traditional crops cultivated in the world. In 2013 buckwheat production worldwide was 2.3 million tons (FAOSTAT, 2013). The main producers of buckwheat grains in 2013 were Russia (35.5%), China (31.2%), Ukraine (7.6%), France (6.6%), Poland (3.9%), the United States (3.5%), Brazil (2.6%), and Japan (1.4%). Each of these countries produced more 10,000 tons (Jacquemart et al., 2012).

Buckwheat is a rich source of valuable chemical compounds such as starch, protein, fat, dietary fiber, vitamins, minerals, *myo*-inositol, D-*chiro*-inositol, fagopyritols, flavonoids, and phytosterols (Zielinski et al., 2001; Bonafaccia et al., 2003a; Krkoskova and Mrazova, 2005; Zhang et al., 2012). Buckwheat grains are of high nutritive value (Jacquemart et al., 2012). Buckwheat proteins contain albumins, globulins, prolamins, and glutelins. They do not contain gluten (Skerritt, 1986; Wieslander and Norbäck, 2001). Consequently, buckwheat food is a common supplement for patients with celiac disease. In comparison to cereals, buckwheat proteins are particularly rich in lysine, arginine, and aspartic acid, but contain less glutamic acid and proline (Zhang et al., 2012). Buckwheat could improve the amino acid balance of products from other species. Buckwheat proteins are characterized by relatively low protein digestibility (Ikeda et al., 1986). Acting in a similar way to dietary fiber, buckwheat proteins prevent constipation and obesity. They have special biological activities regarding cholesterol-lowering and antihypertensive effects. Short length peptides identified in buckwheat are characterized by hypotensive activity (reviewed in Zhou et al., 2015). Buckwheat proteins also may prevent gallstone formation and suppress mammary and colon carcinogenesis by reducing serum estradiol and cell proliferation (Liu et al., 2001). Buckwheat trypsin inhibitors, other proteins,

Molecular Breeding and Nutritional Aspects of Buckwheat. http://dx.doi.org/10.1016/B978-0-12-803692-1.00034-1

and their enzyme hydrolysates demonstrate antibacterial, antifungal, antitumor, hypocholesterol, hypotensive, and antidiabetic activity (Zhou et al., 2015).

Buckwheat groats contain amylose and other resistant starches (Skrabanja and Kreft, 1998; Pomeranz, 1983). Buckwheat lipids contain 79.3% unsaturated fatty acids (Bonafaccia et al., 2003a). Buckwheat soluble fibers reduce total cholesterol content including its LDL fractions (He et al., 1995). Buckwheat flavonoids demonstrate a protective effect against lipid peroxidation as free radical scavengers and metal-chelating agents (Afanasev et al., 1989). Fagopyritols reduce symptoms of noninsulin-dependent diabetes mellitus (Steadman et al., 2000). Phytosterols of buckwheat seeds lower blood cholesterol levels (Li and Zhang, 2001).

Other chemical compounds such as nicotianamine and its hydroxyl derivatives, extracted from buckwheat flour, demonstrate a strong inhibitory effect on angiotensin-I-converting enzyme and lower blood pressure (Suzuki et al., 1983; Aoyagi, 2006).

Different types of buckwheat products such as whole grains, groat, flour, sprouts, shoots, and honey are consumed by humans and animals (Jacquemart et al., 2012). Buckwheat is eaten because of its taste, traditions, and wide range of food products, but mostly because of its importance for human health (Kreft, 2001).

Buckwheat is traditionally used for producing of flour and groat (Ikeda et al., 1999; Varlakhova et al., 2013a). Buckwheat groat is used for preparing porridge in Slovenia, Austria, and Japan, and buckwheat flour is used for producing pasta in Italy and extruded noodles in China, Korea, and Japan (Bonafaccia et al., 2003a). Buckwheat flour alone is not suitable for making bread and is often used in mixtures with other flours in a proportion of 30% (Bojnanská et al., 2009). A mixture of buckwheat and wheat flour is used for preparing noodles, pancakes, girdle cakes, cakes, biscuits, cracknels, and fritters (Levent and Bilgiçli, 2011).

Dehulled (without pericarp) buckwheat grains can be consumed as whole groat and crushed grits. Grits are added to soups, sauces, and salads (Jacquemart et al., 2012). In Russia, milk porridge (kasha) prepared from buckwheat groat is a common breakfast. Crumbly buckwheat porridge is used in dishes with meat, chicken, fish, and vegetables.

Buckwheat breeding was preferably focused on high yields and fineness of grains. But data on biochemical compound content in different tissues of buckwheat grains opened new possibilities for their cultural evolution.

BIOCHEMICAL PROPERTIES OF BUCKWHEAT GRAINS

Common buckwheat is considered an important source of dietary protein. Depending on ecotype, the variety of protein content in buckwheat flour varies from 8.51% to 18.87% (Krkoskova and Mrazova, 2005). Buckwheat flour contains significantly more protein than rice, wheat, millet, sorghum, and maize but less than oat. Buckwheat proteins are composed of 12.5–18.2% albumins,

43.3–64.5% globulins, 0.8–2.9% prolamins, 8.0–22.7% glutelins, and 15.0% residual protein (Javornik and Kreft, 1984; Ikeda et al., 1991). There are a number of differences in the abundance of protein fractions. For example, prolamin content is reported to range from 0.8% to 6.24% (Wei et al., 2003; Petr et al., 2003). As revealed by sodium dodecyl sulfate polyacrylamide gel electrophoresis (SDS-PAGE) and enzyme-linked immunosorbent assay there are no coeliac-active gluten proteins (prolamins, 30 kDa) in buckwheat groats (Petr et al., 2003). This is the main reason for using buckwheat products in the diets of those with coeliac disease.

Buckwheat seed proteins are characterized by high levels of polymorphism as revealed with SDS-PAGE (Lazareva, 2007). Polymorphism of informative protein bands is proposed for identification of buckwheat accessions and cultivated varieties.

The major storage proteins of common buckwheat seeds include 8S, 13S globulins, and 2S albumins (Radovic et al., 1996, 1999; Milisavljevic et al., 2004). Salt-soluble 13S globulins have a hexameric structure with disulfide-bonded subunits composed of acidic and basic polypeptides with molecular masses of 32–43 kDa and 23–25 kDa, but 8S globulins are trimers, composed of subunits of 57–58 kDa (Milisavljevic et al., 2004). Water-soluble 2S albumins contain polypeptides with a molecular weight of 16 and 8–12 kDa (Radovic et al., 1999). Buckwheat globulins and albumins comprise 64.5–70% and 25–30% of total seed proteins, respectively (Elpidina et al., 1990; Radovic et al., 1996, 1999). Buckwheat 13S and 8S globulins, respectively, resemble legumin-like and vicilin-like storage proteins of other species (Milisavljevic et al., 2004). The main storage protein in buckwheat seeds are 13S globulins. The content of 8S globulins is approximately 7% of total seed proteins. Buckwheat seed proteins can be stored for long periods without degradation (Lazareva et al., 2006). This could be explained by the presence of fagopyritols in buckwheat grains (Horbowicz et al., 1998).

Buckwheat flour proteins have a well-balanced amino acid composition (Ikeda, 2002). Buckwheat proteins are rich in lysine, arginine, and aspartic acids in comparison to cereal counterparts (Pomeranz and Robbins, 1972; Wei et al., 2003). About 56% of glutamic and aspartic acids are in the form of amides.

The basic subunit of 26 kDa of 13S legumin is purified from grains of common buckwheat. Its amino acid composition closely matches the World Health Organization's recommended values for a nutritionally balanced protein (Rout et al., 1997). A distinguishing feature of this subunit is the relatively high level of lysine and methionine (Bharali and Chrungoo, 2003).

The high quality of buckwheat proteins is compromised by their low digestibility because of the presence in grains of tannin and protease inhibitors (Javornik et al., 1981; Ikeda et al., 1991; Zhang et al., 2012). True digestibility of buckwheat protein is lower when compared with cereals and reaches 80% (Eggum et al., 1981; Javornik et al., 1981). But protein digestibility-corrected amino acid scores of buckwheat are higher in comparison with wheat and oats because of a relatively low level of lysine in cereal proteins (Wijngaard and Arendt, 2006).

Short length peptides with hypotensive activity have been identified in buckwheat. For example, tripeptide Gly-Pro-Pro isolated from buckwheat (*F. esculentum* Moench) grains inhibit angiotensin-I-converting enzyme and reduce blood pressure (Ma et al., 2006).

Different allergens are identified in common buckwheat (Wieslander, 1996). They are thermostable and remain in food after cooking. Buckwheat allergens that bind IgE have molecular weights from 10 to 30 kDa (Lee et al., 2013). Buckwheat protein BW24KD (24 kDa) is the most frequently recognized allergen, binding to IgE antibodies from 100% of patients' sera (Urisu et al., 1994). This protein is classified as 11S globulin-like legumin (Nair and Adachi, 1999). It is reported that certain proteins of 9–19 kDa are also associated with buckwheat allergy (Park et al., 2000; Yoshimatsu et al., 2000). One of them, buckwheat protein 19 kDa, is specific for patients with clinical symptoms of allergy (Park et al., 2000). Another 16-kDa buckwheat protein is resistant to pepsin digestion and appeared to be responsible for immediate hypersensitivity reaction (IHR) including anaphylaxis, while the pepsin-sensitive 24-kDa protein was responsible for CAP system fluorescein-enzyme immunoassay but not IHR (Tanaka et al., 2002).

Starch is a main component of buckwheat grains (Skrabanja and Kreft, 1998). The content of starch ranges from 60% to 70% (Vojtíšková et al., 2012). Buckwheat starch is composed 25% amylose and 75% amylopectin (Pomeranz, 1983; Qin et al., 2010). Raw buckwheat groats contain 33.5% resistant starch (Skrabanja and Kreft, 1998). Resistant starches and other oligosaccharides could be fermented by intestinal microflora (Krkoskova and Mrazova, 2005). Autoclaving and cooking strongly decrease the level of resistant starch. Generally, uncooked groats comprise 33–38% resistant starch, but after cooking its amount decreases to 7–10% (Skrabanja et al., 1998). In contrast, after autoclaving or boiling, the level of retrograded starch increases from 1 to 4–7%. Rats can completely digest native buckwheat starch but excrete 1–1.6% of starch after hydrothermal processing (Skrabanja et al., 1998).

Fatty acids are an important part of plant grains. Buckwheat grains contain 1.5–3.7% total lipids (Mazza, 1988; Campbell, 1997). Buckwheat lipids are divided into neutral (81–85%), phospholipids (8–11%), and glycolipids (3–5%). Eighteen fatty acids have been identified in buckwheat grains (Dorrell, 1971). Eight main fatty acids (oleic, linoleic, palmitic, linolenic, lignoceric, stearic, behenic, and arachidic) represent 93% of their total content. Palmitic (16:0), oleic (18:1), and linoleic (18:2) are the most abundant (87.3–88%) fatty acids (Mazza, 1988; Horbowicz and Obendorf, 1992). The average content of these fatty acids in total lipids was 14%, 36.3%, and 37%, respectively (Mazza, 1988). Bonafaccia et al. (2003a) reported similar results. In their study the concentration of palmitic (16:0), oleic (18:1), and linoleic acid (18:2) was equal to 15.6%, 37%, and 39%, respectively (Bonafaccia et al., 2003a). The content of other fatty acids such as stearic (18:0), linolenic (18:3), arachidonic (20:0), eicosaenoic (20:1), and behenic (22:0) ranged from 0.7% to 3%. The part of unsaturated fatty acids in buckwheat lipids reaches 79.3%.

Together with starch and proteins, the grains of buckwheat contain minerals, dietary fiber, vitamins, fagopyritols, flavonoids, phytosterols, phytic acid, and tannins.

Buckwheat flour is a good source of essential minerals (Steadman et al., 2001b; Ikeda et al., 2006). Macroelements of phosphorus (P), potassium (K), magnesium (Mg), and calcium (Ca) are presented in concentrations of more than 100 mg/kg dry weight (Steadman et al., 2001b). In buckwheat, large amounts of P are a component of phytic acid. Concentration of microelements is much lower compared to macroelements. Iron (Fe), manganese (Mn), and zinc (Zn) are present in larger quantities than other microelements. In comparison to cereal crops (wheat, rice, and maize), buckwheat flour contains significantly ($P < 0.05$) higher levels of copper, manganese, magnesium, potassium, and phosphorus (Ikeda et al., 2006). About 40% of zinc in buckwheat flour is in water-soluble form (Ikeda et al., 1990). Additionally, major parts of zinc, copper, and potassium are released from buckwheat flour in soluble form after enzymatic (α-amylase, pepsin, and pancreatin) digestion and become available for gastrointestinal absorption (Ikeda et al., 1990, 2006). Buckwheat grains contain other microelements such as Se, Cr, Rb, Co, Sb, Ba, Ni, Ag, Hg, and Sn (Bonafaccia et al., 2003b).

Total dietary fiber (TDF) content in buckwheat groats is comparable to grain crops (Krkoskova and Mrazova, 2005). For nutritional purposes, TDF is classified as either insoluble (IDF) and soluble (SDF) dietary fiber (Cho et al., 1997). Generally, IDF includes cellulose, lignin, and certain noncellulosic polysaccharides but SDF represents pectin and associated noncellulosic polysaccharides.

Buckwheat is a rich source of bioactive compounds. In buckwheat seeds, certain biologically active compounds such as vitamins, fagopyritols, flavonoids, isoprenoids, iminosugars, inositols, and phytosterols are detected (Zhang et al., 2012).

Buckwheat grains contain vitamins A (carotenoids), B1 (thiamine), B2 (riboflavin), B3 (niacin), B5 (pantothenic acid), B5 (pyridoxine), C (ascorbic acid), and E (tocopherols) (Wijngaard and Arendt, 2006). The concentration of vitamins B1, B2, and B6 in buckwheat grains is 0.22, 0.1, and 0.17 mg/100 g, respectively (Bonafaccia et al., 2003a). Buckwheat grains contain vitamins B3 and B5 in concentrations of 1.8 and 1.1 mg/100 g, respectively. Vitamin B1 is strongly adhered to thiamine-binding proteins (TBP) of buckwheat grains (Mitsunaga et al., 1986). TBP enable stability of thiamine during storage and enhance its availability (Li and Zhang, 2001). Vitamin E represents all naturally occurring forms of tocopherols (Zielinski et al., 2001). Total tocopherol concentrations in buckwheat grains range from 1.43 to 5.5 mg/100 g (Zielinski et al., 2001; reviewed in Kalinova et al., 2006).

Fagopyritols are galactosyl derivatives of D-*chiro*-inositol that accumulate in seeds of some species. Fagopyritols are identified as a major soluble carbohydrate (40% of total) in common buckwheat embryos (Horbowicz et al., 1998). Fagopyritol B1 is associated with acquisition of desiccation tolerance during seed development and maturation in planta. It is proposed that fagopyritol B1 facilitates desiccation tolerance and storability of buckwheat seeds (Horbowicz et al., 1998). Seeds of common buckwheat accumulate six fagopyritols. Other soluble carbohydrates analyzed by high-resolution gas chromatography include sucrose (42% of total), D-*chiro*-inositol, *myo*-inositol, galactinol, raffinose, and stachyose (1% of total). Filling of fagopyritols improves when seeds mature at a moderate temperature (18°C) as compared to a more elevated value (25°C) (Horbowicz et al., 1998).

In common buckwheat grains the total content of flavonoids is about 10 mg/g, which is four times less than in *Fagopyrum tataricum* (Li and Zhang, 2001). Six flavonoids including rutin, orientin, vitexin, quercetin, isovitexin, and isoorientin are isolated and identified in buckwheat grains (reviewed in Zhang et al., 2012). All six flavonoids are present in the hull (Dietrych-Szostak and Oleszek, 1999). But in buckwheat groats only rutin and isovitexin are found. Rutin is the main flavonoid component in buckwheat grains. Rutin concentration in buckwheat groats derived from American varieties Mancan and Manor is 0.2–0.3 g/kg (Steadman et al., 2001b). It is similar to the European variety Siva where the rutin content in groats was 0.23 g/kg (Fabjan et al., 2003; Kreft et al., 2006). In comparison to Tartary buckwheat (where the rutin content is up to 81 g/kg) the concentration of rutin in common buckwheat grains is 300-fold less.

Buckwheat grains also contain other kinds of flavonoids such as hyperoside (quercetin 3-O-β-D-galactoside), quercitrin (quercetin 3-O-α-rhamnoside), and catechins (reviewed in Kalinova and Vrchotova, 2009). Mitsuru Watanabe isolated from common buckwheat groats other kinds of flavonoids such as catechins, which showed antioxidant activity higher than rutin (Watanabe, 1998). Catechin distribution in buckwheat plant parts is similar to the distribution of tannins (Kalinova and Vrchotova, 2009). The concentration of catechins in Japanese buckwheat varieties is determined by high-performance liquid chromatography with electrochemical detection. There are 12.7 mg/100 g rutin, 3.3 mg/100 g catechin, 20.5 mg/100 g epicatechin, and 1.27 mg/100 g epicatechin gallate in buckwheat flour (Danila et al., 2007). Catechins and tannins have significant inhibitory effects on peptic and pancreatic digestion of buckwheat globulin (Ikeda et al., 1986). All flavonoids especially rutin demonstrate antioxidant activity in all studied buckwheat varieties (Jiang et al., 2007).

Buckwheat isolated proteins (BPI) contain increased concentrations of polyphenol compounds. For example, in the flour of common buckwheat grown in China, total polyphenol content is about 0.6% rutin equivalent (Tang and Wang, 2010). The content of free polyphenols accounts for 56.7%. BPI has a total polyphenol content of 2.93%. Isolated buckwheat albumin and globulin contain 1.85% and 0.67% polyphenols, respectively. In albumins most of the polyphenols (about 91%) are present in the free form, but in globulins the amount of free polyphenols equals 23%.

Genes encoding flavonoid biosynthesis enzymes in *F. esculentum* such as *FePAL*, *FeC4H*, *Fe4CL*, *FeCHS*, *FeCHI*, *FeF3H*, *FeF30H*, *FeFLS*, and *FeANS* are already cloned and sequenced (Li et al., 2010). The expression level of these genes is examined and the content of various phenolic compounds in different organs, sprouts, and seeds is determined.

Plant sterols (phytosterols) are found in buckwheat seeds at 20 days after pollination (Horbowicz and Obendorf, 1992). The content of total sterols in the embryo and in the endosperm is 2.1 and 0.55 g/kg, respectively. The most abundant sterol is β-sitosterol, which makes up 70% of the total amount. There are 667–753 mg/kg β-sitosterol, 89–97 mg/kg campesterol, and a stigmasterol trace in dehulled groats (Li and Zhang, 2001).

Buckwheat seeds contain higher amounts of phytic acid in comparison to legumes and cereals (Steadman et al., 2001b).

CHEMICAL COMPOUNDS IN PROCESSED MILLING FRACTION AND GRAIN TISSUES

The concentration of chemical components in certain milling fractions has several advantages (Skrabanja et al., 2003b). Milling fractions with high concentrations of certain interesting components (eg, soluble dietary fiber, proteins, and polyphenols) could be used for obtaining desired end-use products. The functionality of such products is very important. Milling fraction with accumulated chemical compounds could stimulate elaboration of new cooking and food technologies.

Buckwheat seed milling fractions are produced either by milling the intact achenes or grains after removing the hull. The number of milling fractions analyzed was 10 (Kreft et al., 1994), 11 (Steadman et al., 2001a), and 23 (Skrabanja et al., 2003a). In these experiments milled products were mainly separated into various fractions of flour and bran. It was proposed that milling processing separated the thicker-walled cells of the outer layer and the thinner-walled cells of central endosperm (Steadman et al., 2001a). For example, 23 milling fractions represented fine flour (FF), coarse flour (CF), small semolina (SS), big semolina (BS), bran (Br), and husk (H) (Skrabanja et al., 2003b). Milling products were differentiated by particle sizes. In FF (7 fractions) the majority of particles were less 129 μm. In CF (3 fractions) 77% of total weight passed a 183-μm mesh sieve. The particle size range of SS (4 fractions) and BS (2 fractions) was 219–656 μm and 505–656 μm (more than 50% of particles), respectively. Six bran fractions were characterized by 219–656 μm milling particles. The milling yield (cultivar Siva) resulted in 55.4% flour, 24.2% bran, 17.4% hull (husk) (Skrabanja et al., 2003a).

Biochemical analysis revealed principal differences in content of biochemical compounds between light flour and bran fractions (Steadman et al., 2001a). In comparison to groats, the concentration of starch in light flour was increased but the content of protein, fat, soluble carbohydrates, dietary fiber, and ash significantly decreased. In contrast, the concentration of starch in bran fractions was decreased but the content of all other compounds greatly increased.

Buckwheat milling fractions had starch contents that varied from 20.3% to 91.7%, and the concentration of protein ranged from 3.7% to 31.3% (Skrabanja et al., 2003b). The highest content of protein was in bran. The lowest content was in fine flour (4.4%) and husk (3.7%). Fat was preferably concentrated in bran (9.7%) and semolina fractions (8.9%). The lowest content of fat was in husk (0.4%) and fine flour fractions (0.5%). Fat content was positively correlated with proteins ($r = 0.93$, $P = 0.0000$) and negatively with starch ($r = 0.92$, $P = 0.0000$).

The concentration of total dietary fiber in milling fractions was 2.7–21.3% (Skrabanja et al., 2003b). In husks, the fiber content was more than 90%, but the proportion of soluble dietary fiber was relatively small (2.9%). Flour fractions contained lower amounts of total dietary fiber (2.7–8.1%) than semolina (6.6–12.9%) or bran fractions (15.4–21.3%). In general, the content of soluble dietary fiber (1.4–6.6%) was weaker than the insoluble counterparts (0.7–15.8%). The percentage of soluble parts in total dietary fiber was much higher in flours than in semolina or bran fractions.

Bran contained 6.4% of total soluble carbohydrates, 55% of which was sucrose and 40% fagopyritols. This ratio was present in all milling fractions. Flour had reduced fagopyritol concentration, which ranged from 0.3% to 0.7%.

In common buckwheat, rutin is concentrated in the hull (0.8–4.4 g/kg). The concentration of rutin in groats was relatively low (0.019–0.3 g/kg) but higher (0.48–0.8 g/kg) in bran containing hull fragments (Steadman et al., 2001b; Skrabanja et al., 2003b).

The highest (6%) concentration of tannins was detected in bran (Skrabanja et al., 2003b). The lowest amount of tannins was in fine flour and husk fractions.

Phytates were present in milling fractions (except husks) at 0.21–0.72% (Skrabanja et al., 2003b). Phytates were the most evenly distributed constituent among milling fractions. The highest (6%) concentration of tannins was found in bran. The lowest content of tannins was found in fine flour (0.1–0.3%) and husk (0.3%).

Buckwheat seed is composed of pericarp (hull, husk), coat, endosperm, embryo with axis, and two cotyledons. The cotyledons are located in major parts close to the surface of the seed (Kreft and Kreft, 1999; Steadman et al., 2001a). Various tissues of seed differ in concentrations of chemical components. The distribution of starch and proteins is different in various tissues of buckwheat seed. Starch preferably accumulates in the endosperm but protein accumulates in the embryo. Whole grains contain 13.8% protein, 16.4% groat, 4% hulls, and more than 50% germ (Pomeranz and Robbins, 1972). Buckwheat seed proteins isolated from embryo and endosperm are separated by SDS-PAGE electrophoresis and characterized with the use of LC-ESI-Q/TOF-MS/MS mass spectrometry (Kamal et al., 2011). In total, 67 proteins are identified. Out of the total, 20 proteins are located in the endosperm, 29 in the embryo, and 18 were shared in both. It is determined that 11S globulins are located in the embryo. The 13S globulins including allergen proteins are found in both of the embryo and the endosperm. Proteins from various seed tissues except the hull have similar amino acid composition (Pomeranz and Robbins, 1972; Bonafaccia et al., 2003a). Dietary fiber mainly concentrates in the hull where the protein content is very low. The concentration of dietary fiber in bran reaches 35.5–40.3% (Steadman et al., 2001a).

Different seed tissues have various level of fat. Lipids are concentrated in the embryo. For example, the content of total fatty acids in the embryo and the endosperm was 123 and 22 g/kg, respectively (Horbowicz and Obendorf, 1992). Phytic acid and phytates in mature seeds are present in vacuole-derived protein bodies located mainly in the embryo and the aleurone layer tissues (Steadman et al., 2001b). Tannins preferably concentrate in the embryo (Skrabanja et al., 2003b). The lowest content of tannins is detected in the hull.

Nonstarch polysaccharides are mainly located in cells of the aleurone layer, seed coat, and hull. Soluble carbohydrate, sucrose, and fagopyritols are concentrated in the embryo and aleurone tissues but not in the pericarp (hull), seed coat, or starchy endosperm (Horbowicz et al., 1998; Steadman et al., 2000). Fagopyritol B1 was 70% of total fagopyritols, with a maximum concentration of 2%. The fagopyritol content in buckwheat grains is influenced by the ratio of embryo to endosperm. Maturation of buckwheat seeds under elevated temperature (25°C vs. 18°C) decreases embryo mass and the content of fagopyritol B1 (Horbowicz et al., 1998).

Polyphenolic compounds are concentrated in the outer layers (seed coat and hull) of seeds. In common buckwheat the flavonoid rutin commonly presents in leaves stems, and flowers but it is detected in buckwheat grains in low concentrations (Kreft et al., 1999). In buckwheat grains, rutin is preferably concentrated (0.8–4.4 g/kg) in the hull of buckwheat grain (Steadman et al., 2001b). The concentration of rutin in groats is 0.2 g/kg. Additional investigations show that each part of a buckwheat grain contains different concentrations of rutin, catechin, epicatechin, and epicatechin gallate (Danila et al., 2007). The embryo proper and cotyledons of buckwheat grain contain higher concentrations of rutin than other parts, while the highest concentrations of epicatechin and epicatechin gallate are present in the embryo proper, cotyledons, and in the testa. The endosperm does not contain any detectable concentrations of these compounds.

Mineral content in buckwheat grains varies from 2.0% to 2.5%; in kernels from 1.8% to 2.0%; in flour from 0.8% to 0.9%; and in hull from 3.4% to 4.2% (Lin, 1994). Minerals P, K, and Mg are stored preferably as phytates in the embryo but less mobile minerals such as Ca, Fe, Zn, Mn, Cu, Mo, Ni, B, I, Pt, and Al are distributed mainly in the coat and hull (Sokolov et al., 1981; Li and Zhang, 2001; Steadman et al., 2001b).

TECHNOLOGICAL PROPERTIES OF BUCKWHEAT GRAINS

Buckwheat breeding in Russia has been conducted since the beginning of the 20th century. In recent times high-yielding buckwheat varieties with large grains and changed architectonics of plants were developed (Bobkov, 1993; Fesenko and Martynenko, 1998). Buckwheat breeding was conducted under the control of grain quality necessary for buckwheat processing. The main product of buckwheat processing is the groat "yadritsa." In recent times buckwheat "yadritsa" is the leading groat produced in Russia (Varlakhova et al., 2013a).

The important technological properties for buckwheat grain processing into groat are fineness of grains (weight of 1000 grains and percentage of grains left on 4.0 mm and 5.0 mm sieves), evenness of size (maximum percentage of grains left on two adjacent sieves among them of 5.0, 4.8, 4.5, 4.2, 4.0, 3.8, 3.6, 3.2 mm), shape of grains, content of kernel, hullness (percentage of hull fraction), and ease of dehulling (Metodicheskie, 1972; Metodika, 1979; Shumilin and Svjatova, 1976; GOST, 1993). Buckwheat groats are produced by sizing and dehulling the grain fractions, air separation of the hulls from the groats, and cleaning of the groats (Steadman et al., 2001a). The quality of groat was evaluated by the following traits: total groat yield, yield of whole groats (yadritsa), yield of crushed groats (grits), and fineness of groats (percentage of groats left on a 3.8 mm sieve).

Buckwheat breeding was aimed at producing varieties with high and stable yields and with a set of value technological traits enabling high yields of whole groats after processing. It was determined that the yield of groats and especially whole groats was positively correlated with fineness of grains (Fesenko, 1970). Grains left on 4.8, 4.5, and 4.2 mm sieves provided the highest total groat yield (Svjatova and Shumilin, 1978). Very large grain (fraction on 5.0 mm sieve) was simply dehulled and gave the maximum yield of whole groats (Shumilin and Svjatova, 1973). Further investigation showed that large grains had enhanced cavities between hull and groat coat, and were characterized by better technological properties (Shumilin et al., 1975). It was demonstrated that differences between size of grains and groats influenced the yield of groat (Svjatova and Shumilin, 1978). The high yield of groats in buckwheat varieties with mean differences between size of grains and groats equaled 0.8–1 mm (Shumilin and Varlahova, 1985).

Buckwheat breeding using high technological properties at the beginning of the 20th century was conducted at the Shatilovskaya Agricultural Experimental Station. Selection of buckwheat genotypes with large grains led to development of the variety Bogatyr. Plants growing from large seeds were characterized by powerful development and gave high yield of grains. If initial local populations of buckwheat had a weight of 100 grains equal to 18 g, the new variety Bogatyr was characterized by the value of 22–25 g (reviewed in Fesenko, 1983). For long periods of time the variety Bogatyr served as etalon of technological properties. Much later, the variety Shatilovskaya 5 was developed. In comparison to the Bogatyr variety, Shatilovskaya 5 had a weight of 1000 grains, a yield of total grains, and a yield of whole grains more than 1.8–4.2 g, 0.3–2.7%, and 6.25%, respectively (Fesenko, 1983). A further increase of grain fineness was observed in buckwheat variety Krasnostreletskaya (Petelina, 1971). This variety was characterized by a weight of 1000 grains of 29–34 g that was 5–6 g more compared with Bogatyr (reviewed in Fesenko, 1983).

Analysis of buckwheat breeding material during the period 1971–2009 revealed the main directions of a number of technological traits alteration (unpublished data). Mean weight of 1000 grains increased from 25.2 to 30.4 g (20.6%), and percentage of grains left on the 5.0 mm sieve changed from 16.6% to 43.2% (60.2%). Fineness of groats (percentage of grains left on the 5.0 mm sieve) increased from 27.9% to 58.3% (109%). Mean values of hull content, total groat yield, and whole groats yield were unchanged for long periods of buckwheat cultural evolution.

The technological properties of grains in buckwheat varieties developed from 1938 to 2010 in various regions of the Russian Federation were studied (Varlakhova et al., 2012, 2013a). Grains of varieties Bogatyr and Shatilovskaya 5 (Shatilovskaya Agricultural Experimental Station, Orel region), Chatyr-Tau and Batyr (Tatar Agricultural Research Institute, Tatarstan), Ilishevskaya (Ulyanovsk Research Institute, Ulyanovsk region), Ballada, Molva, Sumchanka, Dozhdik, Devyatka, and Design (VNIIZBK, Orel region) were investigated. These varieties differed in their normal and changed morphological traits such as limited branching, small leaves, determinant branches, and greenish color of petals. Buckwheat varieties were grown on the experimental fields of the All-Russia Research Institute of Legumes and Groat Crops for periods of 7 and 2 years. The average means of the variety grain properties were analyzed.

According to data obtained the content of hull in modern varieties did not exceed 23% (Table 34.1). Three groups of varieties with respect to fineness of grains (percentage of grains left on the 5.0 mm sieve after separation) were distinguished (Varlakhova et al., 2013b). In the first group (Bogatyr, Shatilovskaya 5, Ballada, Molva, and Dikul) the fraction left on the 5.0 mm sieve did not exceed 20%. In the second group (Sumchanka, Dozhdik, and Devyatka) this value was over 20% but less than 50%. In the third group (Design, Chatyr Tau, Batyr, and Ilishevskaya) the fraction left after separation on the 5.0 mm sieve exceeded 50%.

All studied varieties differed in distribution of grains after separation on a standard set of sieves with holes 5.0, 4.8, 4.5, 4.2, 4.0, 3.8, 3.6, and 3.2 mm (Varlakhova et al., 2013a). Breeding of large grain buckwheat varieties has led to profound changes in fractional distribution of grains. After grain sorting, varieties of Design, Chatyr Tau, and Ilishevskaya were mostly left on the 5.0 mm sieve, and no more than 2% was on the 4.2 mm sieve. Consequently, large grains of modern buckwheat varieties could be dehulled after separation on four fractions (5.0, 4.8, 4.5, and 4.2). Large grains have a considerable pocket between the groat and seed hull, which is why the 5.0 mm fraction of these varieties dehulled rapidly at a 4.9 mm gap of the working zone of the machine (Varlakhova et al., 2013b). The size of groats is less than 4.5 mm and they could be easily separated from products by processing on the sieve with 4.8 mm holes.

Buckwheat varieties with large grains were characterized by high yield of whole groats. Their large whole groats are of high quality. Varieties with small grains, namely, Bogatyr, Shatilovskaya 5, Ballada, and Molva, have little advantage in total groat yield. In terms of technological properties, varieties Sumchanka, Dozhdik, Devyatka, and Batyr are in an intermediate position. Investigation of protein content in grains of various fineness showed that there was no correlation between the size of grains and their protein content (Varlahova et al., 2012).

TABLE 34.1 Grain and Groat Qualities of Buckwheat Varieties

Variety	Year of State Registration	Fineness of Grains		Hull	Processed Fractions (%)		
		Weight of 1000 Grains (g)	Grain Fraction on the 5.0 mm Sieve (%)		Whole Groats	Crushed Groats (grits)	Total Groats
Bogatyr	1938	27.1	12.0	19.3	65.1	9.6	74.7
Shatilovskaya 5	1967	28.9	13.5	18.7	61.2	13.4	74.6
Sumchanka	1985	28.4	22.2	20.2	62.6	11.3	73.9
Ballada	1987	27.0	10.4	19.7	64.4	10.0	74.4
Molva	1997	27.7	9.1	19.2	62.6	12.2	74.8
Dozhdik	1998	30.3	22.4	20.8	64.4	9.4	73.8
Dikul	1999	28.5	20.0	20.4	62.9	10.1	73.0
Devyatka	2004	33.1	47.6	21.6	65.8	7.6	73.4
Chatyr Tau	2005	34.6	78.6	22.2	66.0	7.0	73.0
Batyr	2008	32.3	55.1	21.7	63.7	10.1	73.8
Ilishevskaya	2008	34.0	75.5	22.3	66.6	6.5	73.1
Design	2010	37.4	81.1	21.8	67.1	6.5	73.6

CONCLUSIONS AND PERSPECTIVES

Grains of common buckwheat are a valuable source of starch, protein, fat, dietary fiber, vitamins, minerals, *myo*-inositol, D-*chiro*-inositol, fagopyritols, flavonoids, phytosterols, and other chemical compounds. Processing of buckwheat grains into milling fractions gives some opportunities for producing new kinds of foods and their use in functional diets. Milling fractions with high concentrations of some interesting components (eg, soluble dietary fiber, proteins, and polyphenols) could be used for obtaining desired end-use products of high functionality. Milling fractions with accumulated chemical compounds could stimulate elaboration of new cooking and food technologies. Traditional buckwheat breeding was aimed at developing new varieties with high yield and quality of grains, which could enhance the yield of processed whole groats. Development of varieties with changed composition of biochemical compounds in grains could be an innovative direction of buckwheat breeding. Genes encoding flavonoid biosynthesis could be used in producing transgenic buckwheat plants with high levels of medicinally useful compounds. With the exception of allergenic proteins from proteomes of buckwheat grains, buckwheat varieties could become more attractive for the food industry. The complex composition of buckwheat grains and different distribution of chemical compounds in grain parts such as embryo, endosperm, aleurone layer, and coat could be used in buckwheat breeding. The embryo is the more valuable part of buckwheat grain where protein, fat, dietary fiber, minerals, carbohydrates, and other chemical compounds are concentrated. Breeding of buckwheat varieties with an exchange ratio of embryo/endosperm would enhance the nutrient value of buckwheat grains.

REFERENCES

Afanasev, I.B., Dorozhko, A.I., Brodskii, A.V., Koostyuk, V.A., Potapovitch, A.I., 1989. Chelating and free radical scavenging mechanisms of inhibitory action of rutin and quercetin in lipid peroxidation. Biochem. Pharmacol. 38, 1763–1769.

Aoyagi, Y., 2006. An angiotensin-I converting enzyme inhibitor from buckwheat (*Fagopyrum esculentum* Moench) flour. Phytochemistry 67, 618–621.

Bharali, S., Chrungoo, N.K., 2003. Amino acid sequence of the 26 kDa subunit of legumin-type seed storage protein of common buckwheat (*Fagopyrum esculentum* Moench): molecular characterization and phylogenetic analysis. Phytochemistry 63, 1–5.

Bobkov, S.V., 1993. Sozdanie ishodnogo materiala dlja selekcii intensivnyh sortov grechihi na osnove kompleksnogo ispol'zovanija mutantnyh form. Avtoreferat kandidatskoj dissertacii. Sankt-Peterburg, 16. (Bobkov, S.V., 1993. Producing of initial material for breeding of intensive buckwheat varieties on the basis of mutant form complex use. Thesis of Agricultural Sciences Candidate. Saint-Petersburg, 16) (in Russian).

Bojnanská, T., Chlebo, P., Gažar, R., Horna, A., 2009. Buckwheat-enriched bread production and its nutritional benefits. In: Dobránszki, J. (Ed.), Buckwheat I. Eur J Plant Sci Biotechnol 3 (Special Issue 1), 49–55.

Bonafaccia, G., Marocchini, M., Kreft, I., 2003a. Composition and technological properties of the flour and bran from common and Tartary buckwheat. Food Chem. 80, 9–15.

Bonafaccia, G., Gambelli, L., Fabjan, N., Kreft, I., 2003b. Trace elements in flour and bran from common and Tartary buckwheat. Food Chem. 83, 1–5.

Campbell, C.G., 1997. Buckwheat *Fagopyrum esculentum* Moench. Promoting the Conservation and Use of Underutilized and Neglected Crops. Institute of Plant Genetics and Crop Plant Research; Gatersleben/International Plant Genetic Resources Institute, Rome, Italy.

Cho, S.S., DeVries, J.W., Prosky, L., 1997. Analysis and applications.AOAC International, Gaithersburg, MD, 2–202.

Danila, A.M., Kotani, A., Hakamata, H., Kusu, F., 2007. Determination of rutin, catechin, epicatechin, and epicatechin gallate in buckwheat *Fagopyrum esculentum* Moench by micro-high-performance liquid chromatography with electrochemical detection. J. Agric. Food Chem. 55, 1139–1143.

Dietrych-Szostak, D., Oleszek, W., 1999. Effect of processing on the flavonoid content in buckwheat (*Fagopyrum esculentum* Moench) grain. J. Agric. Food Chem. 47, 4384–4387.

Dorrell, G.G., 1971. Fatty acid composition of buckwheat seed. J. Am. Oil Chem. Soc. 48, 693–696.

Eggum, B.O., Kreft, I., Javornik, B., 1981. Chemical composition and protein quality of buckwheat (*Fagopyrum esculentum* Moench). Qual. Plant—Plant Foods Hum. Nutr. 30, 175–179.

Elpidina, E.N., Dunaevsky, Y.E., Belozersky, M.A., 1990. Protein bodies from buckwheat seed cotyledons: isolation and characteristics. J. Exp. Bot. 41, 969–977.

Fabjan, N., Rode, J., Košir, I.J., Wang, Z.H., Zhang, Z., Kreft, I., 2003. Tartary buckwheat (*Fagopyrum tataricum* Gaertn.) as a source of dietary rutin and quercitrin. J. Agric. Food Chem. 51, 6452–6455.

FAOSTAT data, 2013. Online database. Available from: http://faostat3.fao.org.

Fesenko, N.V., 1970. Metody ocenki tehnologicheskih kachestv zerna na pervichnyh jetapah selekcii grechihi//VNIIZBK. Selekcija i Agrotehnika Grechihi. Orel, 214–224. (Fesenko N.V., 1970. Methods of grain technological properties estimation on early stages of buckwheat breeding. VNIIZBK. Breeding and Agrotechnology of Buckwheat. Orel, 214–224) (in Russian).

Fesenko, N.V., 1983. Selekcija i semenovodstvo grechihi. Moskva, p. 191. (Fesenko, N.V., 1983. Breeding and seed growing of buckwheat. Moscow, 191) (in Russian).

Fesenko, N.V., Martynenko, G.E., 1998. Evolution of buckwheat in the eastern European area and their use in selection. Vestn. Ross. Akad. S-kh. Nauk. 1, 10–13.

GOST 19092-92 (State Standard), 1993. Buckwheat. Requirements for purchases and deliveries.

He, J., Klag, M.J., Whelton, M.J., Mo, J.-P., Chen, J.-Y., Qian, M.-C., Mo, P.-S., He, G.-S., 1995. Oats and buckwheat intakes and cardiovascular disease risk factors in an ethnic minority in China. Am. J. Clin. Nutr. 61, 366–372.

Horbowicz, M., Obendorf, R.L., 1992. Changes in sterols and fatty acids of buckwheat endosperm and embryo during seed development. J. Agric. Food Chem. 40 (5), 742–752.

Horbowicz, M., Brenac, P., Obendorf, R.L., 1998. Fagopyritol B1, *O*-α-D-galactopyranosyl-(1→2)-D-*chiro*-inositol, a galactosyl cyclitol in maturing buckwheat seeds associated with desiccation tolerance. Planta 205, 1–11.

Ikeda, K., 2002. Buckwheat: composition, chemistry, and processing. Adv. Food. Nutr. Res. 44, 395–434.

Ikeda, K., Oku, M., Kusano, T., Yasumoto, K., 1986. Inhibitory potency of plant antinutrients towards the in vitro digestibility of buckwheat protein. J. Food Sci. 51, 1527–1530.

Ikeda, S., Edotani, M., Naito, S., 1990. Zinc in buckwheat. Fagopyrum 10, 51–55.

Ikeda, K., Sakaguchi, T., Kusano, T., Yasumoto, K., 1991. Endogenous factors affecting protein digestibility in buckwheat. Cereal Chem. 68, 424–427.

Ikeda, S., Yamashita, Y., Kreft, I., 1999. Mineral composition in buckwheat by-products and its processing characteristics in konjak preparation. Fagopyrum 17, 57–61.

Ikeda, S., Yamashita, Y., Tomura, K., Kreft, I., 2006. Nutritional comparison in mineral characteristics between buckwheat and cereals. Fagopyrum 23, 61–65.

Jacquemart, A.L., Cawoy, V., Kinet, J.M., Ledent, J.F., Quinet, M., 2012. Is buckwheat (*Fagopyrum esculentum* Moench) still a valuable crop today? Eur. J. Plant Sci. Biotechnol. 6 (2), 1–10, Special Issue.

Javornik, B., Kreft, I., 1984. Characterization of buckwheat proteins. Fagopyrum 4, 30–38.

Javornik, B., Eggum, B.O., Kreft, I., 1981. Studies on protein fractions and protein quality of buckwheat. Genetika 13, 115–118.

Jiang, P., Burczynski, F., Campbell, C., Pierce, G., Austria, J.A., Briggs, C.J., 2007. Rutin and flavonoid contents in three buckwheat species *Fagopyrum esculentum, F. tataricum*, and *F. homotropicum* and their protective effects against lipid peroxidation. Food Res. Int. 40, 356–364.

Kalinova, J., Vrchotova, N., 2009. Level of catechin, myricetin, quercetin and isoquercitrin in buckwheat (*Fagopyrum esculentum* Moench), changes of their levels during vegetation and their effect on the growth of selected weeds. J. Agric. Food Chem. 57, 2719–2725.

Kalinova, J., Triska, J., Vrchotova, N., 2006. Distribution of vitamin E, squalene, epicatechin, and rutin in common buckwheat plants (*Fagopyrum esculentum* Moench). J. Agric. Food Chem. 54, 5330–5335.

Kamal, A.H.M., Jang, I.-D., Kim, D.-E., Suzuki, T., Chung, K.-Y., Choi, J.-S., Lee, M.-S., Park, C.-H., Park, S.-U., Lee, S.H., Jeong, H.S., Woo, S.-H., 2011. Proteomics analysis of embryo and endosperm from mature common buckwheat seeds. J. Plant Biol. 54, 81–91.

Kreft, I., 2001. Buckwheat research, past, present and future perspectives—20 years of internationally coordinated research. Proceedings of the 8th International Symposium on Buckwheat, Chunchon, Korea, pp. 361–366.

Kreft, M., Kreft, S., 1999. Computer aided three-dimensional reconstruction of the buckwheat (*Fagopyrum esculentum* Moench) seed morphology. Res. Rep. Agric. 73, 331–336.

Kreft, I., Bonafaccia, G., Zigo, A., 1994. Secondary metabolites of buckwheat and their importance in human nutrition. Prehrambeno-tehnoloska Biotehnoloska Revija 32, 195–197.

Kreft, S., Knapp, M., Kreft, I., 1999. Extraction of rutin from buckwheat (*Fagopyrum esculentum* Moench) seeds and determination by capillary electrophoresis. J. Agric. Food Chem. 47, 4649–4652.

Kreft, I., Fabjan, N., Yasumoto, K., 2006. Rutin content in buckwheat (*Fagopyrum esculentum* Moench) food materials and products. Food Chem. 98, 508–512.

Krkoskova, B., Mrazova, Z., 2005. Prophylactic components of buckwheat. Food Res. Int. 38, 561–568.

Lazareva, T.N., 2007. Polimorfizm belkov semjan u vidov i sortov grechihi Fagopyrum Mill. Avtoreferat dissertacii na soiskanie uchenoj stepeni kandidata biologicheskih nauk. Sankt-Peterburg, 20 s. (Lazareva, T.N., 2007. Polymorphism of seed proteins in species and varieties of buckwheat *Fagopyrum* Mill. Thesis of Biological Sciences Candidate. Saint-Petersburg, 20) (in Russian).

Lazareva, T.N., Pavlovskaja, N.E., Gor'kova, I.V., 2006. Vlijanie srokov hranenija na soderzhanie zapasnyh belkov v semenah grechihi. Hranenie i Pererabotka Sel'hozsyr'ja 3, 12–15. (Lazareva, T.N., Pavlovskaya, N.E., Gorkova, I.V., 2006. Influence of storage time on storage protein content in buckwheat seeds. Storage and Processing of Agricultural Raw Material 3, 12–15) (in Russian)

Lee, S., Han, Y., Do, J.-R., Oh, S., 2013. Allergenic potential and enzymatic resistance of buckwheat. Nutr. Res. Prac. 7 (1), 3–8.

Levent, H., Bilgiçli, N., 2011. Enrichment of gluten-free cakes with lupin (*Lupinus albus* L.) or buckwheat (*Fagopyrum esculentum* Moench) flours. Int. J. Food Sci. Nutr. 62, 725–728.

Li, S.Q., Zhang, Q.H., 2001. Advances in the development of functional foods from buckwheat. Crit. Rev. Food Sci. Nutr. 41 (6), 451–464.

Li, X., Park, N.I., Xu, H., Woo, S.-H., Park, C.H., Park, S.U., 2010. Differential expression of flavonoid biosynthesis genes and accumulation of phenolic compounds in common buckwheat (*Fagopyrum esculentum*). J. Agric. Food Chem. 58, 12176–12181.

Lin, R.F., 1994. Buckwheat in China. Chinese Agriculture Press House, Beijing, China.

Liu, Z., Ishikawa, W., Huang, X., Tomotake, H., Kayashita, J., Watanabe, H., Nakajoh, M., Kato, N.A., 2001. Buckwheat protein product suppresses 1,2-dimethylhydrazine-induced colon carcinogenesis in rats by reducing cell proliferation. J. Nutr. 131, 1850–1853.

Ma, M.S., In, Y.B., Hyeon, G.L., Yang, C.B., 2006. Purification and identification of angiotensin I-converting enzyme inhibitory peptide from buckwheat (*Fagopyrum esculentum* Moench). Food Chem. 96, 36–42.

Mazza, G., 1988. Lipid content and fatty acid composition of buckwheat seed. Cereal Chem. 65, 122–126.

Metodicheskie materialy Goskomissii po sortoispytaniyu sel'skokhozyaistvennykh kul'tur (Methodological handbook of the state commission on agricultural crop variety testing), 1972. Moscow, 3–4, p. 55 (in Russian).

Metodika otsenki tekhnologicheskikh svoistv grechikhi v processe selektsii (Manual on assessment of the buckwheat technological properties during selection), Orel, 1979, 32 (in Russian).

Milisavljevic, M.D., Timotijevic, G.S., Radovic, S.R., Brkljacic, J.M., Konstantinovic, M.M., Maksimovic, V.R., 2004. Vicilin-like storage globulin from buckwheat (*Fagopyrum esculentum* Moench) seeds. J. Agric. Food Chem. 52, 5258–5262.

Mitsunaga, T., Matsuda, M., Shimizu, M., Iwashima, A., 1986. Isolation and properties of a thiamine-binding protein from buckwheat seed. Cereal Chem. 63, 332–335.

Nair, A., Adachi, T., 1999. Immunodetection and characterization of allergenic proteins in common buckwheat (*Fagopyrum esculentum*). Plant Biotechnol. (Tsukuba) 16, 219–224.

Park, J.W., Kang, D.B., Kim, C.W., Koh, S.H., Yum, H.Y., Kim, K.E., Hong, C.S., Lee, K.Y., 2000. Identification and characterization of the major allergens of buckwheat. Allergy 55, 1035–1041.

Petelina, N.N., 1971. Sozdanie ishodnogo materiala dlja selekcii krupnoplodnyh sortov diploidnoj grechihi. Nauchnye Trudy VNIIZBK 3, 128-135. (Petelina, N.N., 1971. Producing of initial material for breeding of diploid buckwheat varieties with large grains. Scientific Works of VNIIZBK 3, 128–135) (in Russian).

Petr, J., Michalík, I., Tlaskalová, H., Capouchová, I., Faměra, O., Urminská, D., Tučková, L., Knoblochová, H., 2003. Extention of the spectra of plant products for the diet in coeliac disease. Czech J. Food Sci. 21, 59–70.

Pomeranz, Y., 1983. Buckwheat: structure, composition and utilization. Crit. Rev. Food Sci. Nutr. 19, 213–258.

Pomeranz, Y., Robbins, G.S., 1972. Amino acid composition of buckwheat. J. Agric. Food Chem. 20, 270–274.

Qin, P., Wang, Q., Shan, F., Hou, Z., Ren, G., 2010. Nutritional composition and flavonoids content of flour from different buckwheat cultivars. Int. J. Food Sci. Technol. 45, 951–958.

Radovic, S.R., Maksimovic, V.R., Varkonji-Gasic, E.I., 1996. Characterization of buckwheat seed storage proteins. J. Agric. Food Chem. 44, 972–974.

Radovic, R.S., Maksimovic, R.V., Brkljacic, M.J., Varkonji Gasic, I.E., Savic, P.A., 1999. 2S albumin from buckwheat (*Fagopyrum esculentum* Moench). J. Agric. Food Chem. 47, 1467–1470.

Rout, M.K., Chrungoo, N.K., Rao, K.S., 1997. Amino acid sequence of the basic subunit of 13S globulin of buckwheat. Phytochemistry 45 (5), 865–867.

Shumilin, P.I., Svjatova, L.N., 1973. Tehnologicheskaja cennost' grechihi v zavisimosti ot krupnosti i vyravnennosti zerna. Bjulleten' Nauchno-Tehnicheskoj Informacii VNIIZBK 7, 98–97. (Shumilin, P.I., Svyatova, L.N., 1973. Technological value of buckwheat on dependence of fineness and evenness of grain. VNIIZBK Bulletin of Scientific and Technical Information, 7, 98–97) (in Russian)

Shumilin, P.I., Svjatova, L.N., 1976. O metodah ocenki tehnologicheskih svojstv zerna grechihi v processe selekcii. Nauchnye Trudy VNIIZBK. Selekcija, biohimija i agrotehnika zernobobovyh kul'tur 5, 197–205. (Shumilin, P.I., Svyatova, L.N., 1976. About methods of buckwheat technological traits estimation in breeding process. Scientific Works of VNIIZBK 5, 197–205) (in Russian).

Shumilin, P.I., Varlahova, L.N., 1985. Povyshenie urozhajnosti i kachestva krupjanyh kul'tur metodami selekcii i tehnologii vozdelyvanija. Orel, 40–47. (Shumilin, P.I., Varlakhova, L.N., 1985. Increase of yield and quality of groat cultures using methods of breeding and cultivation technologies. Orel, 40–47) (in Russian).

Shumilin, P.I., Svjatova, L.N., Fesenko, N.V., 1975. K metodike ocenki legkosti shelushenija grechihi na rannih jetapah selekcii. Bjulleten' Nauchno-Tehnicheskoj Informacii VNIIZBK 11, 85–88. (Shumilin, P.I., Svyatova, L.N., Fesenko, N.V., 1975. Towards of shelling methods estimation on early stages of buckwheat breeding. VNIIZBK Bulletin of Scientific and Technical Information 11, 85–88) (in Russian).

Skerritt, J.H., 1986. Molecular comparison of alcohol-soluble wheat and buckwheat proteins. Cereal Chem. 63, 365–369.

Skrabanja, V., Kreft, I., 1998. Resistant starch formation following autoclaving of buckwheat (*Fagopyrum esculentum* Moench) groats. An in vitro study. J. Agric. Food Chem. 46, 2020–2023.

Skrabanja, V., Laerke, H.N., Kreft, I., 1998. Effects of hydrothermal processing of buckwheat (*Fagopyrum esculentum* Moench) groats on starch enzymatic availability in vitro and in vivo in rats. J. Cereal Sci. 28, 209–214.

Skrabanja, V., Kreft, I., Modic, M., Ikeda, S., Ikeda, K., Kreft, S., Bonafaccia, G., Knapp, M., Kosmelj, K., 2003a. Nutrient content in buckwheat milling fractions. Cereal Chem. 81 (2), 172–176.

Skrabanja, V., Kreft, I., Golob, T., Modic, M., Ikeda, S., Ikeda, K., Kreft, S., Bonafaccia, G., Knapp, M., Kosmelj, K., 2003b. Nutrient content in buckwheat milling fractions. Cereal Chem. 81 (2), 172–176.

Sokolov, O.A., Timchenko, A.V., Semikhov, B.F., Dunaevsky, Y.E., Belozersky, M.A., 1981. Isolation and chemical composition of aleurone grains from buckwheat. Fisiologiya Rastenii 28, 1166–1173.

Steadman, K.J., Burgoon, M.S., Schuster, R.L., Lewis, B.A., Edwardson, S.E., Obendorf, R.L., 2000. Fagopyritols, D-*chiro*-inositol, and other soluble carbohydrates in buckwheat seed milling fractions. J. Agric. Food Chem. 48, 2843–2847.

Steadman, K.J., Burgoon, M.S., Lewis, B.A., Edvardson, S.E., Obendorf, R.L., 2001a. Buckwheat seed milling fractions: description, macronutrient composition, and dietary fiber. J. Cereal Sci. 33, 271–278.

Steadman, K.J., Burgoon, M.S., Lewis, B.A., Edwardson, S.E., Obendorf, R.L., 2001b. Minerals, phytic acid, tannin and rutin in buckwheat seed milling fractions. J. Sci. Food Agric. 81, 1094–1100.

Suzuki, T., Ishikawa, N., Meguro, H., 1983. Angiotensin-converting enzyme inhibiting activity in foods. Nippon Nogeikagaku Kaishi 57, 1143–1146.

Svjatova, L.N., Shumilin, P.I., 1978. Tehnologicheskoe znachenie nekotoryh pokazatelej kachestva zerna grechihi i vozmozhnosti ispol'zovanija ih pri sozdanii sortov s vysokimi tehnologicheskimi svojstvami. Nauchnye Trudy VNIIZBK 7, 128–134. (Svyatova L.N., Shumilin, P.I., 1978. Technological value of some grain quality traits and possibility its use in developing varieties with high technological properties. Scientific Works of VNIIZBK 7, 128–134) (in Russian).

Tanaka, K., Matsumoto, K., Akasawa, A., Nakajima, T., Nagasu, T., Iikura, Y., Saito, H., 2002. Pepsin-resistant 16-kD buckwheat protein is associated with immediate hypersensitivity reaction in patients with buckwheat allergy. Int. Arch. Allergy Immunol. 129, 49–56.

Tang, C.-H., Wang, X.-Y., 2010. Physicochemical and structural characterisation of globulin and albumin from common buckwheat (*Fagopyrum esculentum* Moench) seeds. Food Chem. 121, 119–126.

Urisu, A., Kondo, Y., Morita, Y., et al.,1994. Identification of a major allergen of buckwheat seeds by immunoblotting methods. Allergy Clin. Immunol. News 6, 151–155.

Varlahova, L.N., Bobkov, S.V., Martynenko, G.E., Mihajlova, I.M., 2012. Tehnologicheskie kachestva zerna novyh krupnoplodnyh sortov grechihi. Zemledelie 5, 40–42. (Varlakhova, L.N., Bobkov, S.V., Martynenko, G.E., Mikhailova, I.M., 2012. Technological properties of novel buckwheat varieties with large grains. Zemledelie 5, 40–42) (in Russian).

Varlakhova, L.N., Bobkov, S.V., Mikhailova, I.M., 2013a. Technological properties of buckwheat varieties grains. Russi. Agric. Sci. 39, 42–45.

Varlakhova, L., Bobkov, S., Zotikov, V., Mikhailova, I., 2013b. Grain qualities of Russian buckwheat (*Fagopyrum esculentum* Moench) varieties. Laško. Proceedings of Papers. 12th International Symposium on Buckwheat. August 21–25, pp. 181–182.

Vojtíšková, P., Kmentová, K., Kubáň, V., Kráčmar, S., 2012. Chemical composition of buckwheat plant (*Fagopyrum esculentum*) and selected buckwheat products. J. Microbiol. Biotechnol. Food Sci. 1, 1011–1019.

Watanabe, M., 1998. Catechins as antioxidants from buckwheat (*Fagopyrum esculentum* Moench) groats. J. Agric. Food Chem. 46, 839–845.

Wei, Y., Hu, X., Zhang, G., Ouyang, S., 2003. Studies on the amino acid and mineral content of buckwheat protein fractions. Nahrung/Food 2, 114–116.

Wieslander, G., 1996. Review on buckwheat allergy. Allergy 51, 661–665.

Wieslander, G., Norbäck, D., 2001. Buckwheat allergy. Allergy 56, 703–704.

Wijngaard, H.H., Arendt, E.K., 2006. Buckwheat. Cereal Chem. 83 (4), 391–401.

Yoshimatsu, M.A., Zhang, J.W., Hatakawa, S., Mine, Y., 2000. Electrophoretic and immuno-chemical characterization of allergenic proteins in buckwheat. Int. Arch. Immunol. 123, 130–136.

Zhang, Z.L., Zhou, M.L., Tang, Y., Li, F.L., Tang, Y.X., Shao, J.R., Xue, W.T., Wu, Y.M., 2012. Bioactive compounds in functional buckwheat food. Food Res. Int. 49, 389–395.

Zhou, X., Wen, L., Li, Z., Zhou, Y., Chen, Y., Lu, Y., 2015. Advance on the benefits of bioactive peptides from buckwheat. Phytochem. Rev. 14 (3), 381–388.

Zielinski, H., Ciska, E., Kozlowska, H., 2001. The cereal grains: focus on vitamin E. Czech J. Food Sci. 19, 182–188.

Subject Index